The Art of Linear Electronics

The Art of Linear Electronics

John Linsley Hood

Butterworth-Heinemann Ltd
Linacre House, Jordan Hill, Oxford OX2 8DP

A member of the Reed Elsevier plc group

OXFORD LONDON BOSTON
MUNICH NEW DELHI SINGAPORE SYDNEY
TOKYO TORONTO WELLINGTON

First published 1993
Reprinted 1994

British Library Cataloguing in Publication Data
A catalogue record for this book is available from the British Library

ISBN 0 7506 0868 4

Library of Congress Cataloguing in Publication Data
A catalogue record for this book is available from the Library of Congress

Printed and bound in Great Britain

Contents

Preface

The past decades have seen a number of changes in the nature of electronics. Of these, few have had a greater impact than the continuous growth of digital circuit technology, with its extensions into electronic computation, data handling and numerical storage.

The enormous potential of digital circuit techniques has fascinated most of those who have become acquainted with them, and it is therefore a matter of little surprise that such a large proportion of studies in the field of electronics at universities and colleges relate almost exclusively to systems based on digital electronic components and circuitry.

However, circuit design in this field is largely restricted to the interconnection of various commercially available circuit building blocks in ways which are largely predetermined by their manufacturers.

While it is obviously essential for the digital circuit engineer to understand the specific functions of the complex circuit blocks he is joining together, it is unnecessary for him to know what actually happens within these blocks of 'integrated circuitry' when these are fed with well defined input pulses at logic level amplitudes, nor to concern himself with the problems which can and do arise in the handling of input signals of minute, indeterminate, or variable size, irregular waveform shape, or uncertain frequency.

This situation is now leading to an increasing shortage of engineers who have any skills in the equally wide field of linear electronic circuit design, and has shut off many competent engineers, quite familiar with the methods of digital circuitry, from the pleasure of creating simple and possibly unique functional circuits from discrete electronic components.

Moreover, while it is increasingly possible for almost any electronic circuit requirement to be met by digital techniques, all the methods which allow contact with the phenomena of nature, normally referred to as transducers, operate with variable signals of a wide range of amplitudes and frequencies, so any digital manipulation of these signals will depend on the availability of some linear 'interface' device to convert the input or output signal to or from its digital form.

While it is far too big a field for any single book to give more than a brief summary of linear electronic circuits and techniques, it does seem to me to be possible to cover the principal aspects of the subject in outline, so that those who would like a rather fuller knowledge of any particular part of the field could then refer to more specialized books for further information, without feeling lost in a totally unfamiliar world.

It is my hope that I will also have provided enough basic information within the chapters of this book for someone, starting with very little previous experience of this subject, to be able, having read it, to do some linear circuit design for himself. For this reason, I have tried not to presume too high a degree of existing knowledge on the part of the reader, and to explain, where I can, the reasons why things are done in the way that they are. I have also tried, particularly in the earlier chapters, to explain the various bits of technical jargon used in this field, since I know, from experience, how frustrating it is to read a text in which all sorts of unknown things are referred to by unfamiliar names or groups of inexplicable initials. I ask the more experienced readers to put up with this, or skip these introductory chapters altogether.

However, in the later chapters, my intention has been, where I can, to include aspects of the subject which I have found, to my regret, are not well covered in the bulk of contemporary textbooks, and which may therefore be interesting, informative, or useful to the experienced engineer as well.

It is not easily possible to sustain a simple and fully explanatory approach through those chapters dealing with moderately complex aspects of the subject, so I have sought to grade the extent of presumed existing knowledge, so that the later chapters lean on those things which I hope the reader will have gathered from the earlier ones.

John Linsley Hood
Taunton
Somerset

Acknowledgements

I would like to thank my many friends for their help and encouragement. In particular, I would like to record my debt to Colin Green, for his great help in ploughing his way through some of the more turgid parts of my prose, to point out obscurities, ambiguities, spelling mistakes and technical errors, and without whose advice the text would have been even less elegant.

1

Electronic component symbols and circuit drawing

Introduction

The normal way by which an electronics engineer will explore the function of an electronic circuit, or will describe his own designs, is by way of a 'theoretical' circuit drawing. In order to be able to understand what he sees or to be able to make his own drawings, it is necessary for him to know what the conventional circuit symbols mean, and which symbols are appropriate for a given device.

It is all too often taken for granted that the reader will understand this, without further explanation. This can be frustrating if some of the symbols used are unfamiliar, or if unexplained conventions are employed as a means of simplifying the drawing.

I have therefore tried, in the following pages, to show the most common graphical forms by which specific components are represented. Although there is a fairly wide agreement on these styles, individual design offices may still employ symbols which are unique to themselves, where some guesswork may be needed. The experienced engineer can skip this chapter without loss.

Basic design philosophy

The purpose of a circuit drawing is to give a rapid visual explanation to the viewer of how the circuit works, and how the individual component parts relate to one another. Unfortunately, in practice, drawings are often made with the sole aims of showing the connections between the components in an accurate manner and of producing a neat looking final result. Whether or not the actual interconnections are easy to follow, or how readily the engineer can discover the way the circuit operates, may not be a particularly high priority at the time the drawing is made.

Certain ground rules will help to keep the drawing simple in appearance and easy to follow. These are:

1. Adopt a consistent policy for the flow direction of signals, such as 'inputs' on the left-hand side of the drawing, moving across to 'outputs' on the right-hand side.
2. Keep positive supply lines at the top of the drawing, and negative supply lines, if present, at the bottom. Also, where polarity sensitive components are employed, try to position these so

that the potential appearing across them has the same orientation as that of the supply lines. If there is an 'earth' or '0V' line this should be positioned between the +ve and –ve lines.

3. Where the proliferation of supply lines will tend to confuse the picture, because of their frequent crossing of signal lines, use conventional symbols, as shown in Figure 1.1, to indicate their destinations. The convention here is that Figure 1.1a denotes a connection to a '0V' supply line, which may or may not be connected to the chassis of the equipment, while that of Figure 1.1b will indicate a direct connection to a metal chassis and Figures 1.1c or 1.1d will denote an earth connection.
The symbols of Figures 1.1e and 1.1f will imply a connection to the positive and negative supply lines. The drawing can frequently be greatly simplified by this technique, and if there are several different power supply or '0V' return lines the appropriate ones can easily be indicated by numbers or letters.

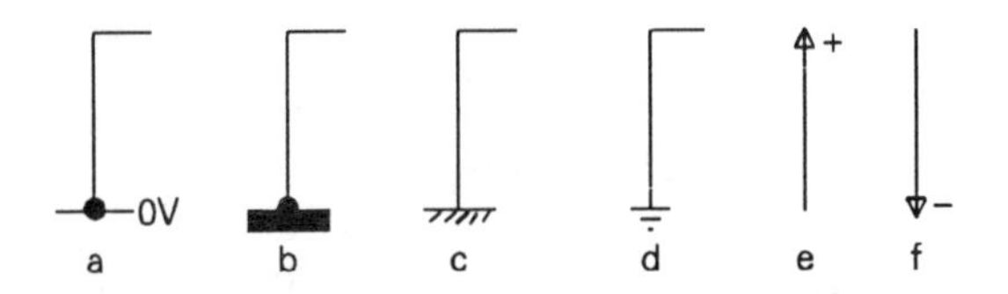

Figure 1.1

4. Avoid groups of connections drawn as closely spaced bundles of parallel lines. Although these can give a drawing a tidy appearance, it is very easy for the reader to lose track of the connection he is tracing through the circuit. It is much better to use labels or supply sign conventions instead.

A comparative example is shown in Figures 1.2a and 1.2b, in which the latter drawing is not only simpler, and easier to follow, but actually gives more information about the operation of the circuit than Figure 1.2a. In view of the complexity which can arise in showing the interconnections of just five components, it is easy to see how confusing the circuit drawings for more complicated systems may become, unless some care is taken to make them easy to follow.

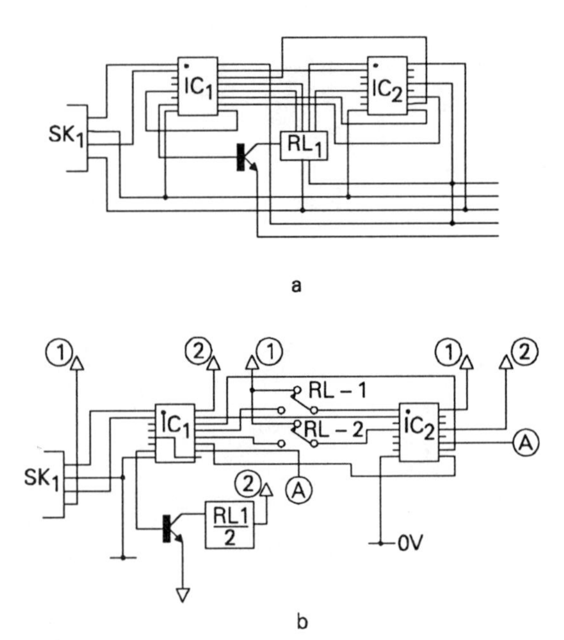

Figure 1.2

Avoidance of ambiguities

The most common source of ambiguity in circuit diagrams concerns wiring connections which join, or which cross without joining. Several conventions exist in this area. Of these the most common is the use of a 'blob' on a junction of two wires, as shown in Figure 1.3a, to distinguish this from a crossing without junction, shown in Figure 1.3b.

Unfortunately, in printing or subsequent reproduction, spurious 'blobs' can appear because of ink accumulations where lines cross, giving a wrong indication of circuit function. To avoid this type of mistake, it is good practice to make sure that junctions are never shown as simple crossings with 'blobs' but as staggered connections as shown in Figures 1.3c or 1.3d. Alternatively, a loop can be inserted where lines cross without connection, as shown in Figure 1.3e.

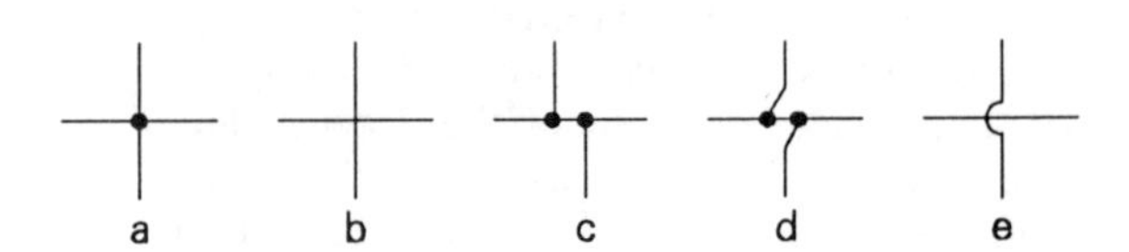

Figure 1.3

The other major uncertainty arises where earth or '0V' line connections are specified. Quite often the performance of the circuit depends critically upon the position of the 0V or earth line returns, and a simple connection to an earthed chassis may not be acceptable. In this case the forms of connection should be distinguished from one another by the use of the symbols shown in Figure 1.1, and where separate 0V or earth line returns are necessary these should be labelled numerically.

Conventional assumptions

Some conventions can cause difficulties to the inexperienced. One of these, as noted below, in Figure 1.13a, in relation to operational amplifiers, is that these ICs are normally powered by stabilized +/–15V DC supply lines – unless otherwise specified – so it may be taken for granted that the IC will be connected to its appropriate supply lines without these connections being shown at all on the circuit diagram.

A similar convention is often assumed with logic ICs, which will usually be connected between the 0V rail and a fixed +5V line. The existence of such a supply line is frequently taken for granted, and not shown in the circuit drawing, as is the presence of a small ceramic 'by-pass' capacitor, to decouple the supply to the 0V line at its point of connection to the IC, as shown in Figure 1.14n. However, there may be exceptions to this rule, and the fact that no connections are shown may not always mean that a +5V rail is used.

Block diagrams

It is often helpful when explaining the function of a relatively complex circuit – or group of circuits – to make use of 'block diagrams', in which the function of each block is described within its outlines, to show how the several parts of the circuit relate to one another. An example based on an audio amplifier and preamplifier is shown in Figure 1.4.

The specific circuit layout of the individual function blocks can then be shown separately at a later stage in the circuit description, when required.

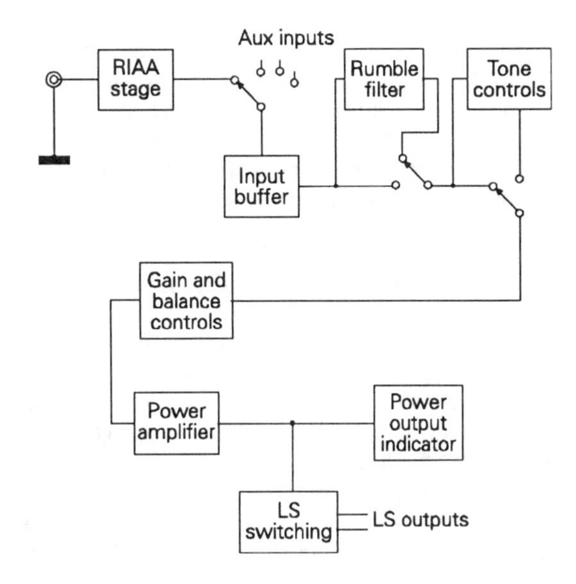

Figure 1.4

The theoretical circuit

In addition to providing an illustration of the way in which the various components of the circuit relate to one another, and an indication of the flow of the signal path and the currents drawn from the power supply lines, the circuit diagram also allows the types and values of the separate components to be shown. It may also be helpful, as an aid to understanding of the circuit function, or subsequent fault diagnosis, if the nominal currents flowing in the various supply line paths are also indicated.

Although designers are traditionally assumed to set down their thoughts on the backs of envelopes, from which a fully-fledged design may spring into existence – and, indeed there may be designers who really do work like this – there is little doubt that the extra time spent, at the design stage, in producing a fully labelled circuit drawing, greatly facilitates the calculation of circuit performance, and allows thermal dissipations and other characteristics of the design to be assessed before the prototype is assembled.

Technical terms

As I explained in the preface, it is my hope that this book will be useful to engineers with little existing knowledge of electronics. To this end, I have tried to explain technical terms as they arise. However there are cases where this would make the text unduly lengthy, so, where I have omitted explanations, I would refer the reader to the chapter which deals with the component or technique in question.

Circuit symbols and useful conventions

In the beginning, most of the circuit symbols were simplified representations of the actual components themselves: such as that for a resistor being shown as a piece of resistive wire laid down in a zigzag path to increase its length, or a capacitor as a nominal pair of parallel conductive plates, in proximity to one another, but separated by an air gap, or a transformer being represented by two coils of wire wound on a laminated iron core.

However, the growth of different forms of these basic component types has led to the adoption of various conventional styles of representation of these sub-species, and it helps greatly in understanding the diagram if these individual variations can be recognized instantly.

In addition, some 'shorthand' forms of symbol have emerged to denote such things as 'current mirrors' or 'constant current sources' – circuit function blocks which will normally be made up from groups of separate components – as well as for the wide variety of integrated circuits which are now available.

The diagrams used to represent the most commonly found types of components, together with some of these 'shorthand' symbols, are illustrated in Figures 1.5 – 1.16.

Capacitors

These are used in various forms. Of these, the simplest is the fixed value component, with either air or some polarity-insensitive dielectric between its plates, for which the circuit symbol is as shown in Figure 1.5a. The actual type of capacitor may not be specified by the drawing used. If this is critical to the function of the circuit, it will then be labelled so that this will be known.

With capacitors, as with most other circuit components, the capacitance value, together with the work-

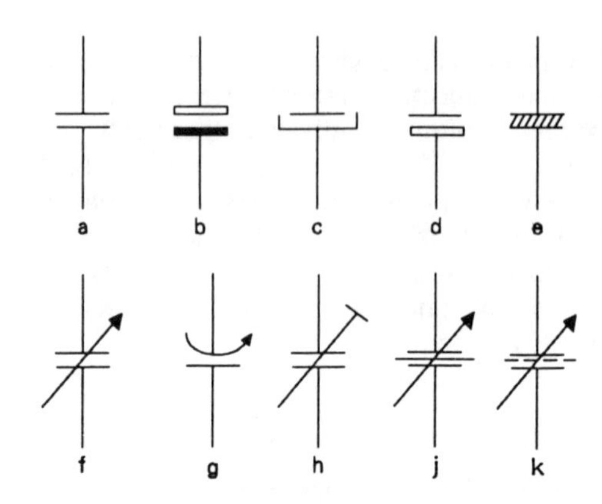

Figure 1.5

ing voltage and the required precision of value, where these are important or non-standard, will be appended as a label, such as 2μ2F/64V/10%. (It is becoming normal practice for the magnifier symbol, such as 'p' or 'μ' or 'k' or 'M' to be used instead of the decimal point, as in the previous example.)

Electrolytic capacitors, that is to say capacitors in which a thin insulating layer of metallic oxide has been formed within the component by electrolytic action, and which must therefore be connected in the circuit so that the applied voltage across its terminals has the correct polarity, are shown in the several forms illustrated in Figures 1.5b, 1.5c, or 1.5d, the actual symbol used depending on design office or national preferences.

Regrettably, the diagram of Figure 1.5c is also sometimes used to denote a non-polar capacitor, as an alternative to that of Figure 1.5a. This source of ambiguity could be avoided if a '+' polarity sign is always appended to the positive terminal of an electrolytic component. In some drawings of Japanese origin, electrolytic capacitors are depicted by the symbol shown in Figure 1.5e.

Capacitors in which the value can be manually adjusted, known as 'variable capacitors', are shown in the forms illustrated in Figures 1.5f – 1.5k. A convention which is worth remembering, because it is used with many other component symbols, is that an arrow with a point on its end usually implies that the component is intended to be adjusted frequently, for some operational or control purpose, whereas an arrow with a square end implies a pre-set component, which might be adjusted once, on setting up the equipment, and thereafter left alone.

Figures 1.5g and 1.5h frequently imply small value 'trimmer' capacitors, whereas Figure 1.5f would be used most commonly to represent a larger value variable capacitor, used as the 'tuning' control in a radio receiver. A special form of tuning capacitor, used in transmitters and short-wave receivers, is the 'split stator' type, in which two insulated plates are separated by a central earthed plate. This is denoted by the diagram of Figure 1.5j.

Where several variable capacitors are 'ganged' together, so that several sections can be adjusted simultaneously from the same control spindle, this fact is usually represented by the use of a dotted line joining the rear ends of the arrows together. The same dotted line convention is used to denote ganged sections of potentiometers and switches.

A similar diagram to that of Figure 1.5j, shown in Figure 1.5k, is used to denote a variable capacitor in which a sheet of some thin insulating material has been inserted between the plates so that these can be operated at much closer spacings, which allows a more compact component to be made, without the likelihood of an inadvertent short circuit occurring between the plates.

Resistors and potentiometers

The normal symbol for a fixed resistor is that shown in Figures 1.6a or 1.6b. The latter is becoming more popular because it is easier to draw neatly, but is slightly more ambiguous in its appearance. A two-terminal continuously variable resistor is represented by either of the Figures 1.6c or 1.6d, while a preset variable resistor is illustrated in Figures 1.6e or 1.6f.

The label applied to a resistor will specify its ohmic value, and may also denote its type, for example wirewound (ww), metal film (mf) or metal oxide film (mo) as well as its power rating and precision in value, giving a description such as 2k2/3W/5%/ww.

A three-terminal potential divider (potentiometer) is shown by either Figure 1.6g or 1.6h, depending on whether it is of a continuously variable or preset form. Note again the 'arrow head' convention. Using the oblong box type of resistor diagram, a continuously variable potentiometer would be shown as in Figure 1.6k.

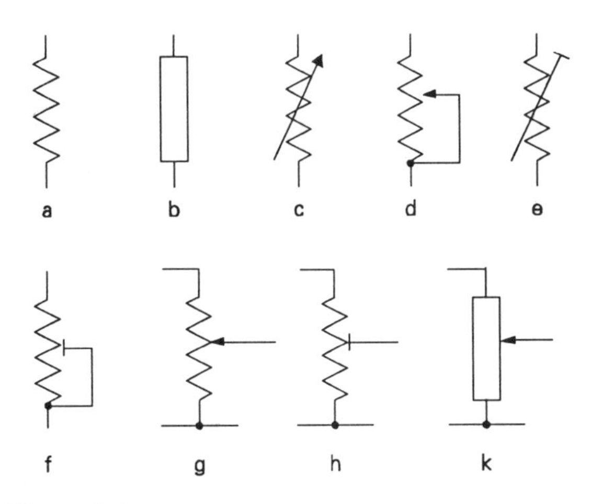

Figure 1.6

Coils and transformers

The simplest form, that of a fixed value air cored coil, is illustrated in Figure 1.7a. For RF use, in order to make the component more compact, a powdered iron core may be used, which will give the coil a higher inductance value for the same number of turns of winding. This is indicated by the use of the dotted line core shown in Figure 1.7b. Figures 1.7c and 1.7d denote dust iron cored, or similar, coils in which the inductance can be adjusted by a preset or continuously variable position for the core.

The use of a laminated sheet iron core, for a low frequency choke, is indicated by Figure 1.7e.

Radio frequency transformers are illustrated in Figures 1.7f, 1.7g and 1.7h, using air cores, and fixed and variable dust iron cores respectively. The diagram used in Figure 1.7j is sometimes used to denote a variable dust iron core position, as an alternative to that of Figure 1.7h. Where the coil or RF transformer is enclosed in a screening can, a box is drawn around it, as in Figure 1.7k.

Where coils are inductively coupled, though not shown close to each other, this is indicated by a curving line drawn through both, as shown in Figure 1.7m, and where the extent of such coupling is adjustable, this would be indicated by the arrow head symbol of Figure 1.7n.

Low frequency power transformers, such as those used for mains power inputs, are depicted by the drawings of Figures 1.7p, 1.7q and 1.7r, of which the last two represent tapped and separated winding types.

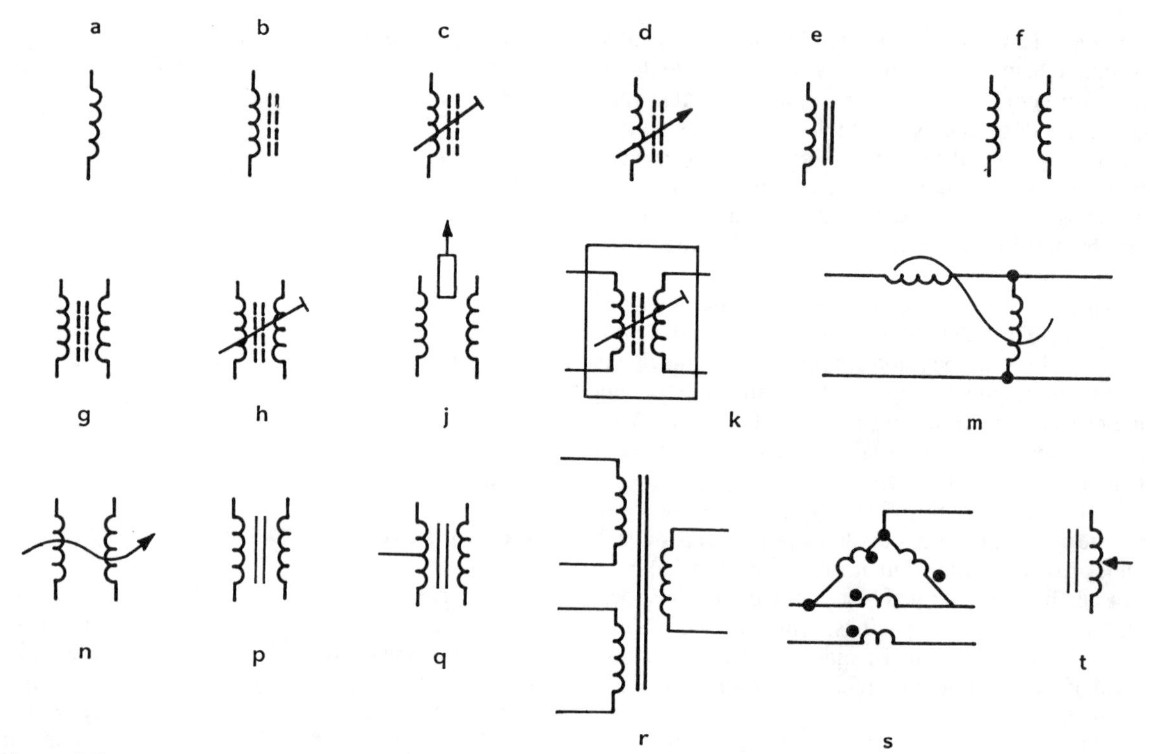

Figure 1.7

If the direction of winding in a coil or transformer is important, the 'top' ends of all windings made in the same direction will be marked by a 'dot', as in Figure 1.7s. By analogy with the symbol for a potentiometer, a 'Variac', or variable transformer, would be drawn as in Figure 1.7t.

Switches

These are mainly drawn in a manner which closely follows their mechanical function, as for example the two-way switch of Figure 1.8a and the five-way rotary switch shown in Figure 1.8b. It is conventional to indicate the moving arm by an arrow head symbol. In normally closed, and normally open forms, this may be indicated by an offset arrow head, as in Figures 1.8c and 1.8d.

Switches with ganged segments, such as the main power on/off switch for a piece of equipment, are commonly denoted by a dotted line linking the moving arms, as in Figure 1.8e.

A momentary contact switch of the kind which makes contact only when the actuating button is depressed is shown as in Figure 1.8f. One in which the operation of the switch momentarily breaks the circuit would be drawn as in Figure 1.8g.

Relays

These are a special category of switches, in which the mechanical operation of the switch arm is performed by an electromagnetic actuator driven by a wound coil. Showing the coil in physical proximity to the switch mechanism, as illustrated in Figure 1.8h – which might depict a 'reed' relay, with an air-cored operating coil – tends to clutter up the circuit diagram, so it is more common to separate these, as shown in Figure 1.8j.

The relay operating coil can then be drawn as a simple box, on which, for example, the legend $\frac{\text{'RL1'}}{2}$ denotes that relay No. 1 has two operating

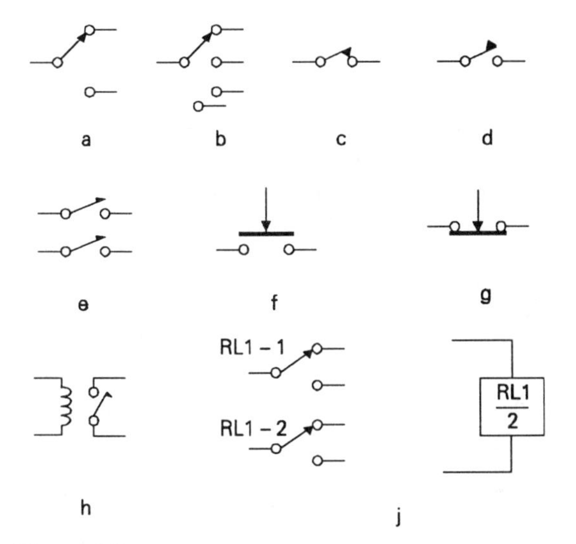

Figure 1.8

switch contacts. This then allows the switch elements which are operated by this relay to be placed anywhere convenient in the circuit drawing, and labelled 'RL1–1' and 'RL1–2' to show how they are operated.

It is good practice to append 'nc' or 'no' signs against such relay switch contacts to show they are 'normally open' or 'normally closed' – when the relay operating coil is unenergized.

Transistors and diodes

The normal semiconductor junction diode is represented by one or other of the equivalent forms shown in Figs 1.9a or 1.9b. No specific notation is used to distinguish between a low-power, 'small signal', diode and a high-power rectifier device, so the engineer must infer this distinction from the part of the circuit in which the device is placed.

'Bridge' rectifiers, or a group of diodes arranged to form a full-wave rectifying block, are conventionally shown as in Figure 1.9c. However, the diode symbols are sometimes omitted from the drawing entirely, giving the picture shown in Figure 1.9d, which relies for its interpretation on the presence of the '+', '–' and '~' (AC input) signs at the corners of the diagonal square. This may be simplified even further to the symbol of Figure 1.9e, in which a diode is shown within the square, and even the '+', '–' and '~' symbols are left out.

'Zener' or voltage regulator diodes are denoted by the symbols of Figure 1.9f or 1.9g. A similar symbol is sometimes used to denote a 'hot-carrier' or 'Schottky' type of point contact diode. A preferred symbol for this device is shown in Figure 1.9h.

Normal NPN junction transistors are drawn as shown in Figures 1.9j – 1.9m, according to the preference of the draughtsman. Enclosing the transistor symbol with a ring, to indicate the physical boundary of the device, as in Figure 1.9m, allows the envelope to be extended, as for example in Figure 1.9n, to denote a dual or multiple device mounted within the same housing. Such a symbol probably implies that both devices are fabricated on the same chip, so that both sections will have closely similar operating characteristics.

In both NPN and PNP junction transistors, the emitter lead is denoted by the line with the arrow head; pointing towards the 'base' in the case of a PNP device, and away from it in the case of the NPN version. The angled line without the arrow head denotes the collector junction. The line joining at a 'T' refers to the base of the transistor. Symbols for PNP junction transistors are depicted in Figures 1.9p, 1.9r and 1.9s.

The line or bar denoting the base area of the transistor should be blocked in. Where this is left as an open rectangle, as in Figure 1.9t, it is usually taken to mean that the device is of the 'super-Beta' type, having a very high current gain but a very low collector–emitter or base–emitter breakdown voltage. These types of transistor are normally only found in integrated circuits.

The drawings shown in Figures 1.9u or 1.9v are sometimes found as a representation of a junction transistor. This is a more logical analogy for the devices now made than the symbols of Figs 1.9j or 1.9p – which typified the early, and now obsolete, point contact devices. However, this more accurate symbol has never been widely adopted, and remains a somewhat eccentric variation.

Field effect devices

These exist as junction types, and as 'MOSFET' or 'insulated gate' (static charge operated) field effect devices. The symbols used to represent junction FETs are as shown in Figures 1.10a and 1.10b, which denote

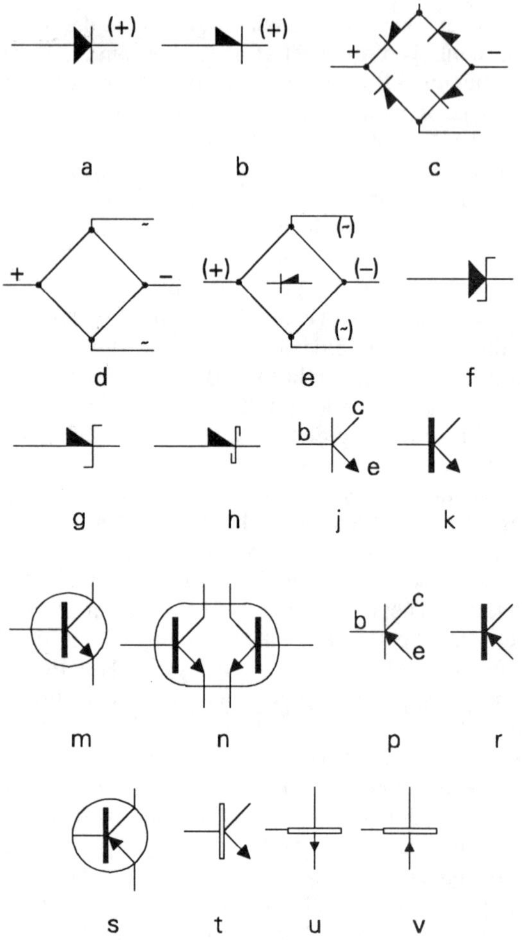

Figure 1.9

'N-channel' and 'P-channel' devices respectively. If these devices are truly symmetrical, so that 'drain' and 'source' connections are interchangeable, the gate junction will be shown centrally, as in Figures 1.10c and 1.10d.

Insulated gate devices are shown by the symbols of Figures 1.10e, (N-ch) and 1.10f, (P-ch). The distinction between an N-channel MOSFET and a P-channel one is denoted by whether the substrate junction contact has an arrow head facing towards or away from the substrate bar symbol.

Sometimes the insulated gate symbol is drawn with the connection at the opposite end of the gate region, as shown in Figure 1.10g, (P-ch), to make this distinction somewhat clearer. MOSFET devices are also available with dual gate regions as shown in Figure 1.10h.

These devices exist in two types, the 'depletion' and 'enhancement' mode forms, distinguished by whether the source/drain channel is normally conducting or whether some forward bias must be applied to the gate to cause conduction. While the 'MOSFET' symbols shown in Figures 1.10e – 1.10h will often be used interchangeably for both of these forms, the drawings of Figures 1.10j and 1.10k, where there are gaps in the channel symbol, may sometimes be used specifically to denote 'enhancement' style devices.

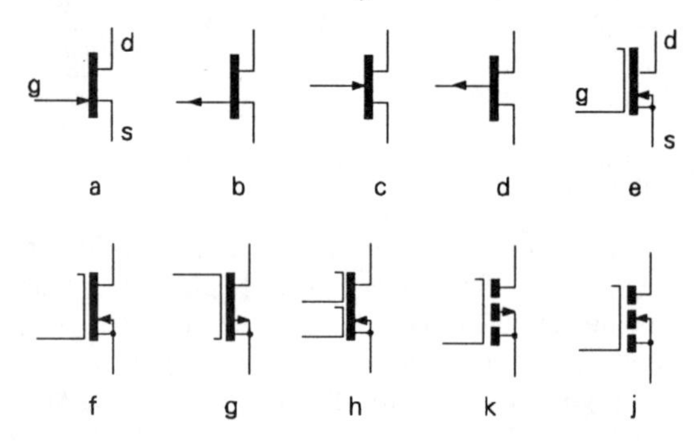

Figure 1.10

Thyristors and triacs

There are a wide range of these specialized current control elements, with a correspondingly wide variety of circuit symbols. Of these the most common are thyristors, also known as silicon controlled rectifiers, and triacs, which are a bi-directional form of thyristor. A modified diode symbol is used to represent these devices, as shown respectively in Figures 1.11a and 1.11b. The device denoted by the symbol of Figure 1.11c is effectively identical in characteristics to the thyristor, but is known as a programmable unijunction transistor.

The device shown in Figure 1.11d is a silicon controlled switch, and those of Figures 1.11e and 1.11f are different forms of 'diac' or bi-directional trigger device. The final commonly found member of this family is the unijunction transistor denoted by Figure 1.11g.

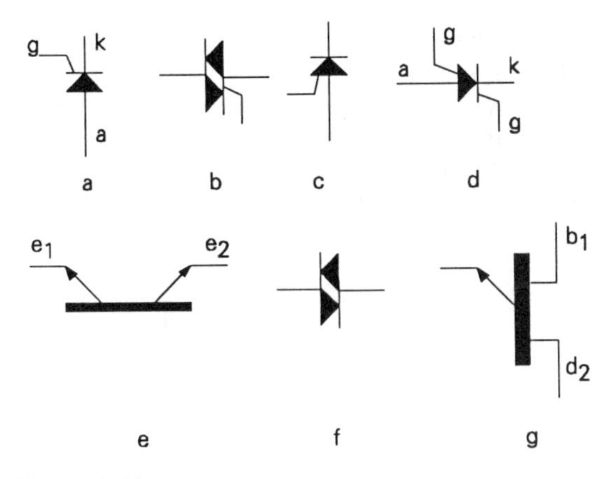

Figure 1.11

Thermionic valves and other vacuum envelope devices

These are also based on representations of the components they depict, with directly heated (filament type) and indirectly heated cathodes drawn as shown in Figures 1.12a and 1.12b. Grids are depicted either as dotted lines or as a continuous horizontal zigzag, interposed between the cathode and the anode, and 'plates' or anodes are shown as rectangular boxes or 'T' shaped symbols, again depending on whether the USA or European convention is used.

For example, in the European style of drawing, a directly heated diode valve would be shown as in Figure 1.12c. while an indirectly heated triode would be shown as in Figure 1.12d. The number of electrodes within the valve – ignoring the heater of an indirectly heated type – is categorized by adding the suffix '-ode' to a prefix based on the classic Greek numerical system, so that a diode means a two electrode valve, a triode means one with three electrodes, and so on with tetrodes, pentodes, hexodes and heptodes, illustrated in Figures 1.12e – 1.12h.

Exceptions are sometimes made to this rule by ignoring the presence of the beam confining plates of the valve shown in Figure 1.12j, which is still classed as a beam-tetrode. Note also that the third or 'suppressor' grid of a pentode, or the beam confining plates of a beam-tetrode, may frequently be internally connected to the cathode. This connection may not always be shown in the circuit symbol.

Combination valves, such as double diode triodes and triode hexodes, are also made. Symbols for these are shown in Figures 1.12k and 1.12m. A valve type which has not been superseded is the cathode ray tube. In this, the electron beam can be focused and deflected either electromagnetically, by means of externally mounted coils, or electrostatically, by the use of internal electrode structures. These differing types give rise to the two different symbols shown in Figures 1.12n and 1.12p.

Some 'cold cathode' or gas-filled valves were made, such as the '0Z4' rectifier or the '0A3' series of voltage stabilizer tubes, shown in Figures 1.12q and 1.12r. In these the presence of a deliberately introduced gas filling is sometimes, but not invariably, denoted by the insertion of a small dot within the envelope symbol.

Some ambiguity exists in the representation of the stabilizer valve, shown in Figure 1.12r, in that this is sometimes drawn inverted, as in Figure 1.12s, with the curved plate as the cathode – since this reflects more truly the actual structure of the component.

Valve symbols for indirectly heated valves are quite frequently drawn with a conventionally shaped cathode, but without showing the heater element at all, as in Figure 1.12t. This omission should not be taken to imply that these are gas filled or cold- cathode devices.

Linear integrated circuits (ICs)

There is an enormous and growing range of these components, of which most will be depicted in a circuit diagram simply as a rectangular box carrying some descriptive label, or perhaps just a type number. A few, however, have become sufficiently widely used that they have acquired their own distinctive symbols. Of these the most common is the 'operational amplifier' IC gain block, shown in Figure 1.13a.

This is normally shown as a triangle, with two inputs on the vertical face, and an output at its vertex. The 'non-inverting' and 'inverting' inputs are indicated by '+' and '–' signs attached to the inputs, within the triangular symbol. Most commonly, the presence of supply connections, from, for example, a pair of regulated +/–15V supply lines is simply assumed, and not shown at all.

Where these are shown, they may often merely be truncated lines terminated by '+' or '–' signs, or by conventional symbols denoting these, as shown in

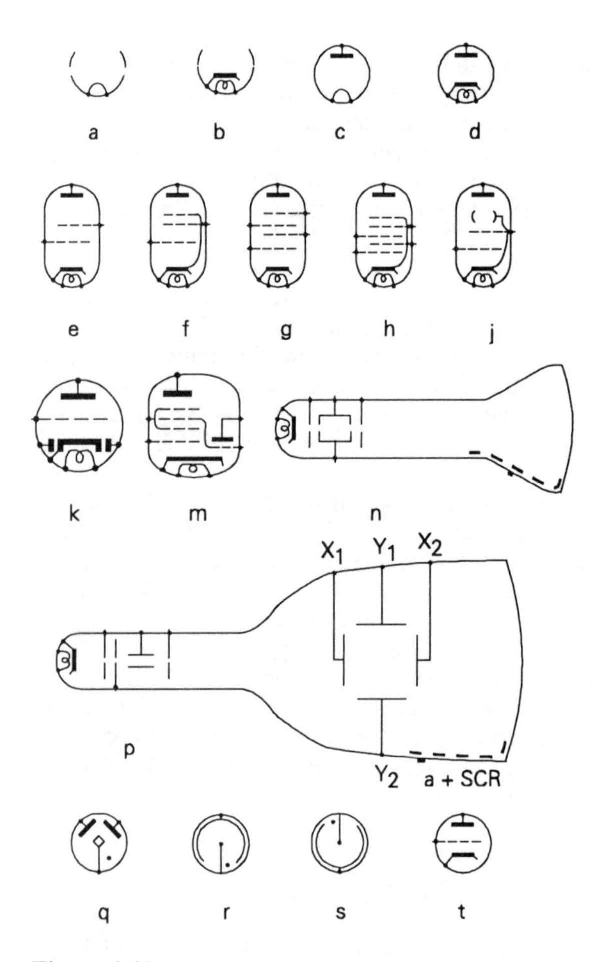

Figure 1.12

Figures 1.13b and 1.13c. Where an external control is used to adjust the output DC offset voltage of an op. amp, this will be depicted as shown in Figure 1.13d. In some RF amplifier gain blocks a pair of antiphase outputs may be provided. This is shown in Figure 1.13e.

A special category of op. amp. ICs, sometimes called 'Norton' amplifiers, is that in which a non-inverting input is added to an inverting gain block by the use of an input inverting circuit, such as a current mirror. To distinguish this type of circuit from a true op. amp., in which both of the inputs would have comparable input impedances and other characteristics, the symbol of Figure 1.13f is used.

Voltage regulator ICs, though very widely used, are still just drawn as rectangular blocks, as in Figure 1.13g. Constant current sources, used both within IC gain blocks, and available as separate ICs, are depicted by the equivalent symbols of Figures 1.13h, and 1.13j.

Current mirrors, or constant current sources whose output current is equal to, or is controlled by, some input reference current, are represented by Figure 1.13k.

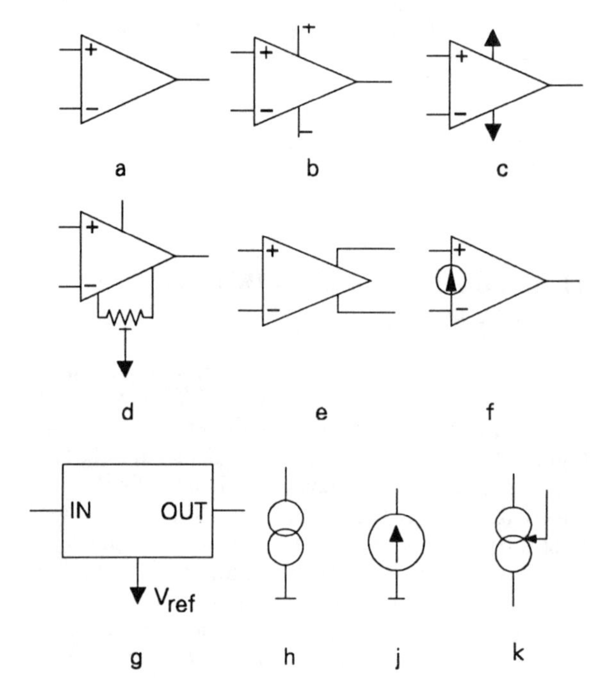

Figure 1.13

Logic ICs

Though digital circuit functions are not within the scope of this book, these devices may often be used as parts of linear circuit designs, and a number of recognized symbols have evolved.

The diagram of Figure 1.14a is that for a simple single-input buffer, with the input and output conventions the same as those for an operational amplifier, i.e. input on the vertical face of the triangular symbol and output at its apex. The small loop at the output point of Figure 1.14b denotes phase inversion, so that this device would be an inverting buffer stage. The internal gain will usually be greater than unity, so these buffers can also be used as low performance linear amplifying stages.

The symbols shown in Figures 1.14c, 1.14d and 1.14e, are 'AND' gates, in which an output requires that appropriate signals are present at all inputs. Two, three and four input AND devices, of the type illustrated, are common. As before, the presence of a small loop on the output implies phase inversion, so that the device of Figure 1.14f will be a 'two input NAND'. Similarly, a loop on an input connection, as shown in Figure 1.14g, would imply a phase inversion of the input signal at this point.

The symbols shown in Figures 1.14h and 1.14j are two and three input 'OR' gates, and that of Figure 1.14k is a two input 'NOR', by the same convention as above. In 'OR' gates, as the name implies, an output will be given if a signal is present at either input. A special case of the 'OR' gate is the 'exclusive OR', shown in Figure 1.14m. This is a device in which an output will be given if a signal is present on either input, but not if it is present on both.

Other logic ICs are usually represented by rectangular blocks, with appropriate symbol markings shown to indicate the connections.

As mentioned above, the DC supply connections to 'logic' ICs, which are typically as shown in Figure 1.14n, are often omitted from the circuit drawing in the interests of simplicity.

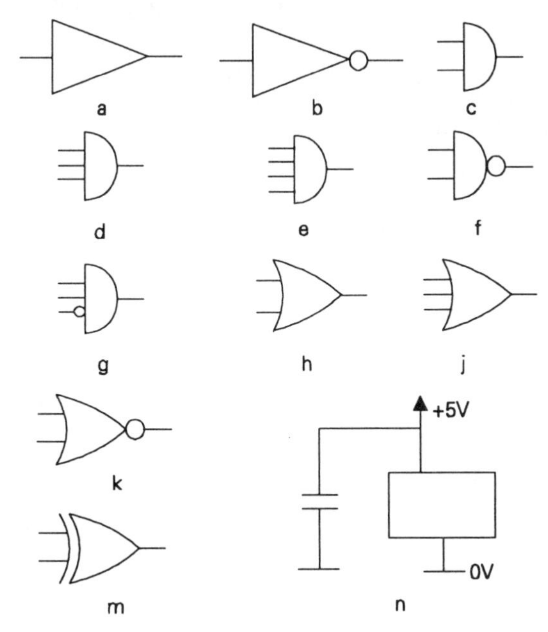

Figure 1.14

Light operated and light emitting devices

This type of function is depicted by the use of a pair of parallel arrows appended to the symbol. If light is emitted, the arrows will point away from the device, as in the light emitting diode of Figure 1.15a. If light can modify the characteristics of the device, the arrows will be shown pointing inwards, as in the wide family of light sensitive or light actuated components such as the photo diode, photo transistor, photo resistor, or light actuated SCR shown in Figures 1.15b – 1.15e.

A different convention is adopted for heat sensitive devices, such as the thermistor, where this is shown by a dot alongside the resistor element as in Figure 1.15f.

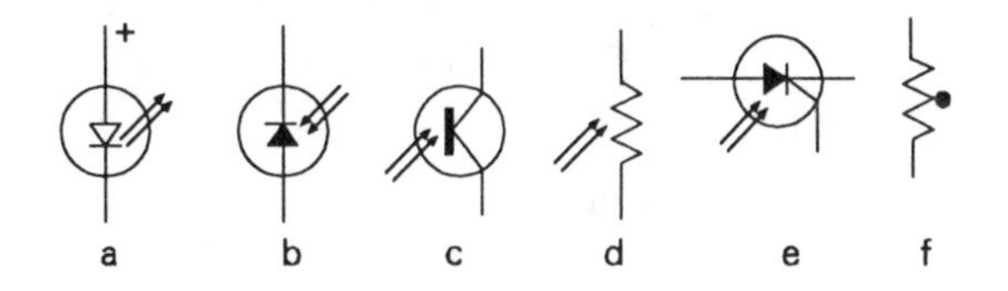

Figure 1.15

Miscellaneous symbols

A DC voltage source, such as a dry cell battery, is drawn as shown in Figures 1.16a and 1.16b, where Figure 1.16a represents a single cell, or perhaps a notional low voltage source, and 16b a multiple cell battery, or a symbolic high voltage DC source. An AC source, of any potential, or an AC signal source, is denoted by Figure 1.16c.

The normal drawing for a quartz crystal resonator is that of Figure 1.16d or 1.16e. The former is preferable because of the possible confusion of the symbol of Figure 1.16e with the Japanese symbol for an electrolytic capacitor. That for a 'surface acoustic wave' filter is as shown in Figure 1.16f. Another common symbol, found in radio circuits, is that of Figure 1.16g, which denotes a 'mixer' or 'modulator' stage. The symbol used in radio circuitry for the aerial connection is as shown in Figure 1.16h, where this is a single wire type. A 'dipole' aerial is shown in Figure 1.16j.

A convenient shorthand way of representing circuit elements – such as a group of coils and capacitors, but possibly made up from other component types – which are used to remove a section of the signal frequency spectrum, known collectively as 'filters', is by the use of the symbols shown in Figures 1.16k, 1.16m, and

1.16n, or 1.16k', 1.16m' and 1.16n', which represent 'low-pass', 'bandpass' and 'high-pass' filters respectively.

The way in which a screened cable is represented is shown in Figure 1.16p, and a coaxial or similar connector as shown in Figure 1.16q. A 'jack' socket is shown in Figure 1.16r.

Output devices such as headphones and moving-coil loudspeaker units are shown in Figure 1.16s and Figure 1.16t respectively. The normal symbol for a microphone is as shown in Figure 1.16u and that for a tape recorder 'pick-up', 'record' or 'erase' coil is as shown in Figure 1.16v. The symbol used for a fuse is as shown in Figure 1.16w.

Various drawings have been used to denote a filament lamp bulb, but I prefer that of Figure 1.16x. A neon or similar gas discharge lamp is shown as in Figure 1.16y, and a 'fluorescent' tube as in Figure 1.16z. A current or voltage indicating meter is represented as in Figure 1.16aa. Where this is part of a more elaborate instrument such as a millivoltmeter or distortion meter, the meter symbol will be enclosed within an appropriately labelled box, as in Figure 1.16ab.

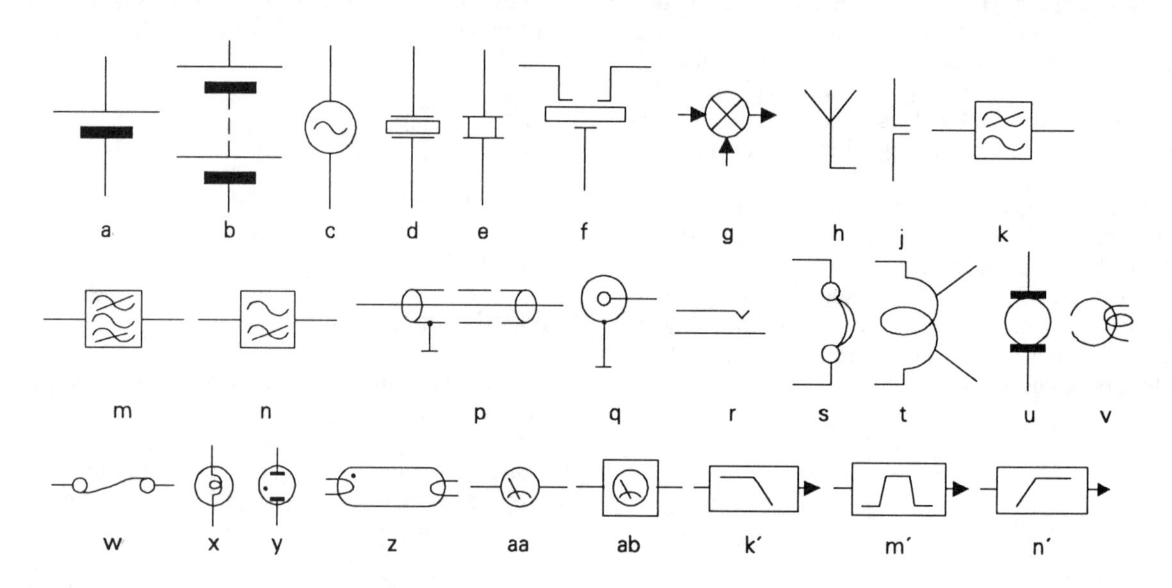

Figure 1.16

2

Passive components

For convenience, the various components used in electronics are usually divided into two descriptive categories, 'active' and 'passive', mainly depending on whether any additional energy is introduced into the circuit by their action. Active components include thermionic valves, together with operational amplifiers, transistors, diodes, and other semiconductor devices, leaving the term passive to describe all the remaining elements of the circuit.

Building any electronic circuit will nearly always require the interconnection of a variety of both these kinds of components, and the way they interrelate will affect the functioning of the design. So, although passive components may seem relatively unexciting in their characteristics the correct choice of these can be important in achieving the intended circuit performance.

Resistors

The purpose of these components is to produce a voltage drop ($V = IR$) which is proportional to the current flow through the device. Ideally, the resistance value of such a component, quoted in ohms, should be constant, independent of time, temperature, current flow or voltage across it – so long as the resultant thermal dissipation ($W = VI$) is within the rating of the device – or the frequency of the applied current.

Most modern components approach fairly closely to these requirements, but earlier, rather less satisfactory component types may still be found in existing equipment. The broad categories are:

Wirewound

These were the earliest form of resistor, and are still widely used where relatively high power dissipation resistor types are needed. As the name suggests, these resistors are made by winding a number of turns of some suitable resistive wire, in a spiral, on a convenient hard, insulating substrate – normally a ceramic rod or tube – in the general form shown in Figure 2.1.

The wire will nearly always be made from a metallic alloy, chosen for its independence of resistance value

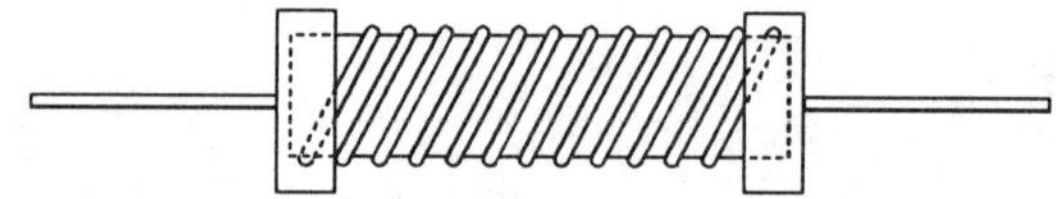

Figure 2.1 *Basic construction of wirewound resistor*

as a function of temperature, over the normal working temperature range: some typical alloys are listed in Table 2.1. Since these alloys are usually difficult to solder, tinned copper wires (or terminals, in the case of high power components) are usually either welded or crimped on to the wirewound element, to assist in connecting the resistor into the circuit.

Table 1.1 *Resistive characteristics of metals and alloys*

Metal or alloy	*Relative resistivity (copper = 1)*	*Temperature coefficient (ppm/°C)*
Aluminium	1.63	+0.0004
Copper	1.0	+0.00039
Constantan (= Eureka)	28.45	+/-0.000022
Karma	77.1	+/-0.000002
Manganin	26.2	+/-0.000002
Nichrome	65.0	+/-0.000017
Silver	0.94	+0.0004

The resistance of any conducting wire is proportional to its length and its specific resistivity, ρ, and inversely proportional to its cross-sectional area. This leads to the equation

$$R=\frac{\rho L}{A} \qquad (1)$$

If the length, L, and the area, A, are measured in metres and square metres, the resistivity, ρ, will have the form ohm–metres.

The resistive characteristics of aluminium, copper and silver are included for reference, but these materials will only be used where a low resistance is necessary. The other materials listed are designed specifically for use in wirewound and other resistor applications.

Since, in a practical wirewound resistor, it is desirable that the resistor should have a stable resistance value, one of the low temperature coefficient alloys will be chosen for its construction, such as 'Constantan' or 'Karma'. Also, since it is preferable not to have to wind more turns of wire on the former than one must, a material with a high value of specific resistivity is also desirable.

A low temperature coefficient of resistivity will usually be associated with a low coefficient of expansion, and the thermal characteristics of the 'former' (the rod on which the resistor is wound) will be matched to that of the wire, so that it neither becomes slack on heating, nor stretched – which would happen if the former expanded more than the wire – since this would bring about a permanent change in the resistance of the wire.

The winding will usually be anchored in place with a coating of some hard, heat and moisture resistant, insulating material, except in the case of variable resistors or potentiometers, where an uncoated track must be left for contact with the slider.

A simple, multiple turn, wirewound resistor will also offer a measure of inductance, and where it is important that this should be kept to a low value, 'non-inductive' wirewound resistors are made by periodically reversing the direction of winding. However, most commercial wirewound resistor types are inductive, and the circuit designer must allow for this characteristic.

Where a component having a high power dissipation, but a low self-inductance is required, it will probably be necessary to find some other resistor type, such as one of 'metal glaze' construction.

Carbon composition

Another early form of resistor, illustrated in Figures 2.2 and 2.3, is made by baking an appropriately chosen blend of powdered graphite and clay to form a solid rod having a suitable value of resistance. Connections to this rod are made by spraying or dipping the ends of the rod, to give a solderable metallic coating to

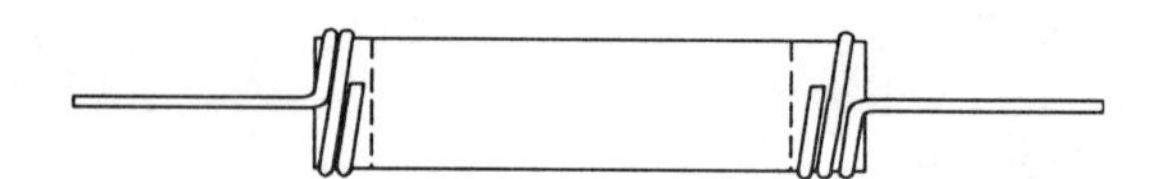

Figure 2.2 *Early form of carbon composition rod resistor*

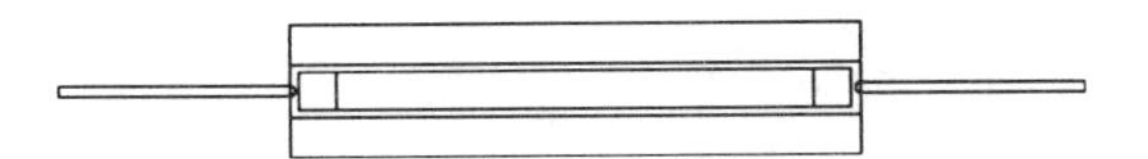

Figure 2.3 *Improved form of carbon composition rod resistor*

which wires can be attached.

The whole rod will then, typically, be impregnated with wax in an attempt to prevent the ingress of moisture which will alter its resistance value, or worsen its already poor long-term stability of value.

Since the only way of controlling the resistance value of the component is by means of its physical dimensions and the composition of the mix, prior to firing, the final value of the resistor will be a matter of guesswork on the part of the manufacturer. This means that the resistors must be sorted, after firing, end-cap metallization and impregnation, in order to select the required values.

In practice, this means that a given value, +/–10%, will be just that, since all those which have a closer approach to the required value will have already been selected out.

In the better forms of this component, the resistive rod will be enclosed in a ceramic insulating tube, to prevent inadvertent short circuits with adjacent components.

As noted above, all resistors of this general type have a poor stability of resistance value, which will change somewhat when they are soldered into a circuit, or if they become hot in use. They also have a relatively high 'excess' noise figure. Fortunately they have now been almost completely replaced by better types.

Carbon film

This is now a very common constructional method for small wattage (0.125 – 2W) resistors, and is based on the use of a ceramic rod on which a hard and durable surface film of carbon has been deposited, in the manner shown in Figure 2.4a. The initial range of values for such a resistor can be controlled by the choice of its physical dimensions, and the thickness of the carbon film.

Resistance values higher than that initially given by the basic manufacturing process can then be obtained by grinding a coarse spiral gap through this film layer,

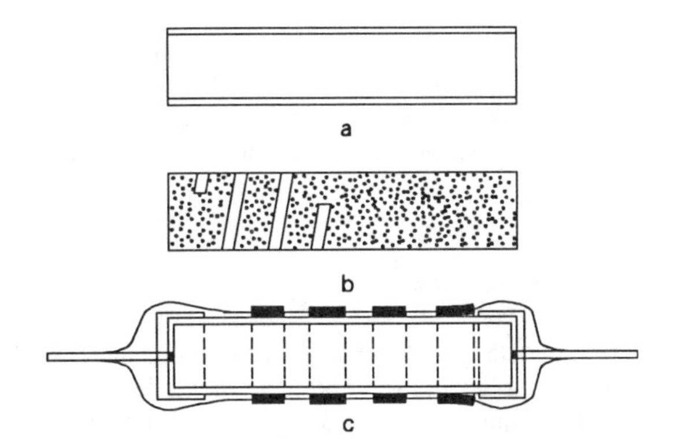

Figure 2.4 *Stages in construction of a tracked carbon film resistor*

as shown in Figure 2.4b. Since it is normal practice to monitor the resistance value of the component while this track is being cut, the accuracy of the final value will be determined mainly by the precision of setting of the track cutting apparatus.

External connections are made by crimping end caps on to the ends of the rod, and connecting wires are normally welded on to these caps. The whole unit is then given a hard protective coating, prior to painting with a colour coded pattern to denote its resistance value, to produce a component of the type shown in Figure 2.4c.

This type of resistor has excellent long term and thermal stability characteristics, and normally offers a high degree of precision in respect of its nominal value. It is, however, being increasingly replaced in the manufacturers' catalogues by resistors having an identical physical structure, but with the resistive film made from a thin vacuum evaporated nickel chromium alloy, known collectively as 'metal film resistors'.

Metal oxide film resistors

These are similar in form to the carbon film types, but with the carbon layer on the ceramic former replaced by a thin layer of tin oxide. This can be fused on to a borosilicate glass or ceramic rod substrate to give a robust and tenacious coating, having very good electrical characteristics.

Since it is somewhat more expensive to make this type of resistor than those based on a metal film layer, these metal oxide resistor types are now becoming rarer, though they are still used where certain military specifications demand very high reliability, since the relatively thick tin oxide layer is more resistant to impact damage or corrosion under high humidity conditions.

Metal film resistors

The past two decades have seen an enormous increase in the use of vacuum evaporated metal films, for applications ranging from packaging materials to integrated circuits, and this growth of technical competence has greatly reduced the cost of this type of process, so that it is now not significantly more expensive to make a metal film resistor than a carbon film one.

Such metal film components have a lower temperature coefficient of resistance (+/– 50 parts per million, per degree celsius), in comparison with the carbon film types (+/– 150ppm), as well as a slightly lower 'excess' noise figure and a rather better ability to survive short term overloads. Since they are now available at a comparable price to their carbon film equivalents, these have now become the most popular low-power resistor type in all new electronic apparatus.

These resistors are available in power ratings from 0.125W – 1W.

Metal glaze resistors

These components are similar to the metal oxide types, except that the resistive coating on the ceramic core is formed from a relatively thick fused layer of glass and metallic powders. The increased thermal inertia of this layer gives the component a better ability to withstand surge conditions, or short term overloads than the metal film types. Typical thermal coefficients for this design lie in the range +/–150 to +/–250ppm. Such resistors are available in power ratings from 0.125W to 3W.

These resistor types are also available as groups of resistors, either entirely separate or with one end of each connected to a common line, mounted together in a single encapsulation, and with their connections brought out in an identical manner to the 'DIL' (dual in-line) integrated circuit package, or the 'SIL' (single in-line) equivalents shown in Figure 2.5.

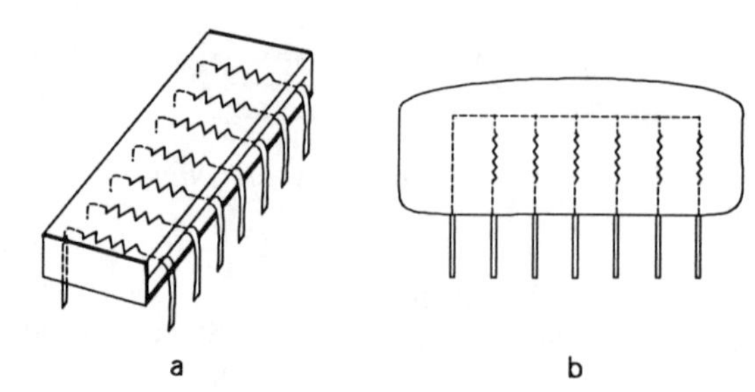

Figure 2.5 *Two of the available forms of DIL and SIL thick-film resistor blocks*

Temperature dependent resistors

These devices are normally called 'thermistors' and are designed so that they will have a very high thermal coefficient of resistance, either positive or negative, for use as temperature sensing elements, or in protective or power control applications.

The earliest forms of negative temperature coefficient devices consisted of fired pellets of clay containing various proportions of silicon carbide, but these components have been the subject of intensive commercial development and a wide range of intermetallic compounds, such as nickel manganite, is now employed, to provide specific electrical characteristics.

Of the negative temperature coefficient (NTC) types, the most common forms are the rod, disc and bead types, together with the vacuum envelope encapsulated bead types used for amplitude control in oscillators and the miniature probe-mounted devices used for temperature probes. These are illustrated in Figure 2.6.

The relationship between resistance and temperature of these devices is shown in Figure 2.7, for some typical NTC thermistors.

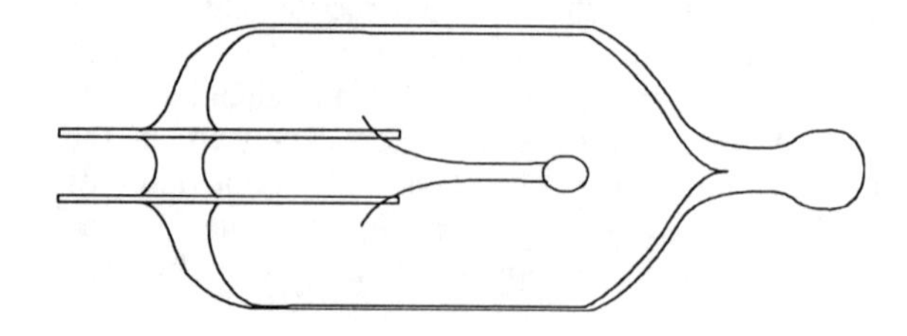

Figure 2.6 *Vacuum envelope miniature bead thermistor*

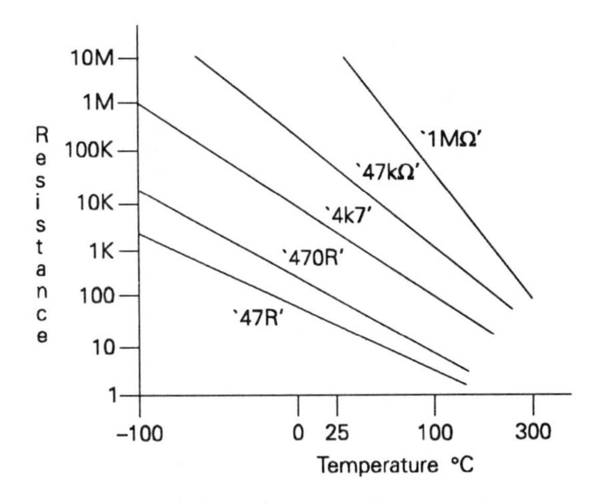

Figure 2.7 *Family of resistance–temperature curves for thermistors of various types and compositions*

Positive temperature coefficient (PTC) components are generally designed so that they have a low value of resistance up to some critical turn-over (or reference) temperature, beyond which the resistance will increase rapidly. They are made from a sintered mixture of lead, barium and strontium titanates, and the turn-over temperature is controlled by adjustment to the composition of the mix, and the firing conditions.

Their physical form is similar to that of the NTC disc thermistor types.

A major application for these PTC devices is in over-temperature protection for components such as transformers and motors – by incorporating the device within the physical structure of the component to be protected, and connecting it, electrically, in series with it, as shown in Figure 2.8. Such PTC devices can also be used as current limiting components.

The relationship between resistance and temperature and current for a given applied voltage is shown in Figures 2.9 and 2.10.

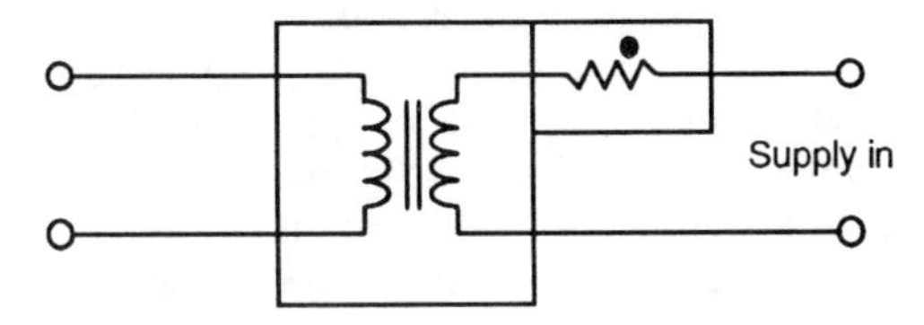

Figure 2.8 *Overload protection by means of a positive temperature coefficient thermistor*

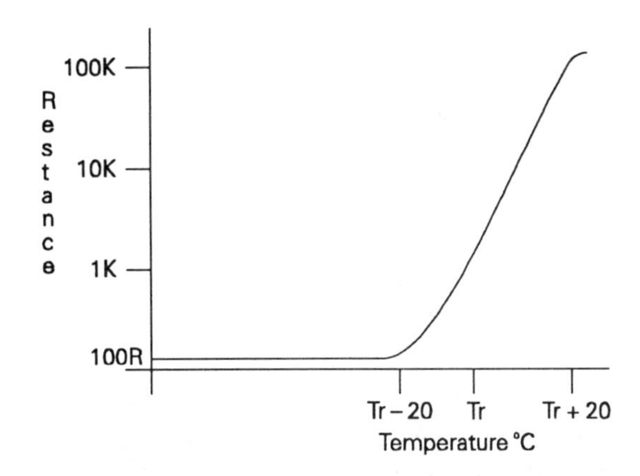

Figure 2.9 *Resistance vs. temperature for a PTC thermistor*

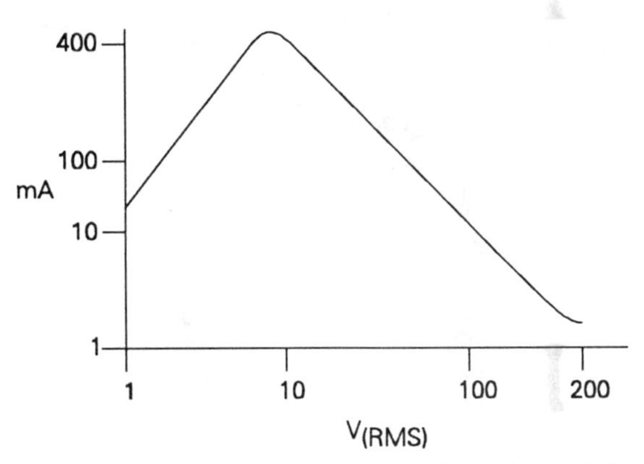

Figure 2.10 *Current vs. applied voltage for a PTC thermistor*

A similar, but rather more restricted, range of positive resistance coefficient, as a function of current, is shown by a vacuum envelope tungsten filament lamp bulb, and this effect can sometimes be employed to provide an inexpensive power or voltage control system.

Variable resistors and potentiometers

The variable resistor is, as its name suggests, a two terminal device, while the 'potentiometer' or potential divider is a three terminal one. These have the circuit forms shown in Figure 2.11a and 2.11b. Because of its greater versatility (in that by ignoring one terminal, a potentiometer can always be used as a variable resis-

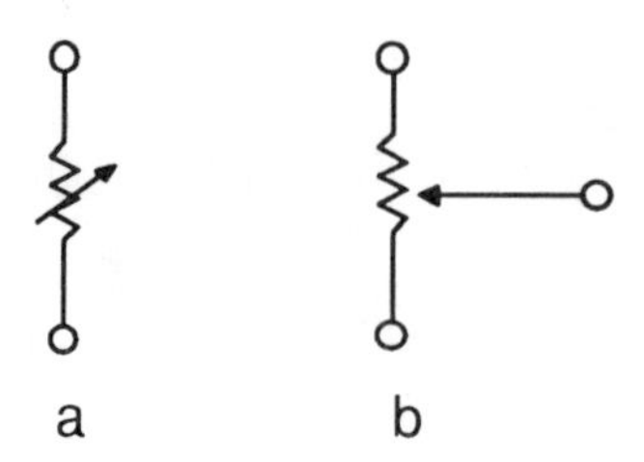

Figure 2.11 *Variable resistor and potentiometer*

tor) the potentiometer is by far the most common form found in electronics.

These components are available in an exceedingly wide range of physical forms, tailored to various practical applications, and a wide range of constructional types, of which the most common is the semi-rotary design, with a part circular track, as shown in Figure 2.12.

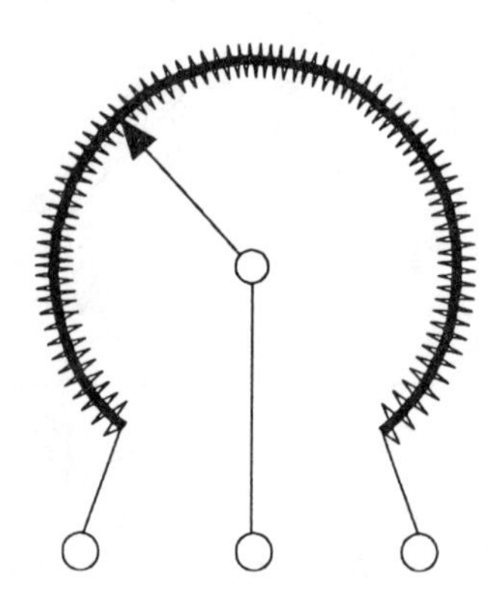

Figure 2.12 *Schematic construction of potentiometer*

The cheapest of these use carbon film tracks, with the sliding contact taking the form of a short graphite rod, to minimize surface wear on the track, since this will lead to electrical noise in operation, and ultimately terminate the useful life of the component.

Better quality potentiometers use conductive plastics, or 'cermet' (i.e. metal glaze thick film) tracks, in the interests of lower noise and surface wear rates, or wirewound construction, with the resistive wire wound on a thin former, as shown in Figure 2.13, to minimize the cross-sectional area of the core, and consequently the winding inductance.

Permissible thermal dissipations will range from 0.2W for a small printed circuit board mounted pre-set

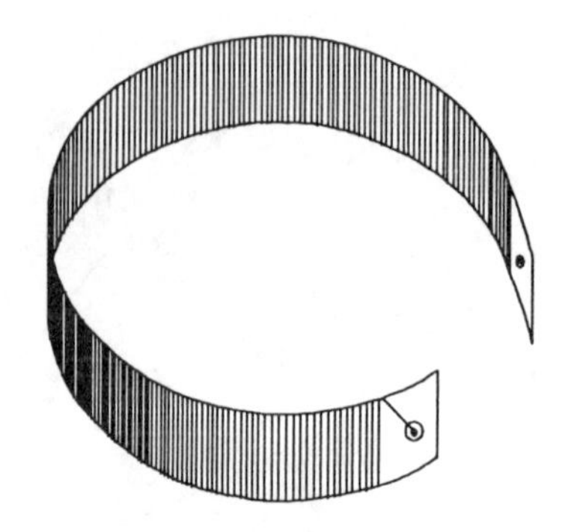

Figure 2.13 *Resistance element for a wirewound potentiometer*

component, up to many tens of watts for a large power control unit. In normal small-signal electronic circuitry, the maximum power ratings for a panel mounted control potentiometer is likely to lie in the range 2–3W. For longevity, it is desirable to keep the total dissipations, and, particularly, the current flow drawn from the slider, to as low a value as practicable.

Two normal 'laws' are used to specify the relationship between the rotation of the shaft of a rotary potentiometer and the resistance between the slider and the beginning of the track.

These are 'linear', in which there is a linear relationship between rotation and resistance value, and 'log' or 'audio', in which the resistance characteristics of the track are chosen so that there will be an approximately constant relationship between the rotation of the control and the sound output level from an audio amplifier in which the potentiometer is used as a 'volume' or 'gain' control.

It is notoriously difficult to produce potentiometers, with such non-linear or graded tracks, so that the actual resistance for a given degree of shaft rotation is the same from one supposedly identical unit to another.

This leads to the problem that, in audio amplifiers, whose gain is controlled by a twin-gang potentiometer, the balance between the output sound levels of a 'stereo' system may vary as the gain control is altered. In high quality audio equipment, the control of output sound level may therefore be achieved by the use of a multiple position switch, between whose contacts selected values of fixed resistor are wired.

Spurious effects

Unwanted characteristics are not likely to be of major importance in the case of these components. Although all resistors are likely to exhibit some degree of 'parasitic' series inductance, this can generally be ignored, except in the case of the larger value wirewound types, or in very high frequency (VHF) applications, where the inductance of the connecting wires must also be considered. There will also be an element of parallel capacitance, but this, again, is unlikely to be troublesome except at VHF.

All resistors will introduce some thermal noise, as a direct function of their resistance value, and this is defined by the equation

$$V = \sqrt{4kT\Delta fR}$$

where V is the mean noise voltage, k is Boltzmann's constant (1.38×10^{-23}), T is the absolute temperature (° Kelvin), R is the resistance value, and Δf is the effective bandwidth of the circuit in which the noise is measured. This means, for example, that a 2k7 resistor would introduce a mean noise voltage of 0.95μV, RMS, into an audio circuit with a measurement bandwidth of 20kHz.

A group of noise voltage curves is shown in Figure 2.14

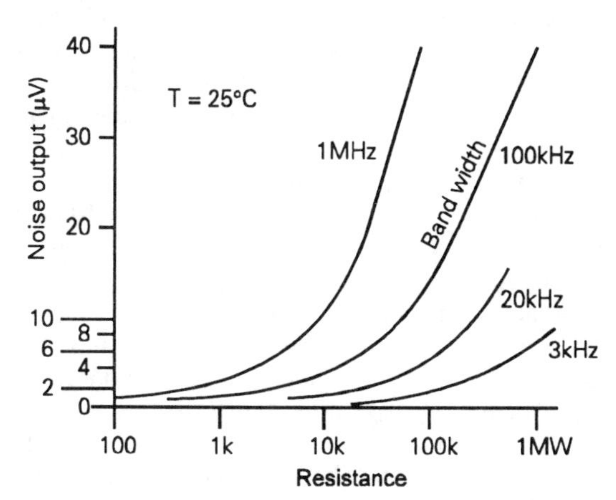

Figure 2.14 *Resistor thermal noise for various measurement bandwidths*

However, in addition to this unavoidable contribution to the circuit noise, purely due to the resistance of the component, there is also the phenomenon of 'excess' noise. This is a function of the way the resistor is made, and of the composition of the materials used. It is due to a variety of causes, including the trapping of electrons by 'holes' introduced by impurities and random electro-chemical potentials. Noise can also be introduced by piezo-electric and tribo-electric effects.

Resistors can also suffer from voltage dependence of resistance value – quite apart from any thermal effects due to current flow, and consequent heating – and also, sometimes, from asymmetry of resistance value, due to imperfections at points of mechanical contact. The better quality modern components are much less prone to these problems than earlier carbon composition types.

Capacitors

If two conducting bodies are brought into proximity with one another, a capacitance will exist between them. This has the effect that, if a potential is applied to one of these bodies, it will cause an electronic current flow into, or out of, the other body until a state of equality of charge, of opposite potentials, is established on both of these bodies, as shown in Figure 2.15.

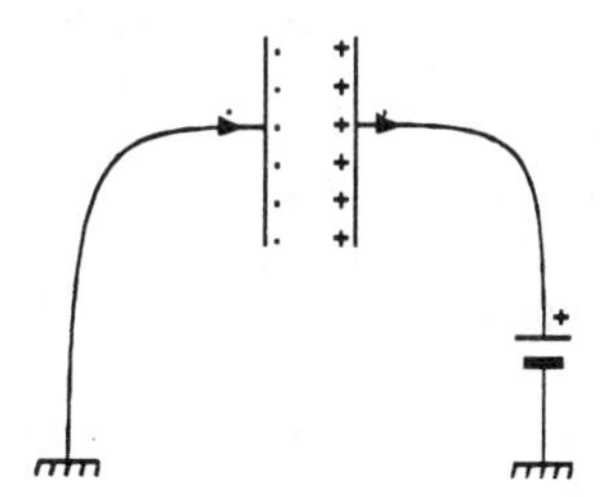

Figure 2.15 *Charge separation between plates of a capacitor*

This has the useful effect that a change in potential may be communicated from one conductor to another without the need for a direct connection between them. For any given applied potential, the quantity of current which will flow in a circuit containing two such separated conductors will depend on the capacitance between them, and this is dependent on their relative areas, their proximity to one another, and the characteristics of the material which separates them.

The presence of any material between the conducting bodies will always have the effect of increasing the capacitance between them, by comparison with that which would exist if the conductors were separated by a vacuum, and this characteristic of the material interposed between the conductors is called the 'dielectric constant' of the material, and is conventionally denoted by the symbol 'k'.

The capacitance of any arrangement of conducting plates separated by an insulating (dielectric) layer is defined by the equation

$$C = M\frac{kA}{d} \qquad (3)$$

where C is the capacitance value (specified in farads or some fraction of a farad), M is some constant, related to the physical form of the capacitor, and the units in which the dimensions are specified, A is the plate area, k is the dielectric constant of the material from which the insulating layer is formed, and d is its thickness.

For a practical capacitor, neglecting end-effects, this formula can be re-written as $C = kA/11.32d$ (pF), for dimensions in centimetres.

From this equation it can be seen that the capacitance of a capacitor increases in direct proportion to the area of the plates, and the dielectric constant of the insulating layer, and in inverse proportion to the thickness of the insulating material between the conductive plates.

On the other hand, the maximum voltage which such a capacitor will withstand is proportional to the thickness of the insulator and to its dielectric strength.

It will be seen from equation (3) that the dielectric constant, otherwise known as the 'specific dielectric permittivity', of the insulating material has a direct effect on the capacitance of the arrangement, and is as important as the thickness of the insulation in this respect.

As a reference standard, the dielectric constant of a vacuum is given a value of unity. Most gases, such as air, also have values which are very close to unity. (For example, dry air has a value of $k = 1.0006$). All solid and liquid materials will have dielectric constant values which are greater than this, and the characteristics of some of the more common dielectric materials are given in Table 2.2.

The dielectric constant should always be quoted at some specific frequency, since in all dielectric materials this property decreases in value as the applied frequency is increased. The k value for all materials

Table 2.2 *Dielectric properties of common insulating materials*

Material	*Dielectric constant (at 1kHz and 20^0 C)*	*Breakdown strength (volts/mil)*
Ceramics (depending on type)	5 – 50,000	100 – 1000
Glass	7 – 8	500 – 2000
Mica	5.5 – 8	600 – 1500
Paper (dry)	4.5 – 4.7	500 – 1000
Polycarbonate	2.8 – 3.0	750 – 1200
Polyester	2.8 – 3.2	2000 – 4000
Polyethylene	2.25	1000
Polypropylene (oriented)	2.20	1500
Polystyrene	2.5	800 – 1000

will also depend on their exact composition and method of manufacture, and will change somewhat with temperature, or applied voltage.

Additionally, all dielectric materials, other than a vacuum, will experience some inter-molecular electronic re-arrangement under the influence of the applied electric field, and this leads to some absorption of energy from the system. This is proportional to frequency, in the case of an alternating electric field, and is known as the 'dielectric loss' or as 'tan δ', where δ is the alteration in phase angle of the resultant current through the component, under AC conditions, due to the reactive and resistive (lossy) components of the current. The effects of this will be examined later.

For practical use, in electronic circuitry, capacitors are made by a variety of methods, described below, with the aim of producing reliable, robust, compact components with precise and reproducible capacitance values, and at a competitive cost.

These can be divided into three broad categories, 'polar' or 'electrolytic', 'non-polar' and variable.

In electrolytic capacitors, the insulating film between the two conductors is formed by electro-chemical action between a conducting electrolyte, and one or a pair of metallic bodies. This can result in a very high capacitance value, but such a capacitor will be sensitive to the polarity of any potential applied across it.

Non-polar capacitors are made from metallic foils or surface coatings, separated by some suitable insulating dielectric, which will normally be either a thin plastics film, a thin layer of mica or a fired ceramic moulding.

Variable capacitors are most commonly built with adjustable metallic plates, mounted parallel to one another, which can be moved into or out of mesh, while preserving an air gap between them. This allows the area of the plates opposite to each other to be continuously varied, which alters the capacitance between them.

Electrolytic

The most common type of electrolytic capacitor is that made from a spiral of aluminium foil, with the turns separated by a layer of some absorbent material, such as cloth or paper, impregnated with some proprietary solution or gel, the whole assembly being enclosed in an aluminium container to give a structure of the general form shown in Figure 2.16.

On the initial passage of current through the capacitor, a thin, continuous (anodized) film of aluminium oxide, which is a good insulator, will form electrolytically over the whole exposed surface of the positively connected foil, as shown in Figure 2.17, and the current flow will fall to a near zero level.

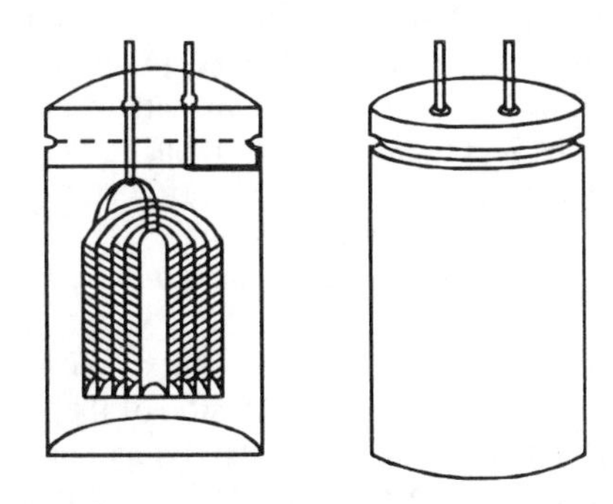

Figure 2.16 *Typical can-type electrolytic capacitor (spiral wound foil/separator)*

This process is known as 'forming', and the voltage applied at this stage determines the maximum voltage at which the capacitor may be operated. It may be done, during assembly, using the electrolyte within the capacitor, or the oxide film may be formed, prior to assembly, using a boric or phosphoric acid anodizing bath.

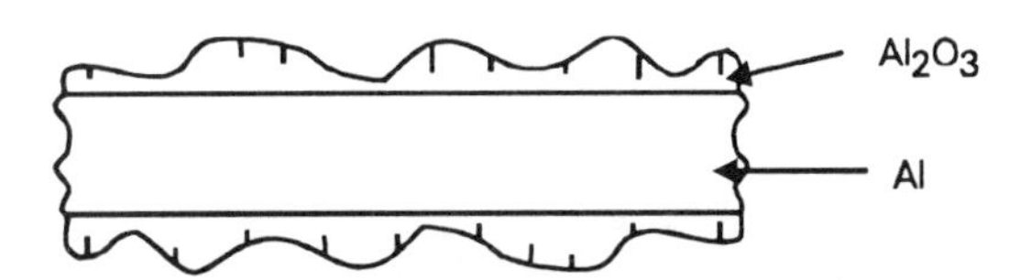

Figure 2.17 *Section through anodized oxide layer on aluminium foil*

Any residual perforations in this oxide film will be automatically healed, by further electrolytic action, during use, and because the anodized oxide film is very thin, such capacitors have a very high capacitance for a given volume.

The capacitance of the units may be increased still further by increasing the effective surface area of the foils by etching the anode foil prior to forming, though such 'etched foil' electrolytics have somewhat lower peak charge/discharge current capability than the plain foil ones.

In early capacitors of this type, the electrolyte, which forms the cathode of the capacitor, would be a water/glycol mix, containing an ionic compound such as boric acid, to increase its conductivity and viscosity, and promote self-healing anodization.

However, internal corrosion could be a problem, so more modern types use organic electrolytes, such as di-methyl acetamide, methyl formamide, or butyrolactone. Some additional gelling agent may also be included to lessen the possibility of leakage from the impregnated paper separators.

Aluminium electrolytic capacitors are available in capacitance ranges from 0.47μF – 200,000μF, and voltage ratings from 2.2V – 470V, though the larger capacitance values will generally only be available at the lower end of the working voltage range.

Such capacitors are widely used where high values of capacitance are needed, such as in supply line decoupling, and power supply applications. Unlike non-polar capacitors, there will always be some leakage current through an electrolytic capacitor, and this will vary with the applied voltage in the manner shown in Figure 2.18.

Recent technology has also led to the development of small 'solid dielectric' aluminium 'bead' capacitors, known as 'SAL', based on a manganese dioxide 'solid electrolyte'. These are very similar in characteristics to their tantalum bead equivalents, but are much less expensive. Their physical construction is shown in Figure 2.19a.

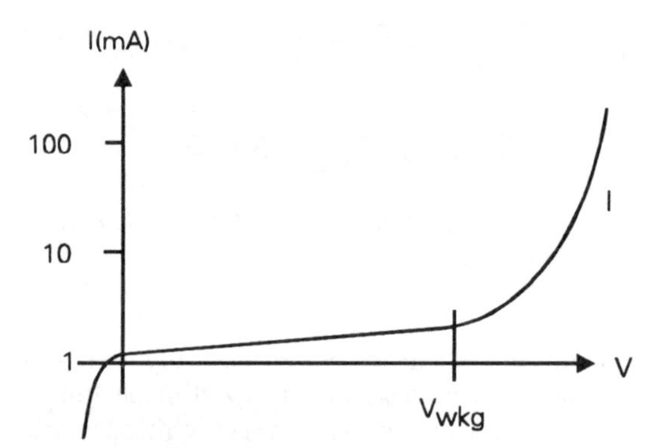

Figure 2.18 *Leakage current in large capacity electrolytic capacitor*

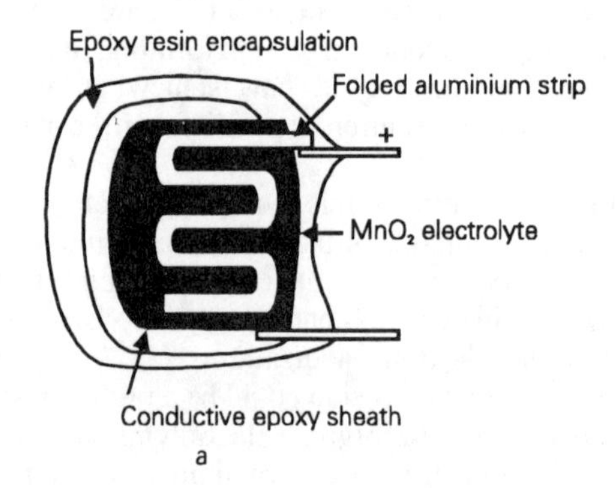

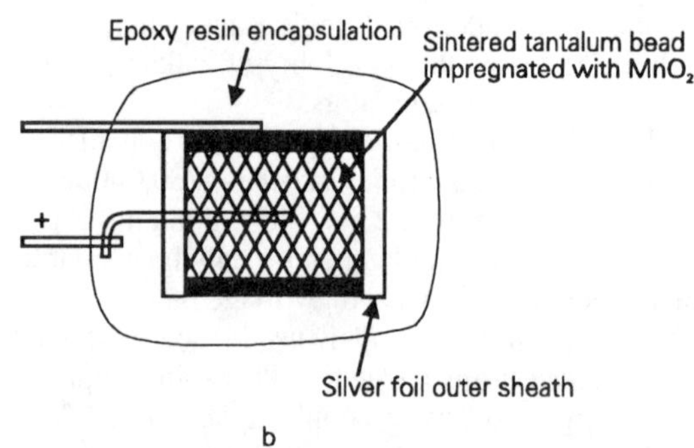

Figure 2.19 *Constructional forms of solid aluminium (SAL) and tantalum bead capacitors*

Like the tantalum bead types, these are physically very small, and offer capacitance ranges of the order of 0.1µF – 100µF, at relatively low working voltages. They are claimed, moreover, to have a much longer life expectancy, prior to failure, than the equivalent tantalum bead components, since the normal tantalum bead capacitor failure mode – that of crystallization of the tantalum pentoxide layer – is absent in aluminium systems.

Since the leakage currents of these particular types of capacitor are usually very low, at voltages below their rated working level, they can be used for inter-stage coupling in electronic amplifier circuitry.

However, although solid dielectric electrolytics are better, in this respect, than the more traditional forms, all aluminium electrolytic capacitors tend to deteriorate somewhat, in respect of their capacitance, and leakage current characteristics, if they are held for long periods without any applied polarizing potential, and this fact must be borne in mind in the design of the circuitry.

The other widely used type of electrolytic capacitor is that in which the anode foil is made from tantalum, on which a thin anodized film of tantalum pentoxide (Ta_2O_5) has been formed electrolytically. These capacitors are made in very similar forms to the aluminium foil equivalents, but at a much greater cost.

In the form most directly equivalent to the aluminium electrolytic, these capacitors take advantage of the chemical inertness of the oxide layer, and use a strongly ionized electrolyte, such as sulphuric acid, or lithium chloride solution, whose high conductivity lowers the equivalent series resistance of the component.

Since the tantalum oxide film is so inert, and physically very robust, tantalum electrolytics will generally offer a lower leakage current than their aluminium equivalents, and are also much more resistant to the deterioration of the oxide layer during storage or use under zero voltage conditions.

Apart from their high cost, such wound-foil tantalum electrolytic capacitors would be very suitable for use in audio circuitry, and this has encouraged the development of a lower cost version of these components, based on a pellet of sintered tantalum powder, impregnated with a solid electrolyte of manganese dioxide, formed in situ by the 'pyrolitic' (i.e. by heat) reduction of manganese nitrate.

This type of electrolyte is similar to that used in solid dielectric aluminium bead capacitors, and serves the same purpose as a liquid electrolyte of maintaining the integrity of the (tantalum) oxide insulating film, but with no possibility of chemical leakage. Since there is no likelihood of gas evolution from this type of component, a simple low cost epoxy resin bead style of encapsulation can be used.

The construction of such a tantalum bead capacitor is shown in Figure 2.19b. They are commonly available in the capacitance range of 0.1μF – 100μF, but mainly at relatively low working voltages, in the range 3 – 30V.

Non-polar

Wound film/foil types

This type of capacitor is made in the manner shown in Figure 2.20, and will usually employ one or other of the available plastics film dielectrics, such as polycarbonate, polystyrene, polypropylene, or polyester (polyethylene terephthalate), wound in a 'Swiss roll' form between a pair of strips of high purity aluminium.

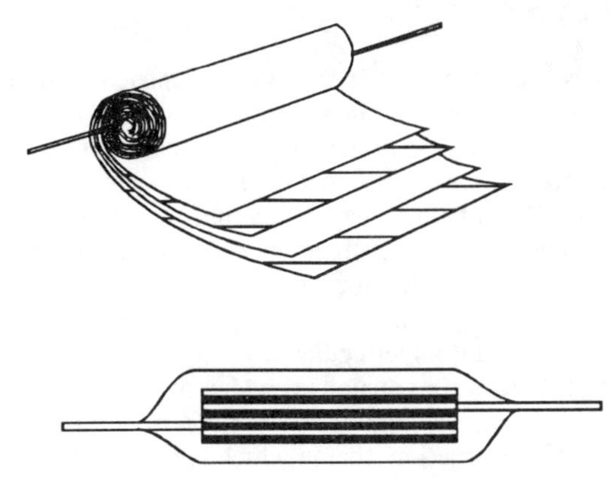

Figure 2.20 *Construction of film-foil capacitor*

In order to attain a high capacitance value the dielectric film should be as thin as is practicable for the required working voltage. Both polystyrene and polycarbonate resin films are made in a similar manner; by 'casting' a thin layer of resin, in a liquid form, in solution in a solvent, on to a continuously moving metal strip, from which the thin film layer is peeled after the solvent has been dried off.

This allows the production of very thin films, but the presence of small imperfections, such as trapped gas bubbles, limits the possible voltage which can be applied between the foil layers. Where high voltage components are required, such solution cast film dielectrics may often be used in double layer form, since this greatly lessens the possibility of two weak spots coinciding at a particular point.

Polypropylene and polyester films are both made by bi-directional (biaxial) stretching of a relatively thick extruded film, which facilitates the production of very thin films, having a high mechanical and electrical strength, together with an almost complete freedom from pin-holes.

In particular, the very high electrical breakdown strength of the polyester films permits high capacitance, compact capacitors to be made at a relatively low cost, and these form the bulk of the commercially manufactured non-polar components for routine use, where special qualities are not required. They are available in the capacitance range 1000pF – 4.7μF, at voltage ratings of 100 – 1000V.

The dielectric loss factor of polyester film capacitors (the extent to which energy is lost during the charge–discharge process) is however, relatively high, so where low loss capacitors are needed, other dielectrics will be preferred.

For low frequency use, polypropylene or polystyrene dielectric capacitors allow the manufacture of very low loss components, but they are relatively bulky where high values of capacitance are desired.

Polystyrene film/foil capacitors are very widely used where high precision capacitors are needed, for timing or filter circuitry.

Because of the low softening point of the polystyrene film, the last stage of the manufacturing process is normally to fuse the finished capacitor, with its insert wire connections, into a solid, stable, block by heating the capacitor briefly in an oven, to give the type of structure shown in Figure 2.21.

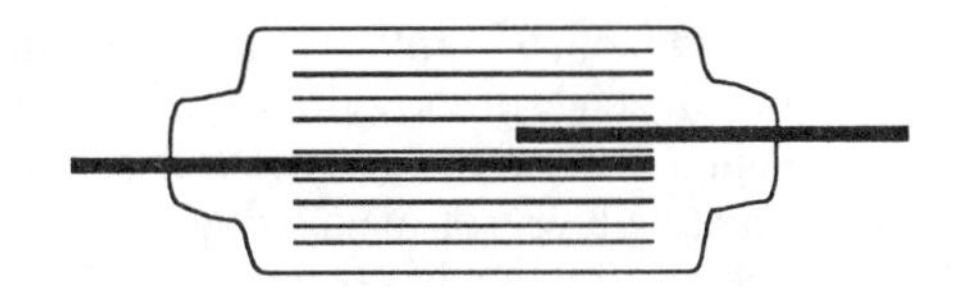

Figure 2.21 *Polystyrene-foil capacitor*

However, the low melting point of this dielectric carries the penalty that, if a soldering iron is applied to the connecting wires for too long, or too close to the body of the capacitor, it may cause internal melting of the film, and lead to a short-circuited component.

Where low working voltages, up to, say, 60V, are adequate, very high capacitance values are possible, in compact forms, with polycarbonate film dielectrics.

Metallized film types

These have much the same physical form as film-foil capacitors, except that the continuous strip foil conducting plates are replaced by thin conducting films of metal, vacuum evaporated on to the surface of the dielectric itself, as shown in Figure 2.22a.

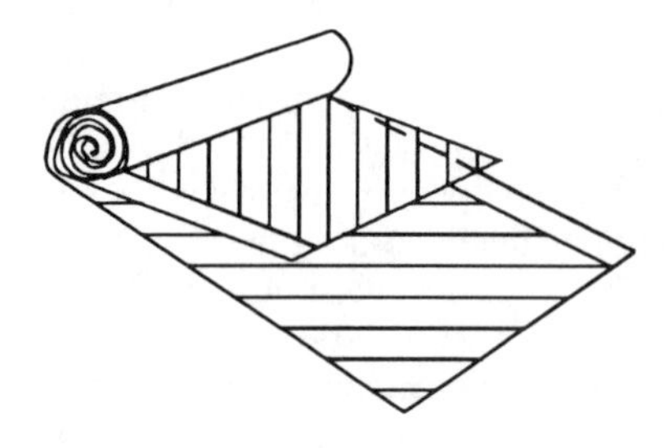

a Single side metallization

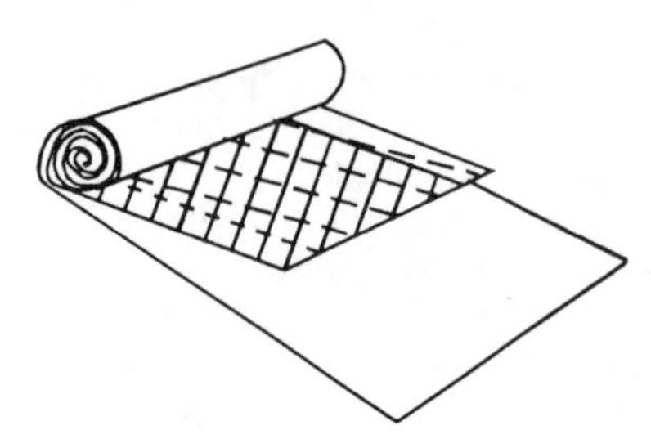

b Film metallization on both sides

Figure 2.22 *Constructional methods of metallized film capacitors*

This is then wound into a roll, as before, and metal is sprayed onto the ends of the cylinder, as shown, in section, in Figure 2.23, to make an electrical contact with the conducting layers of metallizing.

This form of construction has the very great practical advantage that, if an electrical breakdown occurs in the film dielectric layer, this will not destroy the capacitor, but will only burn off a disc shaped area of the metallizing immediately surrounding the puncture, as shown in Figure 2.24.

Figure 2.23 *Cross-section through wound metallized film capacitor*

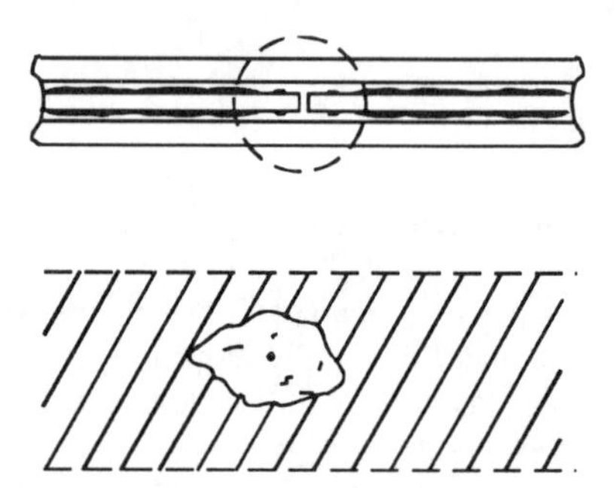

Figure 2.24 *Effect of self-heating of metallized film capacitor*

Also, since the stored energy of the capacitor will be explosively released in such an action, the layer of metallizing will be removed very effectively from the immediate area of the breakdown. The penalty for this self-healing characteristic is that the loss factor of such components is higher than that of the film foil types, and the practicable precision of capacitance value is also much less good.

Such capacitors will generally have a 'tolerance' (a possible error) of capacitance value of +/– 20%.

Two forms of metallized layer can be used in this type of construction: that in which two separate films are metallized, each with a plain strip along one edge, and wound together as shown in Figure 2.22a, or that in which a single strip of film is metallized, on both sides, as shown in Figure 2.22b, and then wound with a plain, un-metallized, strip of film as an interleaving layer.

This construction demands more skill in the original metallizing of the film, but offers the advantages of a higher capacitance per unit volume – because the two conducting layers are immediately in contact with the dielectric layer, without undesired air spaces – and a greater stability of capacitance during use.

Metallized paper, and metal foil/paper dielectric capacitors may still, occasionally, be found. Paper has quite good electrical properties, provided it is dry. However, preventing the long term ingress of moisture is difficult, even when the paper is impregnated with some water repellent material such as oil or wax.

Fortunately such components are now rarely encountered in electronics applications.

Stacked film/foil capacitors

This style of manufacture, which is shown in Figure 2.25, has come into greater use in recent years because it avoids, to a large extent, the inherent parasitic inductance of the wound film/foil or metallized film types.

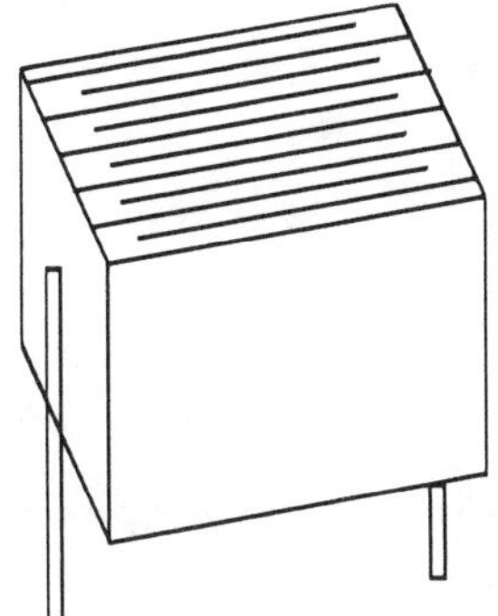

Figure 2.25 *Stacked film-foil capacitor*

It is, understandably, more difficult in manufacture, but it lends itself more readily to the 'radial' style of lead formation, shown in Figure 2.26a, which is more convenient for use in the 'population' of printed circuit boards (PCBs) than the 'axial' style of lead formation shown in Figure 2.26b. Otherwise, the characteristics of the wound foil and stacked foil components are substantially identical.

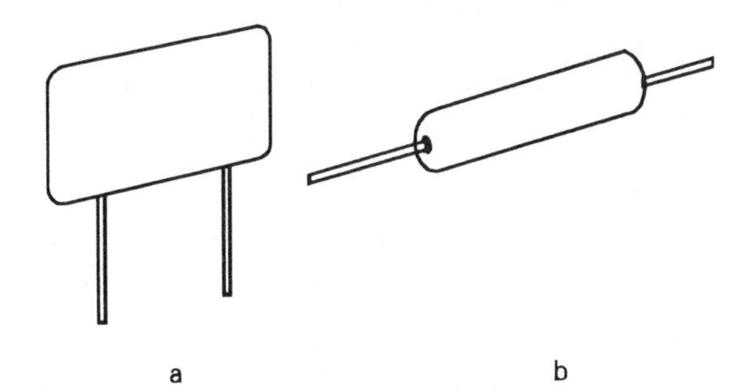

Figure 2.26 *Radial and axial lead configurations*

Mica dielectric capacitors

Mica is, in many ways, an almost ideal dielectric film, in that it can, with care, be split into exceedingly thin, puncture free, layers. It has a dielectric constant which is two or three times greater than that of the common plastics films, it is heat resistant, it has a high electrical strength, and a low dielectric loss, which is maintained up to very high operating frequencies.

Unfortunately, it is expensive, and, because it is rigid, it can only be used in stacked layer constructions. It is used, most commonly, for high power industrial applications, or for the manufacture of high precision, low loss, capacitors having capacitance values in the range 2.2 – 4700pF, at working voltages up to 1000V, for use in radio frequency, or high quality timer applications.

These capacitors usually have their conducting plates formed by opposed layers of silver, applied either chemically or as an electrically conductive paint, though metal foil electrodes are used in higher voltage or power rating components.

Ceramic capacitors

Over the years a very wide range of electrical ceramics has been developed, ranging from relatively low dielectric constant, low loss, materials to those which can give dielectric constant (k) values up to 50,000, depending on their precise composition and production and firing techniques.

These products are based on various mixes of titanium and zirconium dioxides, calcium, barium, strontium and magnesium titanates and zirconates, together with various 'rare earth' additions, the whole being tailored to give a wide range of dielectric properties.

Since there is, understandably, a degree of rivalry between manufacturers, many of their blends and process techniques are not disclosed, and many of their products are unique to themselves. This means that there is now little direct equivalence, except in the case of the more simple materials, or those ceramics made to meet a particular specification, between the products of one manufacturer and another.

Clearly, capacitors made from 'high-k' ceramics can allow high values of capacitance to be offered with units of very small physical size, having values up to, perhaps, 1μF in a pea-sized disc component, at a working voltage of 20–30V. The drawback with these very high-k materials is that their actual dielectric constant may be very strongly temperature dependent, as is also their dielectric loss.

However, where the actual value of the capacitance is relatively unimportant, provided that it is adequate, as would be the case, for example, in RF by-pass applications, such ceramic disc components are widely used.

The lower k materials can offer quite good performance, with low dielectric loss values maintained up to very high frequencies, and reasonable precision of capacitance value. They find wide application in radio frequency (RF) use, in values ranging from 1pF – 2200pF, at 60 – 100V working ratings.

The actual dielectric materials used in ceramic capacitors are usually specified in relation to their temperature coefficient, and the nominal characteristics of the more common ones are shown in Table 2.3.

Table 2.3 *Electrical properties of ceramics*

Material	*Dielectric constant ('k')*	*Temperature coefficient (ppm/°C)*
Alumina	8.1–9.5	–100
Titanium dioxide	70–90	–150
Magnesium silicate ('Steatite')	5.5–7.0	–20 to –50
'NP0'	20–30	+/–30
'N150'	50–200	–150
'X7R'	75–150	–270
'Z5U'	100–500	–750
'N750'	100–500	–750
Barium titanate ('High-k')	1000–10,000	–500 to –1000
Barium titanate zirconate	10,000–50,000	>–1500

While the dielectric constant of most of the common ceramics has a negative temperature characteristic, there are some ceramic blends in which this has a small positive value. These are designated P30 or P100, referring to their temperature coefficient, and they are used for temperature compensation purposes in radio circuitry.

Capacitors are constructed from these ceramic materials in a variety of forms, of which the most common are the disc types shown in Figure 2.27a, the concentric tube type of Figure 2.27b, and the stacked multiple layer type shown in Figure 2.27c. For this purpose, modern manufacturing techniques allow small rectangles of ceramic to be produced in thicknesses in the range 0.025 to 0.1mm, depending on the capacitance values and working voltages required.

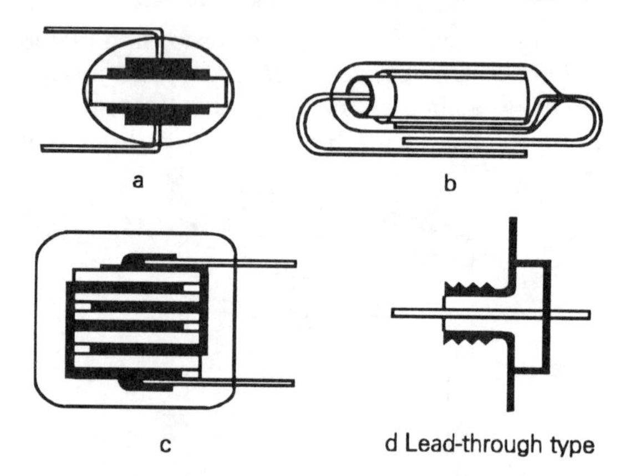

Figure 2.27 *Ceramic capacitor forms*

Originally the conducting plates for these capacitors would be formed by firing on to the surface of the ceramic body a frit composed of glass and silver powders, but higher performance specifications now require platinum or palladium plate electrodes.

As in the case of the other types of fixed value capacitors, the complete unit will be sealed against ingress of moisture by dipping in a suitable synthetic resin encapsulant, although, sometimes, a fired ceramic coating is used.

To produce small, high capacitance value, low working voltage devices, particularly suited for supply line decoupling in computer logic applications, the 'barrier layer' construction is employed. In this technique, a disc of fired high-k ceramic, such as barium titanate, is heated in a reducing atmosphere, so that it is changed into an electrically conducting material.

The disc is then heated in air, to re-oxidize the surface layers to a depth of, say, 0.025mm, to give a very thin effective dielectric layer, and conducting metal electrodes are then applied.

Ceramic tube capacitors are widely employed as 'feed-through' decoupling elements, as shown in Figure 2.27d.

Variable capacitors

It is occasionally necessary to be able to vary the value

of capacitance provided by the component, for example, in the adjustment of the 'tuning' of an inductor/capacitor tuned circuit. The two main classes of component used to provide this facility operate either by moving a pair of parallel plates into or out of mesh, so as to alter the effective area of the opposed conductors, or by altering the degree of compression in a stack of such plates.

Some of the common forms in which such variable value capacitors are made are shown in Figure 2.28.

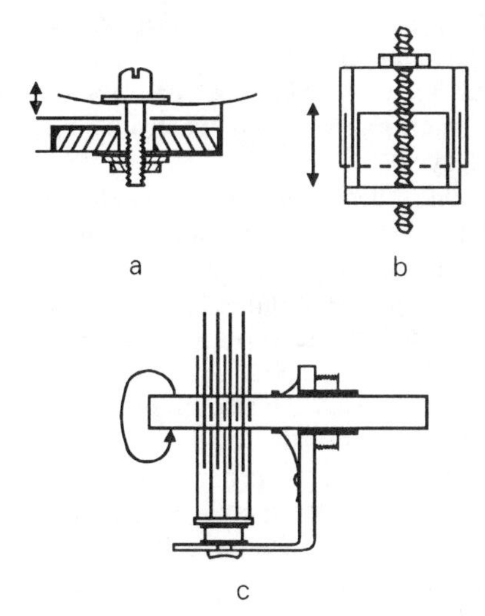

Figure 2.28 *Various forms of variable (trimmer) capacitor*

The variable mesh parallel-plate air-spaced capacitors, generally called tuning capacitors, have the best electrical characteristics of this style of component, but they are bulky, and because of their physical size, suffer from a degree of parasitic inductance, simply due to the necessary length of the conducting path.

They can also be made in multiple ganged forms, as shown in Figure 2.29, which allow electrically isolated capacitors to be simultaneously adjusted by a common shaft, and are commonly used in radio receivers for the parallel adjustment of multiple tuned circuits. They are available in up to four-ganged forms, but variable capacitors with more than two gangs are becoming scarce.

The relationship between the capacitance of a tuning capacitor, and the shaft rotational angle, is seldom linear, except in the case of the smaller 'trimmer' capacitors, since tuning capacitors will normally be

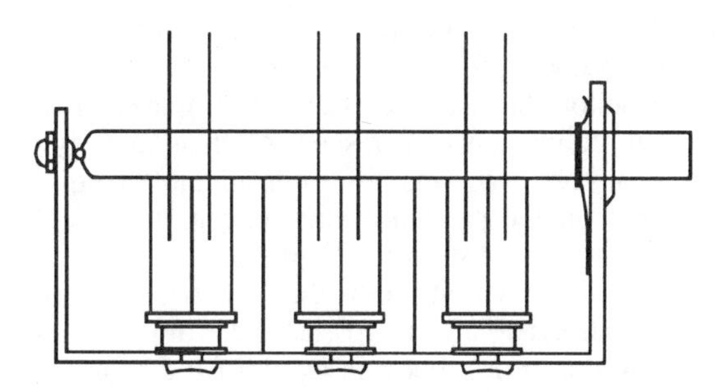

Figure 2.29 *Multiple ganged air-spaced capacitor*

intended to provide a linear change of frequency, as a function of shaft rotation.

This will require a 'square-law' characteristic, because of the way in which the frequency is related to the inductance and capacitance of a tuned circuit, expressed in the equation

$$F = \frac{1}{2\pi\sqrt{LC}} \qquad (4)$$

In the case of the tuning capacitors used in small portable radio sets, the gaps between the meshing plates of the tuning capacitor will frequently be filled with thin sheets of some plastics insulating material, in order to allow the spacing between the plates to be reduced, without the need for a high degree of mechanical precision in manufacture.

These solid dielectric variable capacitors are not suitable for precision purposes because small accidental movements of the plastics dielectric plates may introduce unwanted changes in capacitance, which are unrelated to the degree of shaft rotation.

'Compression trimmers', in which a stack of plates are pressed into greater proximity by a central adjusting screw, usually employ thin layers of a plastics film as a dielectric, although, in earlier times, the material employed was nearly always thin sheet mica.

Spurious effects

Although both resistors and capacitors suffer to some extent from the presence of unwanted characteristics, in the case of resistors the main problem is the presence of inductance, as noted above. However, this is usually swamped by the resistance of the component, and is of little practical effect except at very high frequencies, or with low resistance values, particularly since these will often be of wire wound types.

Capacitors, however, are the most complex of all the passive components, in respect of their underlying physical behaviour, and differ considerably from the notional 'pure' capacitance which one might depict by the circuit symbol shown in Figure 2.30a.

A broad distinction can be drawn between a polar, (i.e. electrolytic) and non-polar (i.e. film, mica, or ceramic dielectric) types, in terms of the equivalent circuit introduced by the component, but, in general, this will more nearly be of the form shown in Figure 2.30b.

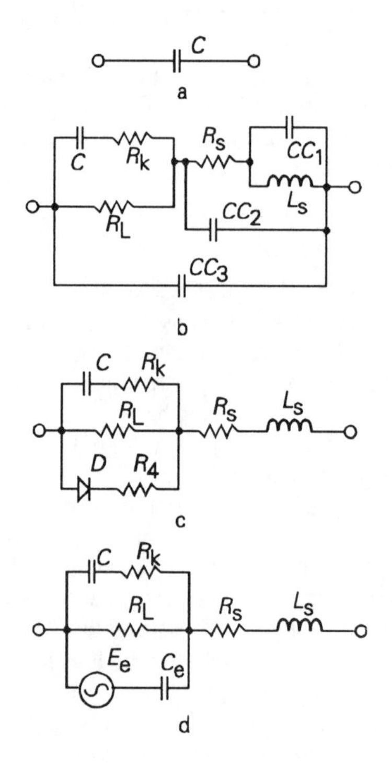

Figure 2.30 *Spurious components inherent in capacitors*

In this, C is the effective capacitance of the unit, which will be somewhat dependent on the operating frequency, voltage and temperature. In series with this element of capacitance is a resistance R_k representing the dielectric loss factor, which is strongly dependent on temperature and operating frequency, and in parallel with C is the leakage resistance, R_l – also very temperature dependent.

In all capacitors, there will be a series element of resistance R_s and a series inductance L_s simply due to the mechanical construction of the component, together with a small amount of distributed parasitic capacitance (C_{C1},C_{C2},C_{C3},) which can probably be ignored, except at radio frequencies, or in the construction of high purity LC oscillators.

In electrolytic types there will also be a unidirectional conductive path, D, in series with a further nonlinear resistance, R_4, as shown in Figure 2.30c, which comes into effect if the polarity is reversed, but can also have an effect under zero polarizing voltage conditions, where these have persisted for long enough to allow deterioration of the electrolytically formed dielectric layer.

The action of the polarizing voltage has a complex electrochemical/ionic effect, and, if reversed polarity conditions are allowed to arise, the nature of the dielectric layer will be modified in ways which will permanently affect the other characteristics of the component.

Although non-polar capacitors avoid some of the undesirable characteristics of the electrolytic types, they can suffer to a much greater extent from dielectric hysteresis and other stored charge effects of the 'electret' type, represented in Figure 2.30d by the generator E_e, and the series capacitor C_e

All of these spurious effects can generally be ignored in most normal applications, but it is prudent to remember that the shortcomings of the capacitor in one's hand may sometimes give rise to unexpected effects, particularly when it is used in some position where the performance of the component is crucial to the behaviour of the circuit.

One must also remember that, for these reasons, the impedance of a practical capacitor does not decrease linearly with frequency, in the manner suggested by equation (4) above, but tends to follow the type of curve shown in Figure 2.31, in which is shown the actual impedances offered by a 0.1μF polyester foil capacitor, and 1μF, 47μF and 4700μF aluminium electrolytic types.

The shape of these impedance curves also illustrates the reason why, if a low impedance is desired over a wide frequency range, it is necessary to connect one or more small capacitors in parallel with any large value electrolytic, to avoid the increase in impedance at higher frequencies which will otherwise occur.

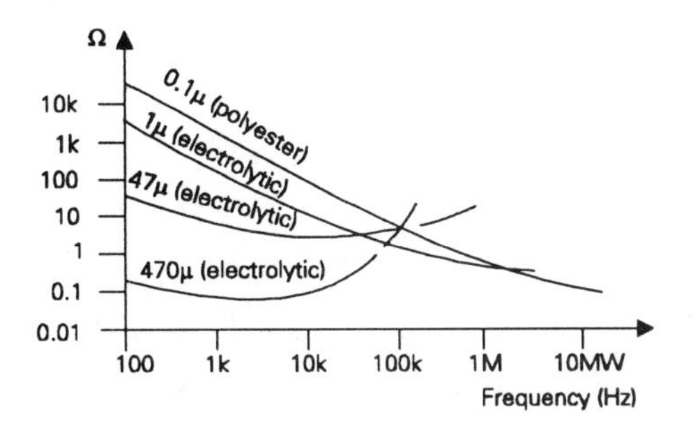

Figure 2.31 *Impedance of capacitors as a function of frequency*

Inductors and transformers

Transformers

'Inductance' is a property possessed by all conductors, and is related to their ability to generate a surrounding magnetic field when current passes through them. This property has a value, expressed in 'henries' (or milli- or micro-henries), which is proportional to their length. However, the inductance of a conductor is increased if it is wound into a coil, and has a maximum value, for a single layer coil, when the diameter of the coil is about five times its length.

If two, or more, coils are placed so that their magnetic fields interact, the passage of a current through one of them will induce a voltage across the other. This arrangement is called a 'transformer', and the efficiency of transformation is dependent upon the completeness of interaction of their associated magnetic fields. This is increased if the coils are wound around a ferro-magnetic core, which lessens the possible loss of magnetic coupling between any one winding and another.

Such a core material will also increase the inductance of the windings, and the extent to which the inductance of the windings is increased by the magnetic characteristics of the core is referred to as its 'permeability' (μ) and is measured on a scale in which a vacuum (or, for all practical purposes, air) is defined as having a value of unity.

The inductance, 'L', of a coil, wound around some magnetizable core, can be calculated from the formula:

$$L = \frac{8.13 \times N^2 \mu a}{10^8 \times l} \qquad (5)$$

where N = number of turns, a = effective cross-sectional area of core in cm^2, l = effective length of magnetic circuit in cm., and μ = effective permeability.

Unfortunately, all magnetic core materials introduce some energy losses into the system due to 'hysteresis' and 'eddy current' effects. These losses will increase with operating frequency, and are due respectively to the internal energy absorption within the crystalline or molecular structure of the material on each reversal of its magnetization, and to the flow of current within the core material itself, constituting a multiplicity of undesirable short-circuited secondary windings.

For low frequency use, as in mains frequency power transformers, eddy current losses can be made adequately small by making the core from a large number of thin 'laminations', typically 0.2 – 0.4mm thick, with one surface coated with some insulating material, and which are stamped out in suitable shapes, as in Figure 2.32, to allow a stack of laminations to be built up to fit the former on which the coils are to be wound.

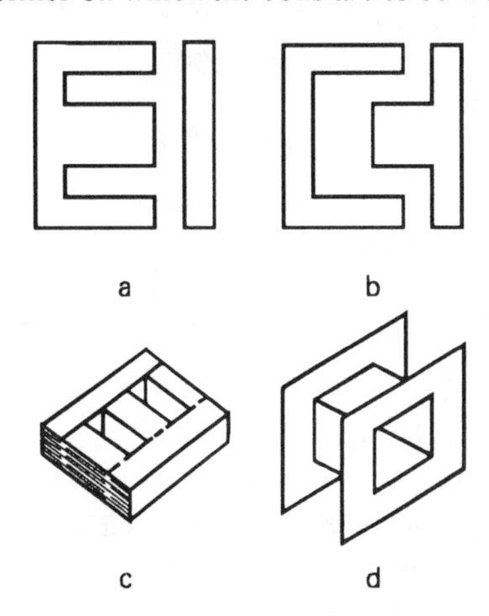

Figure 2.32 *Laminated core assembly in transformers*

At higher frequencies, the possible path length of eddy currents within the core material must be reduced still further, and typical core materials are formed, either from a mixture of magnetic materials and a thermosetting resin binder, in powder form – of which the simplest kind of magnetic material is just is very finely divided iron powder – known as 'iron dust' cores, or from some intrinsically non-conducting material such as a magnetizable ceramic, known as a 'ferrite', moulded into the required shape.

Most of the ferrites are ineffective beyond about 2MHz, though a few formulations can be used beyond this frequency. For use at very high frequencies, as in radio or radar systems, air-cored inductors, or structures having the same effect, are normally used, to avoid the core losses introduced by higher permeability materials.

The permeabilities of a range of magnetic materials used for transformer and inductor cores are listed in Table 2.4.

Table 2.4 *Permeabilities of transformer and inductor core materials*

Material	*Permeability**	*Saturation flux density (Gauss)*
Grain oriented silicon steel (2-5% silicon)	1000–8000	19,700
Ferrites ($Fe.Fe_2O_4$)	2000–5000	2,500
Nickel steel (82% Ni, 2.0% Mo)	5,000–20,000	8,000
'Permalloy' (79% Ni, 0.3% Mn)	20,000–100,000	8,700
'Mumetal' (75% Ni, 18% Fe, 2% Cr, 5% Cu)	20,000–150,000	6,500
'Sendust' (85% Fe, 10%Si, 5% Al)	50,000	c. 7,000

* Value will depend on method of manufacture and on operating flux density

Two things should be noted in relation to this table – that, in general, magnetic alloys which have been designed to give high values of permeability will only allow somewhat lower levels of maximum flux density, and that the actual permeability of a core material will depend on the magnetic flux within it, and will normally increase with increasing flux density – as the magnetic field aligns the magnetic dipoles within the material – up to some saturation value, beyond which the permeability will fall rapidly to a much lower value, as shown in Figure 2.33.

It is important to remember that this effect can occur in any circumstances where there is a DC current flow

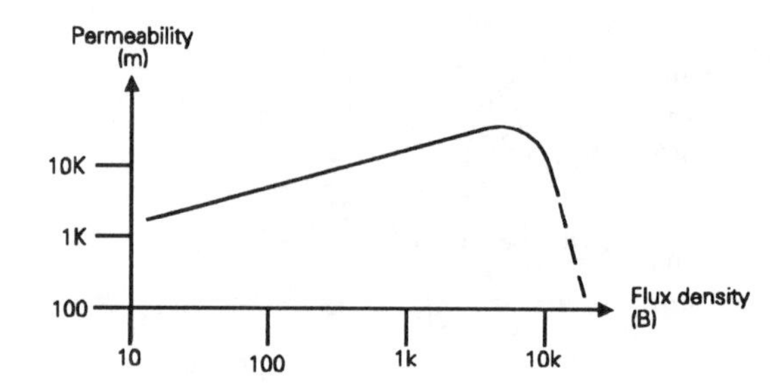

Figure 2.33 *Typical relationship between permeability and flux density in silicon steel core material*

through a coil winding on a ferromagnetic core, as, for example, in the use of an LF choke in a choke–capacitor filter smoothing circuit in a DC power supply, where the flux saturation of the core, due to this current, may substantially reduce the inductance of the choke.

This characteristic is used, deliberately, in 'magnetic amplifiers' where the use of a relatively small DC current, through a third winding on the core, is used to control a much larger output current from a power transformer.

The core losses of a power transformer will depend on the core flux densities used, and will increase rapidly as the flux saturation level is approached. In practical terms, for any given operating frequency, this means that there will be an optimum maximum flux density, which increases as the number of 'turns per volt' is reduced, for any given core material and cross-sectional area.

This is expressed in the equation

$$N = \frac{E \times 10^8}{28.38\, fBA} \qquad (5)$$

where N is the number of primary turns, E is the applied voltage (RMS), f is the operating frequency, B is the maximum flux density, and A is the cross sectional area of the core in cm^2. A fuller treatment of this subject is given by Builder, Hansen, and Langford-Smith in the *Radio Designers Handbook*, 4th edition, Chapter 5.

Grain-oriented silicon steel, in which the crystal structure is oriented during manufacture by cold rolling and annealing, (which also substantially increases its permeability in the direction of crystal alignment), is the most common material used for the cores of LF power transformers and chokes, because of its relatively low cost, and its high possible saturation flux density.

Improvements in LF transformer performance have mainly been obtained by better forms of core construction, to reduce flux leakage, and winding insulation, to allow higher operating temperatures, rather than by the use of more exotic core materials.

Mains power transformers, and other transformers intended for use at relatively low frequencies – say up to 10kHz – are normally made by winding the 'primary' (power input) and 'secondary' (power output) coils on a shaped insulating former within which the core laminations can be stacked, and interleaved to minimize the extent of the inevitable air gaps between the abutting lamination sections.

Typical lamination forms are shown in Figures 2.32a and 2.32b, known as 'E's and 'I's, or 'T's and 'U's from their shapes. ('E's and 'I's are the more commonly found types.) These are assembled into a rectangular stack of the kind shown in Figure 2.32c, and a former of the type shown in Figure 2.32d is made to hold the windings and to fit neatly around the core stack.

Core systems having a better efficiency, mainly in terms of their reduced external magnetic field and higher power handling capability for the same weight of core material, are the 'C' core and 'toroidal' core forms shown in Figures 2.34a and 2.34b. In both these forms, the core is made by winding a continuous strip of thin lamination material, insulated on one or both faces to prevent eddy currents, into a 'toroidal' or doughnut shape, which is then totally impregnated with some synthetic resin to hold it in the required form.

In the C core structure, the core block is then sawn into two segments, and the opposing faces are ground flat so that when the two halves of the core are fitted together the residual air gap will be very small. A steel band is then usually fitted round the core, after the two halves have been inserted into the wound former, to clamp the two halves firmly together.

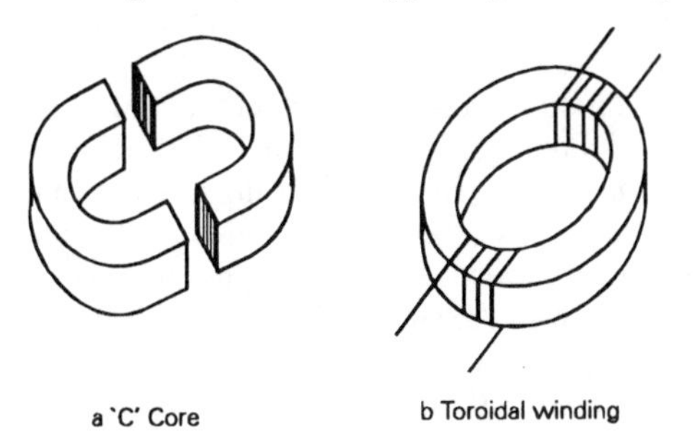

Figure 2.34

This type of core has now largely been supplanted by the toroidal construction, shown schematically in Figure 2.34b, in which the core is left as a complete toroid and the primary and secondary windings are wound within and around it by a specially designed coil winding machine.

Though more costly than the conventional stacked lamination type of transformer, this construction offers the advantage of a very low external magnetic field, which makes it very popular for use in audio amplifiers, where a very low background 'hum' level is a desirable quality.

A useful adaptation of the toroidal type of transformer is the variable transformer, commonly known by the trade name 'Variac'. This takes the form of a cylindrical core on which a single-layer winding has been formed, with the insulating varnish removed from the face of the exposed windings at one end of the core.

An electrical contact is then made to the exposed part of the winding by means of a sliding graphite contact, mounted from a spindle rotated about the central axis of the core. Since the system behaves as an auto-transformer (i.e. one in which a section of the primary winding acts as a secondary), an adjustable proportion of the voltage applied to the primary winding can be tapped off by the sliding contact.

Since one of the possible advantages offered by the use of a multiple winding transformer is that power may be transferred from one circuit to another, while retaining electrical isolation between them, care is taken to ensure that the windings are adequately insulated from one another and from the core.

Sometimes, in high quality units, this is done by dividing up the former, of the type shown in Figure 2.32d, into smaller, separate, isolated sections, but this method of construction may somewhat lessen the electrical efficiency of the transformer by reducing the magnetic coupling. So, more usually, this inter-winding insulation is done by just inserting a layer of flexible insulating material between the windings.

If circuit requirements demand a particularly high degree of magnetic coupling between the windings, the primary and secondary windings may be wound as 'bi-filar' (or 'tri-filar') types, where two (or three) wires are wound simultaneously onto the core. In this case, the inter-winding DC isolation relies solely on the effectiveness of the insulating lacquer coating on the wires.

Electrostatic isolation between the windings can be achieved by the use of an 'electrostatic' or 'Faraday' screen. This is a sheet of electrically conducting foil, of copper or aluminium, insulated on both sides, and wound around the primary winding to give slightly more than a complete single (open circuit) turn, before the secondary winding is applied. A connecting lead, from the electrostatic screen, is led out from within the windings so that it can be taken to an electrical 'earth' point.

For higher frequency use, in low power audio transformers, as, for example, in line matching or as microphone step-up transformers, conventional E and I type laminations are most commonly used, made from 'Mumetal' or one of the other high permeability, and high cost, Nickel–Iron alloys. These laminations may often be very thin, down to some 0.1mm in thickness, to minimize core losses.

Depending on their intended applications, power transformers are made either in open form, with the windings insulated, but otherwise exposed, or in screened forms, with an external metal case surrounding the transformer. However, such a case will not provide electromagnetic screening unless it is made from some ferro-magnetic material.

Circuit diagram symbols for a simple two-winding transformer, a transformer having an internal electrostatic screen, a transformer with a dust-iron core, and a variable transformer, are shown in Figure 2.35. However, as with other components, the actual equivalent circuit of a simple two-winding transformer is more complex, as shown in Figure 2.36.

In this, $L_{l'p}$ is the primary leakage inductance, R_p is the primary winding resistance, C_p is the lumped primary inter-winding capacitance, R'_p represents the primary core losses, L_p is the primary winding inductance, L_s is the secondary winding inductance, R_s is the secondary winding resistance and L_{ls} is the secondary leakage inductance.

The practical significance of these spurious effects is examined in Chapters 11 and 15.

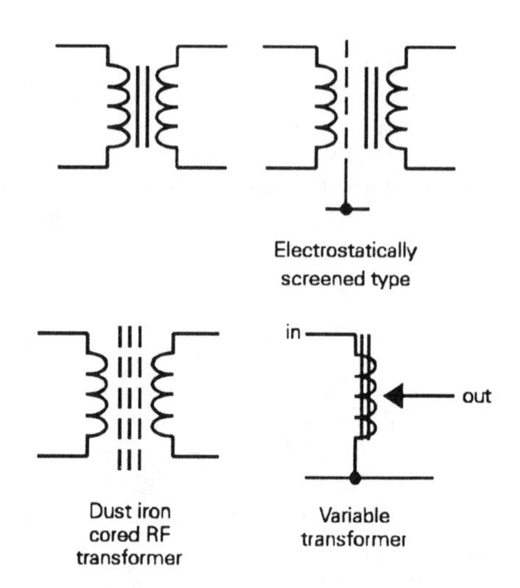

Figure 2.35 *Circuit symbols for transformer types*

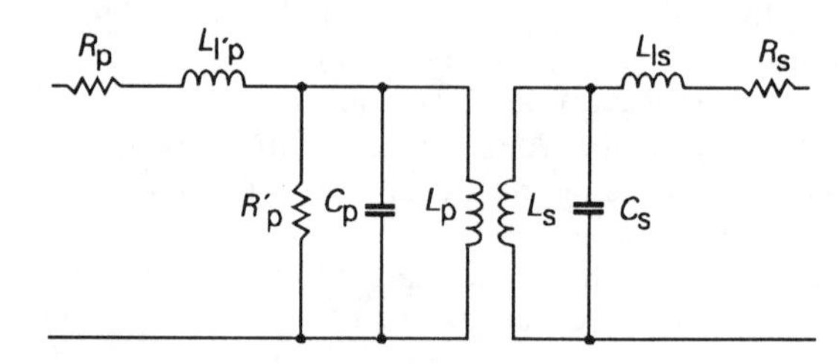

Figure 2.36 *Spurious electrical components within normal two-winding transformer*

Inductors

The inductor is potentially a versatile component in electronic circuit design, but has not been used widely in such applications because it has not been possible, until comparatively recently, to buy, off the shelf, small inductors, having precisely defined values, in as wide a range of values as, for example, wirewound resistors.

The method of construction employed depends upon the inductance value required and upon its intended operating frequency range. Components having inductance values in the range 2–100 henries (H), intended for use in power supplies to smooth out AC 'ripple' from the supply line, in conjunction with a 'smoothing' capacitor, are generally built in a similar manner to a normal mains power transformer, with a multiple layer winding built up inside a moulded former, enclosing a laminated iron alloy core, of the general form shown in Figure 2.32d.

The impedance of an inductor, for which the circuit symbol is shown in Figure 2.37, increases with frequency in a manner defined by the equation

$$Z_L = 2\pi fL \qquad (6)$$

and this characteristic allows it to be used, in AC applications, to perform the function of a resistor, in generating a voltage drop proportional to the (ac) current flow through it.

Figure 2.37

However, it also has the valuable characteristic that, since its impedance is dependent on frequency, networks combining inductors with resistors and capacitors can be used for frequency selective applications, such as filters, or tuned circuits, capable of generating either very high or very low impedance paths at some selected frequency.

Small inductors, intended for low signal-level circuit applications, in the inductance range from a fraction of a microhenry to some hundreds of millihenries, are most commonly made by winding a coil of insulated copper wire upon some rigid insulating former, of ceramic or hard thermoplastic, typically in rod form, to which solderable connecting leads are attached.

It is unusual for such small inductors to employ a magnetic core material, since this would restrict the frequency range over which they could be used, and this, in turn, limits the maximum inductance range which can be provided in a physically small component. Small inductors of this type are often coated with an opaque lacquer, and colour-coded like resistors to show their inductance in microhenries.

Connectors and switches

Contact erosion and contamination

The requirements for both of these types of component are basically similar, in that they are needed to join parts of the electrical circuit together with as low a level of contact resistance as possible. Unfortunately, all metallic surfaces exposed to air will ultimately become contaminated with deposited films of grease or dirt, and will oxidize or corrode because of chemical reactions with atmospheric oxygen or other corrosive gases.

If the contacting surfaces are arranged to slide together, this action will help to scrape the contacts clean, but with the penalty that the greater the contact pressure, the more effort will be required to bring them together and the greater the mechanical wear on the contacts. In both switches and connectors it is good design practice for the component to be constructed to restrict the access of air, to try to preserve contact cleanliness.

Silver is the preferred metal for most plug–socket connectors and switch contacts, in that corrosion processes will normally only form a thin, and partially conducting, layer of silver oxide or sulphide. Thermal diffusion processes will then cause this to be 'doped' with silver atoms in the same manner as occurs in semiconductor junctions, which greatly increases its conductivity. A small 'wetting' current may also be helpful in keeping the contact resistance low, though there may then be some small erosion of the negatively charged contact.

The plating of silver contacts with a thin layer of gold, which is helpful in prolonging the life of sliding contacts, may actually worsen the contact resistance, if silver sulphide growths spread through the minor cracks in the gold surface layer, since the presence of the gold layer may prevent the silver doping effect.

Also, when arcing may occur during the switching of higher power circuits, silver has a better resistance to arc damage than gold, though, if cost is no object, palladium–nickel alloy (85% Pd, 15% Ni) contacts are very arc resistant, and platinum is even better. The presence of atmospheric oxygen helps to lessen arc-induced contact erosion. However, both palladium and platinum suffer from the snag that local surface catalytic action can cause the formation of 'brown powder' – an adhesive non-conducting dust produced by the polymerization of organic gases and other contaminants present in the atmosphere.

At present, the best switch/connector contact construction for low contact resistance is probably gold plated Pd/Ni alloy, since the gold layer acts as a lubricant, and tests have shown such contacts to retain

Table 2.5 *Properties of switch and connector contact metals*

Material	*Melting point (°C)*	*Relative hardness*	*Conductivity (mho/metre)*
Silver	961	80	62
Silver with 0.15% Nickel	960	100 (ref.)	58
Silver with 15% CdO	961	125	42
Gold	1063	60	44
Copper	1083	100	58
Palladium with 40% Silver	1200	100	2.4
Nickel	1453	200	14
Palladium	1552	100	9.0
Platinum	1769	95	9.5
Rhodium	1966	130	22
Indium	2454	220	19
Molybdenum	2610	250	19
Tungsten	3400	450	18

a very low contact resistance, of the order of 2–3 milliohms, after 25,000 operation cycles.

The properties of electrically useful metals and some of the more commonly used contact alloys are listed in Table 2.5.

In general, materials with high melting points will give good arc resistance, and mechanical hardness will assist in prolonging contact life. Alloying almost invariably lowers both melting point and conductivity, but is used where some combination of physical properties is best obtained by this means.

Since the mechanical components of any electronic system are now by far the least reliable, and most trouble prone, part of the equipment, it is preferable to use the most robust and best made plug/socket connectors that physical space or considerations of cost will allow. This advice is even more appropriate to switches, since the area of physical contact is likely to be much smaller.

Switches

These are made in a wide range of styles, for different types of circuit or mechanical applications. Most of the types and descriptions offered are self-explanatory, so little guidance is necessary. However, some points are worth watching.

Rotary switches

These are used as small current wafer switches, and also as higher power stud-type switches. In both of these forms, one contact of the switch is taken to a movable wiper which slides over the other fixed contacts. The wafer switches used largely in small-signal electronics circuitry are offered in three basic switching styles, 'break before make', 'make before break', and 'shorting' types. Their manner of operation is shown schematically in Figure 2.38.

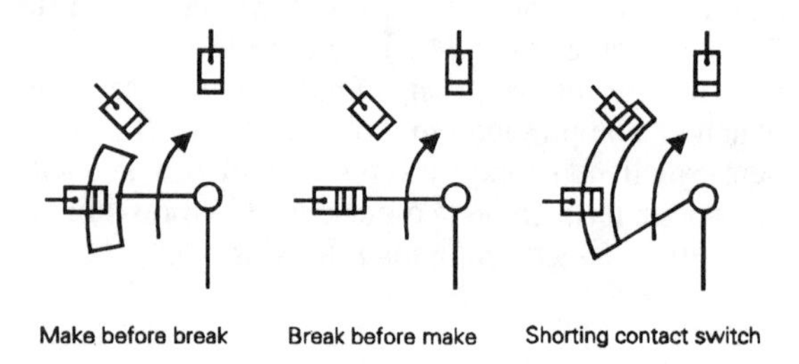

Figure 2.38

A mechanical indent mechanism, usually provided by a spring-loaded ball riding over a notched pathway, locates the rotor at the correct switching positions for the moving contact to register with one or other of the fixed ones. Although there is no specific limit on the range of switching positions, the most common number is twelve, allowing choices, within the range, of 1–pole 12–way, 2–pole 6–way, 3–pole 4–way, 4–pole 3– way and 6–pole 2–way.

The so-called 'miniature' wafer switch, having just a single 12-contact switch section, is an attractive option because of its compact construction and low cost. These switches, however, almost always rely on a single sliding contact, of the kind shown in Figure 2.39a, which is very much less reliable in use than the twin-wiper type of contact, shown in Figure 2.39b, which is almost invariably used in those wafer switches which allow multiple gangs of wafers to be assembled along the same actuating spindle.

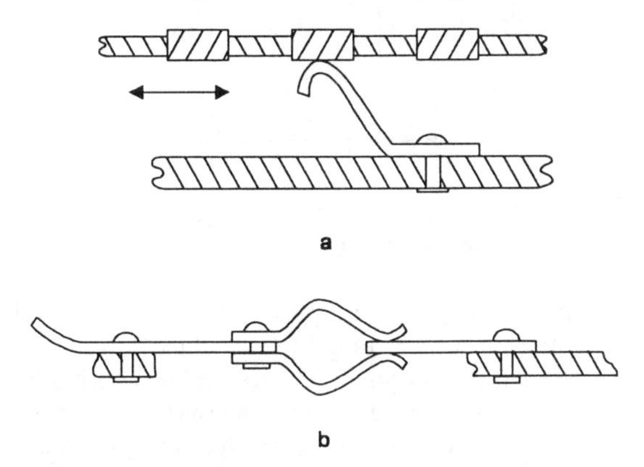

Figure 2.39 *Wafer switch constructional forms*

The thumb-wheel actuated edge switches employ similar constructional forms, and the same comments apply. However, these switches, and the miniature printed circuit board mounting DIL (dual in line) switches are seldom required to perform very many operations during their service life, and do not therefore wear as much as front panel operated units.

Because it is very difficult to exclude the ambient atmosphere from such switches they will inevitably suffer from atmospheric contamination and corrosion. Because of their slow break action, they are unsatisfactory for switching power circuits.

A version of the wafer switch, but assembled on a linear wafer, is used as a small-signal push-button switch. These are particularly valuable in the form in which they latch in position, and where they can be grouped in banks so that the actuation of any one switch of the bank will release the others.

Toggle switches

These are particularly appropriate for switching power supply circuits, because of their rapid make/break action, which minimizes contact damage due to arcing. Such switches will usually be sold with a specific current/voltage rating, which will be for use with AC circuits. Their DC rating will be considerably lower, because arcs may continue for a longer period after the switch contact is broken – especially on inductive loads.

Standard panel mounting toggle switches are available in single, double, triple or quadruple pole forms, in either single or double throw, usually referred to as 'SPST', 'DPDT' and so on. They may also be offered in spring-biased forms, where the toggle will normally return to one or other position when it is released. Centre ('off' position) biasing is also commonly available.

The 'microswitch' is a particular version of a 'rapid make–break' toggle switch, based on an internal beryllium/copper leaf spring, which will trip over from one state to another on a very small movement of the actuating button. These are commonly just make–break types, but change-over forms are available. They can offer very long life expectancies, up to 10^7 operations.

Keyboard switches

These are becoming increasingly widely used because of the spread of calculators, computers, and other similar systems. The main requirements here are for very long life, small size and resistance to the ingress of dust or liquid spillages. Since contact resistance is often not very important, because they will mainly be used to operate very high impedance circuitry, they may simply employ conductive rubber diaphragms, pressed by the key button into contact with one or more conductive tracks on a substrate.

In higher quality equipment, miniaturized toggle switches, based on the microswitch type of mechanism, are more common, because the toggle action gives a better keyboard 'feel'.

Relays

This is the term used to describe the whole class of remotely actuated switches, usually electromagnetically operated. In their classic form, they will have the general type of construction shown in Figure 2.40, in which an electro-magnet is arranged so that, when it is energized, it will pull down a spring-loaded rocker arm, which will push a movable contact away from its rest position so that it comes into connection with another contact.

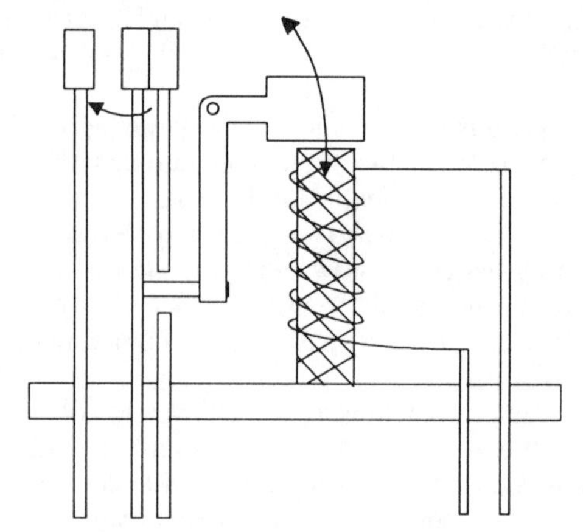

Figure 2.40 *Mechanical toggle relay*

This can give either a single 'make', or a 'change-over' action, and there can be as many such switch sections as required, though four such parallel switches is commonly the maximum available in commercial units.

Since the system can be completely sealed against the ambient atmosphere and other airborne contaminants, good contact life and system reliability is easily obtained, especially on low power circuitry, where contact arcing does not occur.

The major snags of such devices are that contact 'bounce' or 'chatter' will nearly always occur to a minor extent and that the intrusion of the electrical switching spikes, when the coil is energized or de-energized, into low signal level circuitry connected to the relay contacts, is difficult to prevent completely, particularly since the mechanical design of the relay does not normally include any screening between the energizing coil and the actuated contacts.

In addition to the recommended coil operating voltage and coil resistance, the makers will usually also quote the 'pull-in' and 'drop-out' voltages for the relay, and sometimes also the speed of operation, in milli or microseconds, for a given operating voltage. In order to avoid problems when such relays are operated from solid state circuitry, it is customary practice to connect a small power diode across the coil, as shown in Figure 2.41, to 'snub' the switch-off voltage spike which would otherwise occur.

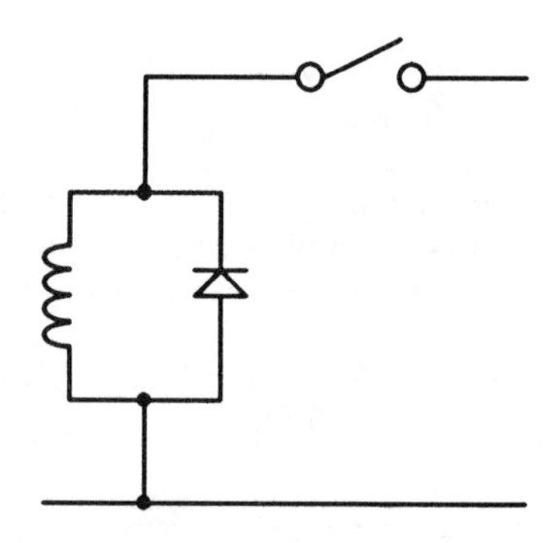

Figure 2.41 *Conventional diode snubber circuit used with relay switching*

A particularly useful type of relay, where high switching speeds and long life expectancy, (up to 10^8 operations), are essential, is the reed relay. This has the form shown in Figure 2.42, and consists of a thin 'reed' of gold-plated nickel or iron, which can be pulled into contact with another such reed when an axial magnetic field is applied, either by some external permanent magnet being moved into proximity to the reed, or by causing current to flow through a coil (sometimes called a 'solenoid') wound around the tube which holds the reed.

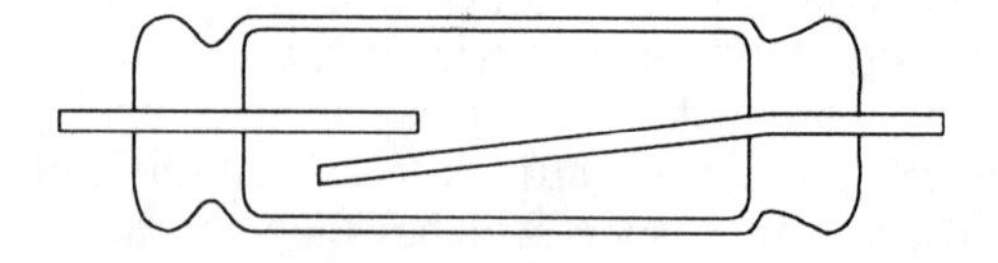

Figure 2.42 *Construction of encapsulated reed relay*

These reed switches are generally sealed inside a thin glass tube containing some inert gas – a vacuum is not used because it would encourage cold welding of the contacts – and are most commonly available as single 'make' or 'change-over' versions. Although not specifically aimed at power switching, reed relays can switch currents of up to 2 amperes.

A convenient modern form of this relay is that in which the whole assembly of reed switch and operating coil is moulded into a small plastics block, having a physical size, and connecting pin layout, which allows it to be plugged into a standard 14-pin dual-in-line integrated circuit holder.

A form of reed relay which has both a low contact resistance and high freedom from contact bounce is the 'mercury wetted' relay, in which the contacting reeds are nickel plated and grooved to assist the flow, by capillary action, of mercury from a small pool at the base of the reed, as shown in Figure 2.43. However, this requires that the reed element shall be mounted vertically to prevent inadvertent shorting of the contacts by the mercury globule.

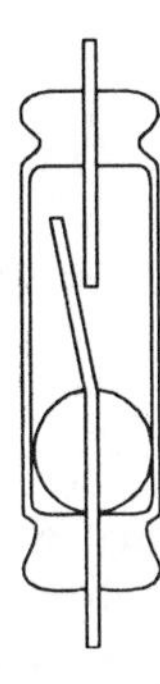

Figure 2.43 *Mercury wetted reed relay*

A similar mechanism is employed in the mercury 'tilt switch', where the capsule is constructed so that the switch can be opened or closed by alteration of its positional angle.

Housings

All electronic systems will require some form of enclosure or housing, to present a neat finished appearance to the unit, to protect the internal components from damage, and to protect the user from possible electrical shock, especially when the equipment is mains voltage operated.

The housing can also serve a number of other useful functions, in assisting in the removal of heat from internal components by conduction from the component, if the housing is made from some thermally conductive material, and in protecting possibly vulnerable components from atmospheric dust, contamination or corrosion. Ventilation of the housing to help keep the circuit components cool may cause problems by increasing the accumulation of airborne dust and grease, so these aspects should be considered in the choice of enclosure.

In very low signal circuitry, or in designs which operate at a high impedance level, the enclosure will normally need to provide some electrical screening for the circuit to prevent the pick up of electrical noise or 'hum' from the surroundings – generally referred to as 'EMI' (electromagnetic interference) – which would worsen the no-signal background noise level of the equipment.

Similar considerations apply, in reverse, if the circuit generates electrical signals, such as high frequency oscillatory voltages, which could cause interference in adjacent equipment. This is particularly the case in radio frequency designs employing local oscillators, and in 'switch-mode' power supplies.

Where electrical screening must offer a high degree of environmental isolation, signal and power supply ports through the metal screening enclosure may require specially designed feed-through insulators, usually combined with a ceramic decoupling capacitor to lessen the extent of RF transmission along the connecting leads.

A particular application in which the design of the enclosure is very important is found in industrial electronics installations where the equipment is to be used in a corrosive, or flame- or explosion-hazard environment. In the latter cases, the equipment may require to be filled with some inert gas, maintained at a positive internal pressure, to ensure that the electrical circuit is never allowed to come into contact with the hazardous environment.

An alternative approach is that defined by the international standards defined in IEC 79-10, BS5501/BS5545 or CENELEC standard EN50-020, in which it is required that the equipment should be designed so that the maximum electrical energy available within the circuitry is below the level at which any spark or thermal effect is capable of initiating a flame or explosion.

3

Active components based on thermionic emission

(thermionic tubes or valves)

Early beginnings

Necessity, it is said, is the mother of invention. This is certainly true of the 'thermionic valve', or 'electron tube' as it is also called, because this invention met an urgent need of the time of its conception, at around the turn of the century, which was for some reliable method for the detection and amplification of very weak radio signals.

This development was a matter of considerable commercial importance, since successful experiments by Marconi had shown the practicability of supplementing or replacing expensive or, as in ship-to-ship, impracticable conductive wire telegraph lines by radio – or 'wireless' – links.

However, the oscillatory frequency of the electromagnetic radiation used for this type of transmission had to be very high, or the efficiency of radiation from the transmitter 'aerial' would be very low. Unfortunately, the use of these very high frequencies meant that the human ear would be quite unable to detect their presence, if the incoming signal current from the receiver aerial was simply connected to a pair of headphones.

The first method which was used by Marconi to detect the presence of this incoming high frequency radiation was the 'Coherer': a device invented in 1890 by the French physicist, Edouard Branly, which consisted of a glass tube with metal end connections, loosely filled with fine metallic powder, which would 'cohere' together if an incoming RF signal was applied to it.

This would cause an immediate increase in the conductivity of the powder filling, and this could be used to cause a pointer deflection in a current meter, but it would then be necessary to tap the glass tube to cause the metal particles to become loose again. In practice, this was arranged by the use of an electric bell-type buzzer to continuously rattle its clapper against the wall of the glass tube.

Though this method did work, and was used successfully as the 'detector' in the first trials of 'Wireless Telegraphy', it was crude and insensitive in action, and could only determine the presence or absence of RF signals. Also, it would obviously have been quite useless in recovering any signal whose transmission was attempted by modulating the amplitude of the signal broadcast by the transmitter.

What was needed was some form of sensitive rectifying system, at the receiver, which would convert the incoming RF signal into a direct current flow, whose size was related to the amplitude of the RF signal. Unfortunately, no such system was known, which would work at radio frequencies, and this need prompted a great deal of experimentation, of which the most important was that concerned with electronic emission.

The initial observation by Edison, the inventor of the electric filament lamp bulb, in 1880, that the glass envelope of the lamp would darken in proximity to the filament, after prolonged use, was correctly interpreted by Sir Ambrose Fleming as being due to the bombardment of the glass by the thermally emitted electrons; which he called 'thermions'; and Fleming exploited this discovery by the construction of a device which consisted of a heated filament, surrounded by a conductive metal plate, or 'anode', housed within an evacuated envelope. An arrangement which he patented in 1904.

Since this device had two elements – the heated filament, or 'cathode' and the anode – he called his device a thermionic 'diode'. The only function which this would perform was to allow current flow in one direction only, and caused him to describe it as a 'valve'. The subsequent introduction, in 1907, of a third electrode – a wire mesh 'grid' between the cathode and the anode – to make a 'triode', was due to an inventor in the United States, called Lee de Forest.

Because this would then allow the flow of electrons between the cathode and the anode to be regulated by the voltage applied to the grid, this device allowed, for the first time, the amplification of small signal voltages, and set in train the whole development of what we now call 'electronics'.

With the introduction of transistors and other semiconductor devices, the thermionic valve has now become largely obsolescent, but there are still a few applications where it has not yet been replaced by more efficient semiconductor components.

These applications include high power, high voltage rectifiers, cathode ray tubes, as used in oscilloscopes and television sets, high power radio transmitting valves, X-ray tubes and the 'magnetrons' used in domestic microwave ovens, as well as a residual use of more traditional small signal amplifying valves in high quality audio amplifier systems. It is still useful, therefore, to possess some understanding of the basic technology.

The modern thermionic valve

The basic structure of a thermionic valve consists, in its essentials, of an evacuated glass or metal envelope, within which is mounted a cathode, as a source of electrons, and an anode (though in some cases there may be more than one) maintained at a positive potential with respect to the cathode, to attract these electrons, and thereby cause a current flow through the valve. Because the electrons are negatively charged, they will be repelled if the anode is itself negatively charged. The current flow through the valve can, therefore, only be in the direction from the cathode to the anode.

A simple valve, having just a cathode and an anode, can be used to 'rectify' an alternating current flow to convert it into a series of unidirectional current pulses, as shown in Figure 3.1a. A typical application, as a rectifier in a power supply circuit, is shown in Figure 3.1b.

In an amplifying valve, one or more grids are interposed between the cathode and the anode, and the potentials applied to these will increase or reduce the flow of electrons reaching the anode. By means of these electrodes, the current flowing through the valve, which can be as large as the design permits, may be directly controlled by some externally applied voltage, and this permits a very high degree of power amplification.

In such an amplifying valve, the grid which is closest to the cathode will normally be negatively charged with respect to it, and this will cause a cloud of free electrons to accumulate between the cathode and the grid. This electron cloud is called the 'space charge' and serves as a reservoir of electrons, to permit high anode currents to be drawn for brief periods.

This grid, if negatively charged, will usually draw no electronic current, and therefore appears – from the point of view of the external circuitry – as an open circuit. This electrode is therefore normally used as the 'control grid' of the valve.

Various different structures of thermionic valves have been evolved for specific purposes, and some of these are discussed in detail below. However, the element which is common to all of these designs is the

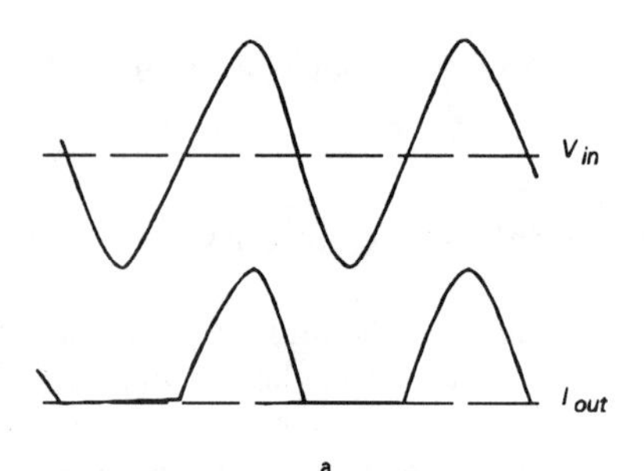

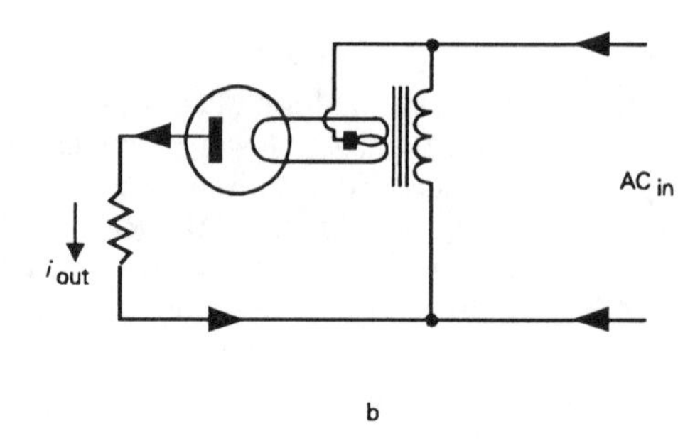

Figure 3.1 *Use of thermionic diode as rectifier*

cathode, as the source of electrons, and the characteristics of this part of the valve and the mechanisms which influence the emission of electrons are of great importance in the operation of the device.

Thermionic emission

The energy due to thermal agitation of an electron within a metallic conductor increases as the square of the absolute temperature, (°K), so that, if a metallic body is heated, in vacua, a temperature will be reached at which the kinetic energy of the electrons will exceed the level, known as the 'work function' of the metal, which is needed for an electron to escape from the electrostatic forces which bind it to the atomic nucleus.

Above this temperature, an increasing number of electrons will escape into the space surrounding the metallic body, until the increasing positive charge appearing on the body, as a result of the loss of electrons, prevents the escape of any more.

The number of 'free' electrons which surround the hot metallic body is thus the result of a condition of dynamic equilibrium between the thermal energy of the electrons and the residual positive charge which their departure has left on the surface from which they have escaped. As noted above, this electron cloud is known as the 'space charge.'

If a positively charged conductor is mounted, in vacuum, in proximity to this space charge, an electronic current will be drawn from this cloud of electrons, which will be replenished from the heated body to maintain the equilibrium. The total current which can be drawn reaches a saturation level when the space charge region is stripped of electrons, and this is dependent on the temperature of the metallic body.

In a thermionic valve, the region emitting the electrons is known as the cathode, or sometimes, loosely, as the 'filament', in those cases where the emitting surface and the heater element are combined in the same component, rather than being formed by a metallic tube, heated by an internal bundle of resistance wire.

The actual thermal energy which is needed to cause electrons to escape from a metallic surface depends on the atomic structure of the metal itself, but very high temperatures may be required, as is shown in Table 3.1, for pure tungsten, in a high vacuum. The figures quoted, as also those of Table 3.2, are due to Williams and Prigmore. (*Electrical Engineering*, 1963, Heinemann.)

Table 3.1 *Electronic emission (J) from metallic tungsten in vacua*

T (°K)	J (A/cm^2)
2000	0.00096
2200	0.012
2400	0.11
2600	0.67

Various techniques have been evolved to reduce the required temperature for electronic emission. Of these, one of the simplest was to incorporate a small amount of thorium oxide in the powder mix from which the tungsten filament was sintered, and then to heat the filament in an atmosphere of acetylene to coat the surface with a layer of tungsten carbide.

This carbide layer performed two functions: to reduce metallic evaporation from the filament – which would, in due course, cause localized thinning and subsequent fracture – and to react chemically with the thorium oxide to produce an atomic thickness layer of metallic thorium on the filament surface. Unfortunately such 'thoriated tungsten' filaments (dull emitters) tend to be more brittle, and more easily broken by mechanical shock, than those from pure tungsten, known as 'bright emitters'.

The second major technique is to coat the filament, or cathode tube in the case of an indirectly heated valve, with a thin layer of nickel, on which is subsequently deposited a coating of mixed barium, strontium and calcium carbonates. These are subsequently reduced to their respective oxides by heating, during the manufacturing process.

In operation, the barium and other oxides in contact with the nickel substrate are reduced to the base metal, and this diffuses outwards through the oxide layer to the cathode surface, where it contributes to the total emission of electrons. Any residual gas in the valve envelope may however combine with the reactive metal surface and render it inactive.

Although these 'oxide coated' cathodes offer much higher efficiencies in terms of output current for a given energy input to the heater, and can provide a copious flow of electrons to sustain the space charge, they have a more limited operating temperature range than either bare tungsten or thoriated tungsten cathodes.

This is because the lower operating temperature is determined by the escape energy required by the electrons, while an upper limit is set by the need to avoid significant evaporation of the relatively volatile cathode metals from the surface. If such emissive metals are allowed to contaminate the grids or the anode, this could lead to secondary emission, and this can lead to a number of problems, as discussed below.

The relative efficiencies of the various practicable cathode types, at various operating temperatures, are shown in Table 3.2

In an oxide coated cathode, the total current drawn from the cathode (space charge) must be substantially less than the possible theoretical maximum if a long operating life is required. If 'saturation current' is drawn, that is if the space charge cloud is entirely stripped of its free electrons, the cathode surface may be damaged.

Such oxide coated cathodes are also much more easily damaged by 'sputtering', which is a term given to the explosive removal of surface particles by bombardment by energetic gaseous ions. These ions will be formed if a rapidly moving electron collides with a molecule of gas, and will be both positively and negatively charged. Those with a positive charge will be accelerated towards the cathode and will cause physical damage on impact if they are sufficiently energetic. This is particularly likely to happen if high anode voltages are employed, since there will always be a small amount of residual gas within the valve envelope, due to outgassing of metal grids and anodes within the valve – especially if these are allowed to overheat – or arising from the reduction of the oxide coatings on the cathode itself.

Table 3.2 *Thermal efficiencies of cathode structures*

	T (°K)	J (A/cm^2)	Relative efficiency
Tungsten	2500	0.4	1
Thoriated tungsten	1950	1.5	10
BaO (as coated cathode tube)	1050	0.2	20
BaO (filament)	1050	0.2	100

To reduce the possibility of sputtering, all normal electronic valves employ an internal gas-absorbing layer, known as a 'getter', made up from highly reactive alkali metals, which is held in a small basin within the envelope of the valve. After the valve envelope has been evacuated and sealed, the cup holding the gettering compound is heated by an eddy current generator, to evaporate the gettering layer onto the inside of the envelope, well away from the active regions of the valve.

If the envelope should crack, and let in air, the loss of the dark mirror-like appearance of the gettering layer is usually the first visible symptom of trouble.

The likelihood of damage due to the bombardment of the cathode, as a result of the ionization of residual traces of gas, limits the voltages which can be applied to the anode, as shown in Table 3.3

This leads to the use of either thoriated tungsten or plain tungsten filament 'bright emitter' valve types for high power transmitter applications, where high anode voltages are needed.

(*Note*. Oxide coated cathodes are, however, usable

Table 3.3 *Anode voltage limits for various cathode types*

Cathode type	Max. anode voltage
Tungsten	100 kV+
Thoriated tungsten	15 kV
Oxide coated	1 – 2 kV (see note in text)

in oscilloscope tubes, even at much higher voltages than this figure, because it is possible to design the internal electrode structure of the tube so that returning positive ions will be physically intercepted before they reach the cathode.)

The superior thermal efficiency of the 'directly heated' oxide coated filament structure leads to this type of cathode being used exclusively where low power consumption is essential, or where, as in power amplifier triodes, the mesh of the grid is too coarse to permit a large space charge to accumulate.

Indirectly heated cathode systems

The major problem with using a heated filament, as the cathode, is that the voltage between the control grid and the filament will vary along its length, and will alter if the voltage applied across the filament changes. If the filament is heated by an AC voltage, this will then cause an AC modulation of the current through the valve in synchronism with the AC filament potential.

Also, unless the filament is substantial in size, its temperature, and its electronic emission, will vary in sympathy with the applied alternating filament voltage, and these effects may introduce a large 'hum' component into the output signal.

Valves intended for use in equipment where the cathode heater power is derived from an AC source therefore employ, with very few exceptions, a cathode structure of the type shown in Figure 3.2, in which a resistive heater, formed either as a twisted helix, or as a simple folded bundle, is given a robust insulating coating of some refractory material – usually alumina – and is then inserted into a separate tubular metal cathode sleeve.

The general construction of typical cathode systems is shown in Figures 3.3a – 3.3c.

Figure 3.2 *Heater system for indirectly heated cathode*

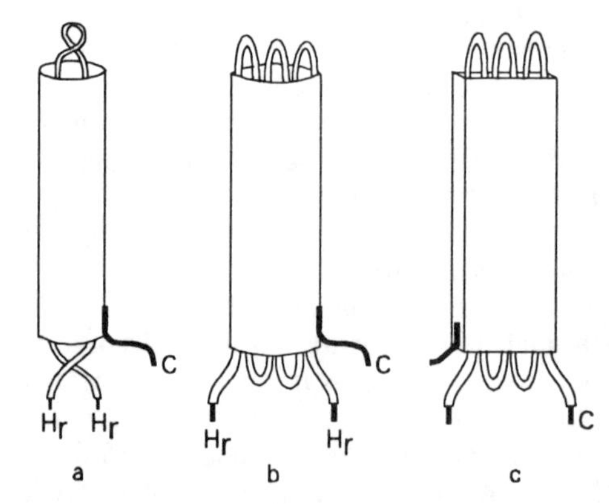

Figure 3.3 *Methods of construction of individually heated cathodes*

To avoid fluctuations in the cathode temperature in sympathy with the heater current, normally supplied at either 50 or 60Hz, the thermal inertia of the cathode system is usually chosen to give a thermal time constant in the range 20–60 seconds. Such valves therefore require a warm-up time before they will operate.

Care is taken in the design of the assembly to keep leakage currents between the cathode tube and the heater wire to a very low level, and also to prevent the emission of electrons from the heater element, particularly as a result of contamination by evaporation from the cathode surface. However, where this factor is critical, it is prudent to arrange that the mean heater voltage is somewhat positive with respect to the cathode.

Anode structures

In its simplest form, the anode will simply be a metal cylinder surrounding the cathode, at a separation determined by the conducting impedance which it is required that the valve should have, and with an area which is adequate to radiate the heat generated by the kinetic energy of the electrons which impinge on it.

The most common metal from which the anodes are made is nickel, and the outer surfaces of the anode will frequently be blackened to assist in the dissipation of radiant heat.

It is also important that the whole internal structure of the valve is held rigidly in place, to prevent mechanical vibration of the electrode assembly from modulating the output current. This can cause audible 'ringing' sounds in audio systems if the envelope of an amplifier valve is struck, and is referred to as 'microphony'.

Because of the need for the maximum rigidity in construction, anode and grids are most commonly made in a rectangular cross-section, supported at their edges or corners by rigid metal rods. These are mounted in the glass seat or 'pinch' at the base, and held in place by one or more stiff mica plates, wedged into the envelope of the valve. Typical construction styles are shown in Figures 3.4a – 3.4d, for diodes and multiple electrode valves.

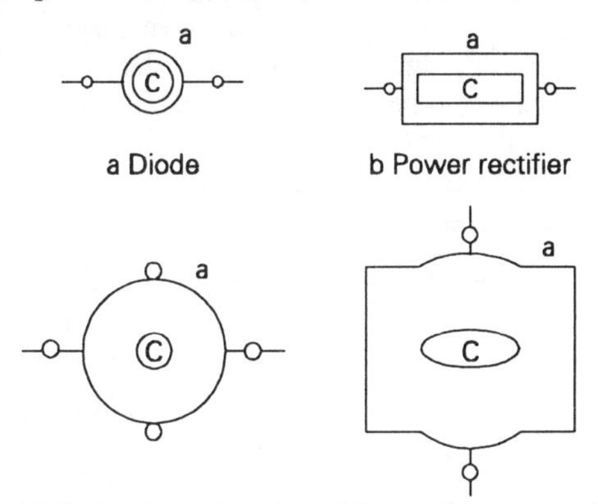

Figure 3.4 *Typical styles of anode construction*

In a rectifier diode, the normal requirement is for as low a conducting impedance as possible, so the anode will be formed in a very tight cylinder around the cathode, at the smallest spacing which is compatible with the avoidance of inadvertent internal short circuits or spark over. The metal of the anode is usually extended on either side of the cylindrical section in order to help radiate any heat evolved.

Valve characteristics

The performance of a thermionic valve in any operating circuit will be determined by its electrical performance characteristics, which will, in turn, depend both on the mechanical construction of the valve and on the conditions in which it is used.

Some of these characteristics will relate solely either to the normal, or to the limiting operating conditions, such as the heater voltage and current requirements – for example 6.3V at 0.3A – or the maximum anode voltage, anode current and thermal dissipation ratings. Others will relate to the specific design of the valve, and the shape and disposition of the internal electrodes.

In the case of a rectifier diode, the main parameter is the ease with which current can flow through the valve, usually specified as its 'anode current resistance', R_a, which is defined as the rate of change of anode current for an increment in anode voltage, as expressed by the equation:

$$R_a = \frac{dV_a}{dI_a} \qquad (1)$$

(This relationship is also valid for amplifying valves, but, in these, assumes that the voltage on the grid is held constant.)

In a rectifier diode, the magnitude of R_a will largely determine the effective forward voltage drop across the valve, in its 'forward' or conducting direction. This is influenced principally, for a given heater temperature, by the area of the cathode, and the cathode–anode spacing.

In triodes, and more complex multiple electrode valves, R_a is also affected by the presence of grid or other electrode structures in the current path, which will restrict the electron flow from cathode to anode. However, in these designs, the structure of the valve also affects the degree of amplification which the component will provide. This is called the 'amplification factor' and refers to the maximum voltage swing which could be produced at the anode for a given change in voltage at the control grid, assuming an infinitely high impedance anode load. The amplification factor is denoted by the symbol 'μ' such that:

$$\mu = -\frac{dV_a}{dV_g} \qquad \text{with } I_a \text{ constant} \qquad (2)$$

where the negative sign takes account of the phase inversion of the applied signal due to the valve.

The third common specification is called the 'slope' or the 'mutual conductance' (g_m) of the valve – or, more correctly, the 'grid-anode transconductance', since it specifies (in amperes/volt or 'Siemens', though, more usually, mA/V or mS) the change in anode current which is brought about by a change in control grid voltage.

This is expressed mathematically by the equation:

$$g_m = -\frac{dI_a}{dV_g} \quad \text{with } V_a \text{ constant} \qquad (3)$$

and these three terms are related by the equations:

$$g_m = \frac{\mu}{R_a} \qquad (4)$$

and

$$\mu = g_m R_a \qquad (5)$$

so that the third can be found if any other two are given in the manufacturer's literature.

Internal grid structures

In a triode, the amplification factor of the valve will be determined by the relative spacings of the grid and the anode in relation to the cathode, and the closeness of the mesh of the grid structure. If the grid is close to the cathode, and the anode is relatively remote, the influence of the grid on the internal current flow will be greater, and the amplification of the valve higher, than if the grid is further away from, and the anode closer to, the cathode.

Similarly, the closeness of the mesh of the grid wire will influence the extent to which an applied negative voltage on the grid will restrict the flow of electrons from cathode to anode, fine meshes giving greater control and vice versa.

However, it will be appreciated that, for a given cathode size, the measures which increase the amplification factor of the valve will also lessen the possible current flow through the valve, and increase its anode current resistance. By and large, therefore, triode valves with high amplification factors will have low anode currents and high impedance values. The comparative structures for grid and anode spacings for high gain, high impedance and low gain, low impedance triodes are shown in Figures 3.5a and 3.5b.

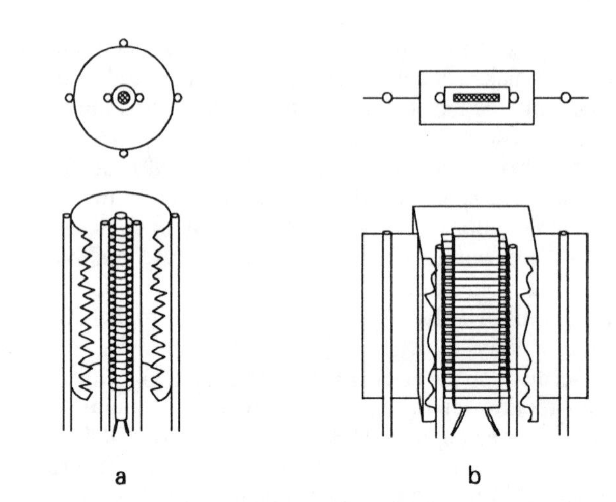

Figure 3.5 *Comparative grid/anode spacings for high impedance and low impedance triodes*

Comparative performance figures for two of the classic small power 'octal' based valve series – the high impedance, high gain, 6SL7 and the equivalent lower impedance type, the 6SN7 – and the 2A3 power triode show this difference.

Valve type	Anode current (mA typical)	Amplification factor (μ)	Anode current impedance (R_a)
6SL7	2.3	70	44.000
6SN7	9.0	20	7,000
2A3	60	6.5	800

Power output triodes, in particular, will have very low amplification factors – typically in the range 5–10. Also, for such valves to operate satisfactorily with very open mesh grids, which do not retain a large space charge, very high cathode efficiencies are necessary, and this usually requires the use of a directly heated system.

Since such an electron source will introduce hum modulation of the signal, output triodes with low voltage filaments, such as the 2.5V operated '2A3', were often used. Such filaments would also be thicker and have greater thermal inertia than those of higher working voltage, which would also help reduce thermal modulation of the electron flow.

In spite of the inconvenience in use of directly heated power output triodes, these enjoyed some vogue in high quality audio power amplifiers because of their favourable distribution of distortion products, which were such as to give a 'rich' or 'warm' quality to the sound.

This preference largely disappeared in favour of triode-connected beam-tetrodes, when these latter valves became widely available, and when the increasing use of negative feedback in audio power amplifier designs reduced the amount of residual harmonic distortion anyway.

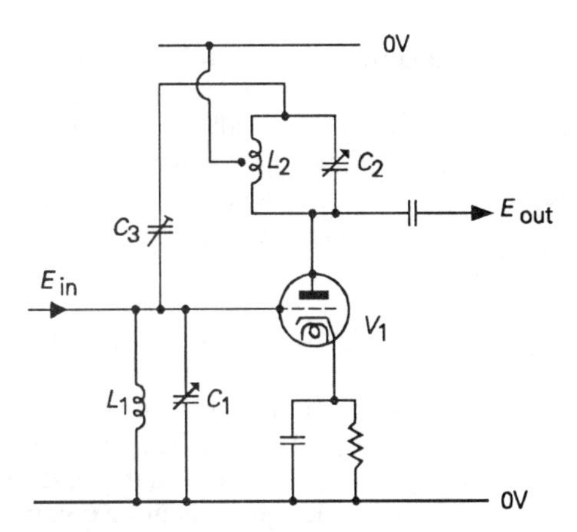

Figure 3.7 *Neutralized triode RF stage*

Screened grid, pentode and beam-tetrode valves

A major problem in the operation of triode valves was discovered as soon as attempts were made to use these valves for RF amplification. In this application, the inevitable inter-electrode capacitance which existed between the grid mesh and the anode plate – typically of the order of 3pF in a small power valve – would cause oscillation, if connected in an RF gain stage circuit layout of the kind shown in Figure 3.6, and stable operation would require some form of 'neutralization' of this capacitance, as shown in Figure 3.7: an inconvenient circuit elaboration, requiring a tapped anode coil and an adjustable 'neutralizing' capacitor.

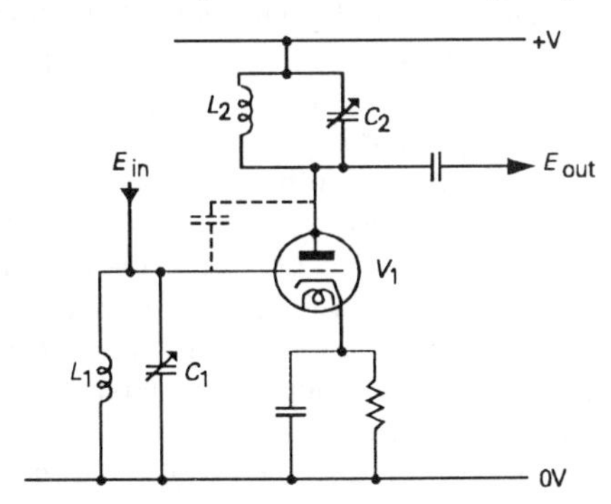

Figure 3.6 *Inter-electrode capacitance in triode RF amplifier slope*

However, if a further fine mesh grid, (G_2), is interposed between the 'control grid', (G_1), and the anode, it is possible to construct valves in which the internal feedback capacitance is as low as 0.005pF, and RF amplifier stages built with these valves will be stable up to very high frequencies.

If this screening grid is connected to the same potential as the cathode, the accelerating field exerted by the positively charged anode upon the electrons within the space charge region will be very small, and very little anode current will flow. Therefore, in order for such a valve to operate, it is necessary to apply a positive voltage to G_2, as well as to the anode.

The actual voltage required for G_2 will depend on the valve design, but will generally lie between 0.3 and 1× the anode potential, and the relationship between G_2 voltage and G_2 current will typically be as shown in Figure 3.8

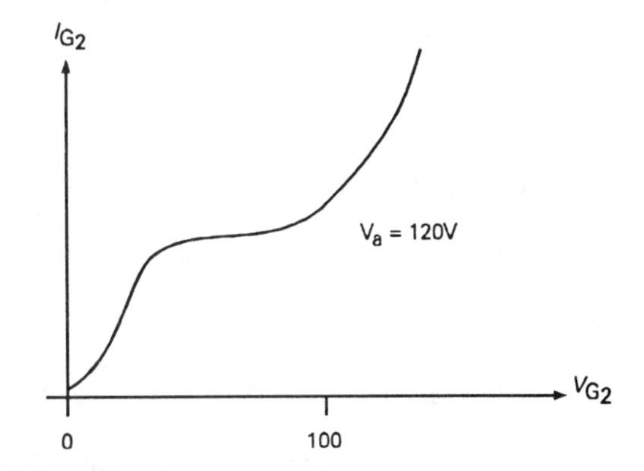

Figure 3.8 *Screen grid current characteristic in tetrode valve*

This current flow is wasteful, in that it contributes to the power consumption and thermal dissipation of the valve, but performs no useful function. It is customary in the design of 'screened grid' or 'tetrode' valves, as they are more usually known, to align the grid wires for G_2 so that they lie immediately behind those of G_1, in order to minimize G_2 current.

An incidental advantage of the tetrode construction is that both the anode current impedance, (R_a), (as defined by the extent to which a change in anode voltage causes a change in anode current) and the amplification factor, (μ), are very high, so that very high stage gains can be obtained from such valves. Typical values are R_a = 1.5 Megohms, $\mu = 4500$.

However, because of the very high values of R_a possible with tetrode designs, it is more common for the g_m of the valve to be quoted in the valve specification, with values in the range 1.5–10 being typical. This allows a rapid estimate of the stage gain (m) to be made, by the use of the relationship:

$$M = -g_m.r_L \qquad (6)$$

so long as the value of the anode load (r_L) is small relative to the AC resistance of the valve.

A further advantage conferred by the very high values of R_a in tetrode and similar valves is that there will be very much less 'damping' of any tuned circuit connected in the anode circuit, as a result of the impedance of the valve appearing effectively in parallel with it.

Secondary emission

There is, however, a practical problem which arises with tetrode valves, which is that high velocity electrons impinging upon the anode may eject lower velocity 'secondary' electrons, which will be captured by the positively charged screening grid, if its voltage approaches or exceeds that of the anode.

Since the velocity of the electrons reaching the anode will depend on the anode voltage, secondary electrons will begin to be emitted only when a certain anode voltage is reached, but will be recaptured by the anode once its voltage becomes higher than that of G_2, this effect can lead to a 'kink' in the anode voltage versus anode current graph of the kind shown in Figure 3.9

While this is relatively unimportant for most RF amplification applications, it would lead to severe

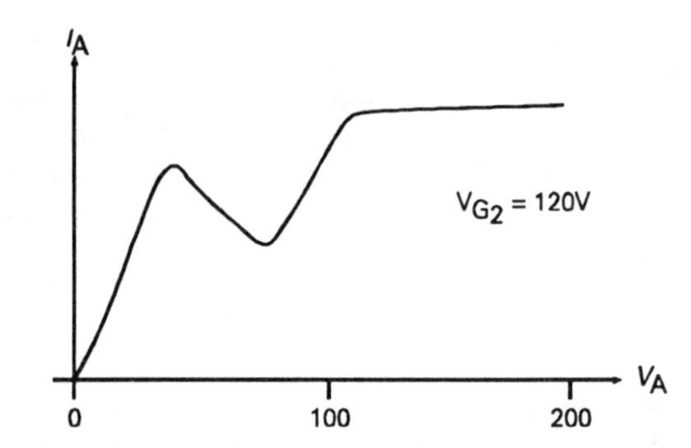

Figure 3.9 *Anode current characteristic of tetrode valve*

output waveform distortions in other uses if large anode (AC) voltage swings were to occur.

Two practical solutions to this problem have been adopted. Of these, the first was to interpose a relatively loose mesh 'suppressor' grid, (G_3), between the screen grid and the anode, and to connect this to a suitably low potential, for example, by an internal connection to the cathode. This would cause a decelerating field between G_2 and the anode, which would have relatively little effect on the high velocity cathode–anode electron stream, but would effectively discourage the flow of the much lower velocity secondary electrons from the anode to G_2.

This was a very successful type of design, which was adopted on a world-wide basis, soon after its introduction, and this layout completely supplanted the normal tetrode RF amplifier valve, except in some battery operated forms.

The high gains and high power efficiencies of the pentode valve also led to its widespread adoption in the period 1935 – 1960 as the output valve in audio amplifiers and radio sets. It did, however, suffer from a substantial amount of third harmonic distortion in its output, and this, by contributing higher frequency components to the output signal, tended to give a somewhat shrill quality to the sound.

The second solution to the problem of the tetrode type anode current kink was to incorporate within the electrode structure a pair of 'beam-confining plates', also typically connected internally to the cathode, of the general form shown in Figure 3.10. These operated by modifying the shape of the accelerating electrostatic fields in the region of the anode in such a way as to discourage the capture of secondary electrons by G_2.

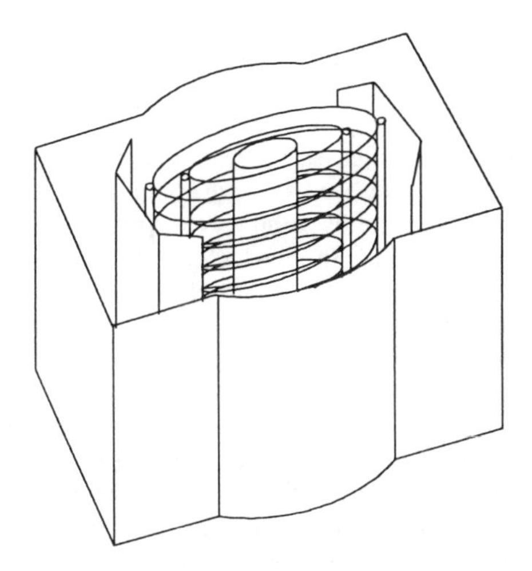

Figure 3.10 *Use of electron beam confining plates in beam tetrode*

The 'beam-tetrode' or 'kinkless tetrode', as this type of valve was known, offered a greater linearity as a power output valve than the pentode, with improved audio quality. It also lent itself readily to triode connection, in which the anode and G_2 were connected together. This allowed even lower output distortion figures, though with a lower stage gain and output efficiency.

Such triode connected beam tetrode output stages soon became the principal type used in high quality audio amplifiers, since they offered all the desirable qualities of the triode output stage, but without the problems associated with the directly heated filament construction.

Other circuit arrangements were also developed for use with beam-tetrodes in audio amplifiers, to allow higher output stage efficiencies, but without a substantial worsening of the distortion figures. These layouts are discussed in Chapter 8.

For RF use, it was not generally felt that the greater large signal linearity of the beam tetrode, in comparison with the normal pentode structure, offered a big enough advantage to justify the extra complexity of the beam-tetrode construction. There were, however, RF beam-tetrodes offered by the Marconi-Osram valve company (the inventors of the design), under the type designations 'KTZ...' and 'KTW...' depending on whether they were sharp cut-off or 'vari-mu' designs.

Variable mu pentodes and tetrodes

With the growing use of 'superhet' type radio receivers, which allowed a very large degree of RF signal amplification, and, consequently, large signal levels at the input to the demodulator, it became practicable to employ 'automatic gain control' (AGC) systems in which the gain of the preceding RF amplifier stages was controlled by the magnitude of the signal at the demodulator.

Special RF amplifier valve types were designed for this use, in which the spiral wire mesh of the control grid (G_1) of the valve was wound so that the spacing between the wires became progressively wider towards one end of the grid structure, as shown in Figure 3.11.

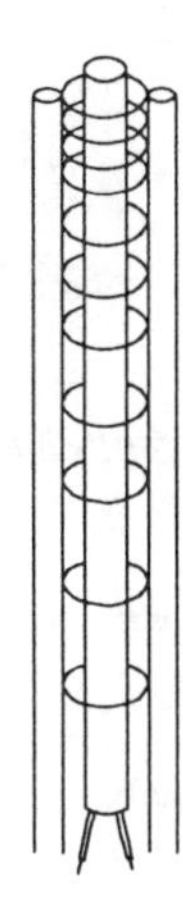

Figure 3.11 *Grid construction of vari-mu valve*

With such a structure, if the negative bias on the grid was progressively increased from its normal operating value, the electron flow would be cut off at the region where the mesh was close together, and progressively restricted to the wider mesh regions of the grid. Since the fineness of the grid spacing determined the amplification factor (μ) of the valve, as the negative bias was increased, so the μ would be reduced. This allowed the stage gain to be controlled by an externally applied bias voltage, as required for 'AGC' purposes. Such valves were called 'vari-mu' types. By contrast, the others were referred to as 'sharp cut-off' types.

Understandably, some proportion of the cathode-anode electron stream in a vari-mu valve would always pass through the wide mesh part of the grid region, even under the normally low bias conditions, and this would mean that the typical performance of such a valve type would be worse than its sharp cut-off equivalent.

Typical anode current vs. control grid characteristics for a small signal triode, a high power triode, a sharp cut-off pentode, a vari-mu pentode, and an output beam tetrode are shown, for comparison, in Figures 3.12a – 3.12e.

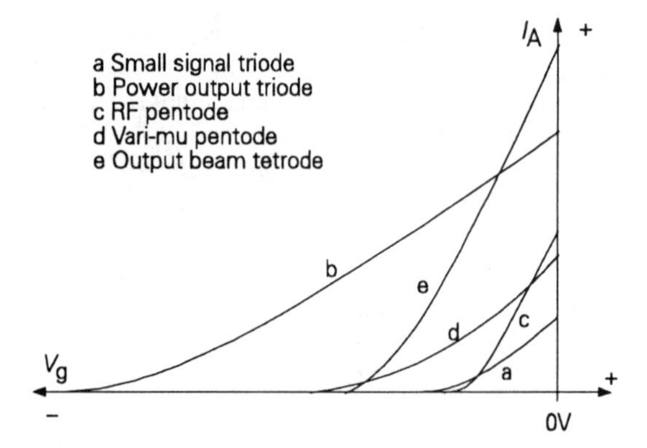

Figure 3.12 *Anode current characteristics of differing valve types*

While vari-mu valves were most commonly found as RF pentodes or beam-tetrodes, the 'hexode' or 'heptode' section of frequency changer types (see below) would also usually be of vari-mu construction.

Other multiple electrode valve types

In the early days of radio set manufacture, a royalty was charged on the use of valves in such apparatus, which tended to encourage the use of as few valves as possible in these designs.

Later, when the relevant patents had expired, the economics of manufacture still tended to encourage the adoption of valves with multiple internal functions, if only because – at a time when the use of a valve meant the punching of a hole through a metal chassis, the fitting of a valve holder, and the wiring up of these valve holders by workers on an assembly line – the fewer valves installed, the lower the manufacturing costs would be.

In a few cases, the use of a multiple electrode structure was the preferred approach, as for example in the frequency changer stages of superhet radio receivers, where it allowed an internal connection between the oscillator and the mixer portions of the valve. In the bulk of cases, however, the main concern was simply to minimize the valve count.

Obviously, the combination valve types which were produced were mainly those which would facilitate the production of mass produced articles, like radios and television sets; using such combinations as double-diode-triodes, or double-diode-pentodes, or twin valves within the same envelope, such as double triodes or triode pentodes; which would provide a substantial market for the valve manufacturers.

Almost all of these multiple valves are now only of academic interest, since, with the exception of the double triode, they are seldom used in any contemporary electronic circuit designs.

Cathode ray tubes

By and large, in low to medium power applications, thermionic emitter systems have been supplanted by various solid state devices, except in very high voltage applications, and even here high voltage MOSFETs are increasingly being used in place of valves. Valves do have advantages, since they are robust and difficult to damage by misuse. However, the major contemporary application for thermionic emitter technology is in the field of cathode ray tubes. These are made in two main types, those intended for electromagnetic and those designed for electrostatic deflection of the electron beam. The former tubes are mainly used for television applications though they are also used in some radar display systems.

In all tubes, an 'electron gun' system is built around the cathode, in order to form an apparent 'point source' of electrons by focusing the emission from the cathode by means of an electrostatic lens. A longer focus electron lens – which can be either electromagnetic, by means of an external 'focus' coil, or electrostatic, by way of a further group of suitably shaped and suitably charged electrodes – is then placed along the barrel of the tube, so that an image of this point source will be formed at the point of impact of the electron beam on the cathode ray tube screen.

For TV use, contemporary practice favours the use of electromagnetic beam deflection but with electrostatic focusing. For use in cathode ray oscillograph tubes electrostatic deflection is invariably used, because of the need for very good high frequency response in the deflection system, which would be impracticable with a magnetic deflection assembly.

Diagrammatic representations of the two types of cathode ray tubes are given in Figures 3.13a and 3.13b.

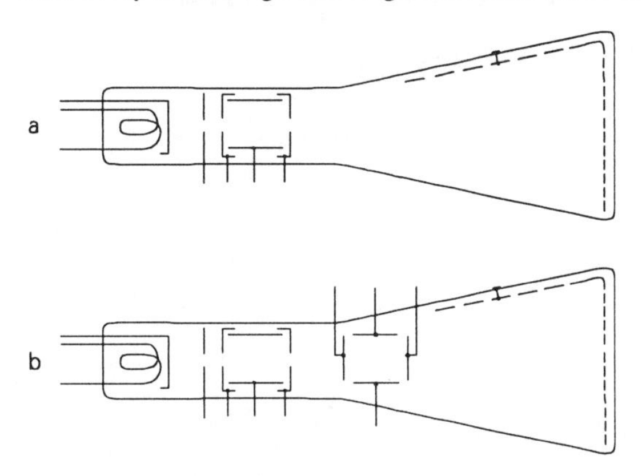

Figure 3.13 *Cathode ray tubes intended for magnetic and electrostatic beam deflection*

Deterioration procesess

One of the drawbacks in the use of thermionic devices is that they gradually deteriorate in use, giving an effective lifetime of some 2,000 – 10,000 hours of service, depending on the conditions of use, and the extent to which their loss of activity impairs the performance of the circuitry in which they are used.

This deterioration comes about as a result of two main causes – the reduction of emissivity of the cathode, and the gradual loss of vacuum within the envelope. Cathode emission deterioration is an inevitable consequence of the chemical reduction of the initial deposit of barium and other oxide coatings, during the use of the valve, and the loss, by evaporation, of the metallic barium, etc., which appears on the surface of the cathode.

This loss of activity is accelerated by higher than normal cathode temperatures, though the life of a flagging cathode may be somewhat extended by an increase in operating temperature. Filament systems will usually outlast indirectly heated cathode types because of their greater relative efficiencies, as shown in Table 3.2.

Deterioration due to loss of vacuum is usually most common in higher power valves, and occurs mainly due to outgassing of the anode and grid structures where these have become hot in use, through excessive current flow. It will also occur, though to a minor extent, due to the reduction of the cathode oxide coatings to base metal; normally the 'getter' will absorb this small gas evolution as it arises.

If a valve becomes 'gassy', it will probably exhibit an internal blue-violet glow, due to the ionization of the gas. This ionization will, however, lead to the production of positive ions which will be accelerated towards the negatively charged cathode, and the resultant ionic bombardment may further damage the cathode emission by stripping off the emissive coating layer.

By comparison, semiconductor devices, if well made and adequately encapsulated, and used within their ratings, will have a virtually indefinite life expectancy, which provided another reason for their use as a 'valve' replacement.

4

Active components based on semiconductors

Basic theory

While the mechanism by which a thermionic valve operates is simple to visualize and easy to understand, the way in which 'semiconductor' devices work is much less easy for the would-be engineer to comprehend, and this has, I think, encouraged the 'black box' approach to circuit design, in which the active devices are treated simply as functional blocks, with appropriate connecting wires and well specified operating conditions.

While this way of working may be entirely adequate for logic systems, in which the required output signal is simply a sequence of pulses alternating between 'zero' and 'logic level' potentials, and the restrictions on the supply line voltages, or device currents, may be irrelevant to the operation of the system, it is too limiting in its constraints to allow much innovative linear circuit design. So, however reluctantly, it is necessary for the designer of 'solid-state' linear circuitry to have a working knowledge of the characteristics and operating mechanisms of the devices he wishes to use. Fortunately, it is possible to possess this understanding at two different levels, which one might call the theoretical and the practical. I will explain the theoretical basis first, with as few departures from the currently accepted view of theoretical physics as are necessary to avoid undue complexity.

With a few exceptions, the electrical characteristics of all pure solid materials may be classified as belonging to one or other of the three categories of insulator, semiconductor or conductor, depending on their electrical behaviour, and this behaviour depends principally on the actions of the electrons in the outer orbital shells.

Although it is convenient to visualize the electrons as occupying planetary type orbits around the nucleus of the atom, this is not really a very accurate representation of their status, which is more correctly defined as a series of energy levels. The distribution of electron energies in these three classes of material is illustrated in Figure 4.1.

It should be remembered that the electrons in low orbits (close to the nucleus) have the lowest energy level, and those electrons in the outer orbital levels

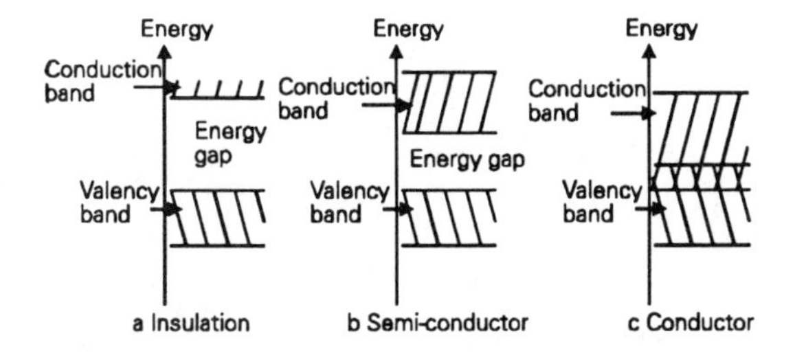

Figure 4.1 *Electron energy distribution in insulators, semiconductors and conductors*

have the highest, so that in order to achieve conduction in insulators or 'undoped' semiconductors some energy input is always required to lift electrons from the outermost normal energy level – known as the 'valency' level – into that state in which the electrons are no longer tightly bound to the nucleus, known as the 'conduction' level. Between these two energy levels is the so-called 'forbidden' band in which electrons may not reside, unless trapped by impurity atoms within the crystal or crystal defects. The forbidden band, also referred to as the 'energy gap', defines the interval between the maximum energy which an electron can possess, and still be bound to the nucleus by attractive forces, and the minimum energy required for an electron to break free from the atom, and migrate from one atom to another within the crystal.

The outermost normal energy levels of the electrons are called the valency levels, because the electronic distribution within these determines the nature of the physical and chemical relationships between adjacent atoms.

In insulators, even the electrons in the highest normal energy states are still very tightly bound to the nucleus of the atom, and the energy gap represented by the forbidden energy levels is wide, so that these valency electrons are not able, except at very high levels of electrical stress, to migrate from these orbital positions into what are termed the 'conduction bands' in which they may move from atom to atom. No electronic current will therefore flow at potentials below those which cause electrical breakdown in the solid.

In conductors, the width of the forbidden band is small, or non-existent, so that the outer orbital levels of the atom (the conduction bands) will be occupied by very many electrons which are loosely bound to the atom and may readily migrate through the solid body, so that electronic current will flow upon the application of a very low accelerating potential.

Those materials within the class known as semiconductors have an atomic structure in which, at temperatures above absolute zero, the thermal energy of the atom may cause some electrons to escape from the tightly bound valency orbits into the more loosely bound conduction band, and this random change of electron orbital position is strongly influenced by the crystal structure in which the semiconductor may exist.

The major class of semiconducting materials of interest to the electronics industry is that in which the atoms have a tetravalent (four electron) outer electron structure, and which crystallize in regular tetrahedral or cubic crystal structures, with the distribution of valency electrons shown in Figure 4.2. Such materials will generally be of those elements which occupy group 4B in the periodic table of the elements, of which the relevant section is shown in Table 4.1, in the order of increasing atomic weight of the elements.

It is also possible, but with rather greater difficulty in manufacture, to utilize materials, having a similar crystalline and orbital electron structure, formed from compounds of elements in Group 3 with those in Group 5, such as gallium arsenide.

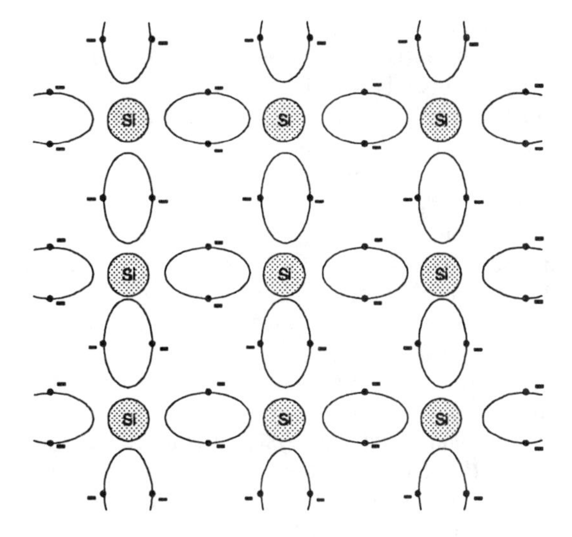

Figure 4.2 *Regular covalent electron bonding in Group IVB element such as pure silicon*

Table 4.1 *Extract from the periodic table of the elements*

Group	3B	4B	5B
Number of outer valence electrons	3	4	5
Element	B Al Ga In	C Si Ge Sn	N P As Sb

The effect of doping

Although some of the materials in Group 4B may be of interest in a completely pure chemical form, because of the phenomena which may occur at the point of contact between a metallic conductor and the surface of the crystal, the principal applications of semiconductor technology concern the effects which arise at a junction between two similar semiconducting materials whose electrical characteristics and crystal structure have been modified by the deliberate introduction of small quantities of impurities – a process known as 'doping'.

If an element with an outer valence level containing five electrons, such as any of those in Group 5B, is introduced into the otherwise regular crystal structure of a Group 4 substance, the interchange between the outer orbits of the valency electrons shown in Figure 4.3 will be disturbed by the presence of the additional orbital electron, and this electron will have a much greater freedom of movement, and can be regarded as occupying a conduction band level.

An impurity atom which has this effect is called a 'donor' atom, and the semiconductor material containing such donor atoms is known as 'N' type, because of its excess of (negatively charged) electrons. Such a 'doped' material will be much more conductive of electricity than the undoped or 'intrinsic' material.

Similarly, if an impurity element having only three outer valence electrons, such as any of those in Group 3B, is introduced into the crystal structure, the disturbance of the shared electron orbits will be such as to leave a vacancy or 'hole' where an electron would otherwise have been, as shown in Figure 4.4.

Such an impurity atom is called an 'acceptor' atom, and material doped in this way is known as 'P' type

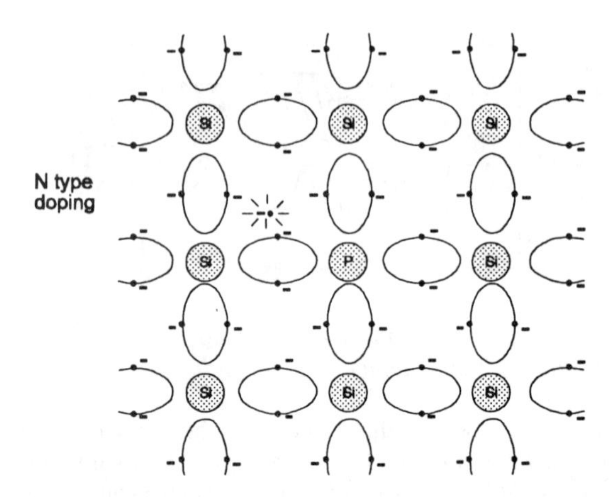

Figure 4.3 *Disturbance of covalent bond structure in silicon crystal due to presence of 'donor' type impurity atom*

material, because of the presence of the (effectively positively charged) 'holes'. This material is also more highly conductive than the intrinsic (undoped) semiconductor crystal because electrons are able, under the influence of an electric field, to migrate through the material by moving to occupy holes, thereby leaving further holes in the place from whence they had come. This gives the effect of the holes themselves moving through the material, simulating the action of microscopic particles having an opposite charge to that of the electron. However, because their movement is, in

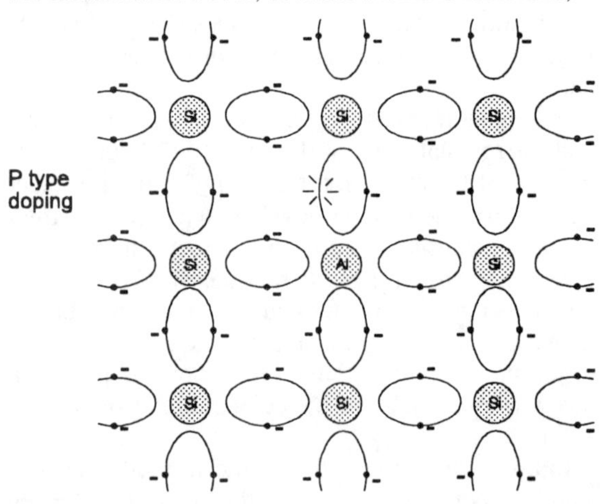

Figure 4.4 *Disturbance of covalent bond structure in silicon crystal due to presence of acceptor type impurity atom*

actuality, built up from a large string of sequential movements of electrons moving small distances to occupy adjacent holes, the effective mobility of hole flow in a P type semiconductor is much lower than that of electron flow in an N type material.

For practical reasons, in order to be able to grow crystals whose structure is not too greatly strained by the presence of the impurity atoms, the impurities chosen are generally those of a similar atomic size, such as aluminium or phosphorus for silicon, and gallium or arsenic for germanium, although, in trace quantities, arsenic can be used as an N type dopant in silicon, and indium as a P type one in germanium.

Sometimes the nomenclature 'N–' or 'N+' and 'P–' or 'P+' is used in explanatory diagrams of semiconductor devices to denote materials having a lesser or greater degree of doping than the standard N or P material.

The semiconductor junction and the junction diode

The currently accepted theoretical model of the energy structure within a multi-electron atom is that proposed by the Italian physicist Enrico Fermi, and, because of this, the notional 'average' electron energy level in such an atom is known as the 'Fermi level', or the 'Fermi energy' (E_f). This concept is of importance at the region of contact between dissimilar materials, and particularly so in the case of junctions in semiconductor materials.

In an 'intrinsic' (undoped) semiconductor material, at a temperature of absolute zero, (0°K), the energy distribution of the electrons will be as shown in Figure 4.5a, in which the maximum energy level of any of the electrons will be that of the valence band. If the material is at a higher temperature, say room temperature (300°K), the thermal energy of the electrons, even in an intrinsic semiconductor, will allow some of them to escape from the pull of the nucleus, and attain energy levels appropriate to that of the conduction bands, causing the Fermi level to lie between the conduction and valence levels, as shown in Figure 4.5b.

Where the material is of N type doping, in which the impurity material has led to an excess of electrons, there will also be electrons in the conduction bands – in addition to those present solely due to thermal excitation – simply because they have been excluded from their normal valence position because of the effects of the crystal structure on the inter-atomic relationships, as shown in Figure 4.3. This leads to an electron energy distribution, even at low temperatures, which is more like that shown in Figure 4.5b.

On the other hand, if the material is of P type, with an orbital electron distribution of the type shown in Figure 4.4, then very few electrons, escaping from the valency levels, will survive being trapped by holes and arrive in the mobile conduction band zone, so that the energy distribution will be more akin to that shown in Figure 4.5a. Because of this, the average electron energy level (E_f) in a P type material will be lower than that in an N type, simply because fewer electrons will normally be found in the higher energy conduction bands. The effect of this is that when an N region and a P region are brought into contact, the reciprocal flow of electrons at different energy levels across the junction, due to thermal agitation, will bring the Fermi levels on the two sides of the junction into an energy equivalence, and this will have the effect of displacing the relative energies of the conduction bands of the P and the N type materials. For this reason, electrons cannot flow from the N to the P side of the junction without some external source of energy, and this leads to what is termed a 'potential barrier' (E_b), between the two zones. Forward conduction will not occur until the applied potential exceeds this value.

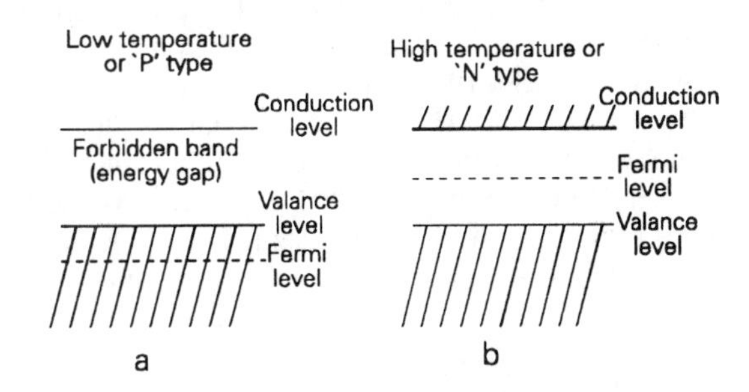

Figure 4.5 *Average electron energy levels in semiconductors as a function of temperature or doping level*

Increasing the temperature of the material will increase the thermal energy of the electrons, and will increase the proportion which have sufficient energy to make the transition from valence level to conduction level, even in P type materials where there would normally be few conduction band electrons. This will have the effect of displacing the Fermi energy to a higher level, particularly in the P regions, and this will reduce the potential barrier presented by the junction.

The actual potential barrier will therefore depend both on the nature of the material and on the temperature of the junction. For germanium junctions at room temperature (20° C) this is about 0.15V, while for silicon junctions it is of the order of 0.58V. (Note that the diagrams of Figures 4.5 and 4.6 relate to electron energy, and so have a polarity sense which is opposite to the normal convention of +ve at the top.)

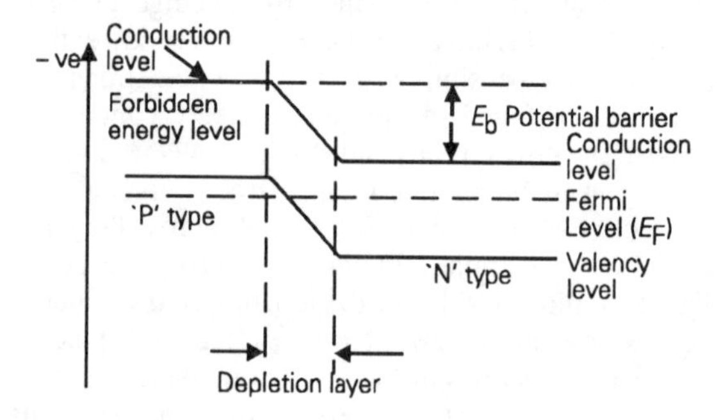

Figure 4.6 *Mechanism by which potential barrier arises at P–N junction*

The depletion zone

There is a further important, and related, phenomenon in P–N junctions which is that a depletion zone will arise on either side of the junction, as shown symbolically in Figure 4.7. This is an additional effect of the migration of electrons across the junction from the N type (electron surplus) into the P type (electron deficient) regions of the semiconductor, in that electrons which diffuse into the adjacent P type region will

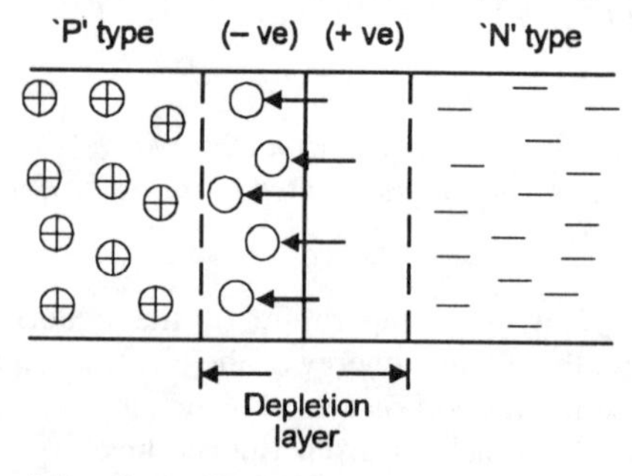

Figure 4.7 *Growth of depletion layer on either side of a P–N junction*

occupy the holes which are characteristic of this material.

The effect of this migration is therefore to remove both electrons and holes from a band on either side of the junction, and leave a region which is effectively depleted of available current carriers. However, the loss of an electron from an atom – even if it has more than will fit into the shared orbital electron structure of the crystal – will leave a positively charged ion on the N type side of the junction. Conversely, the filling of a hole on the P type side of the junction will also create a negatively charged ion in this region, and the accumulating potential of these '+' and '–' charges will eventually shut off further electron migration across the junction.

In the absence of any externally applied voltage across the junction, the thickness of the depletion layer will depend on the population density, 'ε' (epsilon), of free electrons or holes within the junction materials, and also on the size of the potential gap arising at the junction (which in turn depends on the nature of the materials in contact), in that this will determine the extent to which a potential difference across the junction can arise before electronic migration is cut off. Therefore, due to this movement of electrons, even in the case of a junction with no externally applied potential, there will be an effective potential gradient across the depletion zone, which will be +ve on the N doped side of the junction, and –ve on the P doped side.

However, if an external potential is applied, the effect of the additional electrostatic force field on the electrons and holes will either widen the depletion zone – if a reverse bias is applied, as shown in Figure 4.8a – or make it more narrow – if a forward bias is applied, as shown in Figure 4.8c, by comparison with the width, in the absence of any applied potential, as shown in Figure 4.8b. When a sufficient forward bias is applied to overcome the inherent potential barrier at the junction, the depletion zone will vanish and a current will flow, as shown in Figure 4.8d. This gives rise to the type of forward and reverse current flow effects shown in Figure 4.9a. Because of the differences in their electron energy characteristics, and in the size of their potential barriers, different semiconductor materials will have dissimilar forward and reverse conduction characteristics, as shown in the diagram for silicon and germanium P–N junctions.

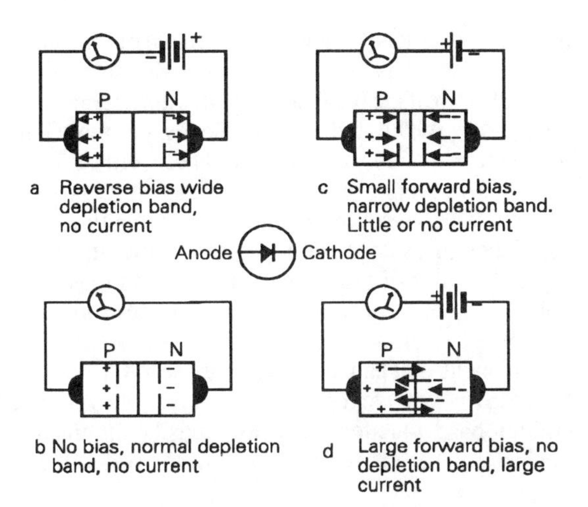

Figure 4.8 *Influence of external potential on width of depletion zone in junction 'diode'*

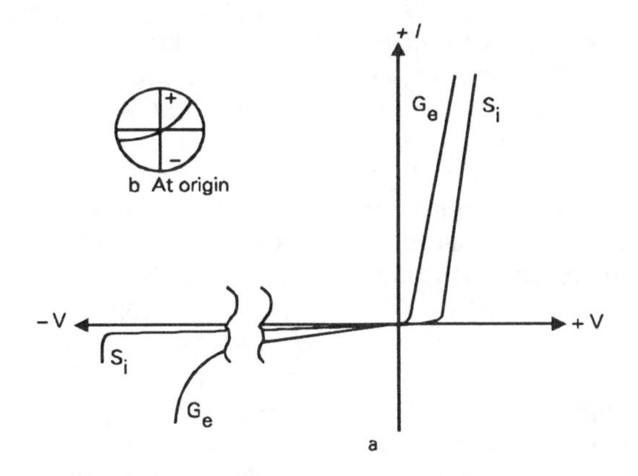

Figure 4.9 *Normal forward and reverse characteristics of silicon and germanium junction diodes*

Reverse breakdown effects

Ideally, there would be no current flow in the reverse direction until the junction breakdown voltage was attained, nor would there be any forward conduction until the voltage defined by the potential barrier was reached. However, in practice, due to crystal imperfections and residual contaminants, there will always be some small leakage current at low applied potentials, as shown in the expanded view of the region around the origin of the graph, Figure 4.9b.

There will also be an increasing reverse leakage current, at increasing reverse voltages, due both to contaminants and imperfections, and also to ionization of the semiconductor due to electrostatic stress effects, related to the geometry of the device. This last effect is very important in high voltage semiconductor rectifiers.

If the reverse voltage applied across the junction is high enough, the thermal dissipation in the junction, due to the small reverse current flow, may be high enough to increase the leakage current still further, so once some critical reverse voltage is reached, the reverse leakage current will begin to increase rapidly.

In addition, the population density of carriers within a semiconductor depends on the doping level of the materials, so that junctions between highly doped materials will give rise to thin depletion zones, while those between lightly doped materials will have relatively wide ones. This has the effect of modifying the gradient of the P–N junction energy diagram of Figure 4.7, as is shown in Figures 4.10a and 4.10b. This has two practical consequences in very highly doped materials, of which the first is that if the boundaries of the N and P regions become too close together, and the energy levels which electrons can occupy are the same on both sides of the depletion layer, the electrons may 'tunnel' in both forward and reverse directions through the depletion layer, at very low applied potentials, in spite of the apparent existence of a substantial potential barrier.

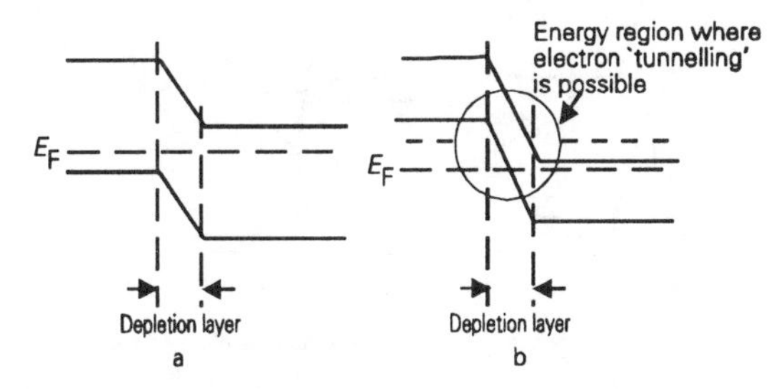

Figure 4.10 *Effect of doping on potential gradient across P–N junction*

This effect is employed in the manufacture of 'tunnel diodes', which have the characteristic voltage/current characteristic shown in Figure 4.11a, where there is very little forward or reverse voltage drop across the junction. The downward kink in the conduction curve is due to the fact that if a forward bias is applied, this will have the effect of displacing the possible electron energy levels, and preventing further tunnelling, which can only occur if the electron can cross the junction without gain or loss of energy. This causes the forward current to drop once more, up to the voltage at which normal forward conduction resumes.

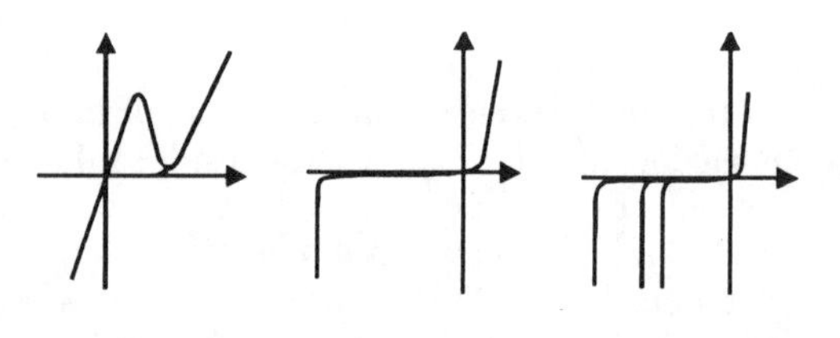

Figure 4.11 *Forward pulse characteristics of tunnel, zener, and avalanche diode*

The same mechanism is used in 'zener' diodes, whose forward and reverse conduction characteristics are shown in Figure 4.11b, except that the doping levels are arranged so that electron tunnelling does not take place until a few volts of reverse junction potential is applied. The energy distributions across the junction boundary at zero, and at breakdown potentials, are shown in Figures 4.12a and 4.12b. Typical zener diode reverse voltage breakdown potentials lie in the range 2.5–5V.

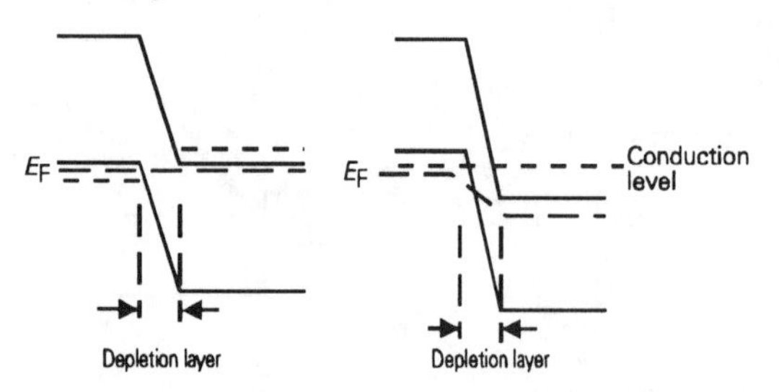

Figure 4.12 *Fermi levels and conductor 'mechanisms' in zener diodes*

The second effect is that in highly doped junctions, the increase in width of the (very thin) depletion layer as a reverse potential is applied will not be adequate to prevent an electrical breakdown of this layer once the kinetic energy of any leakage electrons, accelerated by the applied field and colliding with other atoms, is sufficient to cause an avalanche of ionization in the material. This mechanism gives rise to the abrupt low-voltage breakdown diodes also commonly, but incorrectly, known as zener diodes, but more correctly termed 'avalanche' diodes. These will be found usually in reverse breakdown voltage ratings from 5–100V. The type of forward/reverse conduction characteristics of these diode types are shown in Figure 4.11c. Because of the predictable reverse breakdown potentials, both avalanche and zener diodes form useful voltage reference components.

Avalanche mechanisms also give rise to the reduced breakdown voltage of the relatively highly doped emitter–base junction of a normal silicon junction transistor, which is typically only of the order of 5–6 volts.

(It should be noted that the actual distribution of the depletion zone on either side of the junction is not necessarily symmetrical, unless the P and N materials have an identical concentration of donor or acceptor atoms, but will depend on the relative doping levels of the two contacting regions, so that the width of the depletion layer, other things being equal, will be less on the more highly doped and greater on the less highly doped side of the P–N interface.)

Although high electrostatic stress levels at the edges of junctions can cause premature breakdown effects, the reverse breakdown voltage of a P–N junction is principally a function of the extent of the doping of the materials, and will be greater in those materials which have a lower concentration of impurities (lightly doped, e.g. N– or P–) than in more heavily doped ones. On the other hand, the lightly doped materials will have a lower conductivity, so, for a given junction area, there must always be some degree of compromise in the design of semiconductor rectifiers between conductivity and breakdown voltage.

Junction capacitance

An important inherent characteristic of a non-conducting (e.g. reverse biased) P–N junction is that there will be a significant capacitance between the two halves of the junction, and this will depend on their area and the thickness of the depletion layer by which they are separated. If the bias is low, or the dopant level is high, this depletion zone will be thin, and the junction capacitance will be high.

If, on the other hand, the reverse voltage is increased the depletion zone will widen and the junction capacitance will decrease, and this allows the construction of voltage controlled variable capacitance diodes, which can be made in forms which offer a wide range of initial capacitances, typically from 3–300pF, depending on the junction area and doping levels used, and with a variety of voltage/capacitance characteristics.

The junction capacitance of a reverse biased diode is approximately defined by the relationship:

$$C \approx \frac{\varepsilon A}{w} \approx \frac{k}{\sqrt{V}} \qquad (1)$$

where C is the junction capacitance, ε is the mean population density of holes and electrons on either side of the depletion zone, and depends on the doping level, A is the junction area, w is the width of the depletion region, and V is the (reverse) voltage applied across the junction.

To summarize the practical aspects of semiconductor action, certain elements within the chemical group 4B will, when in crystal form, show electrical characteristics which are intermediate between insulators and conductors, and the normally low conductivity of these materials may be increased by doping them with certain types of impurities. Depending on the type of impurity present, this will cause them to have an excess of electrons, (N type), or conversely a deficiency of electrons (P type) giving rise to holes where these electrons should normally have occurred.

When a junction is formed between an N type and a P type semiconductor, electronic current can normally flow only in one direction, from the N type region to the P type one, but only when a sufficient voltage is applied in the forward direction, and the size of this voltage will depend on the junction temperature and on the materials from which the junction is formed.

The conductivity of semiconductor materials is increased by increasing the concentration of the doping impurity, but this also reduces the reverse breakdown voltage of the junction.

Although it is convenient to visualize the carriers in P type materials as mobile holes, having the same characteristics as electrons but with a +ve rather than a –ve charge, in reality, the movement of a hole is due to the movement of a sequence of electrons, each leaving a hole behind when it move to fill another adjacent hole. For this reason, while the movement of holes, in response to an applied electric field, appears to be that of small positively charged bodies, the actual speed of movement of holes is much lower than that of electrons.

Majority and minority carriers

A P type material is defined as a material having a population of acceptor atoms, and an N type one as one with a population of donor ones. However, because of crystal structure irregularities or unwanted impurities, some free electrons will always be present in a P type material, and some holes will always occur in an N type one, especially during the passage of current through the junction.

These 'alien' bodies will be less effective as current carriers than either electrons in an N type or holes in a P type, because of their proneness to recombination with the principal carriers within the material. They are therefore termed 'minority' carriers to distinguish them from the holes in a P type or electrons in an N type material, which are known as 'majority' carriers.

Ohmic contacts

In order to convert a piece of semiconductor material containing one or more P–N junctions into a practical component, wires must be attached to it. If such wires are simply placed in contact with the material, rectifying action will probably occur, so the method of making such a contact must be chosen so that this does not occur.

A common method is to arrange that the material is very highly doped at the point of contact, with a dopant material which gives the same effect as that already present: for example, a 'P++' layer on a normally P type material, or an 'N++' layer on an N region. A thin layer of metal, such as gold or aluminium, which would cause a similar type of doping action, can then be evaporated on to the surface of the semiconductor material to form a contact pad.

An even easier method is simply to rely on the fact that even when an element such as aluminium is applied, say, to an N type silicon substrate, on which the aluminium will cause the formation of a P type region – because Al is a donor type impurity – the depletion region of such a P–N junction will be so thin that electrons can tunnel through at very small potential differences. So, for all practical purposes, such a junction is also a non-rectifying one.

Point contact devices

Although nearly all modern semiconductor technology is based on the use of inter-crystalline junctions, it is worth remembering that, historically, the first application of solid state electronics was in the 'crystal and cat's whisker' point contact diode, based on the use of a thin metal wire pressed into contact with the surface of a piece of crystalline semiconducting material, most commonly galena (lead sulphide), but including other substances, such as household coke, at the whim of the experimenter.

Such metal/semiconductor rectifiers were usable at the input signal frequencies of early radio transmissions and, in the early days of radio, allowed the construction of simple home built receivers. However, devices of this kind were very fragile and uncertain in operation, because of the proneness of the wire contact to become dislodged from the surface of the crystal.

With the advent of VHF and UHF 'radar' systems in World War 2, there was a need for a rectifier diode, for frequency changer applications, which would operate beyond 500MHz. Thermionic diodes became increasingly inefficient as the operating frequency was increased, and were of little value at these frequencies.

Use was therefore made of an updated version of the original 'crystal and cat's whisker' system, but using a small slice of monocrystalline silicon as the crystal, and with a tungsten whisker held in firm contact with its surface. This type of point-contact diode was subsequently supplemented by similar diodes based on germanium, of which the lower melting point facilitated the production of larger lumps of single crystal material. In manufacture, it is customary to pass a brief pulse of high current through the diode which will 'spot weld' the wire contact – which will generally be either of tungsten or of gold – onto the surface of the crystal, which makes the whole structure physically robust.

Silicon point contact diodes have a higher forward voltage drop (typically 0.3 – 0.45V) than germanium ones (which will typically be in the range 0.1 – 0.2V), but have a better high frequency characteristic, and remain usable up to the GHz range, whereas the germanium diode loses efficiency above some 50–100MHz.

Because of the high electrostatic stresses which arise at a point, the reverse breakdown voltages in point-contact diodes will always be relatively low, in the range 5 – 15V.

Germanium point contact diodes are still commercially available, and find use in relatively low frequency applications where their low forward conducting voltage drop is of advantage. They do, however, have a relatively low reverse resistance and reverse breakdown voltage in comparison with the silicon junction devices, which are now by far the most common commercial type.

A useful, and now, happily, almost universal convention is that the cathode (N type region) of a rectifier diode will always be marked on its case with a black band or ring. As a mnemonic, if the end of the diode marked with a black ring is connected to the negative end of a battery, current will flow.

The point contact transistor

Theoretical studies of the action of the metal/semiconductor rectifier had led to the concept of a depletion zone within the semiconductor material in proximity to the metal contact, which would lead to a modification of the distribution of carriers as shown for a point metallic contact on a P doped substrate (base) in Figure 4.13a.

The idea of introducing a further metallic point contact into this depletion zone, so that electrons emitted by this contact would be collected by the other positively charged wire, instead of disappearing into the P type base, was conceived, in December 1947, by John Bardeen and Walter Brattain, at the Bell Telephone Laboratories, in the USA, using the structure illustrated in Figure 4.13b.

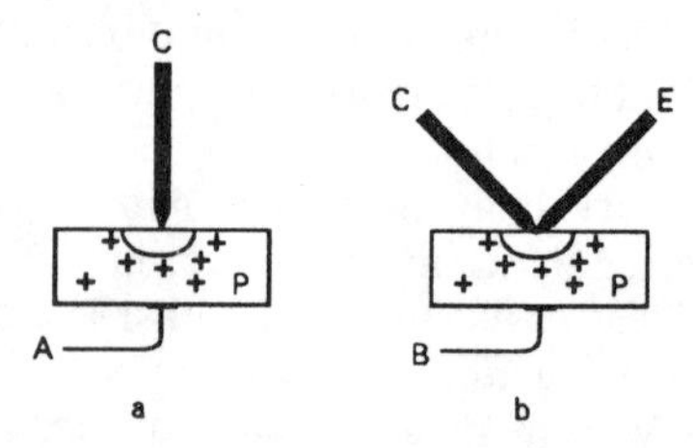

Figure 4.13 *Point contact diode and point contact transistor*

The major practical problem, that of maintaining the two contacts sufficiently close together to allow effective collection of the emitted electrons was solved by Bardeen and Brattain, in their first successful experimental model, by the use of a 'V' shaped wedge of polystyrene, which had been coated on its angled faces with a thin film of metallic gold. The continuity between the two faces was then broken at the apex of the 'V' by cutting off the tip of the wedge.

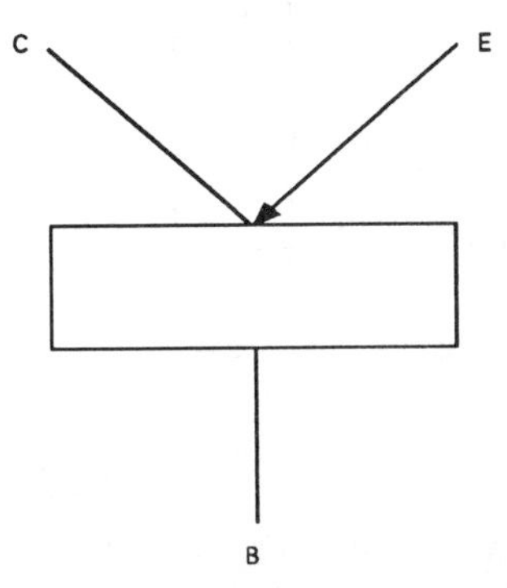

Figure 4.14 *Symbol for PNP transistor – derived from structure of early experimental model*

While this early experiment is of little practical importance, apart from establishing the feasibility of such a device, it has given both the 'emitter', 'base' and 'collector' terminology to the transistor, and also the graphic symbol which is now used to represent the device, as shown in Figure 4.14 – in spite of the fact that the physical structure of contemporary transistors is a long way removed from this layout.

In early point-contact transistors, the only practical method of operation was to inject a current into the low impedance emitter circuit, so that a substantially identical current could be drawn from the collector. Amplification was possible because the output impedance, at 10,000–20,000 ohms, was very much greater than the input impedance, which might be typically in the range 25–200 ohms.

Because of the possibility of voltage gain arising from this characteristic, the term 'transfer resistor' or 'transistor' was coined.

While, ideally, the collector current would be identical to that of the emitter, too great a spacing between emitter and collector points would reduce this figure, while any ionization due to the electrostatic stress at the collector point might even increase the current transfer ratio to rather greater than unity.

All in all, the point contact transistor was never much more than a laboratory curiosity, since its mechanical fragility and the general unpredictability of its electrical characteristics made it a difficult component to use. However, the possibility of achieving the required proximity between collector and emitter regions, by diffusion of suitable impurities within the crystal body, allowed the construction of fully usable and reliable components.

The bipolar junction transistor

In the previous sections we have considered the nature of the semiconductor P–N junction, and the way in which such a P–N junction can be used to make a robust solid state rectifier diode. While such semiconductor diodes are valuable components, in that they are much more compact, much more efficient and simple to use than the thermionic equivalents, and have both greater reliability and much greater life expectancy, it is the fact that this mechanism can also be used – in various different arrangements – to make an amplifying device which has caused semiconductor technology to have such an enormous impact on the electronics field.

The normal 'bipolar' junction transistor is made by combining a pair of P–N junctions, so as to form a composite three-segment N–P–N or P–N–P element, within a single crystal block, to which connecting wires are attached by 'ohmic' (i.e. non-rectifying) contacts, in the manner which is shown schematically in Figure 4.15. Because of the way in which these regions of the semiconductor material operate when the device is used, the upper section is called the

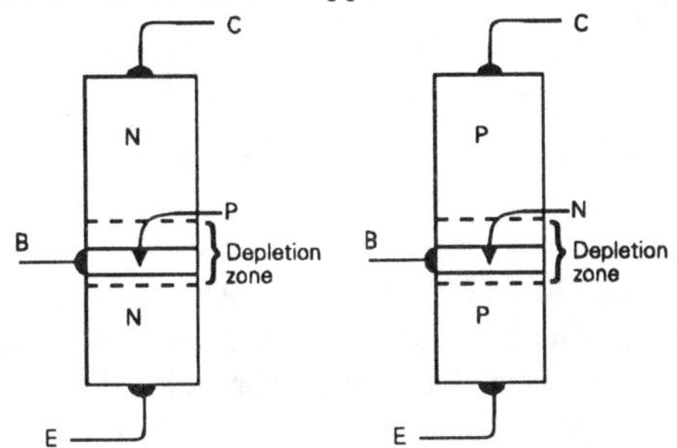

Figure 4.15 *Single crystal junction transistor structures*

'collector', the central one the 'base', and the lower one the 'emitter'.

For practical reasons, the P type base region in the NPN slice – or the N type base region in the PNP one – will have a relatively low impurity level, as will the collector region. The emitter region will have a relatively high doping level.

As explained above, when a P–N junction is formed within a single crystal block a depletion band arises on either side of the junction, in which all the normal carriers are removed by recombination as a result of electron diffusion across the junction. In the case of a thin, lightly doped, base layer, sandwiched between two oppositely doped regions, the whole of the base region will be depleted of carriers. Therefore, when the collector–base junction is reverse biased, as shown in Figure 4.16, no significant amount of current will flow from the base region to the collector at any potential below the collector–base junction reverse breakdown voltage. If, on the other hand, the switch S_1 is closed, forward biasing the emitter–base junction, a substantial flow of electrons will pass from the emitter into the base region, and, since this region is almost entirely depleted of carriers, these electrons will be swept up by the accelerating effect of the base–collector electrostatic field.

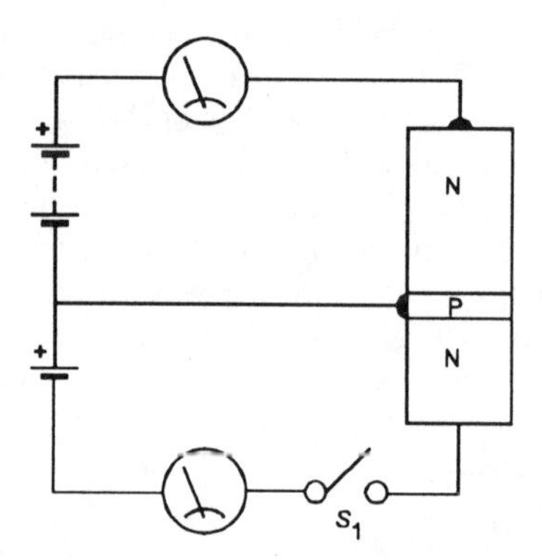

Figure 4.16

The only loss of emitter current on its way to the collector will be due to recombination with the residual holes present in the base, and provided that the base region is thin (of small volume) and lightly doped, this loss will be small.

The same circuit can be rearranged as shown in Figure 4.17, when similar considerations will apply, in that when S_2 connects the base region to the emitter there will be no current flow from the emitter to the base, so that the only collector current will be that due to the reverse leakage current of the base–collector junction.

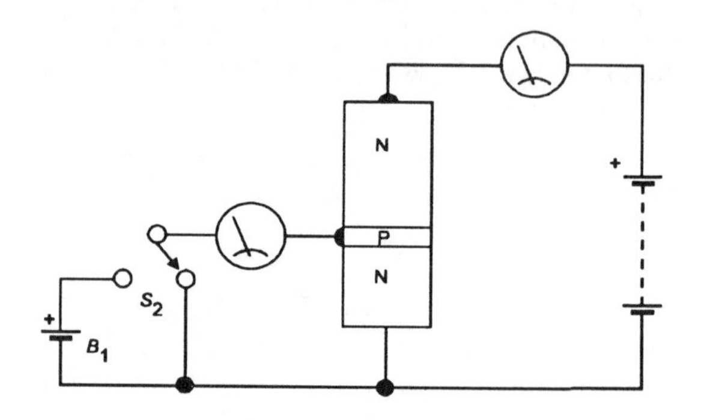

Figure 4.17

If, however, S_2 connects the base to the battery B_1, the emitter–base junction will now be forward biased, and virtually the whole emitter current will be swept up by the collector. The only base current which will flow into B_1 will be due to those parts within the base region where the accelerating base–emitter field is locally stronger than the collector–emitter one. The internal layout of modern junction transistors is chosen to minimize this.

The base current flow therefore effectively controls the emitter collector current flow, and the ratio of the collector to base currents is known as the (common emitter) 'current gain', and denoted by the Greek symbol 'beta'. This can have a value in the range 15–50, for high current power transistors, to 100–1000 for low current 'small signal' devices. The less commonly specified ratio of collector to emitter current, as would be found in the circuit layout of Figure 4.16, is denoted by the Greek symbol 'alpha'.

Since the emitter current in a transistor is the sum of the base and collector currents, the relationship between these two ratios can be derived, and is:

$$\alpha = \frac{\beta}{1+\beta} \text{ and } \beta = \frac{\alpha}{1-\alpha}$$

Although the example of an NPN transistor has been chosen to explain junction transistor action, the same arguments will apply equally to a PNP one if the applied potentials are reversed in polarity and the action of 'holes' is substituted for that of electrons.

Modern manufacturing techniques allow the production of PNP transistors which are substantially identical, in their current gain and other characteristics, to NPN ones. However, because the motion of holes is slower than that of electrons, PNP transistors, in which the P regions are relatively thick, are slower in response speed than the NPN ones where the only P region is the very narrow base section.

The collector–base breakdown characteristics of PNP type silicon junction transistors are also worse than NPN ones, due, at least in part, to the differences in the effectiveness of the technology available for producing N doped silicon, as compared with P doped material, so that high voltage silicon transistors will almost invariably be NPN ones.

The converse is true for germanium devices, where the performance of PNP devices is superior to that of the NPN ones. Because of their inferior reverse leakage characteristics, due mainly to their lower maximum operating temperature, germanium devices have now been supplanted almost completely by those based on silicon.

Moreover, because manufacturing skills and techniques depend on research, and, for commercial reasons, this is mainly devoted to those aspects of the business which offer the greatest market return, a very substantial gulf in manufacturing and applications technology now exists between germanium and silicon materials, and this disparity is entirely in favour of silicon based components. The semiconductor circuit designs, and related comments, offered in this book will therefore apply to silicon based devices, unless otherwise stated.

To summarize, in a P–N junction, electronic current can flow easily from an N type (electron surplus) region into a P type (electron deficient) one, provided that the forward potential is enough to overcome the potential barrier which will exist at the junction. This potential barrier will depend on the nature of the materials forming the junction, on the junction temperature, and, to a lesser extent, on the doping levels used.

Electronic current will not flow from a P type junction to an N type one, at potentials below those high enough to cause reverse junction breakdown, apart from a leakage current mainly due to minority carriers, and this leakage current is strongly dependent on junction temperature, reverse junction potential, impurity (doping) level and the presence of defects in the crystal structure.

If, however, within a single crystal body, two PN junctions are formed to make an NPN or PNP three layer sandwich, and if the outer regions of this sandwich are sufficiently close together that the depletion zones which will arise in the central layer can overlap, then electrons (or holes) drawn into the intermediate region (known as the base) from a forward biased one (known as the emitter) can readily cross this depleted base region, without loss due to recombination.

If the other P–N junction is reverse biased with respect to the base, this means that it will have a forward bias in respect to any electrons or holes injected by the emitter, so that these current carriers (electrons or holes) will be swept up by the accelerating potential present on the third (or collector) region in this NPN or PNP sandwich.

In practice, this allows a relatively small current injected into, or drawn from, the base region, at a relatively low potential – of the order of 0.5– 0.6V – to control a much larger current at a much higher collector potential, and this allows a large degree of signal amplification.

The current gain or beta of such a three-layer device, known as a junction or bipolar transistor, will depend on the thinness of the base region, as will the high frequency response, since it will take a finite time for an electron or hole to cross the base region.

The maximum operating voltage at which such a transistor may be used will also depend on the width of the base layer, and will increase as the base region thickness is increased. For this reason, high levels of current gain, or good high frequency response characteristics, are more readily obtained in transistors intended to operate at relatively low collector voltage levels.

5

Practical semiconductor components

Junction transistor manufacturing techniques

Having established in the early 1950s that it was possible to make 'solid state' semiconductor devices which would perform many of the same circuit functions as the existing thermionic tubes or 'valves', but with many advantages in size and power consumption, progress since that time has largely been in the development of constructional techniques giving lower manufacturing costs, improved performance, and greater reproducibility and reliability.

Early junction transistors endeavoured to simulate, though with rather more fully controlled manufacturing processes, the basic structure of the early point-contact devices, but with the points brought towards each other on the opposite sides of a semiconductor slice, as shown in Figure 5.1a, into cavities which had been etched to cause a local reduction in the effective thickness of the slice.

Since the positional stability of point contacts is not very good under conditions of mechanical vibration, a manufacturing improvement was made by replacing these by infilling the cavities with a material (indium in the case of a germanium transistor) which would produce, by local surface diffusion, a pair of P regions on either side of the N type base, as shown in Figure 5.1b.

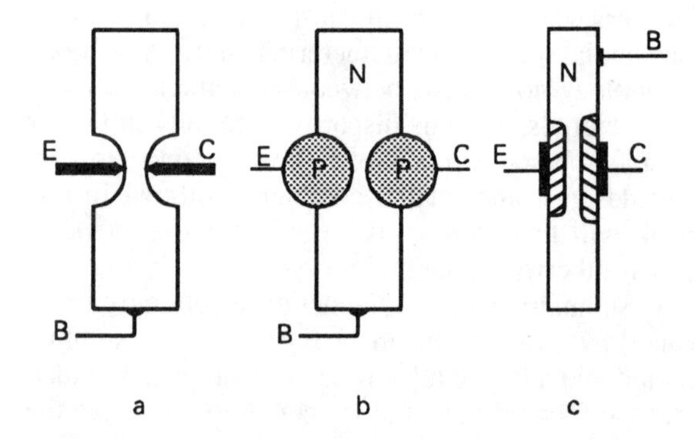

Figure 5.1 *Early PNP junction transistor manufacturing processes*

There are obvious difficulties in controlling the

depth of an etched cavity so that the residual base thickness – which will determine both the current gain and the HF performance of the device – is precisely constant from one device to another during a production run. A further improvement was therefore evolved, in which the required P type impurities were simply deposited on opposed areas of a thin flat slice, which was then heated to cause these dopants to diffuse inwards to the required extent, as shown in Figure 5.1c.

The problems with this approach were that the silicon slice needed to be very thin to start with, which meant that it would also be fragile, and that if the device ran hot in use, the P type junction regions would continue to diffuse towards each other, causing the current gain of the device to increase, and its breakdown voltage to reduce, until the two regions eventually met in the middle, and the device became defunct.

The other shortcoming of all these approaches was that such devices were made on a 'one off' basis, whereas what was needed was a technique which allowed a large number of devices to be made, simultaneously, on the same semiconductor crystal slice.

This was accomplished by the introduction, in 1960, of the 'planar' manufacturing technique, by the Fairchild Instrument Corporation, in the USA. In this process, by the use of photo-lithography, selected areas of the slice were exposed through apertures in a protective mask, and dopants were allowed to diffuse inwards, to produce, for example, on an N type silicon wafer, a series of flat bottomed P doped base 'wells', into which a further series of N+ type dopants could be diffused to produce the emitter regions, as shown in Figure 5.2. The silicon wafer can then be sliced into individual transistor segments, known as 'dies', by a diamond cutting wheel.

This masking and diffusion process could be carried out to produce, side-by-side, as many transistors as the size of the slice or the skill of mask production would allow, and this greatly reduced the cost of the individual devices.

The planar system is particularly suited to use with silicon, since it can easily be oxidized to silica, which is an excellent insulator, and is also an effective resistor to limit the areas affected by the added vapour phase dopants. The planar process allows the production of both PNP and NPN devices, of which the current gains and breakdown voltages are readily controllable.

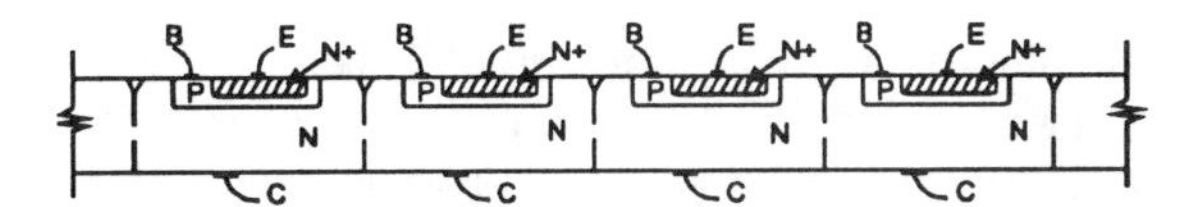

Figure 5.2 *The planar manufacturing process for multiple transistors on a single slice of silicon*

An improvement on this process, introduced in 1962, was the so-called 'epitaxial' system, in which the silicon wafer, after grinding and polishing, is exposed to an atmosphere of hydrogen and silicon tetrachloride, at a temperature of some 1200° C, so that a thin monocrystalline layer of silicon, containing appropriate traces of dopant impurity, will be grown on the surface of, and with an identical crystal structure to, the single crystal silicon substrate. The masking process is then only required to define the regions through which the N+ impurities would be diffused into, say, a P type epitaxial layer on an N type substrate, to produce a series of NPN transistors.

It will be appreciated that the performance of a junction transistor is greatly affected by the characteristics of the base region. For this reason, the 'planar epitaxial' technique is now by far the most commonly used manufacturing process, because it avoids the need for the base region to contain both N and P type impurities, which arise in the planar process when, for example, a P type dopant is diffused into a region which already has an existing N type impurity.

In order to lower the conducting resistance of the relatively thick collector region, it is customary to diffuse a more heavily doped layer into the reverse side of the slice, to produce a P+ region in a PNP device, and an N+ layer in an NPN one, as illustrated in Figure 5.3.

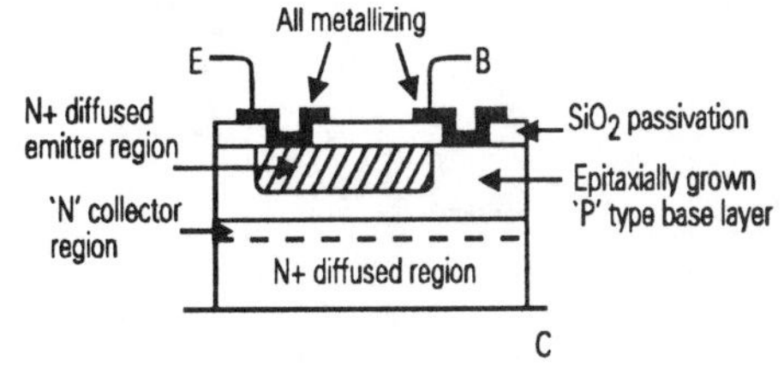

Figure 5.3 *Epitaxial planar transistor structure*

The 'ohmic' connection to the collector will normally be made by soldering the individual dies onto metal 'header' tabs and connections to the emitter and base regions will be made by vacuum deposition of aluminium through apertures in the upper 'passivating' layer of silica, so that they make contact with the appropriate exposed semiconductor regions, as is also shown in Figure 5.3.

The typical dimensions of the die used for a small-signal transistor are 0.5mm square, ranging up to about 4mm square for a high power device. In order to conduct heat away from the junction regions, the slice thickness is kept as small as possible compatible with ease of handling and freedom from breakage. Values of 0.15 – 0.5mm are typical.

Junction transistor characteristics

The major difference in use between a thermionic valve and a junction transistor is that the transistor is a current operated device, with a relatively low input (base–emitter) impedance, whereas the valve is a voltage operated component, with a very high input impedance. Moreover, while for quite a large part of its usable input voltage range, the valve has a quite linear relationship between input (G_1) voltage and anode current – as shown in Figure 3.12 – this is not true for the junction transistor, which has very nonlinear input voltage vs. output current transfer characteristics. This arises partly because the relationship between base-emitter voltage and base current is nonlinear, of the general form shown in Figure 5.4, and partly because the common-emitter current gain (beta) is influenced by collector current, being lower at very small and large values of collector current than at some intermediate current levels, as shown in Figure 5.5.

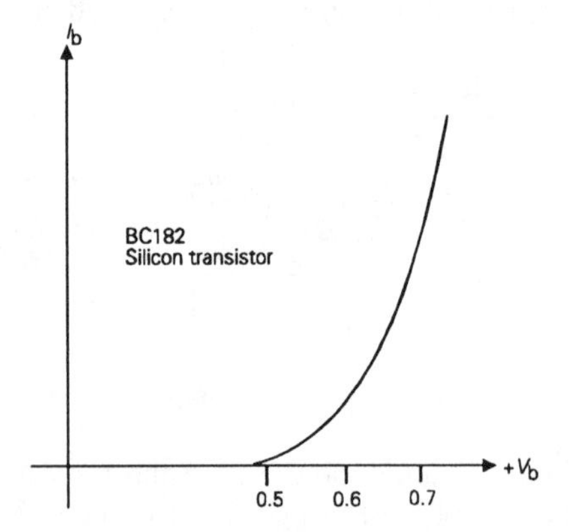

Figure 5.4 *Relationship between base current and base voltage in junction transistor*

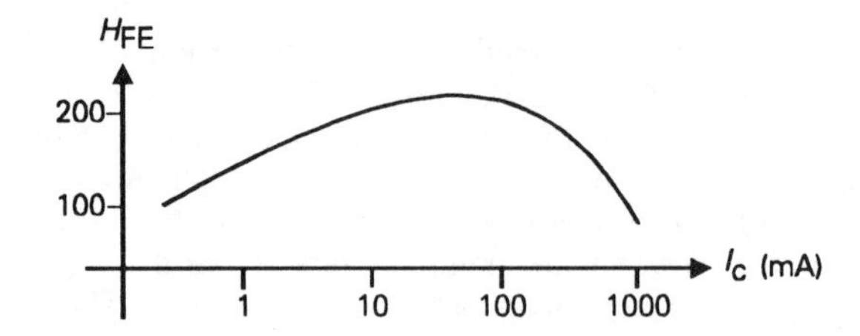

Figure 5.5 *Relationship between current–gain and collector current in silicon junction transistor*

For these reasons, the use of a transistor as a voltage amplifier, in the circuit shown in Figure 5.6, would lead to substantial distortion in the output waveform. Care must therefore be taken in the design of transistor voltage amplifiers to minimize the undesirable effects of the device characteristics. To some extent, the linearity of the relationship between current gain and collector current can be helped by care in the design of the internal structure of the device, and recent device types are better than earlier ones.

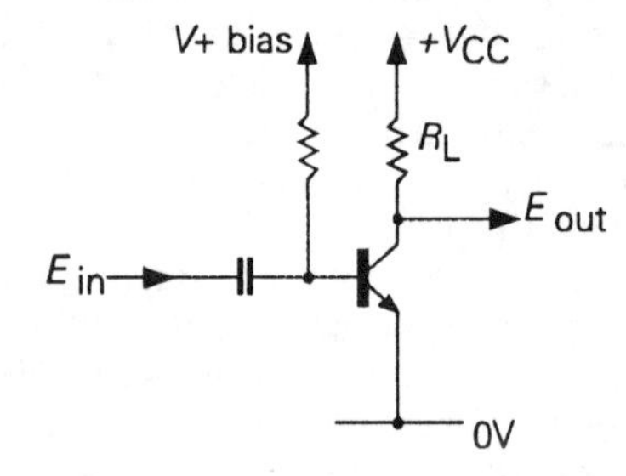

Figure 5.6 *Simple transistor voltage amplifier circuit*

The output impedance of the bipolar transistor is high, so that the relationship between collector current and collector voltage is typically as shown in Figure 5.7, for a range of potentials applied to the base. This high output impedance characteristic is often of considerable benefit in bipolar transistor circuit design.

Under small-signal use, the input impedance of a junction transistor is dependent on the value of collector current, as shown in Figure 5.8, as is also the 'noise figure' (the ratio, in dB, between the random noise introduced by the transistor, as an amplifier, and the thermal noise which would be expected from a perfect amplifier having the same input impedance), as discussed in greater detail in Chapter 16.

With thermionic valves, an important performance

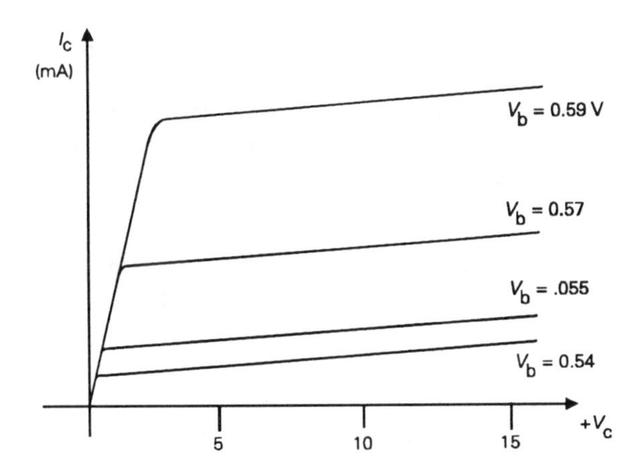

Figure 5.7 *Relationship between collector current and collector voltage for small signal silicon transistor*

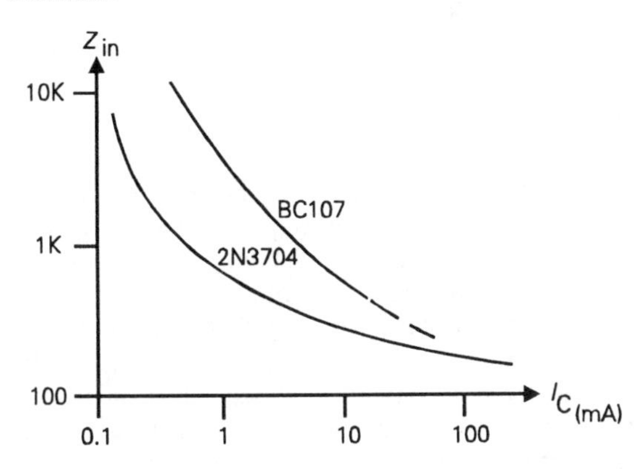

Figure 5.8 *Relationship between input impedance and collector current in silicon junction transistors*

characteristic is the 'slope' of the input voltage vs. anode current curve. This is usually called the 'mutual conductance', g_m, and expressed as mA/V, or, in contemporary terminology, milliSiemens (mS).

Because of their very steep collector current turn-on characteristics, bipolar transistors have very high values of mutual conductance (I_c/V_b), measured in Siemens (A/V), but this figure is not usually specified because it depends so strongly on collector current, (I_c). This is shown for a small signal and a power transistor in the graphs of Figures 5.9a and 5.9b. The theoretical relationship between mutual conductance and collector current is

$$g_m = I_c \frac{q}{kT} \qquad (1)$$

where q is the charge on the electron, (1.60×10^{-19}), k is Boltzmann's constant, (1.38×10^{-23}), and T is the absolute temperature. For a junction temperature of 25° C (298° K), this equation gives a calculated value of 38.9S/A. In practice, the actual value for g_m in a transistor will be influenced by its construction. Typical values of 'mutual conductance' for silicon bipolar transistors, as a function of collector current, lie in the range 15–45 S/Ampere.

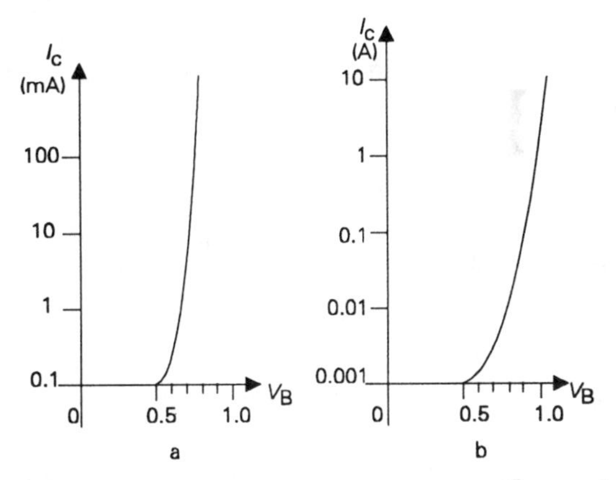

Figure 5.9 *Relationship between base voltage and collector current in junction transistors*

h parameters

Because the junction transistor is a conductive element, in which the output and input circuits will interact with one another, the relatively simple methods for calculating circuit performance which were used with thermionic valves were not of use with transistors.

A more fundamental approach was therefore adopted in which these devices were treated simply as four terminal 'black boxes', as shown in Figure 5.10, in which the relationships between the input and output circuits could be specified in terms of their mutual impedance (Z) or admittance (Y) parameters.

With growing familiarity in the use of these by circuit engineers it was found that some of these parameters were more useful than others, and so a combination of these, known as 'hybrid' or h parameters, has become almost universally adopted.

In these, h_i is the input impedance, and is defined as

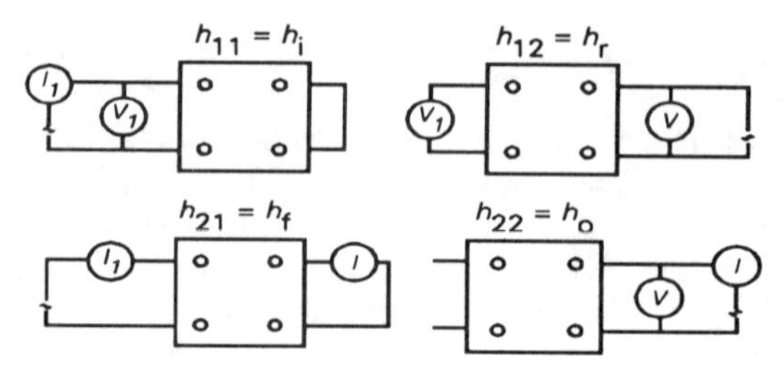

Figure 5.10 *Black box approach to derivation of h parameters*

$$h_i = \frac{V_i}{I_i} \quad (2)$$

h_f is the forward current transfer ratio, and is defined as

$$h_f = \frac{I_o}{I_i} \quad (3)$$

both of these measured with the output voltage held constant, h_o is the output admittance, and is defined as

$$h_o = \frac{I_o}{V_o} \quad (4)$$

and h_r is the voltage feedback ratio, defined as

$$h_r = \frac{V_i}{V_o} \quad (5)$$

both of these as measured with the input circuit open circuit (o/c) to AC.

The numerical value of these parameters will vary, depending on whether the transistor is connected in the 'common emitter' (emitter returned to a neutral no-signal line), 'common collector', or 'common base' configurations, and the condition for which the parameter is defined will be denoted by the subscript 'e', 'c' or 'b'.

For example, if the 'forward current transfer ratio' – commonly known as the 'current gain' – were to be specified for a transistor connected, as it would normally be, in the common emitter mode, the symbol h_{fe} would be employed.

Another convention is that an upper case (capital) subscript is used to denote the large signal, or DC value, of this characteristic. If the small signal value is intended, as, again, would normally be the case, the subscript to the h symbol is written in the lower case, so a numerical value quoted for h_{fe} means the 'small signal current gain, for a transistor connected in the common emitter configuration'.

From the published values for these parameters, the circuit performance may be calculated. For example, the stage gain given by a junction transistor in the circuit of Figure 5.11 can be derived from the equation

$$\frac{V_{out}}{V_{in}} = \frac{h_{fe} R_l}{R_s + r_i} \quad (6)$$

where R_s is the source resistance, R_l is the load, and r_i is the internal (base/emitter) impedance of the transistor.

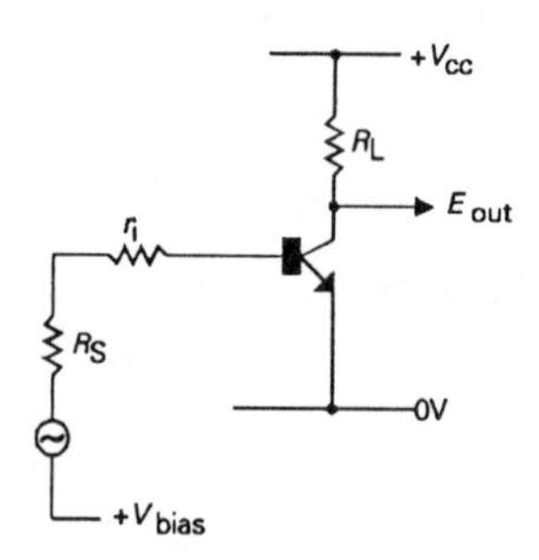

Figure 5.11 *Transistor stage gain model*

The last of the commonly published performance parameters relates to the high frequency performance of the device. Because of the finite time which it will take an electron or a hole to cross the base region of the transistor, and because of the effect of base–emitter and base–collector junction capacitance, in combination with the finite base input impedance of the device, the effective current gain will decrease with increasing frequency. This is specified as the 'transition frequency', or 'common emitter gain–bandwidth product', f_t, which simply means the frequency by which the small-signal current gain of a transistor amplifier stage, using a circuit layout in which its emitter is connected to an AC neutral line, will have decreased to unity.

This method of specification tends to give an over-generous impression of the HF performance of the device, since a quoted value of f_t = 300MHz, for a small signal transistor with an h_{fe} of 300, will imply that the current gain of the transistor will have fallen to 150, at 2MHz, to 30 at 10MHz, and to 5 at 60MHz.

Thermal effects

The other characteristic of the bipolar junction device

which must be considered is that the junction transistor is a temperature dependent device, in that both the forward base–emitter voltage and the collector–emitter breakdown voltage decrease with temperature. Not only does this determine the maximum working temperature which is permissible with the device, but it also controls the permitted thermal dissipation of the junction, for any given die size, and for any given thermal conductivity through the mounting substrate to free air, or to some external 'heat sink'.

A specific problem which arises with all these devices is that the emitter current (and hence the collector current) for any input voltage, will depend on the temperature of the base/emitter junction. If this becomes hotter, more current will flow, and the device dissipation will increase, making the base/emitter junction still hotter.

This problem is particularly acute in power transistors with a large junction area, in that, if one small region of the base/emitter junction should become hotter than the rest, its forward voltage will decrease, in comparison with the remainder of the junction, and this will cause the total device current to be funnelled through this small section of the device, heating this small local region still further.

To avoid the possibility of device failure due to this cause, manufacturers specify the limits of the 'safe operating area' (or SOA) for all power devices, in the form shown in Figure 5.12. Meanwhile, care should be taken in all transistor circuit design to ensure that the dissipation of the devices stays within the specified limits. If long device life expectancies are required, the thermal safety margins should be generous.

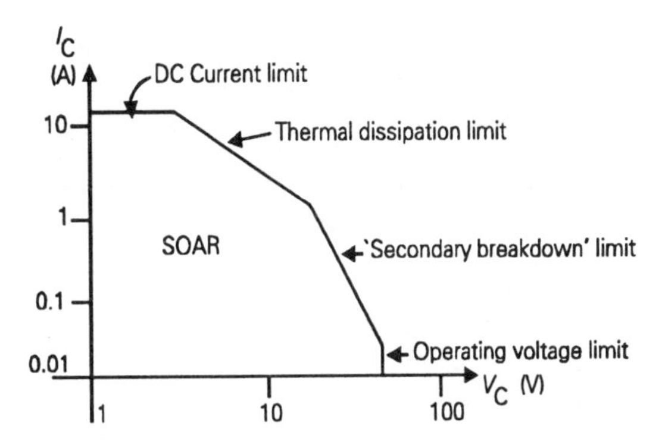

Figure 5.12 *Typical safe operating area curve for power transistor*

Junction field effect transistors (FETs)

These devices utilize semiconductor junction technology to construct a type of transistor which has very similar characteristics to those of a thermionic valve, but without the need for a heated cathode or a relatively high supply line voltage.

The method of construction is shown in Figure 5.13, and consists, in its essentials, of a thin slice of silicon into which a pair of gate regions, with a different type of doping to that of the substrate, have been diffused from opposite sides of the slice. If the slice is of N type silicon, and the diffused gate regions are of P type, as shown in Figure 5.13a, the device is described as an N-channel FET. If the slice is P type, and the gate regions are N type, as shown in Figure 5.13b, it will be called a P-channel FET. These two types of FET are substantially identical in characteristics, except that one operates from a +ve supply line, and requires a negative bias on its gate to cut off the channel current, while the other operates from a –ve supply, and requires a +ve gate bias.

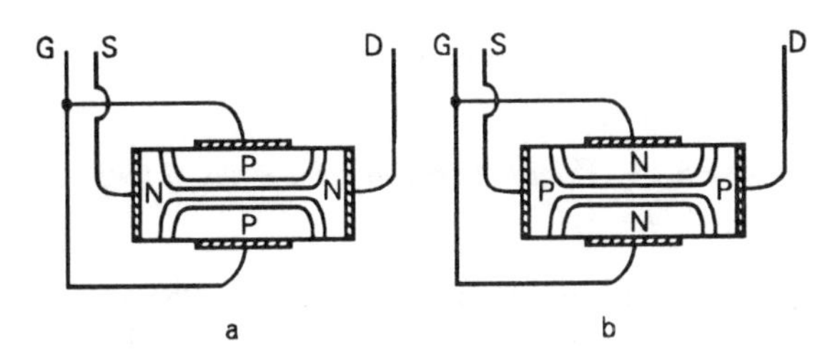

Figure 5.13 *Schematic construction of N- and P-channel junction FET*

The method of operation of the FET is basically very simple. If the slice of semiconductor material has some suitable impurity content it will conduct electricity by electron or hole flow between ohmic contacts at either end of the slice. These contact regions are called the 'source' and the 'drain' respectively, and the conducting path between them is called the 'channel'. However, for reasons given in Chapter 4, when oppositely doped regions are formed on the slice, depletion zones will occur as a result of electron and hole diffusion, both within these regions, and within the slice, and in these regions, which have been depleted of carriers, no current will flow. That part of the slice which can carry the current is therefore reduced in thickness and the resistance of the channel is increased. If the reverse bias on the gate regions is increased, the channel will be further narrowed. On the other hand, if it is biased in a forward direction, the depletion zones will decrease, up to the point at which forward conduction of the gate occurs, as a simple junction diode.

This leads to the type of relationship between gate voltage and drain current shown in Figure 5.14 for a typical N-channel junction FET. The curve for a P-channel FET would be similar, but with reversed polarities.

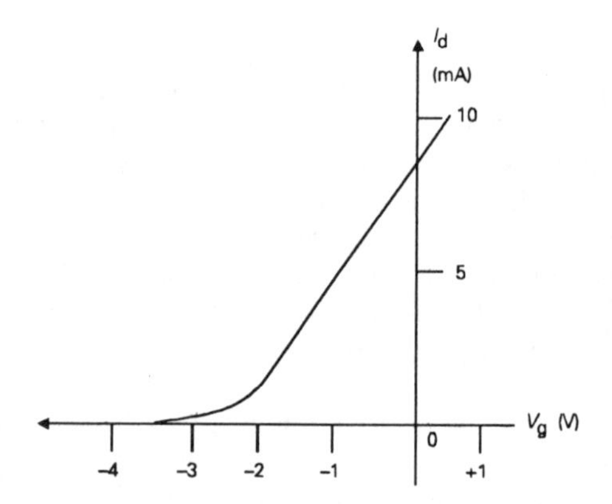

Figure 5.14 *Conduction characteristics of typical junction FET*

A further feature of the FET which must be noted is that, because there is a potential drop along the conducting channel, from source to drain, the actual gate–channel potential will vary across its width, leading to an unsymmetrical depletion region, as shown in Figure 5.15. As the drain–gate potential is increased, the depletion regions extend further towards each other until they meet, and 'pinch off' all current.

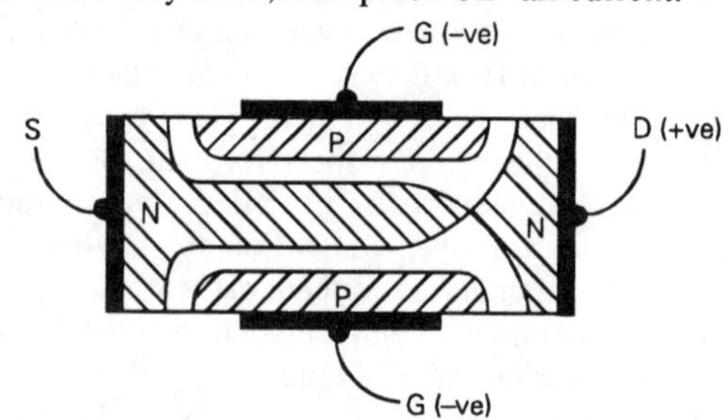

Figure 5.15 *Effect of pinch off in junction FET*

When this happens, the potential drop along the remainder of the channel, towards the source, disappears – with a consequent widening of the conducting channel, apart from the narrow neck at the drain end, which becomes sufficiently short that electrons can still tunnel through it.

In this 'pinched-off' condition, the current through the FET is substantially constant, for any given gate–source bias, over a wide range of source–drain potentials, as shown in Figure 5.16. For a typical small-signal FET, pinch-off will occur at a relatively low drain–source potential (typically of the order of 2–3V), and because the FET has better performance characteristics when operated in this condition, especially because of its very high output impedance, FETs are usually employed at drain potentials which are greater than the pinch-off value.

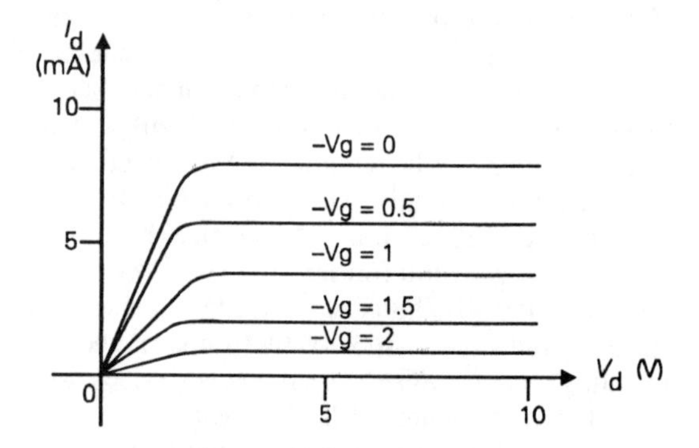

Figure 5.16 *Junction FET drain current characteristics at voltages above pinch off*

The very flat I_d/V_d characteristic of the junction FET, when used in this mode, is much more characteristic of a pentode valve than of a triode. Junction FETs do, however, have a fairly high gate–drain capacitance – varying with gate-drain voltage, but in the range of several pF – so they are not very useful, in the common source mode, as RF amplifiers, unless some form of 'neutralization' is employed.

Because of the much more linear character of the gate voltage/drain current curve, the distortion introduced by junction FET gain stages is much lower than for bipolar transistors. On the other hand, because the mutual conductance (g_m) of the FET is very much lower – values in the range 2 – 10mS (mA/V) being typical – the stage gain for the same value of load impedance will also be much lower. Also, because the

FET has a very high input impedance, (up to a million megohms), in its normal reverse biased state, the thermal noise associated with this is higher than for a bipolar device. An exception to this is the case where the device is used in an application in which the source impedance is comparably high. In this case the noise figure of the FET will be better than that of a bipolar transistor in a similar application, as explained in Chapter 16.

Junction FETs are normally only found in small-signal forms, at drain–source voltage ratings below 40V, and with free air (maximum) dissipation figures of the order of 350mW. Although power FETs have been made, these have largely been superseded by MOSFET types, and are now seldom found.

Insulated gate field effect transistors

These devices, most commonly known as 'MOSFETs' (metal oxide semiconductor field effect transistors) or sometimes as 'IGFETs' (insulated gate FETs), represent a relatively recent development of semiconductor technology, if only because the technical problems in their manufacture are severe, and satisfactory methods for overcoming these problems have taken time to evolve. Commercial devices of this type are almost exclusively fabricated from silicon, though a few VHF devices have been made using gallium arsenide.

The basic idea behind the IGFET is a very simple one, and is illustrated in Figure 5.17. It is that if an electrostatic charge is brought into proximity to a body in which charge migration is possible, polarization will occur so that charges, of opposite sign to that inducing this polarization, will be drawn towards that face of the body which is closest to it, while charges of the same sign will be repelled away from it. This effect can be utilized in a practical device if the type of layout shown in Figure 5.18a is adopted. In this a strip of P type material is formed on a P+ substrate, and a pair of N type zones are diffused into it at either end, to form source and drain connections. At the source contact, the metallization is extended to connect to the base/substrate as well, and a further metal contact, referred to as the gate, is deposited on a thin insulating layer formed between the source and drain regions.

If a positive voltage is applied to the drain electrode, with respect to the source and substrate, no current will

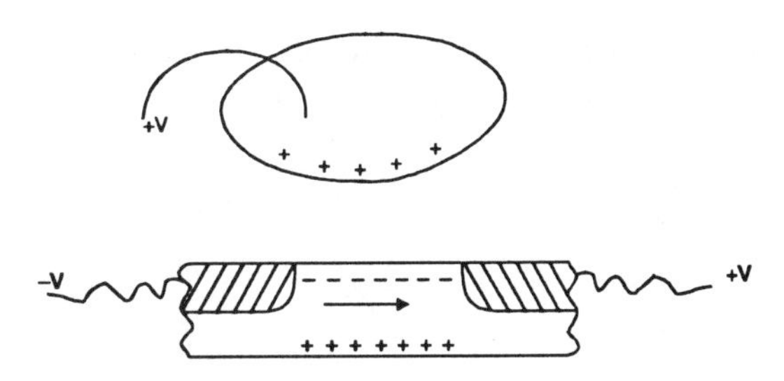

Figure 5.17 *Mechanism of electrostatic induction of mobile charges*

flow, because the substrate/drain P–N junction is a reverse biased diode, and there is no conducting path through the P– layer, between the source and drain, underneath the gate region, which forms, in effect, an NPN transistor with its base shorted to its emitter.

If, however, a positive charge is also applied to the gate metallizing, a layer of negative charges will be induced into the P– substrate, as shown in Figure 5.18b, to form a link between the source and drain regions, and current will flow. Because this current flow is parallel to the surface of the silicon die, transistors of this type are called 'lateral' MOSFETs, to distinguish them from devices, such as bipolar transistors, where the current flow through the silicon die is 'vertical', that is to say, in a direction normal to the device surface.

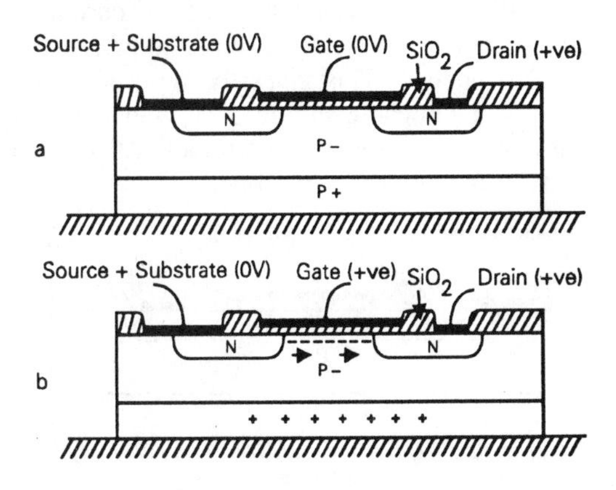

Figure 5.18 *Simple insulated gate FET*

This source/drain current will be directly proportional to the magnitude of the positive potential applied to the gate electrode, apart from minor effects due to the imperfect geometry of the system, and this leads to a very linear input voltage vs. output current transfer characteristic, as shown in Figure 5.19.

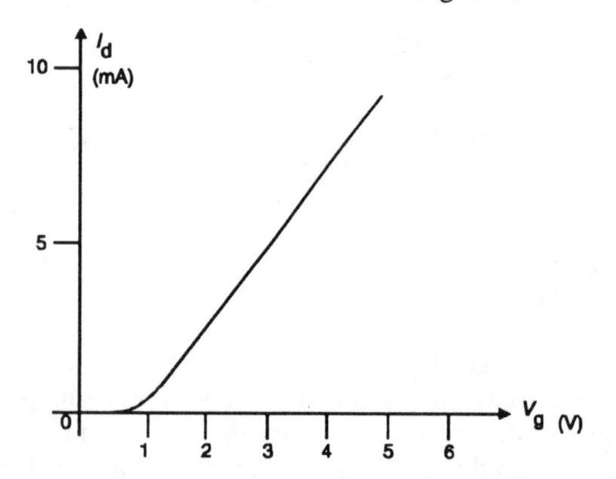

Figure 5.19 *Conduction characteristics of small signal MOSFET*

Because the current flow is due to induced negative charges (electrons) and there is no current flow at zero gate voltage, such a device is known as an 'N-channel enhancement type MOSFET'.

If a thin layer of N type silicon is formed underneath the gate region during manufacture, as shown in Figure 5.20, so that there is an already existing conductive path between the source and the drain, a negative charge applied to the gate will drive the N type carriers away, leading to an eventual cut-off of current altogether when a sufficiently large negative gate potential is applied. This gives the type of transfer characteristic shown in Figure 5.21, which is very similar in its form to that of a thermionic valve, and is probably even more linear in slope.

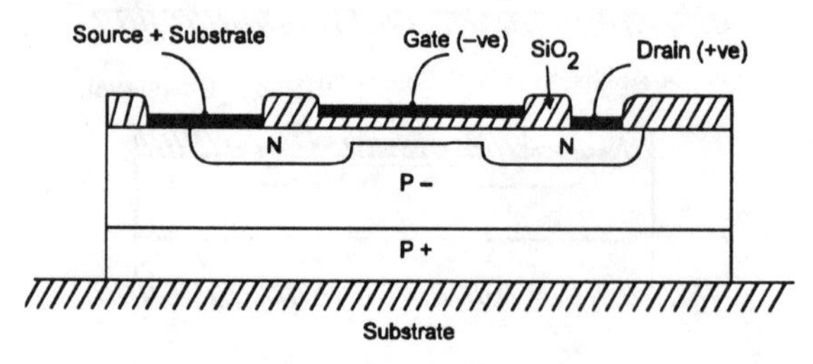

Figure 5.20 *Depletion type MOSFET*

This kind of IGFET is called a 'lateral N-channel depletion MOSFET', and is typical of the small signal MOSFETs used particularly for HF amplification purposes.

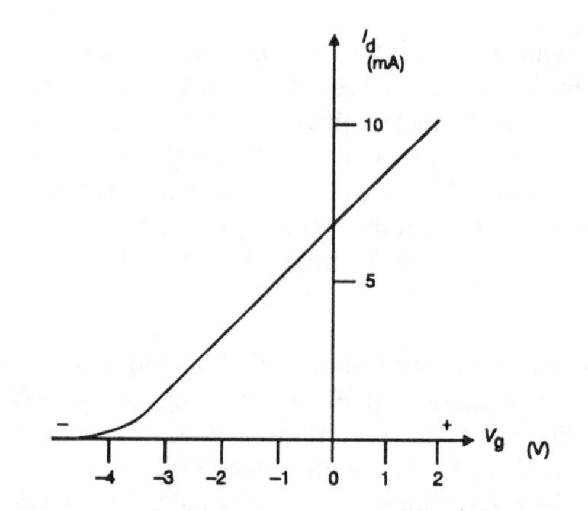

Figure 5.21 *Conduction characteristics of N-channel depletion MOSFET*

By using a complementary symmetry in the N and P type regions, P-channel MOSFETs can also be made, which have very similar electrical characteristics to the N-channel versions, and use the type of structure shown in Figure 5.22. The circuit symbols used for these devices are shown in Figure 5.23.

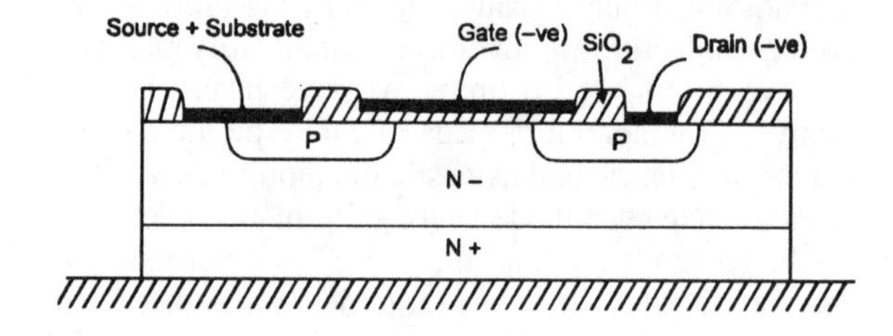

Figure 5.22 *Construction of lateral P-channel enhancement MOSFET*

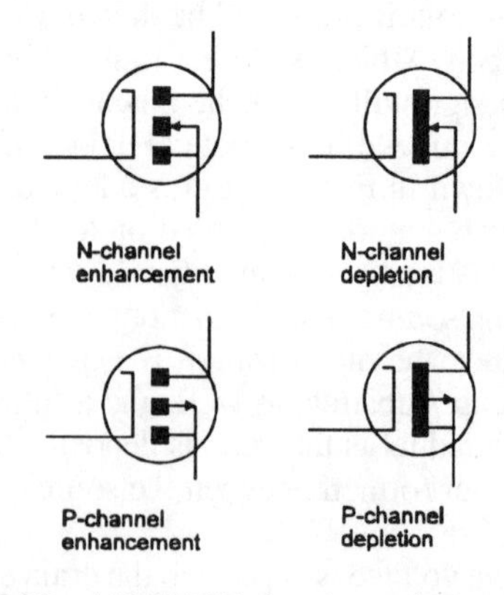

Figure 5.23 *MOSFET circuit symbols*

The practical problems which delayed the realization of this type of transistor were mainly concerned with the production of a sufficiently thin and uniform layer of insulation between the gate electrode and the silicon die. The requirement being that a useful amount of current carrying charges could be induced into the channel at a sensibly low operating voltage, without this insulation being so fragile that inadvertently applied potentials would destroy it.

However, it was also necessary to maintain a very high degree of control over the chemical purity, and physical structure of the channel region, since any inadvertent contamination or crystal structure irregularity could lead to uncontrolled leakage currents.

Developments in manufacturing techniques, mainly evolved to allow the routine production of integrated circuits, have provided process systems which are well able to meet these needs so that reliable insulated gate FETs with repeatable device characteristics are now available at a modest cost.

Dual-gate MOSFETs

As with triode valves, there is a significant inter-electrode capacitance between drain and gate, and this would cause instability, and possibly continuous oscillation, if such devices were used as simple RF amplifier stages. So, by analogy with the triode to tetrode elaboration, dual-gate MOSFETs are made, as shown in Figure 5.24, with a second gate region between the signal gate (G_1) and the drain. This effectively isolates G_1 from the drain, and allows feedback capacitances as low as 0.02pF to be attained, so that such transistors can be used as stable high gain RF amplifiers. In straightforward RF amplifier use, such a transistor would normally be operated with a small negative potential (2–3V) applied to G_1, since such dual-gate MOSFETs are usually depletion type devices, while G_2 might be taken to a small positive potential, in the range 0–3V.

However, the availability of this second control electrode allows the use of such a transistor in other circuit applications, such as mixers or frequency changers. Since dual-gate MOSFETs are principally used for HF/VHF applications, and the flow of electrons is very much more rapid than that of holes, these transistors are made exclusively as N-channel devices.

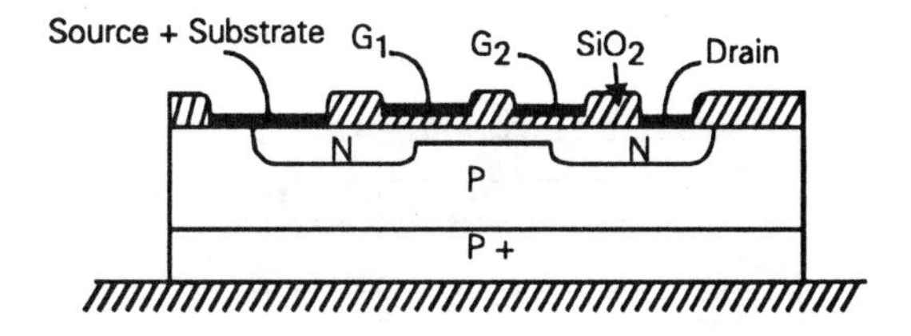

Figure 5.24 *Lateral dual gate N-channel depletion MOSFET*

Power MOSFETs

Small signal transistors based on MOS technology give a very good HF response, since, in general, the rate of change of the conductivity of the channel is determined solely by how fast the input resistance and capacitance of the gate circuit will allow the gate potential to change.

Since this type of transistor ought, in principle, to work just as well in larger power versions, considerable technical effort has been directed, over the past decade, towards the production of power MOSFET devices which can be used as a replacement for the more sluggish bipolar power transistors.

The use of the 'lateral' MOSFET structure, shown in Figures 5.18 and 5.20, is unsuitable for high current applications because the length and the shallowness of the current carrying channel leads to too large a path resistance. Early attempts to reduce the length of the channel, by restricting the width of the gate region, led to problems due to the limited precision of the photolithographic processes by which the etching patterns are defined.

However, it was noted that in bipolar junction transistors the 'vertical' current flow path, from the emitter to the collector, is defined by the thickness of the diffused layers, as shown in Figure 5.25a, and this thickness can be controlled with great accuracy.

So-called vertical MOSFET devices have therefore been produced, as shown in Figure 5.25b, in which the source, channel and drain regions are formed as horizontal layers, as in a bipolar device, and a 'V' shaped notch is then etched through these layers to allow an insulated gate electrode to be applied across the exposed channel region. This gave rise to the description 'VMOS' for this type of device, though, because of the relatively high electrostatic stresses which arise at sharp points, the bottom of the notch is more usually

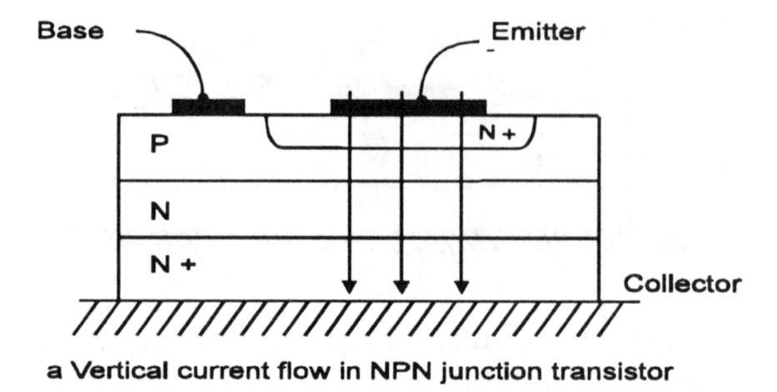

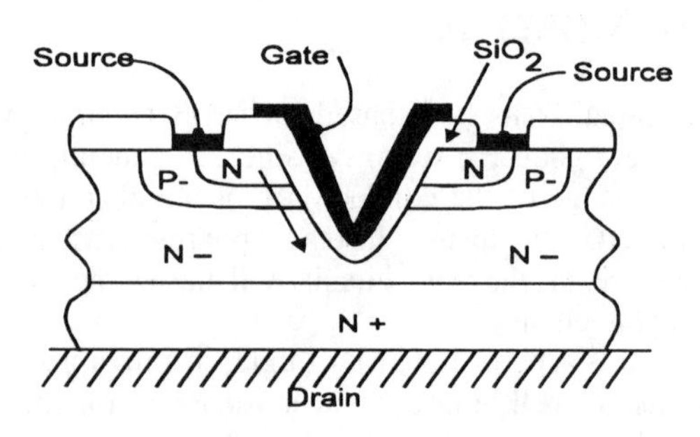

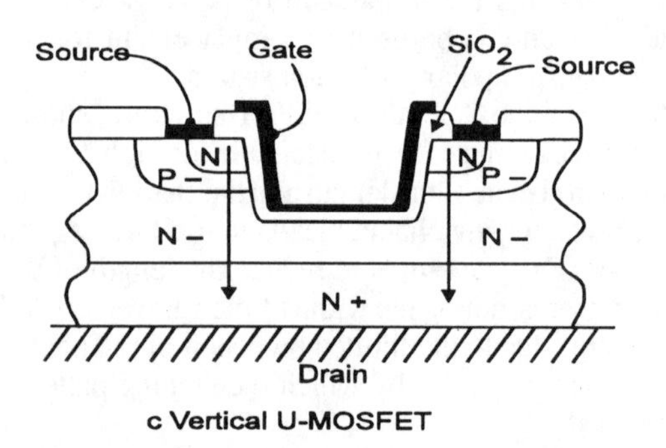

Figure 5.25 *Structures of various vertical MOSFET designs*

flattened out, as shown in Figure 5.25c. These devices are usually called 'U- MOSFETs', with reference to the 'notch' shape. Although this arrangement allows a much shorter effective channel length, the conducting resistance of the channel, even over this short vertical path, will still be relatively high, so, in order to achieve 'on' resistances equivalent to those of a bipolar power transistor, many such conducting channels need to be formed – electrically in parallel – on the same chip.

Various geometrical layouts have been adopted in order to provide a large number of parallel channel paths within a limited chip area, and one of the most elegant of these is that introduced by International Rectifier, Inc., in their 'Hexfet' range of power MOSFETs, although this is employed in a 'D-MOS' rather than a 'U-MOS' structure. In this design the masking, oxidation and etching pattern, used to define the diffusion structure of the source and substrate regions, takes the form of a series of hexagonal pits, arranged in honeycomb fashion, as sketched in Figure 5.26.

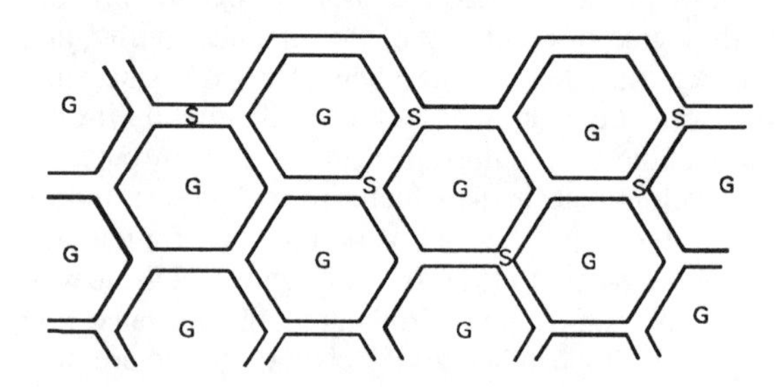

Figure 5.26 *Honeycomb structure used to provide multiple parallel gate channels*

Channel resistances, specified as 'R_{ds} (on)', as low as 0.02 ohms are now readily obtainable with power MOSFETs, as are operating voltages as high as 1200V, though at rather higher values of R_{ds} (on). The 'U-MOS' or 'V-MOS' type of construction lends itself more readily to N-channel MOSFETs, so the P-channel versions of these transistors are relatively more expensive, and offer a more limited range of working voltages and channel conductivities.

There are a number of practical difficulties in the use of these power MOSFET devices, of which the most immediately apparent is that the parallel connection of a number of gate regions leads to a high gate/source capacitance, and values for this in excess of 1nF are typical. This requires that the device must be driven from a low source impedance if good high frequency performance is required, in spite of the fact that the gate circuit is effectively an open circuit to DC.

Another problem is that, because of the exceedingly fast internal electrical response of the transistor, high frequency parasitic oscillation can readily arise if the gate, source or drain connecting wires are long, or run parallel to one another. A 'gate stopper' resistor, in the range 220–470 ohms, will usually serve to prevent this type of malfunction, which can often cause rapid destruction of the device.

Care must also be taken to ensure that the maximum gate/source voltage rating is not exceeded, since, if the gate insulating layer should break down, the device will be defunct. It is common practice in small signal MOSFETs, such as the dual-gate types used for RF amplification, or in CMOS logic devices, for one or two Zener diodes to be diffused onto the MOSFET chip to act to limit excessive gate/source or gate/drain potentials. Unfortunately, this simple protective measure is inapplicable with power MOSFETs since it will cause the same layout of junction areas as is found in a thyristor, so that an applied gate voltage high enough to activate the protective diode could cause the whole device to trip into a short-circuit condition. Externally connected Zener diodes can usually be employed as protective components without problems.

T-MOS devices

An alternative structure used for MOSFETs is that shown in Figure 5.27, called 'T-MOS', from the conducting channel path shape, – or also sometimes 'D-MOS', as a contraction of 'double-diffused, vertical current flow MOS'.

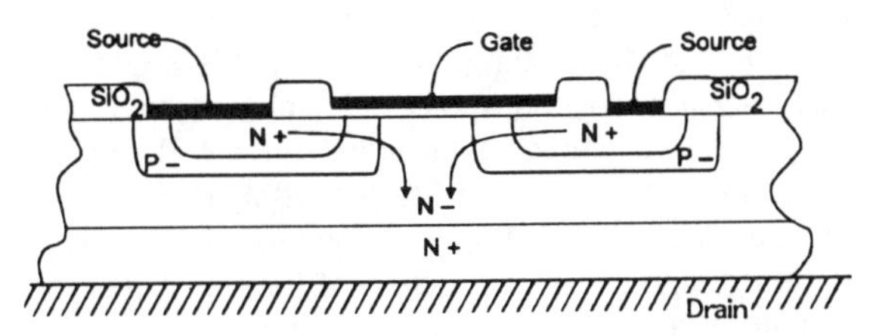

Figure 5.27 *Structures of T-MOS or D-MOS FET*

These devices reflect the advances which have been made in manufacturing technology, in that they now use masking and etching processes to define very close tolerance channel lengths in a combination of lateral and vertical current flow, and offer the advantage that both N-channel and P-channel versions can be made, with less difficulty, although, even so, the N-channel versions are both more rugged and more suitable for high voltage use.

In principle, T-MOS devices are rather slower than the U-MOS versions, but the use of polycrystalline silicon gate electrodes, rather than deposited aluminium ones, has allowed the HF response to be improved to a value which is now not much lower than that of the V-MOS and U-MOS types.

With both U-MOS and T-MOS transistors there is a possibility of an excessive gate–drain voltage arising, at high drain potentials, since part of the gate insulation serves to separate the gate from the drain as well as from the channel. This danger is minimized, in the diffusion process, by arranging that that part of the drain region which is in proximity to the gate is of a very lightly doped (N– or P–) material. This is known as the 'drift' region, and, because of its relatively low conductivity, permits a relatively large voltage drop between the end of the channel and the highly doped (N+ or P+) substrate. The relative distribution of potential drops between the channel and this drift region is, however, influenced by drain current flow so that gate–drain breakdown can occur more easily at high device currents.

The drain current vs. drain voltage characteristics of MOSFET devices are similar to those of junction FETs, and are shown in the comparative graphs of Figures 5.28 and 5.29.

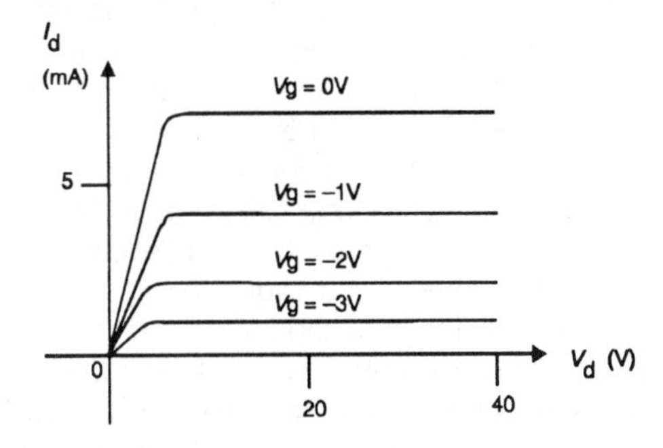

Figure 5.28 *Drain current characteristics of N-channel FET*

With the major exception that some care must be exercised in circuit design and applications to ensure that the gate insulation layer breakdown voltage is not exceeded, MOSFETs offer a very valuable and versatile type of active component. In particular, the very good transfer, (V_{in}/I_{out}), and high frequency characteristics of these devices make them ideally suited for use in audio amplifier designs, and there are few such applications for which thermionic valves or bipolar transistors have been used, in the past, which could not be performed better by one or other of the MOS type devices.

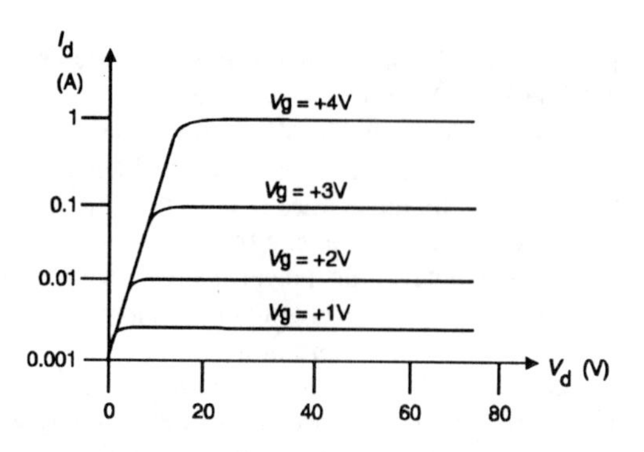

Figure 5.29 *Drain current characteristics of N-channel D-MOSFET*

Moreover, with the growing availability of P–channel devices, in addition to the well established N–channel versions, considerable scope exists for innovations in circuit design, using complementary polarities, which could never have been implemented with thermionic devices.

Also, whereas with both thermionic valves and junction FETs, in which both the grid and the gate electrodes will form a conductive path with the cathode or source, if it should become forward biased, the gate electrode of the MOSFET remains a high impedance point over its whole allowable voltage range. Again, the gate/source capacitance of the MOSFET is largely independent of the gate/source voltage.

A practical point which should be noted with all MOSFETs is that their method of manufacture, which employs a common connection to both an N and a P region at the source contact, as shown in Figures 5.20, 5.22, 5.25 and 5.27, leads also to the formation of a parallel connected diode: a P–N diode in the case of an N-channel MOSFET, and an N–P diode in the case of an P-channel device. In normal use this diode is non-conducting.

Thyristors and triacs

While power transistors exist which can handle large currents and voltages, in many cases all that is required is a semiconductor device which will operate as a low impedance switch. Indeed, a term which is sometimes used for one form of these components is 'silicon controlled switch'. In its simplest form, such a bistable semiconductor switch can be contrived by connecting a pair of PNP and NPN bipolar junction transistors, as shown in Figure 5.30.

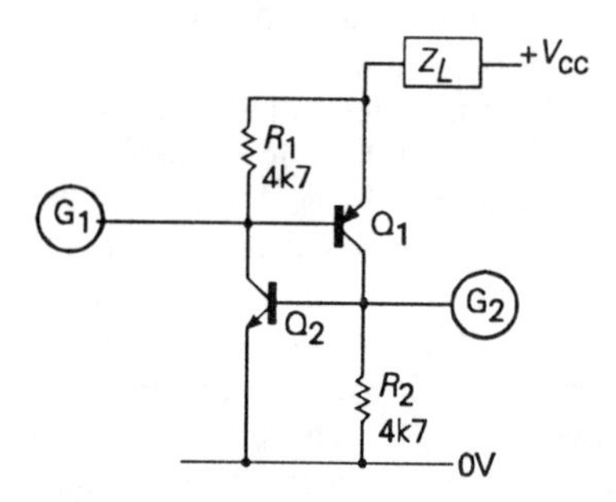

Figure 5.30 *Two transistor embodiment of thyristor*

If Q_1 is initially non-conducting, then no collector current due to Q_1 will flow into Q_2 base, so Q_2 also will be non-conducting, and no collector current from Q_2 will flow into Q_1 base; a state of affairs which will continue as a conditionally stable state.

However, if a small current is injected into the bases of either Q_1 or Q_2, of such a polarity as to cause either of these to conduct momentarily, then both transistors will 'latch' into a stable conducting mode, with both devices being turned hard on. Some external current limiting components must exist, of course, to prevent the transistors from being damaged by this action.

So long as an adequate potential remains across such a regenerative pair of transistors, they will remain conducting, and can only be turned off again by either reducing the applied supply voltage to a level below that needed to cause conduction of the transistors, or by putting a short circuit across one or other of the base–emitter connections.

The use of a circuit of this type constructed from separate transistors is quite feasible, but since the current gains of both transistors will be much larger than is necessary to cause regenerative action, it is prudent to connect a pair of resistors, as I have shown at R_1 and R_2, to prevent leakage currents from causing the circuit to trip inadvertently.

Such a PNP/NPN combination can be merged into a four-layer PNPN device, as shown in Figure 5.31, in which the external connections are called the anode, cathode and gate terminals.

With a practical four-layer switch, called a 'silicon controlled rectifier' (SCR), or 'reverse blocking triode thyristor', or simply a 'thyristor', the current gains of

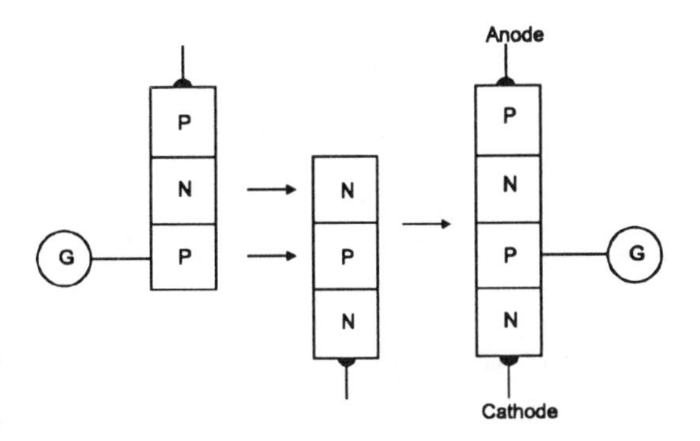

Figure 5.31 *Evolution of 4-layer SCR/thyristor from two transistors*

both of the incorporated transistors are chosen to be only slightly greater than unity, which helps both stability in use, and the ability of the device to operate at high applied voltages.

It is an established convention that the gate connection will be that to the lower P type base region, (equivalent to Q_2 base). Those SCRs with the gate connection made to the upper N type region (equivalent to Q_1 base) are called 'complementary' SCRs.

The forward and reverse characteristics of the SCR are shown in Figure 5.32, and a typical schematic cross-section, showing the disposition of the diffusion layers, is shown in Figure 5.33. The circuit symbol used is that shown in Figure 5.34.

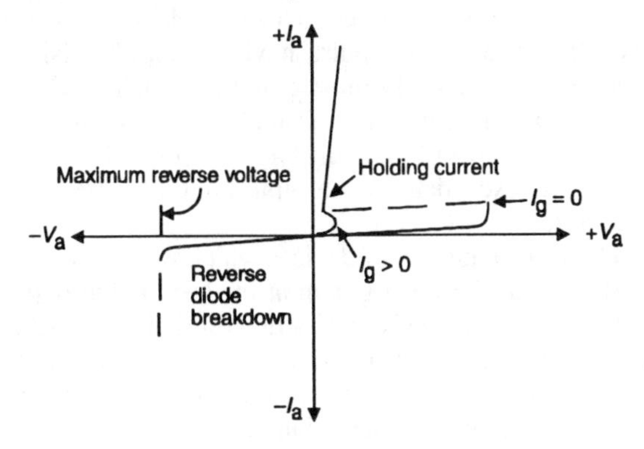

Figure 5.32 *Forward and reverse thyristor conduction characteristics*

Although many thyristor applications may be for relatively low power use, thyristors are intrinsically very robust devices. Depending on the die size, and

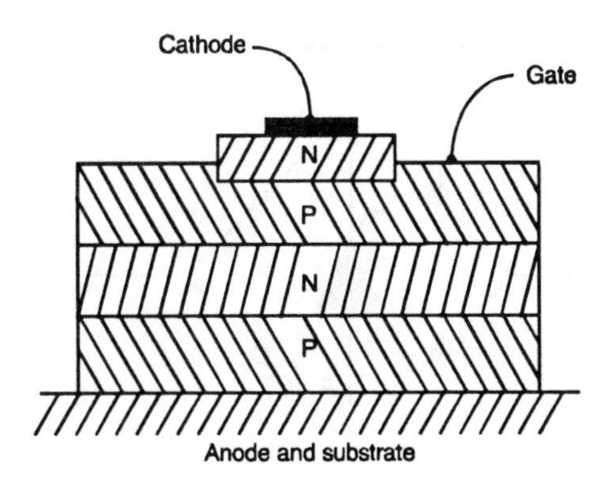

Figure 5.33 *Construction of practical reverse blocking thyristor*

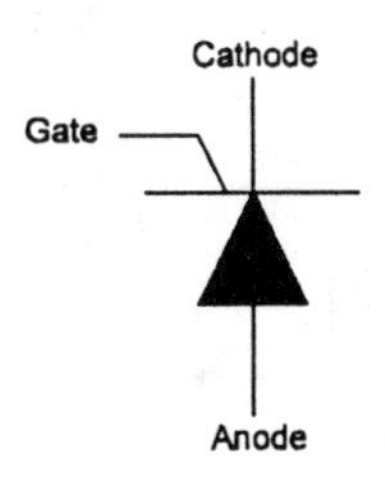

Figure 5.34 *Circuit symbol of thyristor*

the diffusion and heat-sinking arrangements, thyristors are available with peak current ratings of the order of thousands of amperes, and capable, if required, of holding off potentials in excess of two thousand volts.

Since many of the uses of the SCR lie in the field of alternating current control, it would normally be used within a bridge rectifier layout, as shown in Figure 5.35, in order to allow it always to switch in the correct polarity. However, a version of this device, known popularly as a 'triac', (triode AC switch), has been evolved to meet the specific need for an AC control component.

These devices can be visualized as a pair of PNPN and NPNP four-layer devices, joined in parallel, as shown in Figure 5.36, with a gate electrode so arranged that it will be able to 'fire' either of these, as required. This leads to the type of internal structure shown in Figure 5.37 and the bi-directional turn-on characteristics shown in Figure 5.38.

The circuit symbol for this AC switching device is shown in Figure 5.39.

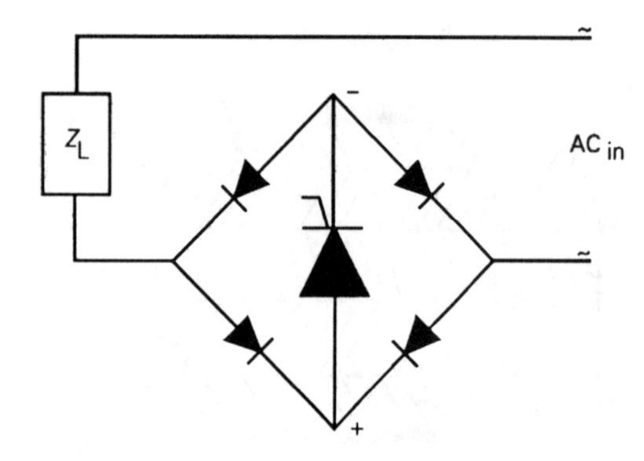

Figure 5.35 *Use of rectifier bridge to allow thyristor control of AC*

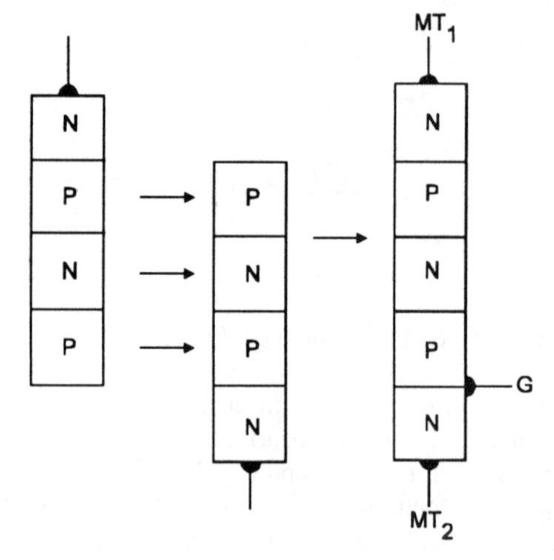

Figure 5.36 *Evolution of TRIAC from parallel connected thyristors*

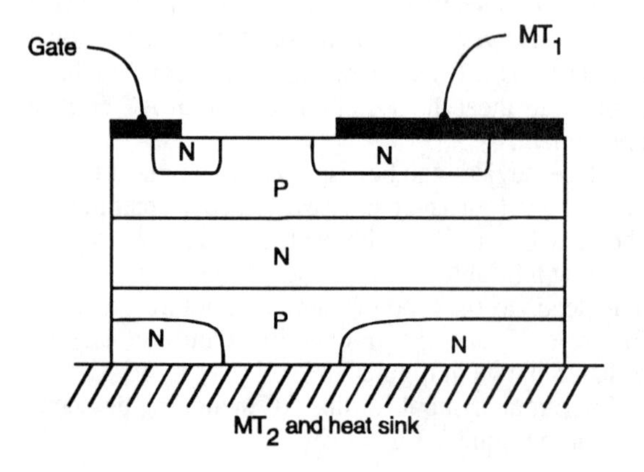

Figure 5.37 *Practical construction of triac*

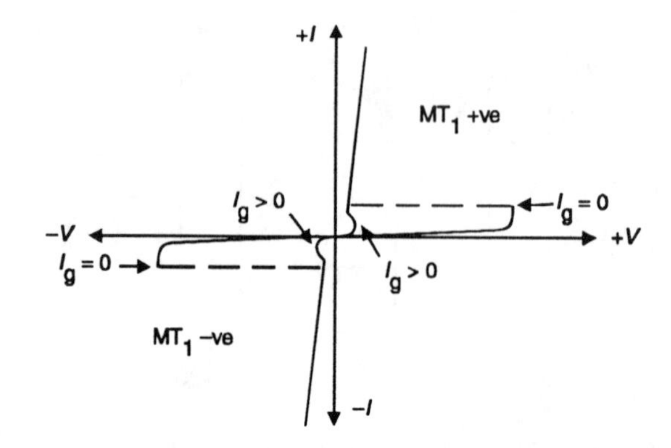

Figure 5.38 *Conduction characteristics of triac*

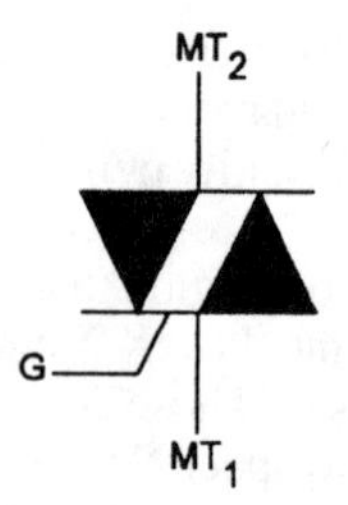

Figure 5.39 *Circuit symbol for triac*

Silicon bilateral trigger diodes (diacs)

Proper operation of both SCRs and triacs is assisted if the turn-on voltage applied to the gate electrodes is in the form of a voltage pulse having a rapidly rising leading edge. This is most commonly achieved in practice by incorporating a small power switching element, which will suddenly break-over if its hold-off voltage is exceeded, in the input circuit to the gate electrode.

There are a number of devices which will serve this purpose, but the most common of these is the diac, which is basically an NPN junction transistor in which two identical N+ doped 'emitter type' junctions have been diffused on either side of the P type base layer. When whichever of these junctions happens to be reverse biased is triggered into avalanche type break-down, transistor action multiplies the resultant current flow, to give the type of conduction characteristics shown in Figure 5.40.

Other trigger devices are also available, as are a wide range of variations of the basic thyristor/triac devices, to fit specific application needs. In general, all of these

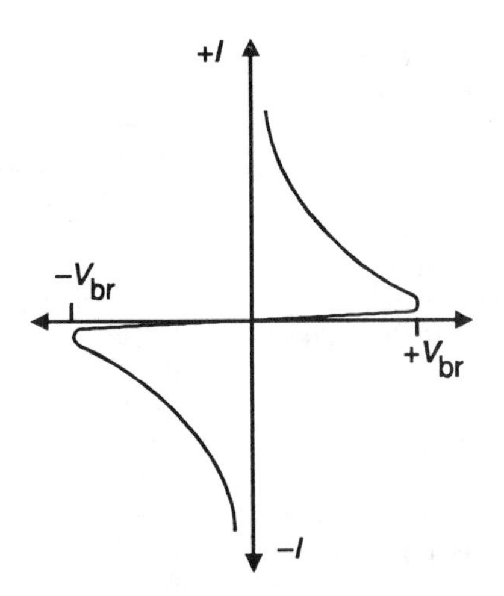

Figure 5.40 *Breakover characteristics of diac*

high power devices are relatively sluggish in operation, because of the time which is required for the current carriers to pass through the relatively thick N and P layers which are employed in their construction, and for the decay of the many minority carriers generated by the high conduction currents.

Because of the great utility of these switching devices in power control applications, this has become an important branch of semiconductor technology, and there are many specialist manuals which give further information on this topic.

Linear integrated circuits

If it can be said that, with the invention of the transistor, thermionic valves became obsolescent as amplifying devices, apart from certain specialized usages, then it can be said, with equal truth, that the invention and successful development of the 'integrated circuit' has largely done the same for the discrete semiconductor. Indeed, it can be argued that it would usually be a wasteful exercise to employ many separate components and semiconductor devices in order to achieve the same result that could be obtained, more cheaply, more predictably, more reliably, and in less space by a single IC block. It may often also be that the ready-made integrated circuit will offer a higher degree of optimization in design than the average engineer is likely to pursue for any one-off application.

This philosophy has encouraged the growth of 'application specific ICs' (ASICs), in which the design skills of the development laboratories of the semiconductor manufacturer are placed at the disposal of the engineer, even for applications in which the likely volume of sales is relatively small. However, the main forms of linear ICs remain the 'op. amp.' gain blocks and voltage stabilizer circuits, discussed below.

Operational amplifiers

These are the universal low power gain stage units, commonly represented by the symbol shown in Figure 5.41. These will typically require, for operation, a pair of positive and negative supply lines whose potentials will bracket the input signal voltage range. The recommended potential for these supply lines will normally be +/–15V, but the IC may operate, with some restrictions on possible output voltage swing and HF response, on supply line potentials as low as +/–3V.

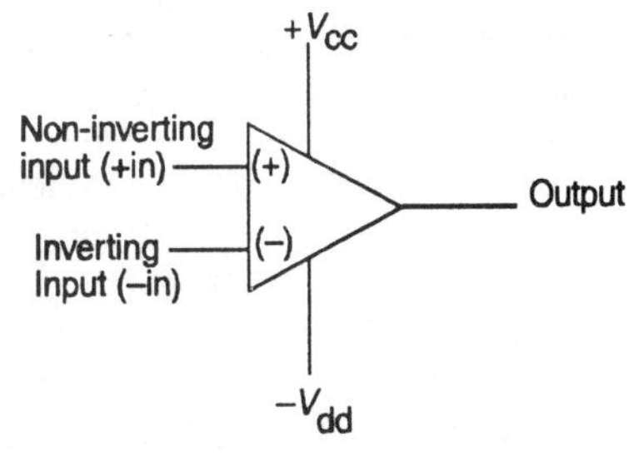

Figure 5.41 *Conventional circuit symbol for 'op.amp.'*

The output (peak-to-peak) voltage swing, obtainable without clipping, will usually be some 0.5V less than the supply line potentials, provided that the output frequency or current demand requirements are modest. The circuit design is normally such that up to 100% negative feedback can be applied across the gain block without loop instability, usually manifest as HF oscillation, but this will depend on the specific IC type and the output load characteristics.

It is taken for granted in modern op. amp. ICs that internal protection circuitry will be incorporated which will prevent inadvertent damage through short circuits from the output connection to the '0V' line, or through excessive output current demands.

Developments in these devices are continually

introduced, as a result of competition between manufacturers, so, provided that equipment is designed to accept ICs having conventional pin configurations, of the types shown in Figure 5.42, equipment performance can often be upgraded simply by replacement of the existing ICs by those of more recent design. Examples of the supply line requirements, and the performance specifications for two such units, representing 'second generation' and 'third generation' designs are shown in Tables 5.1 and 5.2.

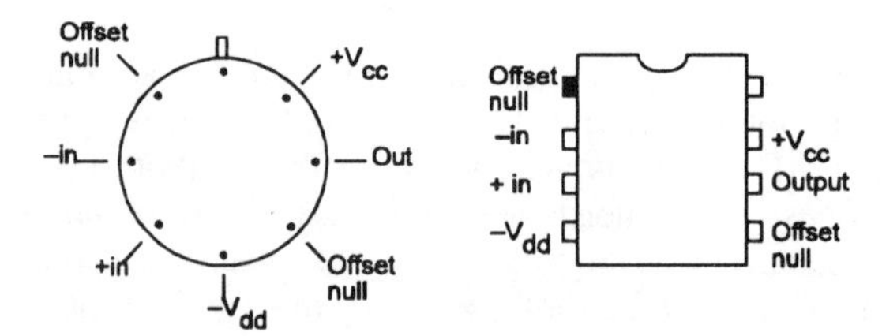

Figure 5.42 *Normal pin connections for single IC 'op.amps.' – shown from top of package*

Table 5.1 *Performance specification for '741' type IC amplifier unit*

Supply voltage	+/–15V typical, +/–18V max.
Max. power dissipation	500mW
Differential input voltage	+/–30V, max.
Input offset voltage	2mV (typical)
Gain/bandwidth product	1 MHz
Input impedance	2 Megohms
Slewing rate	0.5V/µS (at unity gain)
Voltage gain, DC	200,000×
Noise figure	200nV/√Hz
Common mode rejection ratio	90dB
Supply line rejection ratio	90dB
Output voltage swing	+/–14V
Maximum output current	25mA
Power consumption	60mW (typical)

Other operational amplifier designs have been evolved for specific end uses, such as the 'OP07', which is intended particularly for low frequency instrumentation applications, in which the input DC offset voltage has been reduced to 10µV, and the input voltage drift with time is as low as 0.2µV/month.

In the field of low-noise, low-distortion amplifiers, the 'OP27' offers a typical input noise figure of 3nV/√Hz, coupled with an LF voltage gain figure of 1,000,000×, which allows harmonic distortion figures of the order of 0.002% at 1kHz, and 10V RMS output, at circuit gain values up to 100×.

Table 5.2 *Performance specification for 'LF351' type IC amplifier unit*

Supply voltage	+/–15V typical, +/–18V max
Max. power dissipation	500mW
Differential input voltage	+/–30V
Input offset voltage	5mV (typical)
Gain/bandwidth product	4MHz
Input impedance	10^{12} ohms
Slewing rate	13V/µS
Voltage gain, DC	300,000×
Noise figure	15nV/√Hz
Common mode rejection ratio	110dB
Supply line rejection ratio	110dB (average)
Output voltage swing	+/–13.5V
Maximum output current	13mA
Power consumption	55mW (typical)

For audio applications, the 'LM837' offers a slewing rate of 10V/µS, a gain bandwidth product of 25MHz, and a power bandwidth of 200kHz, typical. The input noise voltage of this device is 4.5nV/√Hz, coupled with a THD figure, at up to 10kHz/6V RMS output of 0.0015%.

There are also ICs of this general type which are designed for use in high frequency applications, such as the 'LM6165' family, which offer a gain/bandwidth product of 725MHz, coupled with a slew rate of 300V/V S, although many of the very high speed devices are designed as unity gain buffer (impedance converter) stages or 'voltage followers'. Devices of this type, such as the 'LH0063', offer a slew rate of 2400V/µS, and a working bandwidth of DC–200MHz.

Voltage regulator units

This is the other major field in which ICs have established a very wide presence in linear circuitry, because of the almost universal desirability of supply lines which are stable, and noise and ripple free, from which this circuitry can be operated.

These ICs are made in two main types, the so-called

'3-terminal regulators' (available in both positive and negative supply line versions) which are designed to provide highly stable, overload proof, fixed output voltage sources, at a wide range of output voltage and current ratings, covering the range +/–5V to +/–24V, at current outputs of 100mA to 10A, and 'programmable voltage regulators'.

The programmable, or adjustable types are intended to confer the same performance benefits as the fixed voltage types, though with the added advantage of being able to vary the actual output voltage. If a small reduction in the performance of the voltage regulator is acceptable, it is possible to achieve some adjustment of the output voltage by the simple expedient of inserting a variable resistor in the '0V' return path of the IC.

The normal ways of using these ICs are shown in Figure 5.43. The circuitry employed for this type of application is discussed in greater detail in Chapter 15.

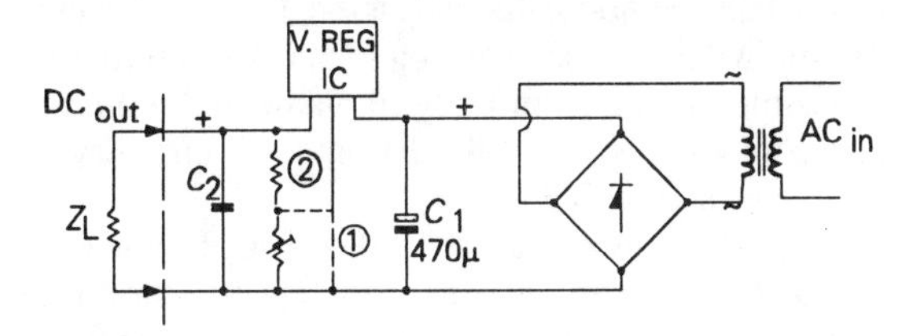

Figure 5.43 *1 normal method of use of voltage regulator IC, and 2 method of adjusting output potential*

Voltage reference devices

In addition to the voltage regulator ICs, there are also integrated circuit blocks which are intended to provide a high precision replacement for the zener or avalanche diodes. These will be used in similar circuitry to the normal 2-terminal regulator diode, but will offer a much higher degree of output voltage precision, and with much lower noise and output voltage drift as a function of time or ambient temperature.

Transistor and diode arrays

These are a simple assembly of conventional components, fabricated at the same time on a single semiconductor chip, as shown in Figure 5.44, with the intention of allowing precise matching of the device characteristics, in applications such as 'long-tailed pair' configurations – discussed in Chapter 6 – where any difference in device characteristics could cause unwanted drifts in the circuit performance or DC output potential.

Because of some necessary compromises which must be adopted in the manufacturing techniques adopted for the production of multiple device arrays, the performance of, say, the individual transistors in a transistor array, of the kind shown in Figure 5.44, may differ from, and may be somewhat less good than, a single small signal transistor of equivalent type.

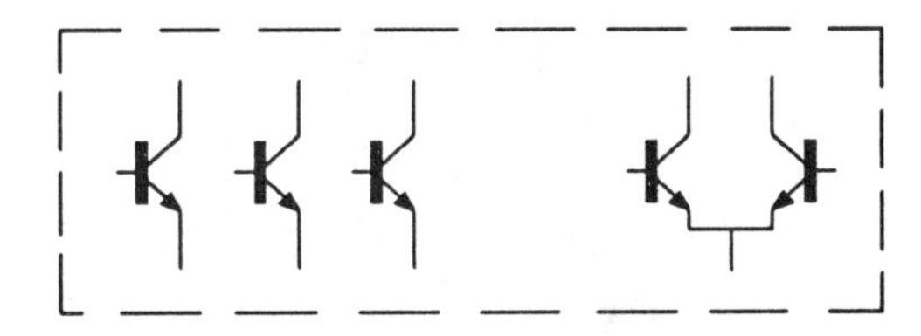

Figure 5.44 *Multiple individual transistor array*

Current mirrors

These devices are a specialized application of the facility offered, by simultaneous manufacture on the same semiconductor die, for the production of separate devices having identical electrical characteristics.

In the case of the 'current mirror' the function of the circuit is to generate an exit current from one limb of the circuit which is identical to, or bears a constant, fixed, proportional relationship to, the current drawn from another limb. This is a very useful circuit device for many circuit applications, and is widely used in IC operational amplifiers.

The circuits used are normally one or other of the forms shown in Figure 5.45, and an appropriate circuit symbol for the general class of these devices is shown in Figure 5.46.

Constant current sources

These are basically junction FETs, constructed so that their gate and source electrodes are joined together, and fabricated in a two-terminal package to take advantage of the intrinsically high independence of drain current upon drain voltage in FETs. They are available with output currents in the range 0.1 – 10mA, and with working voltages up to some 50V.

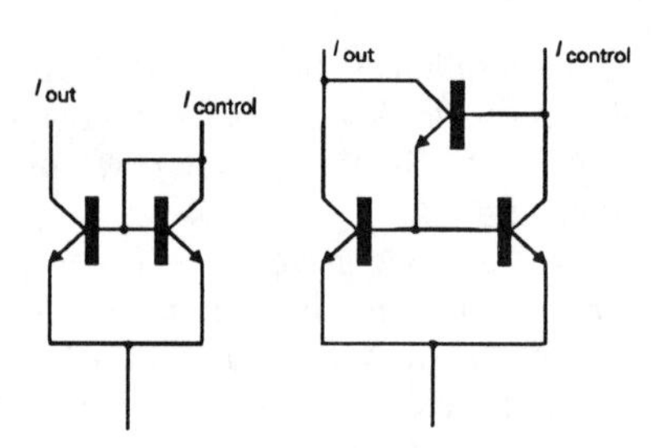

Figure 5.45 *Typical internal circuitry of current mirrors*

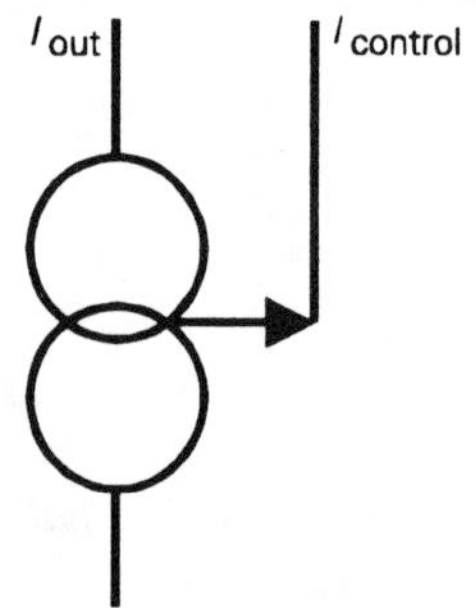

Figure 5.46 *Circuit symbol for current mirror*

Instrumentation amplifiers

These are a specific category of IC amplifier intended to offer a very high input impedance, together with a very stable value of gain and DC offset, when used in DC amplifier applications.

These devices will either be straightforward, but high quality, operational amplifiers, in which the circuit design and constructional process has been optimized, with low noise, and high DC and gain stability as the major design requirements, or will be integrated circuit blocks containing the circuit layout shown in Figure 5.47. The operation of this type of circuit is considered in greater detail in Chapter 6.

Other useful characteristics such as a logarithmic relationship between input and output signal voltages may also be offered, in this type of IC, depending on the manufacturer.

Specialized circuitry

This covers such a wide, and expanding, field that it is impracticable to discuss it in detail, and the catalogues of IC manufacturers should be consulted to provide

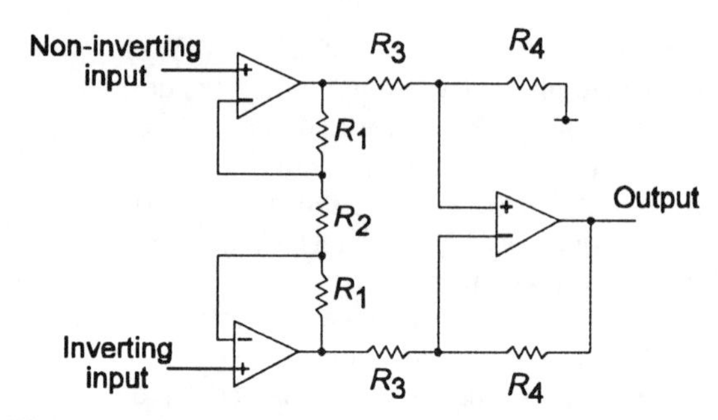

Figure 5.47 *Internal circuit configuration of instrumentation op.amp.*

detailed information on the available devices. As a rule, such ICs are offered either to allow the use of circuits whose complexity would prevent their satisfactory implementation using discrete components (resistors, capacitors, transistors), such as, for example, the demodulator and stereo decoder circuitry used in FM tuners, or to offer a lower cost alternative to existing circuitry normally implemented with discrete components, as in the RF and AF circuitry of portable radio receivers.

The continuing reduction in the cost of consumer electronics hardware, and its growing incursion into other fields, has only been possible because such specialized circuit blocks have been developed, and, in general, their use has contributed to the reliability of this equipment.

Two cautions should be noted. Of these the first is that special circuit ICs may not be designed to offer as high a standard of performance as the discrete component circuit they are designed to replace, particularly when their main purpose was to allow a reduction in manufacturing costs. Secondly, experience suggests that such special application ICs may be discontinued without warning, as a result of a marketing or sales decision, and this can leave any equipment, in which these ICs have been incorporated, in difficulties in respect of future maintenance. It is prudent, therefore, to look to see how many directly interchangeable replacements are available for the IC in question, before making a decision to incorporate it into equipment.

Linear uses of digital ICs

The fact that an IC has been designed for use in digital

circuitry does not necessarily preclude its use in linear applications, although some sacrifice of linearity or noise figure may have to be accepted in comparison with a comparable IC specifically designed for linear applications. On the other hand, the low cost of such digital ICs, and their availability for use with single 5V line supplies may make their use very attractive in combined linear/logic circuit designs.

A particularly versatile type of logic device for linear applications is the CMOS inverting amplifier/buffer, which can be self biased into a linear mode of operation using the circuit shown in Figure 5.48. This can either be used as a single stage amplifier, giving a gain in the range 20–50, or, in cascade, as a means of converting a small signal input into a 'logic level' switching waveform.

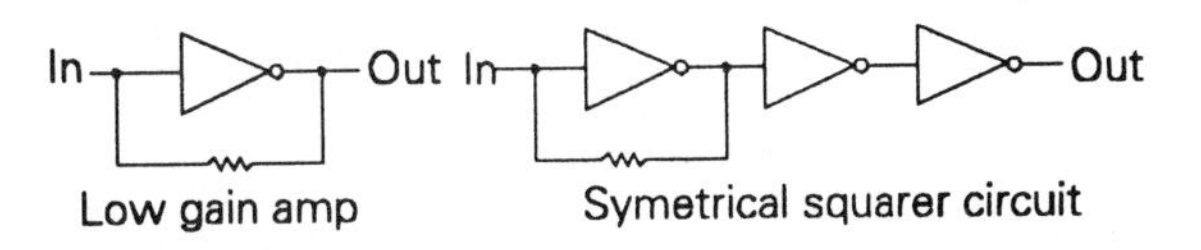

Figure 5.48 *Linear application of CMOS (digital) inverter stage*

6

DC and low frequency amplifiers

Introduction

Although it is quite feasible to use thermionic valves for many of the applications where electronic signal amplification is required, it is, in general, very much simpler, and easier, and more cost-effective, to design using one or other of the available range of semiconductor devices, and I propose, therefore, to confine this chapter to 'solid-state' amplifying systems. However, for those who are interested in thermionic valve amplifiers, a few representative valve circuit designs are shown in Chapter 9.

The basic types of discrete solid-state device which are currently available to the linear circuit designer are bipolar junction transistors, in both NPN (positive collector supply line) and PNP (–ve supply line) types: junction FETs, in both N-channel (+ve supply line), and P-channel (–ve supply line) forms: and 'insulated gate' field effect transistors – sometimes called 'IG-FETs', but more commonly referred to as 'MOS-FETs', because of their physical construction – again sometimes available in both N-channel and P-channel versions.

Among the MOSFETs, there are, again, two quite distinct types: the 'lateral' forms, in which the structure of the junctions is laid down on the surface by masking and diffusion, and the current flow is parallel to the surface of the device, and the 'vertical' ('V' or 'T'), MOSFETs in which the semiconductor junctions are formed one on top of another, by in-depth diffusion, and the current flow is perpendicular to the surface.

Lateral MOSFETs are mainly intended for use in relatively low voltage, small signal applications, and are mostly only available in N-channel forms. They are more commonly found in dual-gate designs, intended for use in RF amplifier and mixer applications. Vertical MOSFETs are available in a wide range of operating voltages and power levels, and in both N-channel and P-channel types.

All of these devices have their individual advantages and drawbacks, and part of the skill of the circuit designer is to be able to select which of these device types is most appropriate to his particular application, or whether, indeed, it would be more sensible and cost-effective to use one or other of the many readily available linear integrated circuit packages.

Although many practical electronic circuits will employ a combination of the various device types available to the designer, because the circuit technology which will be used will differ from any one device type to another, it is simpler to consider each kind of semiconductor component individually.

Circuitry based on bipolar junction transistors

Small-signal (silicon planar) transistors

General characteristics

Bipolar junction transistors can be considered either as current or as voltage amplifying components. In the first of these applications, shown in Figure 6.1, current which is injected into the emitter junction, which offers a low input impedance (typically in the range 50-500 ohms for small-signal devices), will pass with very little loss through the device into the collector circuit, which has a very much higher output impedance (typically in the range 20k–200k ohms). Only a minute proportion of this current will be taken by the base junction, giving an output current, 'I_c', which is probably about 0.995 I_e, where 'I_e' is the current injected into the emitter. This emitter/collector current transfer factor is denoted by the Greek symbol 'α' (alpha). This arrangement will only give a voltage gain if some advantage can be derived from the increase in impedance from input to output, as, for example, in the RF amplifier stage shown in Figure 6.2.

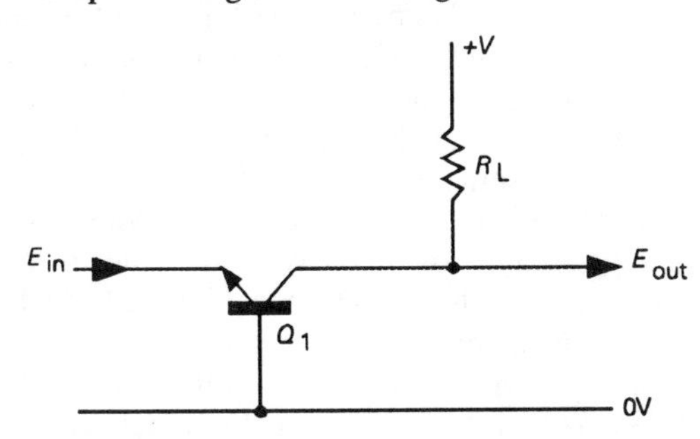

Figure 6.1 *Simple grounded base transistor amplifier circuit*

The most common way of using a junction transistor as an amplifier is in the circuit layout shown in Figure 6.3, where the emitter is returned to a low impedance point, and current is injected into the base. In this

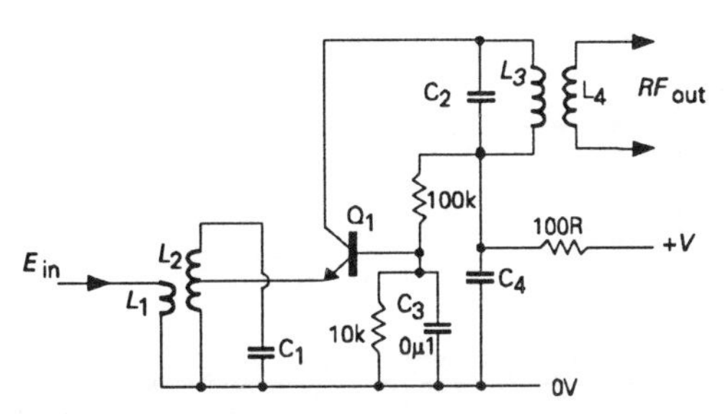

Figure 6.2 *Typical grounded base RF amplifier stage*

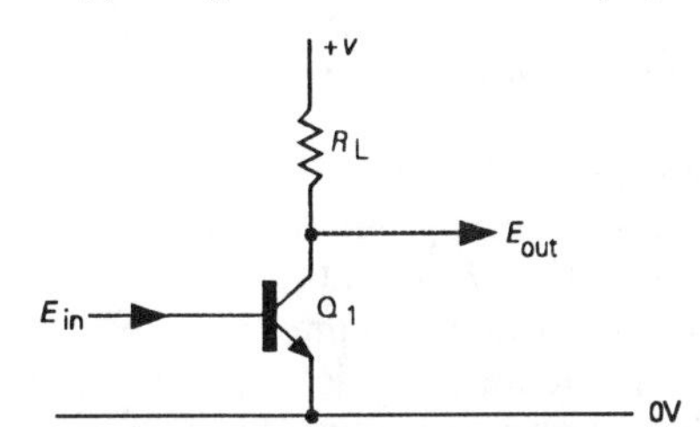

Figure 6.3 *Typical grounded emitter amplifier layout*

method of use, a much increased current will then flow in the collector circuit, so that I_c will typically be 50–500 times greater than I_b. This current increase, from base to collector, is called the 'current gain' of the transistor, and is known by the Greek symbol 'β' (beta). [Please note that this is an entirely separate and distinct use of this symbol from that in feedback systems, described in Chapter 7.] To avoid this confusion, the 'common emitter current gain' of a transistor is more properly described as the h_{FE}, where the quality described is the low frequency, or DC, value of this term, or h_{fe} – using lower case letters – where the current gain is measured at some (specified) higher frequency. The 'e' in this subscript refers to the fact that it is the 'common emitter' characteristic which is measured. For example, h_{fb} would be the term used to describe the 'common–base' figure, α (alpha).

The current gain of a transistor will decrease with increasing operating frequency, as shown in Figure 6.4, and the frequency at which it decreases to unity is known as the 'transition frequency', or f_T, sometimes also described as the 'gain–bandwidth product'. This tends to give a somewhat over-optimistic impression of the HF performance of a transistor, in that a device with a current gain of 200, and a 'transition frequency' of 300MHz, would, in reality, have a current gain value which would decrease, at a slope of 6dB/octave,

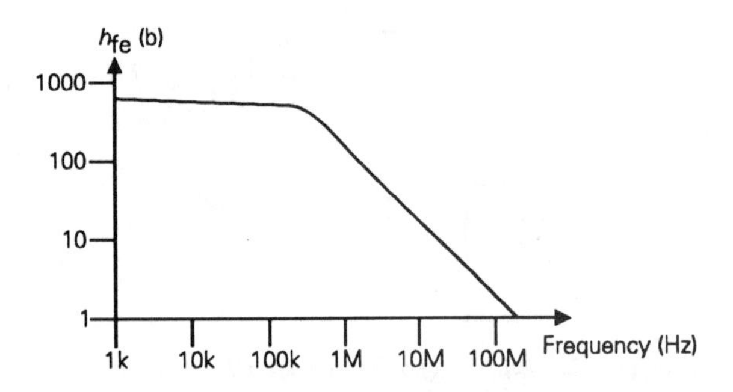

Figure 6.4 *Variation of h_{fe} with frequency*

beyond 1.5MHz.

The low frequency current gain also depends somewhat on collector current, in the manner shown in Figure 6.5, but used as a current amplifier the transistor is a relatively linear device, and one of the ways by which the transistor's performance, as a voltage amplifier, can be made more linear is by inserting a large value of resistor between the signal source and the base of the device, as shown in Figure 6.6. The problem with this method of use is, of course, that the gain of the stage is reduced, and its input noise figure worsened, by the presence of this added resistor. However, it does mean that a transistor voltage amplifying stage will be more linear in its operation – other things being equal – if its base is driven from a high impedance source, such as the collector circuit of a preceding transistor amplifying stage.

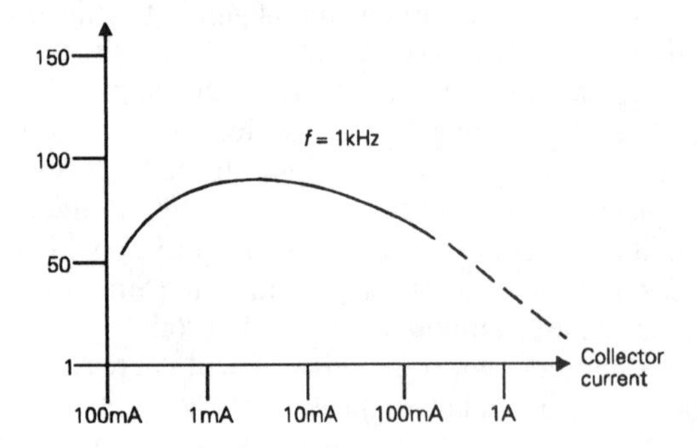

Figure 6.5 *Variation of h_{fe} with collector current*

Junction transistor voltage amplifiers

The simple circuit layout shown in Figure 6.7a will act as a voltage amplifier, provided that the transistor is forward biased – by a suitable current injected into its

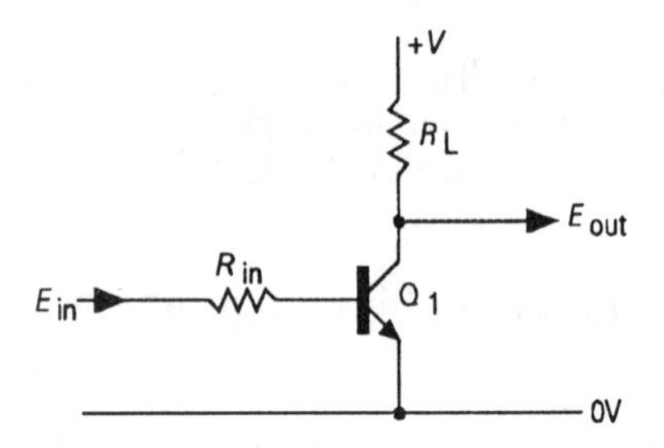

Figure 6.6 *Use of input swamp resistor to linearize current gain*

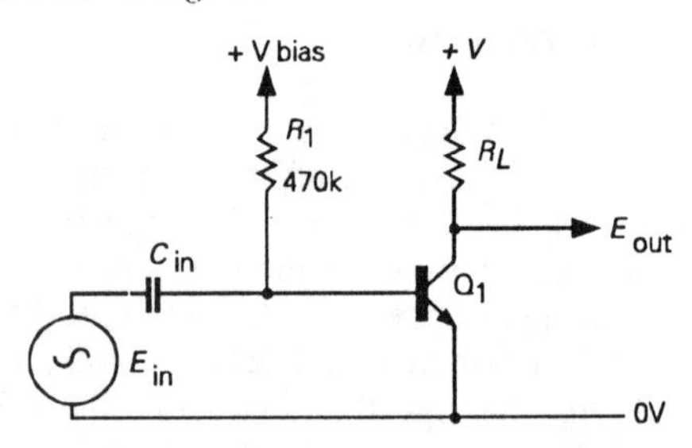

Figure 6.7a *Simple and unsatisfactory input biasing arrangement*

base, perhaps by the simple expedient of connecting a resistor, R_1, between the base junction and some suitable voltage source – into a condition where the collector DC voltage is somewhere between that of the collector DC supply rail and the collector saturation voltage of the transistor, which will probably be of the order of 0.2 – 0.5V.

Because the actual current gain of a transistor used in any given circuit design cannot be predicted in advance, and it is poor design practice to offer a circuit for which the semiconductors must be pre-selected, the injection of some fixed value of base bias current will not be a satisfactory answer to this requirement.

The repositioning of R_1 between the collector and base junctions, as shown in Figure 6.7b, will provide a simple, though crude, means of controlling the base bias current within fairly broad limits, since if the current gain of the device turns out to be high, and this leads to a high collector current – so that the collector potential falls towards 0.5 volts – the bias current through R_1 will similarly fall, while if the current gain is low, and the collector potential rises toward the supply rail voltage, because of inadequate collector current, the base bias current through R_1 will also rise. This is, however, a clumsy method of controlling the working conditions of the transistor, and suffers from the problem that, because of the effect of the internal negative feedback through R_1 (see Chapter 7), the gain of such a stage can never exceed R_1/Z_{in}. In the circuit shown in Figure 6.7b, this gain limit will be 47.

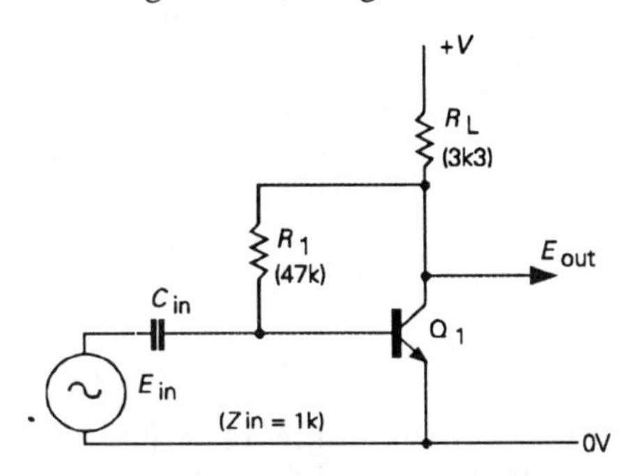

Figure 6.7b *Improved input biasing arrangement*

A much more elegant method of biasing such a transistor amplifying stage is that shown in Figure 6.8a, where the base is fed from a relatively low impedance potential divider, and the current through the transistor is controlled by the value chosen for the emitter resistor, R_4.

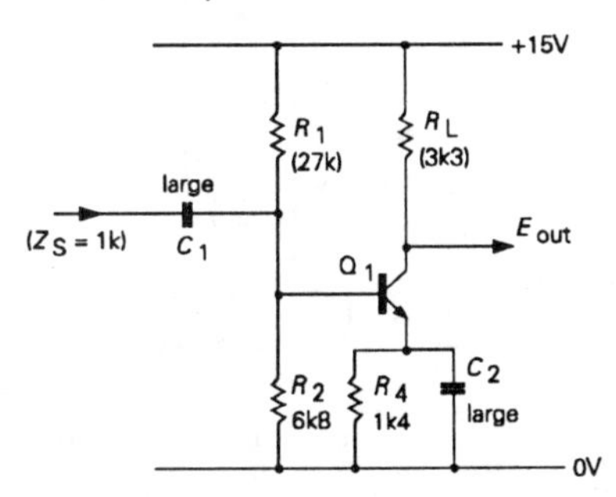

Figure 6.8a *More satisfactory method of biasing amplifier transistor*

For the circuit shown, ignoring the relatively small base bias current, the values shown in the drawing will set the collector current fairly precisely to 2mA, and the collector potential to a voltage about half way between the potential of the supply rail and the emitter voltage. Moreover, this set of DC conditions will be largely unaffected by the actual current gain of the transistor, over quite a wide range of likely values. This arrangement has the drawback that there can be unwanted signal breakthrough into the input from the supply line. This problem can be avoided by the layout shown in Figure 6.8b, where C_2 is used to decouple the input bias circuit, at the cost of a somewhat greater component count. This input bias system can also be used with other circuits where an input bias voltage must be derived from the DC supply rail.

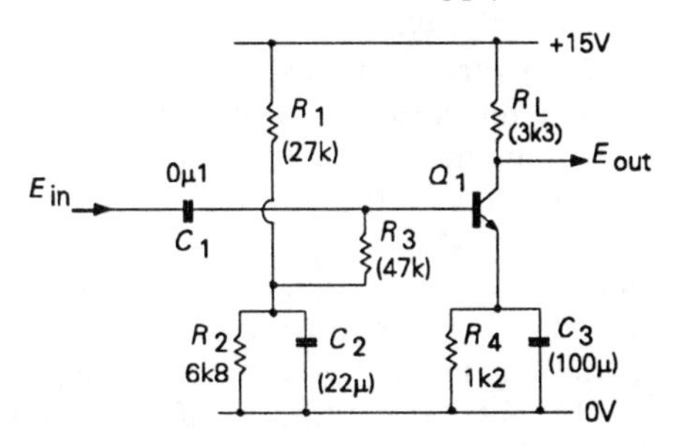

Figure 6.8b *Means of improving circuit loading and supply line ripple rejection*

If a two stage amplifier circuit can be used, for example, as shown in Figure 6.9, the use of DC negative feedback, through R_4, can be used to stabilize the working conditions quite precisely, and it will again be almost independent of the actual current gains of the transistors used. For example, for the resistor values quoted, the base potential of Q_1 will be 8.25V, its collector current will be 0.18mA, the collector current of Q_2 will be 6.5 mA, and its collector potential will be 6.5V DC, allowing an output voltage swing of some 14V peak to peak. If the current gains of the transistors are low, the output DC potential will fall, though only by 0.1–0.2V, and this will increase the current through R_4, and cause the collector currents of both Q_1 and Q_2 to increase, and conversely.

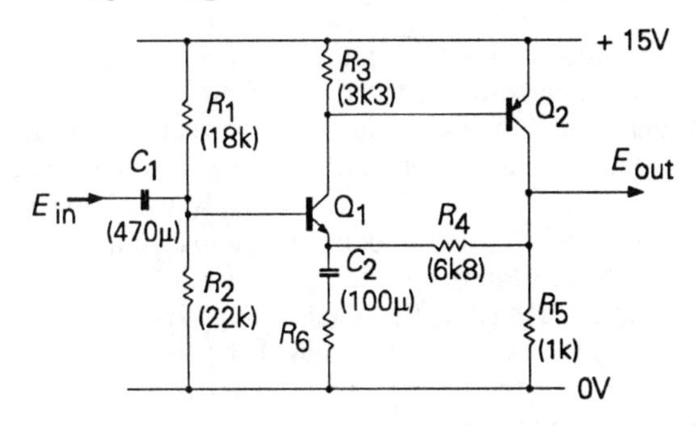

Figure 6.9 *Two stage amplifier with NFB used to stabilize both gain and working point*

An amplifier stage of this type will probably have a voltage gain of some 2000, provided that C_2 has a large enough value. A more precise control of stage gain can be obtained by inserting a resistor, R_6, in series with C_2. The gain, for reasons explained in the next chapter, will then approximate to $(R_6 + R_4)/R_6$, for gain values which are small relative to 2000. This particular circuit layout makes a very useful and versatile gain block, and shows the convenience of being able to use a PNP transistor in combination with an NPN type, to be able to avoid the need for DC blocking inter-stage coupling capacitors.

A comparable circuit layout, using only NPN transistors, is shown in Figure 6.10. This has a somewhat higher intrinsic gain, though with a rather less good stability of DC operating conditions. If C_2 is removed, the stage gain is then controlled by the values of R_4 and R_5, and is, approximately, $(R_4 + R_5)/R_5$, over a wide frequency range. Both of the circuit layouts of Figures 6.9 and 6.10, give very much lower levels of harmonic distortion than either of the earlier single transistor circuit layouts.

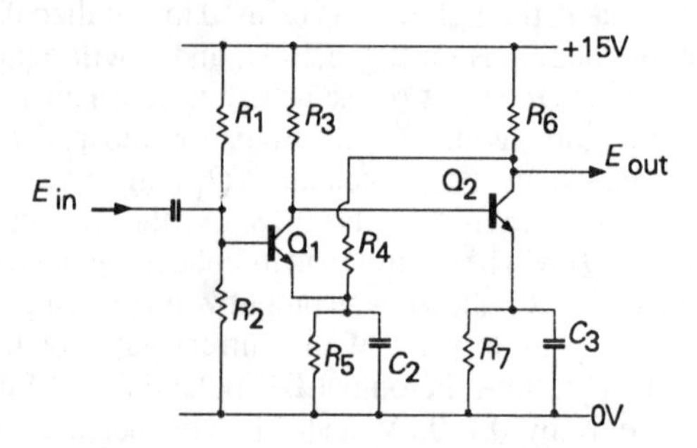

Figure 6.10 *Two stage feedback amplifier using only NPN transistors*

The circuit layout of Figure 6.10 can be elaborated to make a three-transistor layout of the kind shown in Figure 6.11, in which the two transistor amplifier stages, Q_1 and Q_2, are followed by an impedance converting, 'emitter follower' stage, Q_3, which makes the gain of Q_2 independent of the load impedance applied to the output of the circuit. Both DC and AC negative feedback are applied through R_6 and R_4, to stabilize the (AC) working gain of the circuit, as well as its DC operating conditions. If R_4 is partially, or wholly bypassed by a capacitor (C_2), the gain can be adjusted, up to a possible figure of 10,000, with high gain transistors.

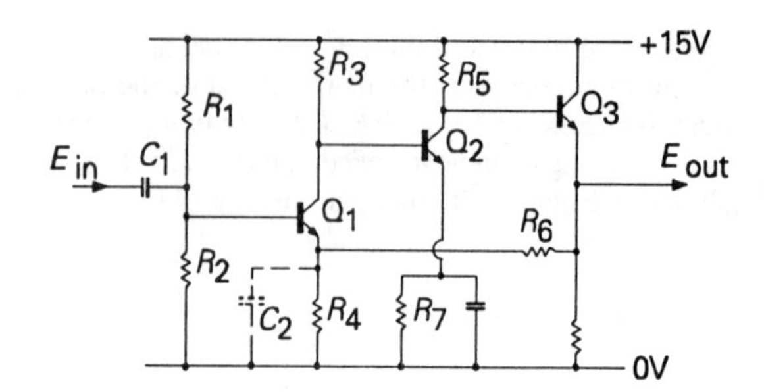

Figure 6.11 *Three stage feedback amplifier with emitter-follower output*

Transistor stage gain and distortion

One of the minor inconveniences of the junction transistor, as a voltage amplifier, is that its base voltage/collector current relationship is very non-linear, as shown in Figure 6.12a, by comparison, for example, with a typical grid voltage/anode current characteristics of a thermionic valve, shown, for comparison, in Figure 6.12b.

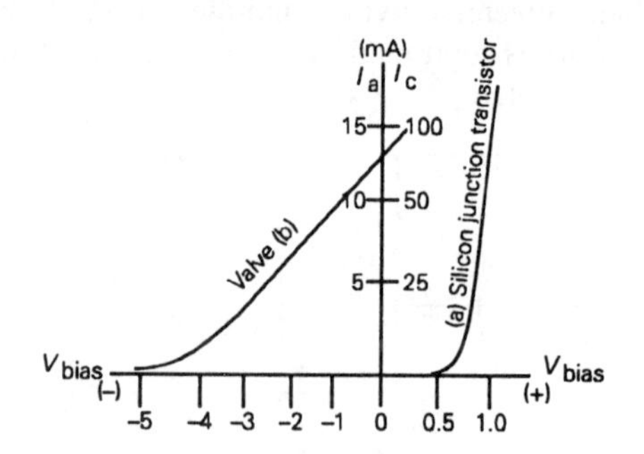

Figure 6.12 *Relative input transfer characteristics of junction transistor versus valve*

Several aspects of transistor behaviour can be seen from Figure 6.12a, of which the first is that the rate of change of collector current, for an increment in base voltage – known as the slope or mutual conductance, and denoted by the symbol g_m, measured in amperes (or milliamperes)/volt, now known as 'Siemens' (or mS) – increases rapidly with increasing collector current. The theoretical formula, from which the mutual conductance of a junction transistor can be derived, is

$$g_m = I_c(q/kT) \qquad (1)$$

where I_c is the collector current, q is the electronic charge in coulombs, (1.6×10^{-19}), k is Boltzmann's constant, (1.38×10^{-23}), and T is the absolute temperature of the junction, in °Kelvin. For most modern silicon planar small signal transistors, this gives values of g_m, in the range 25–40mA/V, per milliampere of collector current.

A simple approximation to the low frequency voltage gain of a transistor amplifier stage, of the type shown in Figure 6.8, is given by the relationship

$$\text{Stage gain} = g_m R_l \quad (2)$$

where R_l is the effective load impedance of the circuit, which would be nearly 3k3 in the case of Figure 6.8a. So, for a collector current of 2mA, this circuit should give a voltage gain of about 150–250, if it is driven from a low enough source impedance and if both C_1 and C_2 are adequately low in impedance at the chosen operating frequency.

The second characteristic which can be deduced from Figure 6.12a is that there will be a substantial amount of distortion in the amplified output signal, because of the curvature of the V_b/I_c relationship. However, any curve will approximate to a straight line if a small enough segment of it is taken. This means that if the input signal is small enough, the distortion given by the stage will also be small. This is more or less independent of the actual gain, and the consequent output voltage, given by the stage, since the relationship between collector voltage and collector current, for any given base voltage setting, as shown in Figure 6.13, is, in fact, quite flat. However, the simple expedient of increasing the collector load resistor, to increase the stage gain, suffers from the drawback that the collector current would also have to be reduced – unless the supply rail voltage can be proportionally increased – to maintain the chosen DC collector potential, and this would reduce the effective 'slope' of the transistor, which would lessen the expected increase in stage gain. If, however, some arrangement can be found which would provide a high collector load impedance, while still allowing a moderately high working collector current, without the need for impracticably high supply rail voltages, a very much higher stage gain could be obtained from a single transistor amplifier stage. Moreover, because this would require only a very small input signal voltage, it would also give a very low level of signal distortion.

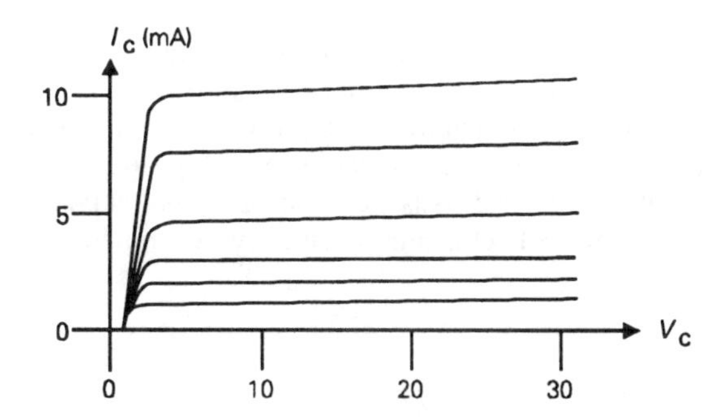

Figure 6.13 *Collector current/voltage relationship*

Circuits of this type have been designed, (JLH, *Wireless World*, Sept. 1971, pp. 437–441) and are offered commercially, in IC versions, such as the LM359. The operation of these circuits will be examined later in this chapter.

DC amplifier layouts

All of the circuit layouts so far described have been aimed specifically at the amplification of alternating signal voltages, in which any direct voltage component has been blocked off by the inclusion of a DC blocking capacitor in the signal line. If, however, a +/– pair of supply voltage rails is available, so that the emitter of the transistor can be made about 0.6V negative, in the case of an NPN device, the transistor can also be made to act as a DC amplifying stage, as shown in Figure 6.14a. This would give a DC output voltage which would not normally sit at 0V, DC, for a 0V DC input potential.

A simple elaboration of this circuit, shown in Figure

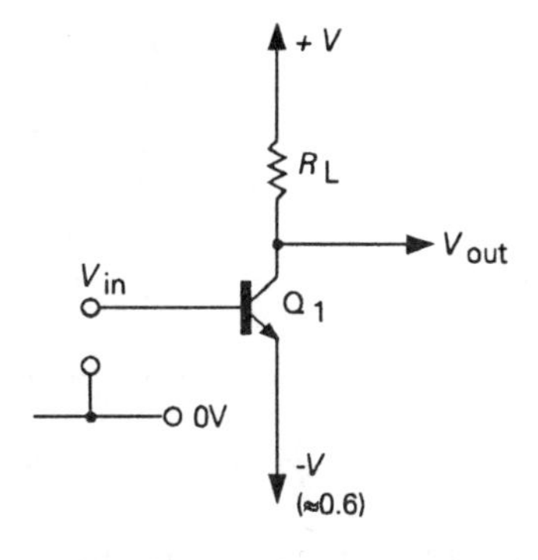

Figure 6.14a *Very simple DC amplifying stage*

6.14b, to include a zener diode (ZD_1), in the output circuit could offset the normal +ve quiescent collector voltage of Q_1, and allow an output voltage which was at 0V DC, but the circuit would be highly susceptible to output voltage drift as the static base–emitter potential, and the collector current, varied with temperature.

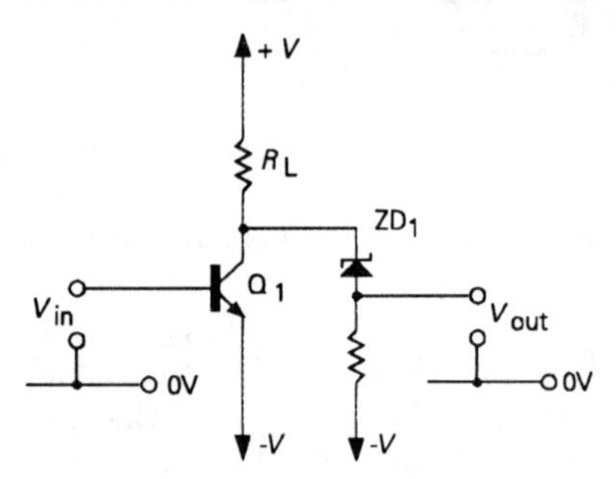

Figure 6.14b *Simple amplifying stage with DC offset minimized by zener diode*

A useful circuit layout which allows automatic compensation for the temperature dependance of the base–emitter potential of Q_1 is that known as the long-tailed pair, shown in Figure 6.15a, where if a matched pair of transistors is used, the current through Q_2 will be the same as that through Q_1, and any changes in base–emitter voltage will apply equally to both transistors, and will merely result in a change in emitter voltage to sustain the existing collector currents through Q_1 and Q_2.

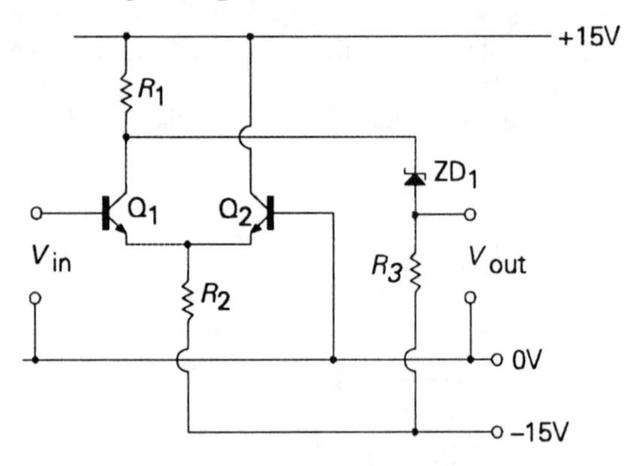

Figure 6.15a *Input long-tailed pair arrangement*

Once again, a zener diode could be used to offset the static DC potential on Q_1 collector, to give the required zero output level for zero DC input potential. However, a much neater arrangement is that shown in Figure 6.15b, where a PNP transistor is used to give additional amplification, and invert the output current flow, so that the potential drop across R_5 can give a true 0V DC output. The AC and DC gain is then set by the choice of the feedback resistors R_3 and R_4, to give a figure which will be very close to $(R_3+R_4)/R_3$ in value. For high values of R_1, so long as the current gain of Q_3 remains constant, the actual base-emitter potential of Q_3 is relatively unimportant since Q_3 is current rather than voltage driven.

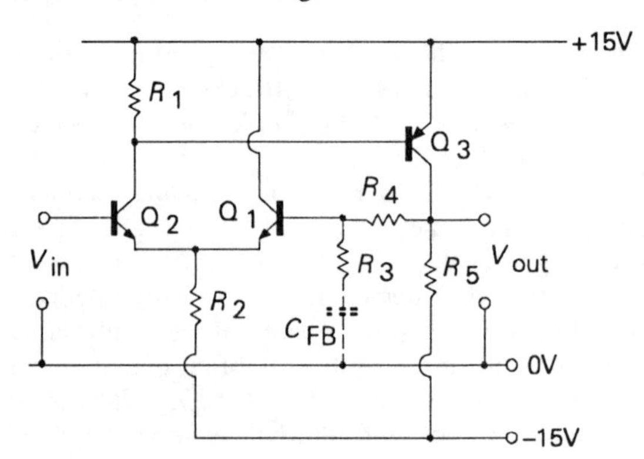

Figure 6.15b *Improved DC amplifier circuit based on an input long-tailed pair*

Although this circuit is capable of use as a DC amplifying stage, and, indeed, forms the basic type of input circuit used in almost all the op. amp. integrated circuits, some of which are specified especially for DC amplification, its other qualities, such as low distortion and stable performance characteristics, commend it for use as an AC amplifier, in which use, a capacitor, C_{fb}, will be inserted in series with R_3. This has the effect of making the circuit have a unity gain, at DC, so that the quiescent output voltage will be accurately equal to the quiescent input voltage, while giving a stage gain of $(R_3+R_4)/R_3$ at all frequencies where the circuit 'open-loop' gain is high enough, and the impedance of C_{fb} is small enough to be ignored.

Constant current sources and current mirrors

As mentioned earlier, it is very useful to be able to operate a transistor with a high load impedance without also needing to limit the actual value of collector current employed, since reducing the collector current would reduce the operating 'slope' of the transistor, and thereby lose some of the extra stage gain which the use of a higher load impedance was intended to provide. The simplest answer to this need is to use a circuit arrangement known as a 'constant current source'. As its name implies, such a circuit would provide a constant output current over a wide range of applied voltages, and, in its simplest form would just be the collector circuit of a transistor operated at a fixed forward bias.

A practical version of this is as shown in Figure 6.16, where, for the reason shown in Figure 6.13, the collector current is largely unaffected by collector voltage, so that the circuit behaves as if it were a very high resistance connected to a sufficiently large negative potential to cause the measured current flow, but with, in reality, only a modest supply voltage requirement. The circuit of Figure 6.16 works best with a fairly large value of both V_{ref} and R_2, but this limits the minimum voltage drop at which the circuit can be used.

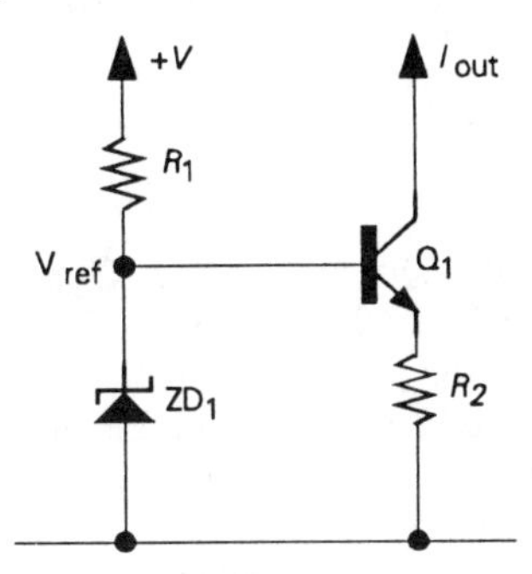

Figure 6.16

Because of the usefulness of such constant current sources, a wide range of circuits has been evolved for this purpose, of which one of the most effective is the layout shown in Figure 6.17. In this, if the voltage drop across R_2 exceeds the voltage required to force Q_1 into conduction, the collector current of Q_1 will 'steal' the current which would otherwise flow through R_1 into Q_2 base – which will, in turn, reduce the current flow through Q_2 and R_2. This circuit has a high dynamic impedance, and will also operate at a relatively low voltage drop across Q_2/R_2 – down to a little above 1.2V.

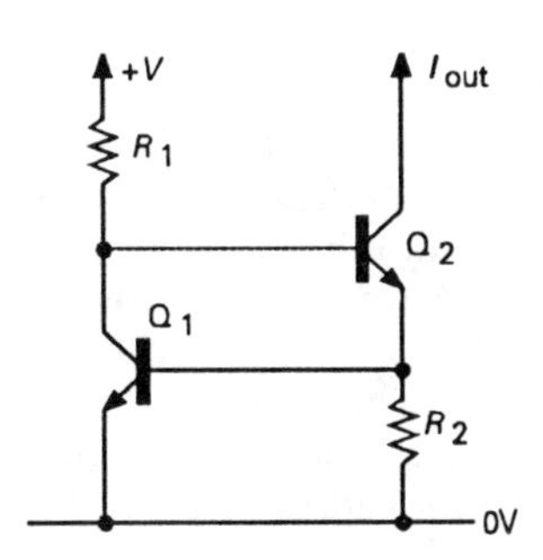

Figure 6.17

A very common, contemporary form of constant current source is the two-terminal version based on a field effect transistor, and shown in Figure 6.18. The method of operation of this type of device is explained later in this chapter, under the heading 'Junction FETs'. This type of device offers a very high dynamic impedance, at a range of specified operating currents, but has, usually, a more restricted maximum working voltage than a layout of the types shown in Figures 6.16 and 6.17, based on small signal bipolar transistors.

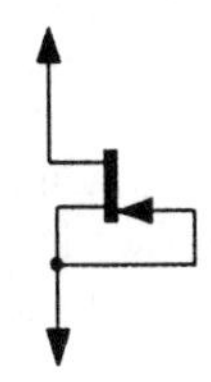

Figure 6.18 *FET constant current source*

(*Note.* It is customary for all of these circuits to be called constant current 'sources', though this would only be true where they were operated from a more negative supply rail than their load – where one defines 'current' as the flow of electrons, as I have done throughout this book. For accuracy in terminology, therefore, where such circuits are connected between the operating system and the positive supply rail, they should be referred to as constant current 'drains'. Luckily, the term 'constant current source' is acceptable in both –ve and +ve supply rail applications.)

'Current mirrors' are a special class of high dynamic impedance constant current sources, in that their output current is dependent upon some arbitrarily chosen input current, and in which the output current usually is chosen to 'mirror' the input one.

The simplest circuit arrangement of this type is that shown in Figure 6.19, in which Q_1 and Q_2 are closely matched transistors, so that the current drawn from Q_1 establishes a forward base bias for both Q_1 and Q_2 which will be, ignoring the small base currents drawn from both Q_1 and Q_2, of the correct value for the two collector currents to be identical. This circuit configuration is also available as a 'three terminal' IC package, for which types are available in which $I_{out} = I_{in}$, or $= 2I_{in}$, or $= 0.5I_{in}$, and so on, depending on user requirements.

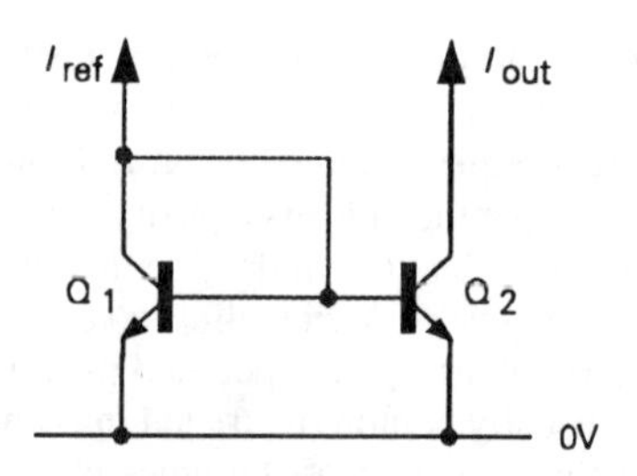

Figure 6.19 *Current mirror*

A more sophisticated discrete component current mirror circuit is shown in Figure 6.20, in which Q_3 is added as a buffer to remove the base currents of Q_1 and Q_2 from the current drawn from the input. Where matched transistors are not available, the degree of current matching can be improved by inserting small value resistors between the emitters of Q_1 and Q_2 and the supply rail. As with 'constant current sources', these circuits are employed with both –ve and +ve supply rails.

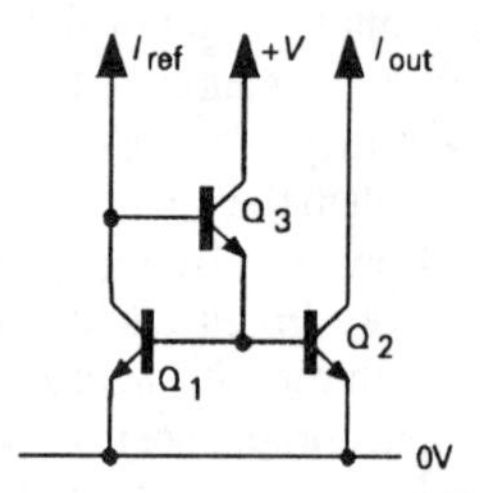

Figure 6.20 *Improved current mirror*

A typical application of a current mirror is as a high impedance load for a long-tailed pair input stage, such as that shown in Figure 6.15b, redrawn, in Figure 6.21, to include, in addition, a pair of constant current sources, of the type shown in Figure 6.16, in place of the load resistors R_2 and R_5. The different currents required by the input and output transistors are arranged by the choice of different values for R_1 and R_5. The higher dynamic impedance of Q_3 than R_2, in Figure 6.15b, gives very much better rejection of any unwanted supply line ripple on the –ve DC rail, and the higher impedance of Q_7 than R_5, in Figure 6.15b, greatly increases the gain, and linearity of the output transistor (Q_6).

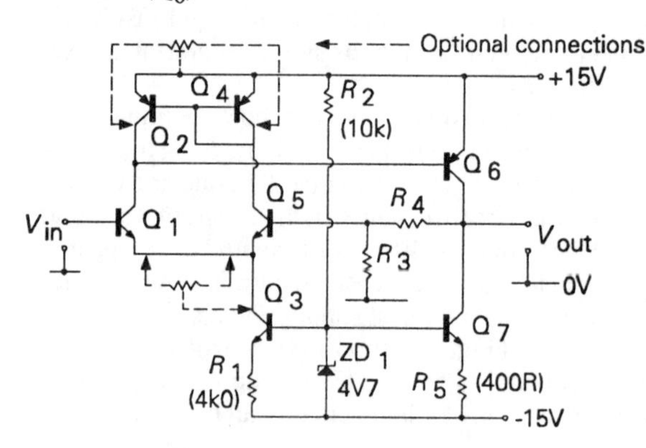

Figure 6.21 *Circuit arrangement using constant current source to supply a long-tailed pair and a current mirror as its collector load*

The use of the current mirror Q_2/Q_4 in place of the simple load resistor (R_1 in Figure 6.15b) has a number of beneficial effects. Firstly, it allows the collector current of Q_1 to be chosen at a high enough value to obtain a good stage gain from Q_1. Secondly, it improves the linearity of Q_1 as an amplifier stage. Thirdly, it combines the output currents of both Q_1 and Q_5, and thereby improves the symmetry of the input stage – as well as effectively doubling its gain – and finally, it also improves the rejection of unwanted supply line ripple.

An additional feature of this type of circuitry is that it greatly minimizes the usage of resistors in the design, and this is of great value to the manufacturers of ICs, in which transistors and diodes are easy to fabricate, and not very demanding of 'chip' area, whereas resistors are more difficult to make with any degree of precision, and are relatively extravagant in their use of chip space.

Circuitry of the kind shown in Figure 6.21 is representative of much contemporary low-frequency amplifier design, though, in order to increase the available output voltage swing from Q_6, it is more usual for the constant current source, Q_7/R_5, to be replaced by a circuit of the kind shown in Figure 6.17.

Where the circuit layout shown in Figure 6.21 is to be used for DC amplification, as could be the case with IC op. amps, provision is usually made for some form of output DC 'offset' zero adjustment, either by an external potentiometer connected between the collectors of Q_2 and Q_4, or, less commonly, by an offset adjustment potentiometer in the emitter circuit of Q_1 and Q_5, as shown in dotted lines in the diagram.

It will be appreciated that the whole circuit layout shown in Figure 6.21 can be inverted, by the use of PNP transistors in place of NPN types, and vice versa, and by altering the polarities of the supply rails. This could have certain advantages in practice, in that Q_6 might, more advantageously, be an NPN device, in that these devices are better suited as higher power output transistors.

Similarly, the use of PNP transistors for the input long-tailed pair (Q_1/Q_5), would give a somewhat lower input noise level, because the N type base region of a PNP transistor suffers less from recombination noise, and its 'base spreading resistance' will usually be lower, with lower associated thermal noise.

Since current mirrors and constant current sources are so common a feature of circuit design, the symbols of Figures 6.22a, and 6.22b are widely used as a draughtsman's 'shorthand' for the circuit layouts of Figures 6.16–6.18, and of Figures 6.19 and 6.20 respectively. Using this notation, a very high gain single transistor stage can be formed from the simple combination shown in Figure 6.23, of an amplifying transistor and a high impedance constant current load. This circuit arrangement would, however, suffer from two major practical drawbacks – it would only give the expected high gain figure if any output load, Z_L, had an impedance which was very large in relation to both the dynamic impedance of the constant current source and the output impedance of the amplifying device, and if the operating frequency was so low that the shunt impedances of any stray capacitance, both from the collector of Q_1 to ground, and, more importantly, to Q_1 base, could be neglected. In practice this would impose severe constraints on the performance of such a circuit.

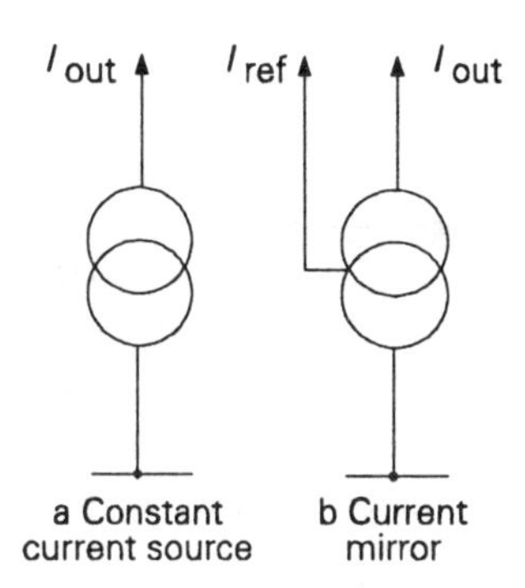

Figure 6.22

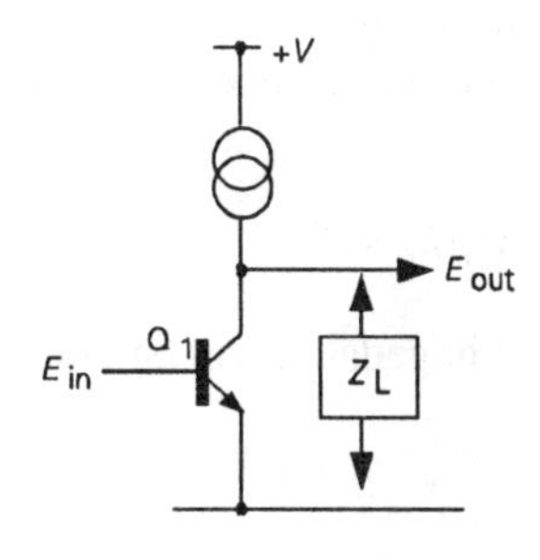

Figure 6.23 *Very high gain stage using constant current load for a transistor amplifier*

Impedance conversion stages

The simplest form of impedance conversion stage is the emitter follower, shown in Figure 6.24. The method of operation of this circuit is that, on switch on, an increasing current will flow through R_1 until a potential is reached at which the base–emitter potential is just sufficient to sustain that value of emitter current. If the base voltage is increased, this equilibrium state will be disturbed, and the emitter current and emitter potential will also increase to restore the equilibrium. Similarly, if the base voltage is caused to fall, the emitter voltage will also fall, and this change in output voltage will occur, in either direction, as rapidly as the internal carrier creation or recombination processes within the transistor, or the external circuit capacitances, will allow.

From a DC point of view, in the case of the NPN transistor shown, the output voltage will be the input potential less the forward base bias potential required to make the transistor conduct – typically 0.55 – 0.6V, depending on emitter current. Similarly, for a PNP

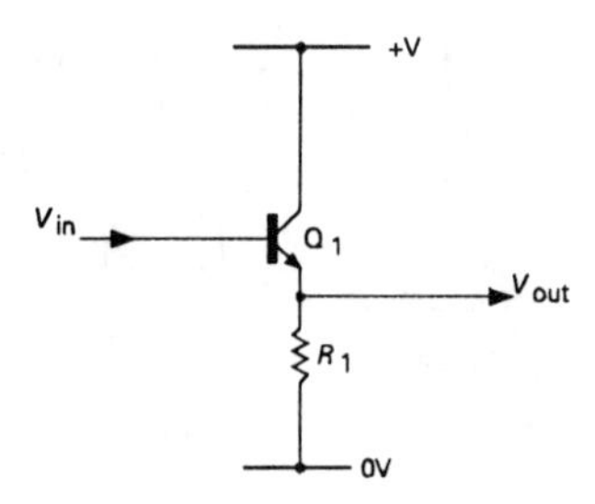

Figure 6.24 *Simple emitter follower*

transistor, the emitter voltage will be more positive than that present on the base.

From the AC, or 'dynamic' point of view, for small signals, the input impedance of such a circuit will be high

$$[Z_{in} = R_1(h_{fe} + 1)] \quad (3)$$

and the output impedance will be low

$$[Z_{out} = R_{in}/(h_{fe} + 1)] \quad (4)$$

The performance of such a circuit can be improved still further by the use of the circuit elaborations shown in Figures 6.25a and 6.25b, known respectively as a 'Darlington pair' and a 'compound emitter follower', both of which arrangements can, of course, be used in their 'complementary' – PNP vs. NPN – forms. In both of these cases, the effective current gain, for the purposes of equations (3) and (4) approaches the product of the current gains of the two transistors. The circuit layout of Figure 6.25b is superior from the standpoint of DC operation because it has only one base-emitter junction potential between the input and the output lines, and so suffers from only half the potential DC thermal drift.

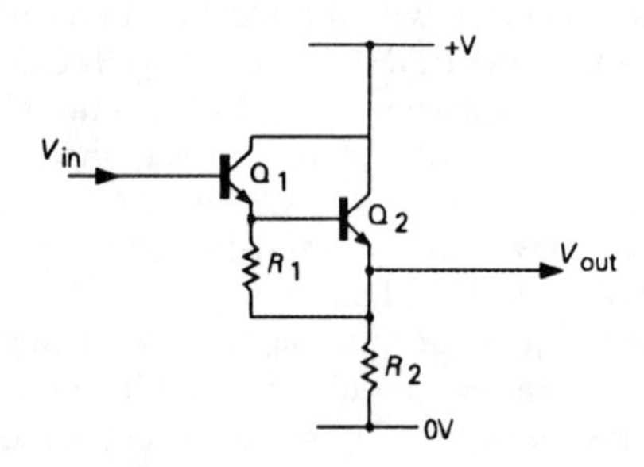

Figure 6.25a *Darlington pair emitter follower*

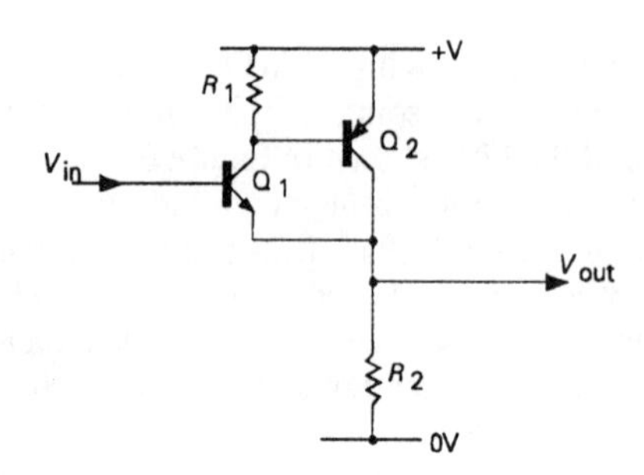

Figure 6.25b *Compound emitter follower*

In both of the arrangements shown, there is a degree of asymmetry, in that the ability of the circuit to source current (through R_2), is usually very much less than its ability to sink it (through Q_1 and Q_2). This asymmetry in single-ended emitter followers is particularly apparent where such an emittter follower stage is required to drive a capacitative load, where, if there is a rapid change in input voltage, the output waveform could show very different 'rise' and 'fall' times, as shown in Figure 6.26, for the case of a square-wave input waveform.

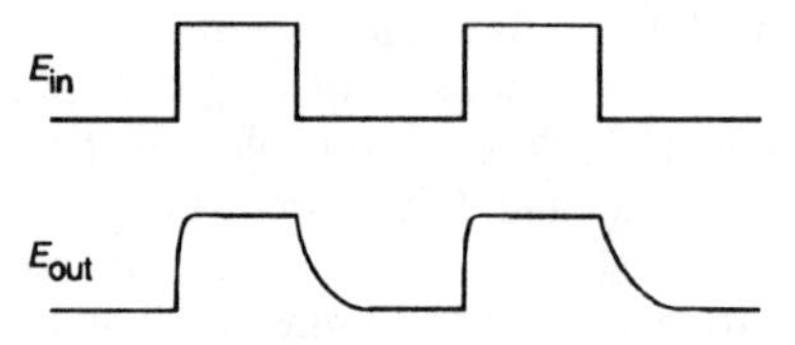

Figure 6.26 *Effect of asymmetry of circuits Figure 6.25 when used to drive a capacitive load*

For audio amplifiers, where it is normally required to provide a symmetrical output impedance conversion stage between the voltage amplification circuitry and the low impedance loudspeaker load, a balanced pair of emitter followers will normally be used, such as that shown schematically in Figure 6.27, and this will also avoid the problem of output waveform asymmetry.

Some external bias source, such as the notional pair of batteries, V_1 and V_2, will be required to cause the transistors to pass an adequate quiescent current to avoid conspicuous 'crossover' phenomena at the point where one pair of transistors ceases to conduct, and the other pair takes over. Combining such an impedance converter stage with the voltage amplifier shown in

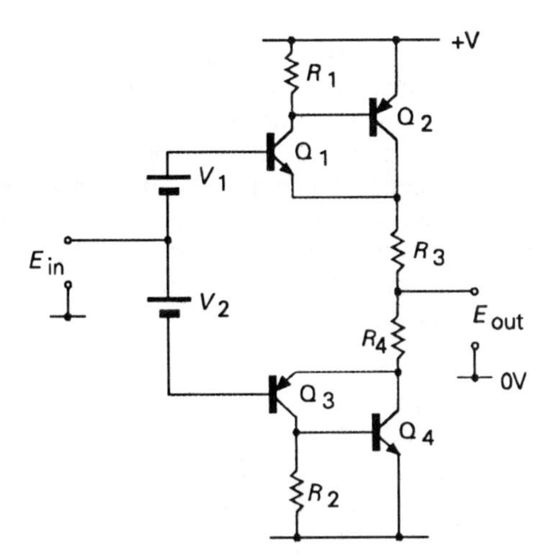

Figure 6.27 *Typical symmetrical emitter follower circuit used in output stage of audio amplifier*

Figure 23 will greatly improve its performance in respect of the stray capacitances, and the shunting effect of the load impedance itself, but will still leave the problems, in respect of HF gain, associated with the 'Miller effect' of the collector–base internal capacitance of Q_1. This arises because any change in collector (or base) potential will require the collector–base capacitance to charge or discharge through either the base or the collector circuit. Moreover, the higher the stage gain, M, the greater the effect will be. So, if a capacitance, C_{fb}, exists between collector and base of the input device, the stage gain will cause the input capacitance, C_{in}, to look like

$$C_{in} = C_{fb}(M + 1) \tag{5}$$

and symmetry would argue that the capacitance, seen at the collector, when the base was returned to a low impedance source, would appear to be the same.

A further, incidental, advantage of the long tailed pair circuit shown in Figure 6.15a, if the output signal is taken from the collector circuit of Q_2, as shown in Figure 6.28, is that the dynamic input capacitance of Q_1 is not increased by the Miller effect, and this allows this circuit to be used for HF amplifier stages.

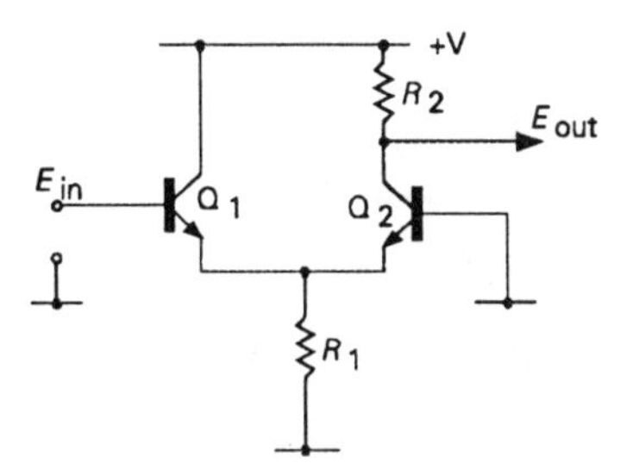

Figure 6.28 *Use of long-tailed pair input circuit to avoid loss of HF gain due to Miller effect*

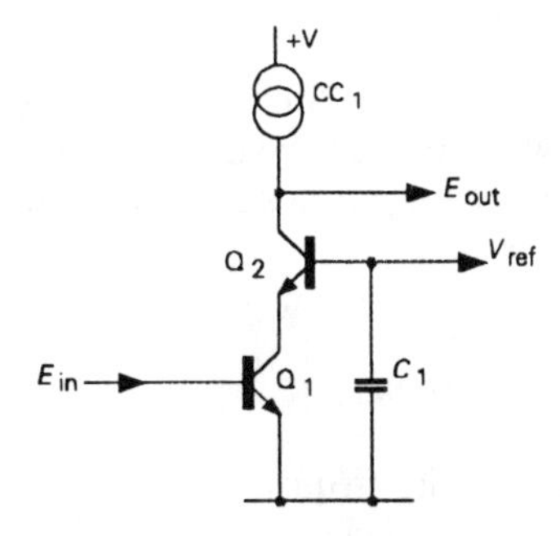

Figure 6.29 *Cascode amplifier connection*

Cascode connection

The problem of the unwanted effects of collector–base capacitance is most satisfactorily remedied by connecting a pair of transistors in cascade, referred to, in electronic engineers jargon, as a 'cascode' connection. Taking the simple circuit of Figure 6.29, the input impedance offered by the emitter of Q_2 is so low that its potential will vary by very little over quite a significant change in emitter–collector current. The input capacitance seen at the input of Q_1 will therefore just be the sum of the static base–emitter and base-collector capacitances, not modified by the voltage gain of Q_1, because, effectively, this is nearly zero. In the case of Q_2, if the base junction is returned to a point offering a low impedance to AC, it will act as an electrostatic screen between its collector and emitter, and the only reactance seen at Q_2 collector will be that of its normal internal collector–base capacitance. So, if the circuit of Figure 6.23 is rearranged to use an input cascode stage, combined with an output emitter follower, as shown in Figure 6.30, a very high stage gain can be obtained, up to perhaps 20,000 at 1kHz, coupled with an extremely low level of harmonic distortion – even in the absence of external distortion reducing 'negative feedback' loops – due to the fact that exceedingly small base–emitter voltage excur-

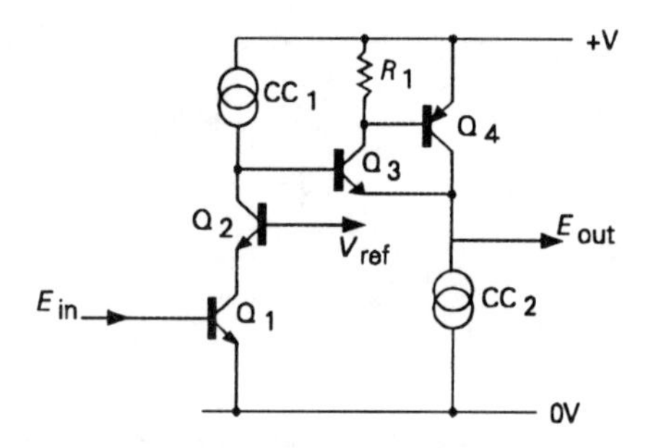

Figure 6.30 *Very high gain, low distortion amplifier circuit*

sions are required to provide the desired output voltage swing, and over such very small excursions the base voltage/current characteristics of the junction transistor are quite linear.

Phase-splitter systems

The need arises quite frequently to convert a single alternating input signal into a pair of identical 'mirror-image' waveforms, as illustrated in Figure 6.31, perhaps for driving a so-called 'push-pull' amplifier stage, as shown, schematically, in Figure 6.32, where a pair of transistors are driven with opposed signal voltages so that their outputs will combine to give a symmetrical drive.

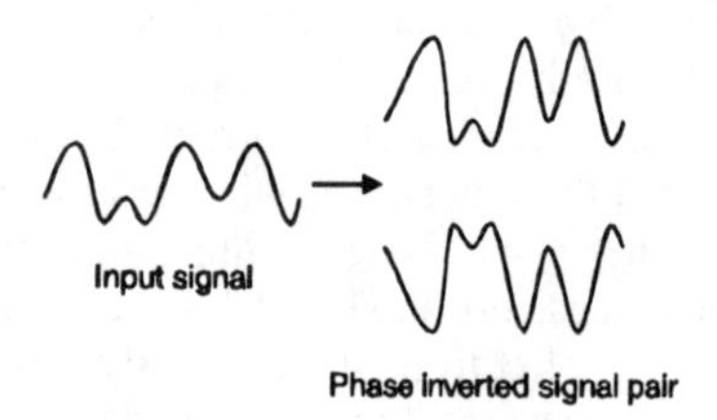

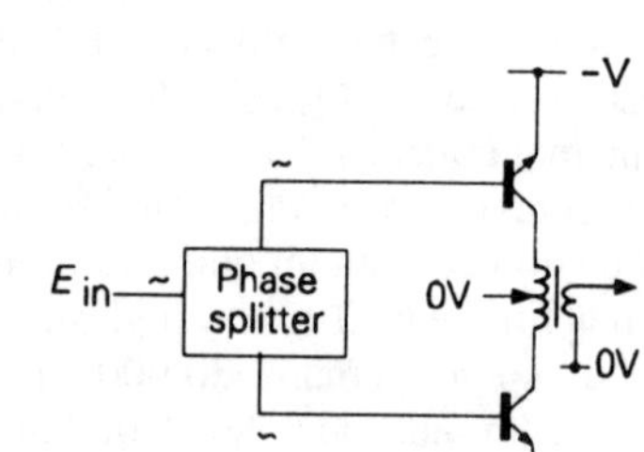

Figure 6.31

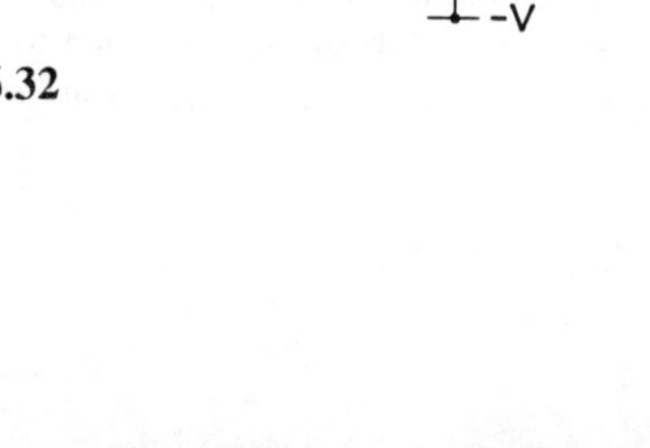

Figure 6.32

One of the easiest ways of achieving this requirement is to use a single high-gain transistor with identical resistive loads in its emitter and collector circuits, as shown in Figure 6.33a. The pair of 'anti-phase' output voltages generated by this circuit will then be virtually identical, since the collector current will only differ from the emitter current by the loss of the base current, which could well be less than one two hundredth of the whole.

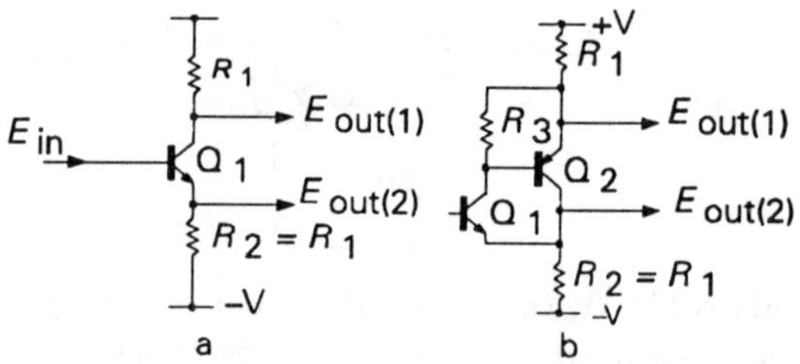

Figure 6.33

Since Q_1 is acting as an emitter follower, with unity gain, the voltage gains of this system will be very nearly +1 and –1. Also, since Q_1 is an emitter follower, the circuit can be elaborated by the use of any one of the improved forms shown in Figure 6.25, as for example, as shown in Figure 6.33b.

While this arrangement would be quite satisfactory for AC coupled systems, in which there were DC blocking capacitors in the two output limbs to avoid problems due to the dissimilar DC potentials at the two output points, it would be inconvenient for use in DC amplifying circuits.

An alternative arrangement which is sometimes used, in spite of its lack of elegance in design, is that shown (neglecting DC potentials) in Figure 6.34a, where a second amplifier stage, Q_2, is added simply to invert the phase of the output signal at Q_2 collector, and the degree of signal attenuation introduced by the potential divider R_2/R_3 is chosen to remove the additional gain provided by Q_2. This does, however, allow the two output signals to be delivered at the same static DC level.

A greatly improved form of this circuit, known as a 'floating paraphase', is shown – again neglecting DC coupling potentials – in Figure 6.34b. In this, the resistors R_2/R_3 are part of a negative feedback network which will allow the output signal voltages from Q_1 and Q_2 to be maintained accurately equal, though in

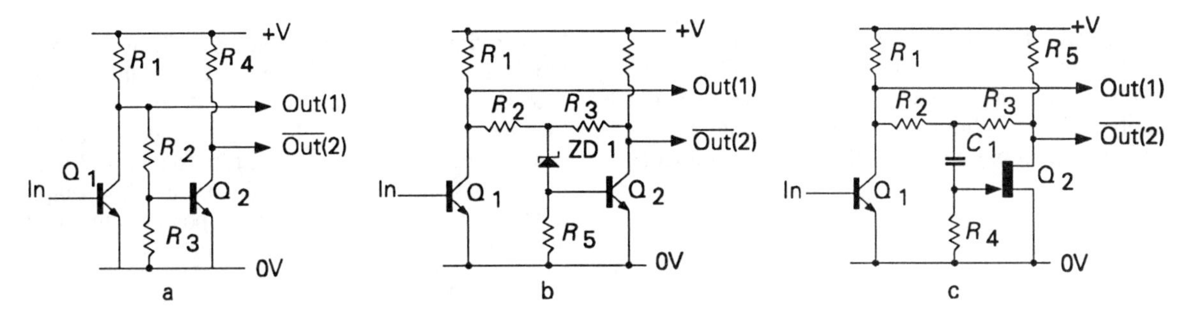

Figure 6.34 *Phase-splitter circuit arrangements*

antiphase, over a wide frequency range, and independently of variations in the gain of Q_2. The difficulty with this circuit, in DC operation, lies in the provision of a proper DC bias for Q_2, though a zener diode between the junctions of R_2 and R_3 and Q_2 base could provide a somewhat restricted answer. For circuits using thermionic valves, or various field effect devices, intended for use only in AC systems, the use of a DC blocking capacitor, C_1, illustrated, in the case of a junction FET, in Figure 6.34c, provides a very satisfactory answer to this problem.

However, by far the best answer to this requirement is the use of a long-tailed pair circuit, shown in Figure 6.35, where, if Q_1 and Q_2 are matched in their characteristics, and the common emitter circuit resistance, shown in Figure 6.15a, is replaced by a constant current source having a high dynamic impedance, the two output potentials will remain closely similar over a very wide frequency (and temperature) range.

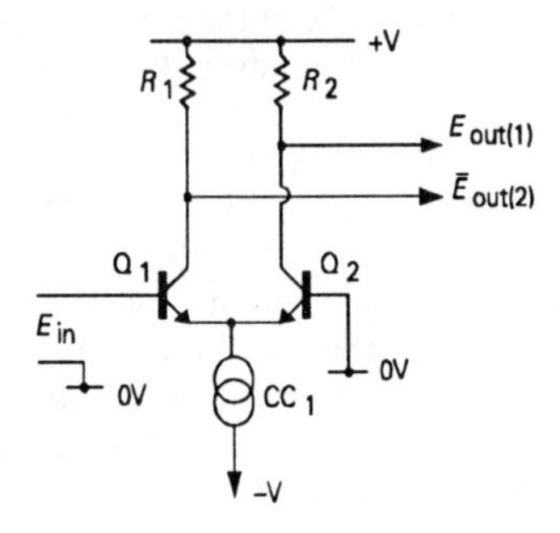

Figure 6.35

If more amplification is required, the input long-tailed pair can be followed by a further pair, Q_3 and Q_4, as shown in Figure 6.36. If the two push–pull outputs are then recombined by a current mirror, this layout forms very linear symmetrical high gain amplifier stage, an arrangement which has formed the basic gain stage in both operational amplifier, and high quality audio amplifier circuitry.

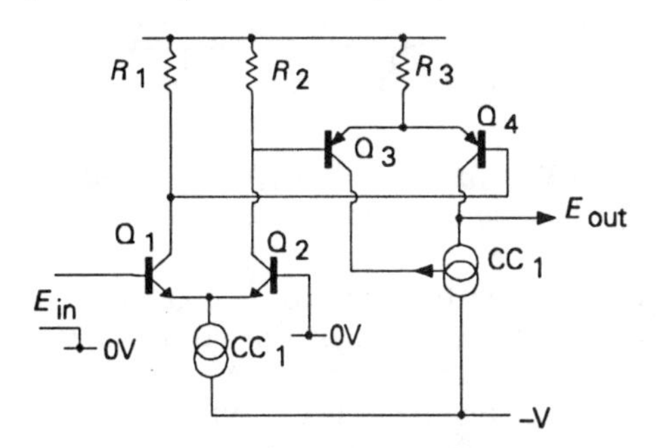

Figure 6.36

Other small-signal junction transistors

The designs shown above have been intended, almost exclusively, for use with silicon planar transistors, because these are by far the most commonly found in practice, and are, generally, the least expensive in cost.

There are, however, other junction semiconductor types based on germanium and gallium arsenide – the latter being an artificial type of semiconductor made, expensively, by the co-crystallization of a trivalent element, gallium, with a pentavalent one, arsenic, a so-called 3–5 semiconductor – which have uses in specific applications. Of these, germanium junction transistors are currently made by similar manufacturing techniques to those used for silicon devices, and differ from them, principally, in having higher leakage currents and lower maximum working temperatures, which leads, in turn, to lower permissible power dissipations for any given 'chip' size. They do, however, have a lower 'turn-on' point in their base voltage/collector current curve, as shown in Figure 6.37, and this can be useful in circuitry for which the available supply voltage is limited. In general, though, they represent an obsolescent area of semiconductor technology.

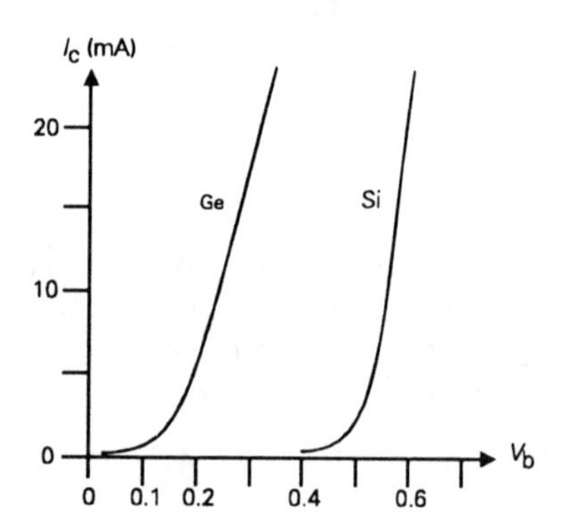

Figure 6.37 *Conduction characteristics of silicon and germanium transistors*

This is far from being the case with gallium arsenide, which has, in several ways, superior performance to silicon – for example, lower noise and a higher possible operating frequency, due to the greater electron mobility within the crystal lattice – but limited in use by the higher material cost. Its principal application, to date, is in the manufacture of VHF (microwave) field effect transistors, operational up to beyond 12GHz.

Bipolar junction power transistors

These, also, are mainly made from silicon, though some germanium types are still available. They are, in essence, substantially identical to small signal devices, except that the method of construction is specifically adapted to allow high working voltages as well as large collector/emitter currents and power dissipations.

The penalty incurred by the use of the larger junction areas, and greater junction thicknesses, required for this type of use, is that the current gain will usually be much less than that of small-signal junction transistors, and will be more strongly dependent on collector current, as shown in Figure 6.38. Because, as shown in equation (1), the mutual conductance, g_m, of a bipolar transistor depends on its collector current, power transistors can offer very high values of g_m indeed, up to 200 Siemens (A/V). Also, because for low input impedances, the output impedance of a simple emitter follower, of the type shown in Figure 6.24, approximates to $1/g_m$ ohms, very low output impedances can be provided by power transistors, when operated at large quiescent currents.

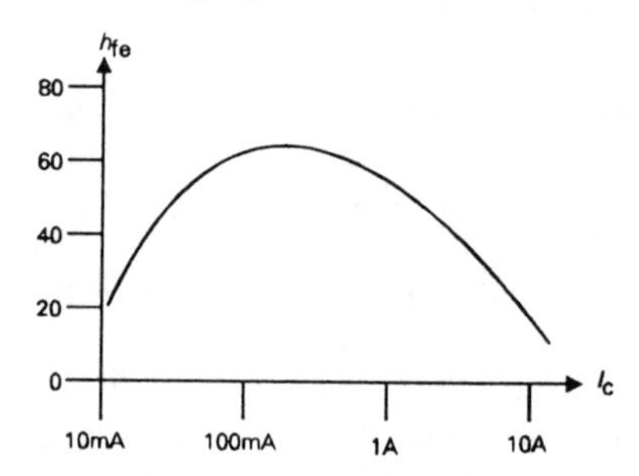

Figure 6.38

There is a restriction in the use of silicon junction power transistors in respect of the permissible collector current which may pass for any given value of collector voltage. This then leads to the need to define usable operating conditions, referred to as the 'safe operating area rating' (SOAR). This problem arises from the fact that, owing to the large emitter junction areas necessary to permit high collector currents, the thermal conductivity across the junction may not be high enough to ensure that the whole area of the junction will always be at the same temperature.

Also, because of manufacturing imperfections, there are likely to be some variations in junction thickness and dopant concentration across this junction, so there is a probability that the curent flow through the transistor, even if the junction temperature is constant across its width, will not be uniformly distributed across this junction. This means that, in use, some small areas of the base/emitter junction may become hotter than others, and since the current flow for a given base voltage will increase as the junction temperature increases, this can lead to a situation in which current will tend to flow increasingly through these areas of the junction – which will cause these regions to get still hotter, so that more current will tend to flow through them, and so on. In the absence of any external restriction on the collector current flow, this effect could result in a destructively high current through some small region or regions of the junction, a condition known as 'thermal runaway'.

To avoid this problem, the manufacturers of power transistors publish safe operating area curves, which define safe values of collector current at specified values of collector voltage, of the kind shown in Figure 6.39. In this, that part of the graph between points 'A' and 'B' relates simply to the maximum current which can be carried by the wires used for the internal connections to the transistor before these wires fuse. The part between points 'B' and 'C' defines the maximum thermal dissipation of the transistor, (the simple product of current and voltage) when the transistor mounting pad is held at a fixed temperature, usually 20°C.

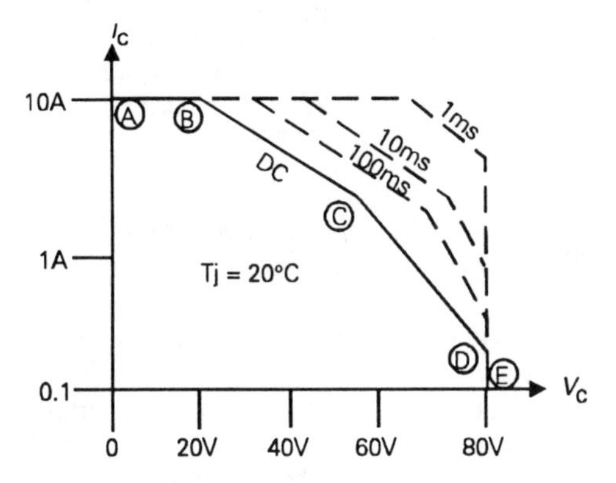

Figure 6.39 *Safe operating area rating (SOAR) for typical power transistor*

It should be noted that this quoted value is usually unrealistically high, since even the best transistor heat sinks will not keep the transistor at room temperature, so it is wise not to approach this value too closely in practical designs.

That portion of the curve between points 'C' and 'D' indicates the area of likely thermal runaway, if either the current or voltage levels are exceeded. Finally, the portion between points 'D' and 'E' simply defines the value of collector/emitter reverse breakdown voltage of the transistor at relatively low current and dissipation values.

The actual boundaries of the thermal limits and the secondary breakdown threshold are time dependent, because of the thermal inertia of the junction, so an 'overrun', if of very short duration, and not repeated at too high a rate, will not necessarily lead to device failure.

It should also be noted that the actual voltage at which this breakdown will occur, for any transistor, will depend on the base/emitter circuit conditions, in that the collector/base breakdown voltage, V_{cbo}, at zero collector current, particularly when the base/emitter junction is reverse biased, is greater than the collector/emitter breakdown voltage, V_{ceo}, at zero base bias. In addition, the value for V_{ceo} will depend on the base/emitter circuit resistance, and will be higher for low values of base/emitter external resistance.

In general, for reliable operation, the designer should ensure that the operating conditions of the transistor keep well within the SOAR curve.

Improvements in manufacturing technology, principally in respect of the purity of material and uniformity of thickness of the junctions, and in keeping the total thickness of the 'die' (the tiny slice of single crystal semiconductor on which the diffused regions are formed) as small as possible consistent with its required operating voltage – so that proximity to the thermally conductive 'header' will hold the junction temperature more uniform – mean that recent designs of power transistor will tend to have a better performance, and may be more reliable under arduous operating conditions.

So, in the case of US 'JAN' '2N...', and the Japanese '2SK/2SA...' series devices, as well as for the European 'Pro- Electron' 'BD...' type nomenclature, the higher type numbers may indicate improved designs.

A widely available form of power transistor is the so-called Darlington device, in which an input transistor, Q_1, is fabricated, as an emitter follower driver stage, on the same silicon chip as the power transistor, Q_2, as shown in Figures 6.40a and 6.40b, for the NPN and PNP versions. These devices are widely used in 'Hi-Fi' audio amplifiers, and the 'economies of scale'

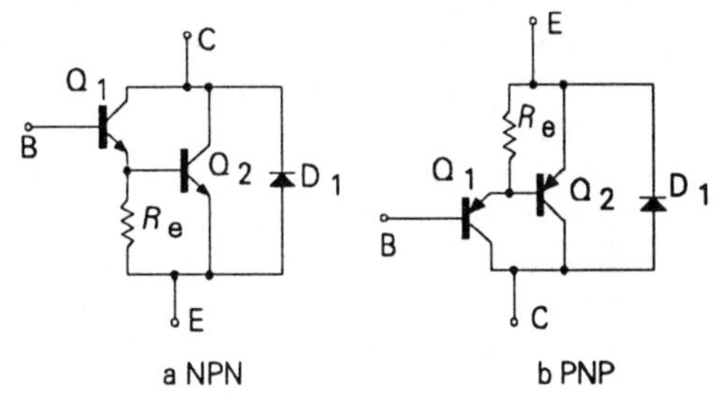

Figure 6.40 *Monolithic Darlington transistor construction*

therefore means that they are relatively inexpensive for their operating voltage and power ratings.

Darlington output transistors offer very high values of current gain, up to, say, 2500, which means that they require relatively low values of input drive current for full output. As output transistor pairs in push–pull Hi-Fi amplifiers they suffer from the small drawback that, since the driver transistor is on the same chip as the power device, it will also heat up at the same time, which makes maintaining the correct value of forward bias for the transistor pair – which will, of course, alter with chip temperature – somewhat more difficult.

Circuit characteristics of Field effect devices

Junction field effect transistors (FETs)

General characteristics

The construction of these transistors was discussed in Chapter 5, but, in general, they rely on the fact that any P–N junction, especially if reverse biased, will generate a depletion zone (a region denuded of current carriers) on either side of the junction, and that this depletion zone can be widened by increasing the reverse bias on the junction, or narrowed by reducing the bias, to provide a means for controlling current flow through any semiconductor region in proximity to this junction. This action is illustrated, schematically, for an N-channel FET, in Figure 6.41. Although this summary of the method of operation of an FET, and the circuit applications which I have shown, refer to N-channel devices, (i.e. those in which the conducting channel is an N type material, and the gate electrodes are P type), the method of operation and the circuitry shown, will apply equally to a P-channel FET, except that it could be a little slower in operational speed.

In Figure 6.41a, a thin strip of N type semiconductor has been formed with ohmic (ie. non-rectifying) contacts, 'X' and 'Y', at either end, and a pair of P type junctions has been diffused into either side of the strip. With no negative bias applied to the P junctions, the depletion zones will extend uniformly on either side of the P–N junction leaving a channel through which current can flow from 'X' to 'Y'. If the P type regions

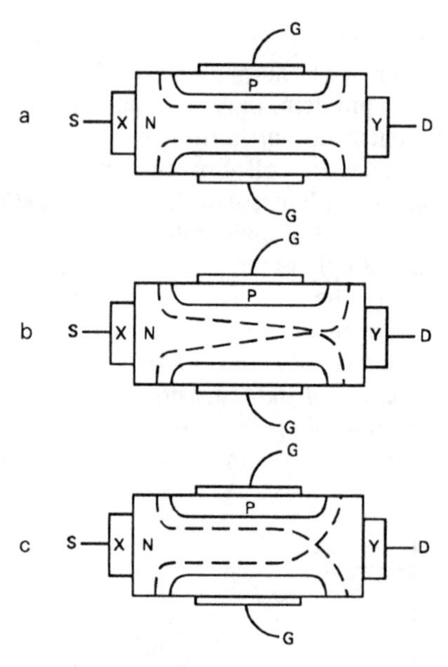

Figure 6.41 *Effect of increasing drain voltage on FET with constant source–gate bias*

are at the same potential as the current source, 'X', and the drain, 'Y', is then made more positive, the potential drop along the strip of semiconductor material, due to current flow, will reverse bias the gate junction towards the drain end of the slice, and will cause the boundaries of the depletion regions to become deformed, as shown in Figure 6.41b, restricting the thickness of the open channel through which current carriers can flow.

Finally, when the drain voltage, on 'Y', is high enough, depending on the doping level and thickness of the channel strip, the two depletion zones will meet, which will cut off the free flow of carriers, as shown in Figure 641c. When this happens, there will be no residual potential gradient along the rest of the channel – all of which, up to the point where the depletion zones join, will be at the same potential with respect to the P type gate region. With normal small-signal junction FETs this effect occurs at a drain voltage of 2–3V, with respect to the source, and, in this condition, the flow of current along the channel is restricted by the ability of the carriers to tunnel through the depleted neck of the channel at the end nearest to the drain: a condition in which the drain current flow is largely independent of the source–drain potential.

This condition is known as pinch-off – a term which

should not be confused with cut-off, where the negative gate bias has been increased to the point where all current flow along the channel has been cut off – and is the normal operating condition for a junction FET, when used as an amplifier.

All normal bipolar junction FETs are of depletion type, which is to say that the drain current will be at a maximum value when the gate voltage is zero with respect to the source and will decrease as the gate is made more negative (in the case of an N-channel FET). When operated in the pinch-off condition, the drain current depends almost entirely on the ability of the current carriers – electrons in the case of an N-channel FET – to tunnel through the depletion zone, and this is largely unaffected by drain voltage. However, the breadth of this pinched-off region is still affected by gate bias, giving the characteristic drain current vs drain voltage curves shown in Figure 6.42, for various values of gate voltage.

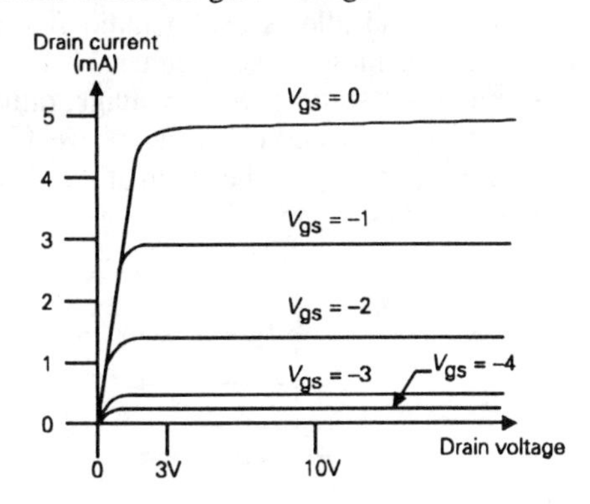

Figure 6.42

Circuit applications

Because the drain current is almost completely independent of drain voltage, this gives the junction FET a very high dynamic impedance, as measured at its drain, and also allows the FET to be used as a very simple and efficient 'two- terminal' constant current source, in either of the circuits shown in Figure 6.43, up to the drain/gate breakdown voltage of the device – usually in the range 25–40V for a small signal FET.

In the case of the circuit shown in Figure 6.43a, the output current will be predetermined by the characteristics of the FET; a fact which allows the manufacture of two-terminal constant current sources which

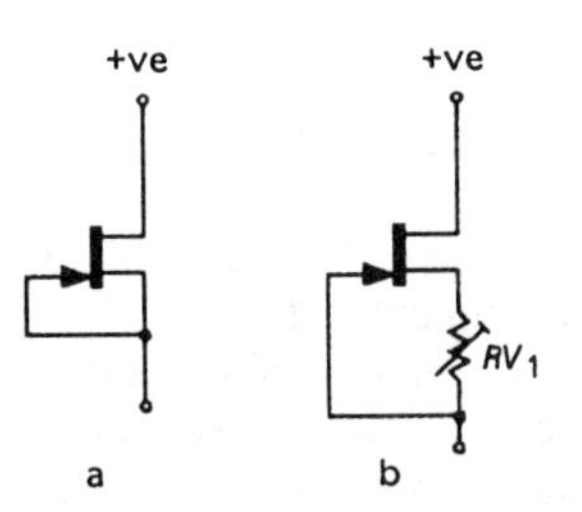

Figure 6.43 *FET constant current sources*

can, by suitable choice of channel thickness and doping level, be offered with a range of output currents, usually within the range 2–20mA. The circuit shown in Figure 6.43b is usable with most junction FETs, and allows the current through any given FET to be controlled by the chosen value of source resistor, RV_1, which can be adjustable.

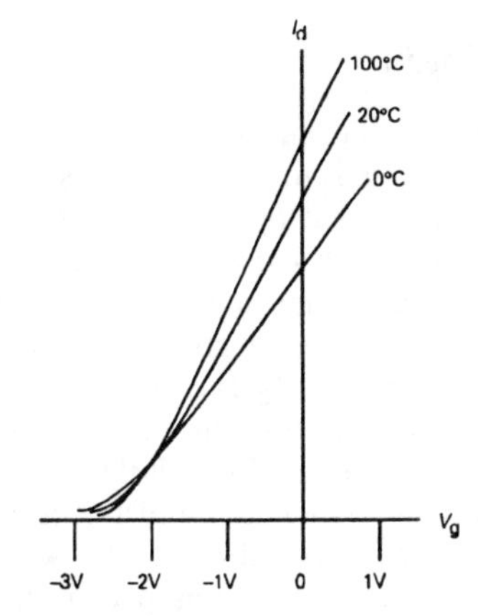

Figure 6.44 *Influence of temperature on drain current characteristics*

The drain current vs. gate voltage relationship of a typical small-signal N-channel junction FET is shown in Figure 6.44. Because the energy of the carriers is dependent on temperature, and their energy influences the boundaries of the depletion zone, the drain current for any gate voltage is still somewhat temperature dependent. It will be noted also that the usable gate voltage extends only up to some +0.5V, beyond which the gate/channel junction will become just a conducting junction diode, and FET action will cease until the gate is once more reverse biased.

The slope of the I_d/V_g curve is governed, for an ideal device, by the relationship

$$I_d = I_{dss}[1 - (V_{gs}/V_{gc})]^2 \qquad (7)$$

where I_d is the drain current, I_{dss} is the drain current at zero gate voltage (gate shorted to source), V_{gs} is the gate–source voltage, and V_{gc} is the gate (cut-off) voltage for zero drain current.

Although this relationship specifies a curved I_d/V_g characteristic, it is still significantly more linear than the comparable base voltage/collector current relationship for a typical silicon junction transistor. However, for typical small-signal FETs it is also much less steep, with g_m values of only 2–10mS (mA/V) as compared with, say, 45mS for the bipolar transistor. This means that a small-signal amplifier stage based on a junction FET will have a lower stage gain, by a factor of some 5–20 times, than the equivalent stage using a bipolar transistor, as can be seen from the relationship

$$\text{Gain} \approx g_m R_l \qquad (8)$$

where R_l is the load impedance.

When the FET gate is in a reversed bias condition – which is that of normal operation – the gate input impedance is very high indeed, up to a million megohms, and the drain impedance can be as high as 2 megohms. This makes the junction FET very suitable for high impedance circuitry. It is also a very convenient device for circuit design, since it is not necessary to derive a specific forward bias for the gate from the supply line (which can lead to problems of unwanted signal breakthrough), instead of which, 'source bias' – similar to the 'cathode bias' used with thermionic valves – can be used, as shown in the simple amplifier stage of Figure 6.45. Where the input signal excursions are less than +/– 0.5V p–p, the bias circuit, R_2/C_2, can even be dispensed with entirely, with suitable FETs.

The junction FET has similar characteristics in use to a low power 'pentode' valve, and will give similar levels of stage gain and distortion, if used at the same supply voltage levels. However, operation at valve type supply voltages is not usually practicable, since junction FETs are mainly limited to drain voltages in the range 15–50V, so with usable values of drain load impedance, the stage gain will not usually be very high.

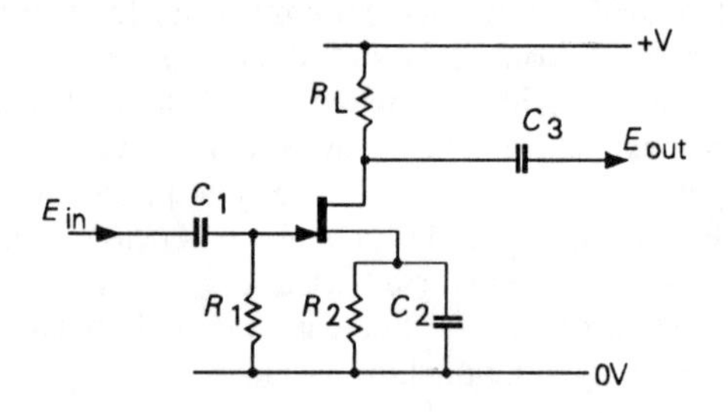

Figure 6.45 *Simple FET amplifier*

'Cascode' operation

The junction FET can, within the limitations indicated above, be used in almost all the circuit designs shown above for bipolar junction transistors. However, there are some applications where the fact that a small-signal FET will operate at a 2–3V negative gate bias is very convenient, such as the high output impedance cascode circuit shown in Figure 6.46. This layout gives very good isolation between the input and the output circuits, and allows use at radio frequencies without 'neutralization' – see Chapter 14. If the gate of Q_2 is returned to some positive voltage, rather than the zero volt line, a junction FET can be used in place of Q_1, to allow advantage to be taken of the FETs very high input impedance.

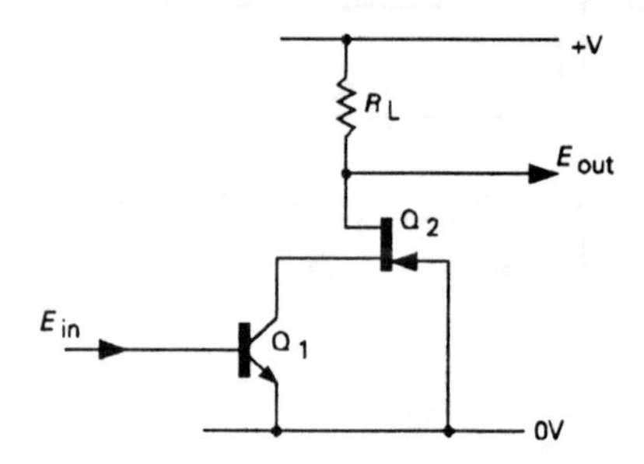

Figure 6.46 *Bipolar/FET cascode amplifier*

The circuit of Figure 6.46 can be adapted, as shown in Figure 6.47 (JLH *Wireless World*, September 1971, pp. 437–441), to give an extremely high stage gain, probably in excess of 10,000, by using another junction FET, Q_3, as a high impedance load, and then interposing a high input impedance emitter follower circuit, Q_4/Q_5, between the output of Q_2 and the load. Moreover, because the stage gain is so high, the input signal voltage required by this circuit, even for the maximum possible output voltage swing, is very small, so that the distortion introduced by the

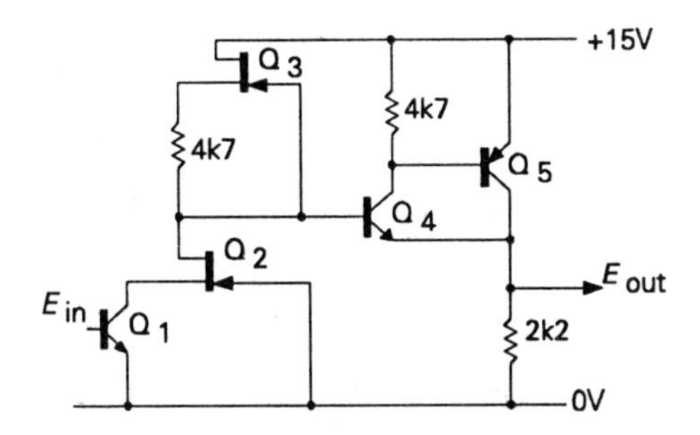

Figure 6.47 *High gain amplifier using FET – bipolar cascode input*

curvature of the base voltage characteristics will also be kept to a very low level.

Very similar circuit arrangements, but based on bipolar transistors rather than junction FETs, are available in IC form, as, for example, the National Semiconductors LM3900 and LM3301, or the later LM359 IC. This particular device offers a typical gain, at frequencies up to 100kHz, of some 4000, into a 10kΩ load, a very low input noise level, when used as an inverting amplifier, and an HF gain/bandwidth product of up to 400MHz. For reasons noted above, it will also offer a very low level of harmonic distortion.

However, the problem with all circuit arrangements which obtain high stage gain values by the use of very high values of load impedance is that they are dependent on keeping the value of stray capacitance at the collector or drain of the amplifying device to a very low level, which can pose problems in discrete circuit layouts. So, if high gain values are required it is usually better to use several conventional amplifying stages in cascade than to seek to get it in one go.

MOSFET devices

General characteristics of low power types

There is a wide class of semiconductor devices, known variously as insulated gate FETs, or MOSFETs (named after their usual method of construction, metal-oxide-semiconductor FETs), usually based on a slice of very high purity single crystal silicon (though, in some very high frequency devices, gallium arsenide is also used), which rely on the ability of a potential applied to a gate electrode to electrostatically induce negative or positive electric charges, electrons or holes, in an otherwise non-conducting semiconductor layer

Historically, this was one of the earliest forms of semiconductor device whose manufacture was attempted, and is illustrated schematically in Figure 6.48. In this, a strip of some semiconductor material having a very low impurity level, and therefore having a very low level of intrinsic conductivity, is formed with two more highly doped regions (the source and drain electrodes), diffused into it – the regions designated N+ in the case of the N-channel MOSFET illustrated. A thin layer of some non-conducting material, usually silicon dioxide or silicon nitride in the case of a silicon substrate, is then formed on its surface. (In practice, this would normally be done as a first step, and the N+ regions diffused through apertures etched through it). On top of this insulating layer a conducting electrode, the gate, is deposited, whose geometry is chosen so that it overlaps both the source and drain N+ regions. If the insulating layer is very thin, when a positive potential is applied to the gate metallizing a layer of electrons will be formed, by electrostatic induction, under the surface of the insulating layer, in the very low conductivity (P–) material joining the two N+ regions. These electrons are mobile, and will move from source to drain, or vice versa, in response to an applied positive potential, with the vacancies left by their removal being instantaneously replaced by electrostatic induction.

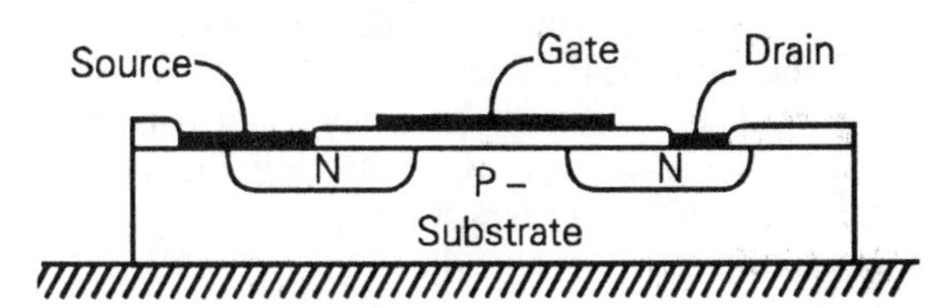

Figure 6.48 *Enhancement type*

This type of device, in which little or no current will flow from source to drain in the absence of a gate voltage, is described as an N-channel 'enhancement mode' MOSFET, and it will have a gate voltage vs. drain current characteristic of the kind shown in Figure 6.49.

As in the case of junction FETs, 'complementary polarity' MOSFETs (ie, P-channel rather than N-channel) can be made by diffusing highly doped P regions,

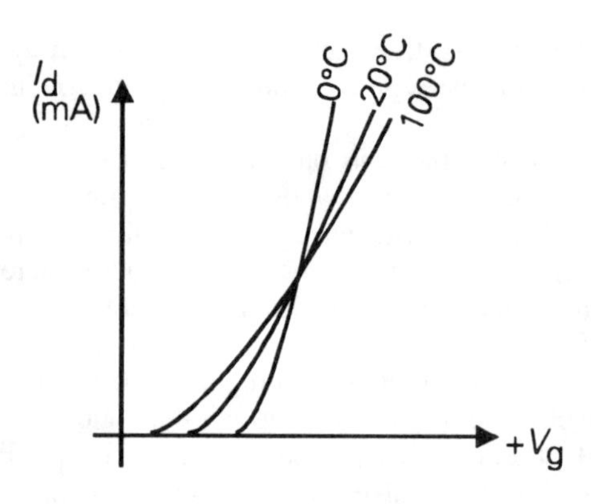

Figure 6.49 *Characteristics of enhancement type MOSFET*

usually known as P+, into the silicon strip, in place of N+ ones, and these devices will have very similar characteristics, though of opposite working polarity, to the N-channel types.

If, in the case of an N-channel MOSFET, a thin layer of N–doped silicon is formed underneath the gate region by surface diffusion, before the oxide or nitride insulating layer is formed, as shown in Figure 6.50, some source–drain current will flow, even in the absence of an applied gate voltage, to give the kind of gate voltage vs. drain current relationship shown in Figure 6.51. This type of device would be described as an N-channel depletion mode MOSFET, since the residual n-doped conducting layer can be 'depleted' of carriers by the application of a negative gate voltage. This kind of MOSFET is usually only found in small-signal N-channel devices, especially where two separate gate regions are formed, as shown in Figure 6.52, a dual-gate MOSFET, to provide electrostatic screening between gate 1 and the drain, for use in RF applications.

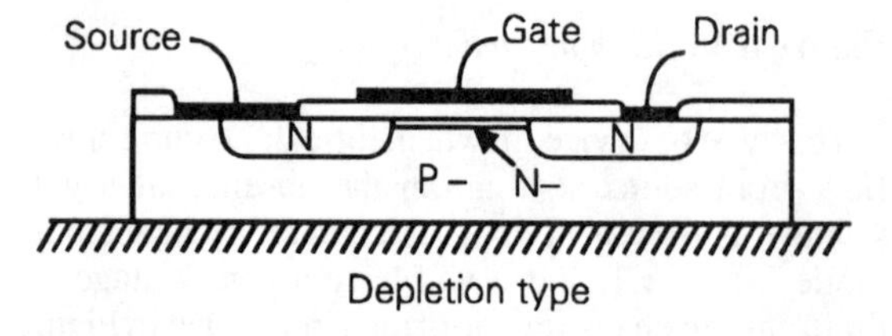

Figure 6.50 *Construction of N-channel MOSFETs*

As shown in Figures 6.49 and 6.51, there is a degree of temperature dependence of the drain current for a

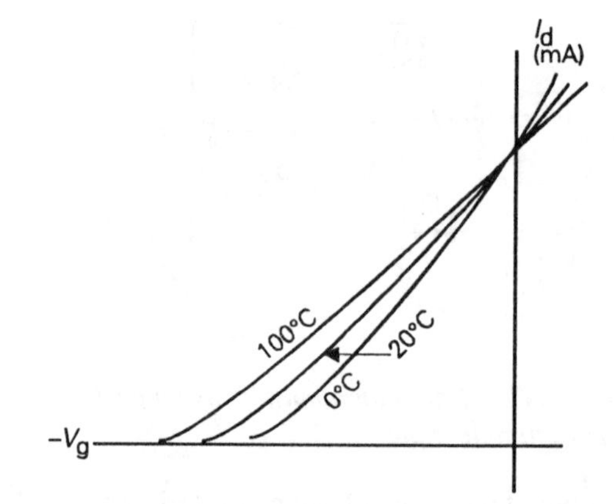

Figure 6.51 *Characteristics of depletion-type MOSFET*

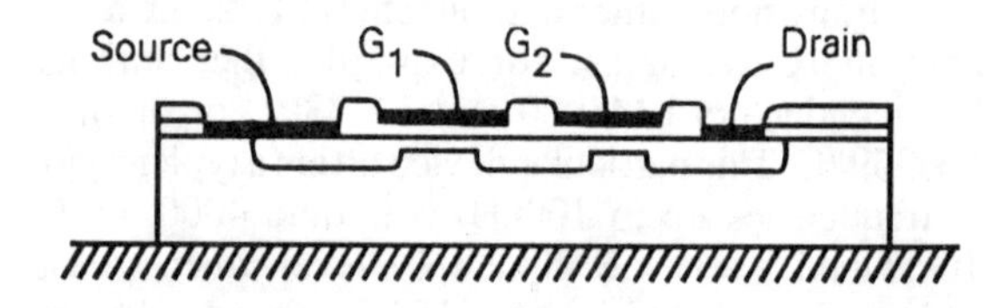

Figure 6.52 *Construction of dual-gate MOSFET*

given gate voltage, though, because these curves intersect, there will be some point on the I_d/V_g curve where the effect of temperature will fall to near zero.

Circuit applications

MOSFETs can be used in almost all of the circuit layouts applicable to junction FETs, or even to bipolar junction transistors, with the added advantage that the operating gate voltage range can extend on either side of 0V, up to the level at which the gate/substrate insulation breaks down, when the device is destroyed.

The characteristics of MOSFETs cannot be predicted from any theoretical model, since the relationship between gate voltage and drain current will depend to a large extent on the thickness and dielectric constant of the gate insulating layer, in relation to which the skill of the manufacturer lies in producing an insulating barrier which is very thin, so that a voltage as low as 2–3V will allow a usable (mobile) charge layer to be induced in the channel, while being able to sustain voltages of the order of 30–40V without breakdown.

However, apart from the minor stray field effects due to errors in the manufacturing geometry, which mainly affect the shape of the I_d/V_g curve at the onset of conduction, the linearity of the I_d/V_g curve is very high, since it is largely controlled by the physical relationship between applied voltage and induced charge, which is very linear. Also, because of the nearly instantaneous induction of charge in the channel, following on application of a potential to the gate, and the absence of 'hole storage' effects, MOSFETs are extremely fast in operation – limited mainly by the relatively high gate–source capacitance, which forms a capacitor which must be charged or discharged through the gate/source circuit impedance before the gate-source voltage can increase or decrease. This encourages the use of MOSFETs in RF applications, where they can outperform both bipolar transistors and junction FETs. Otherwise, they can be used in most of the circuits shown above, with the reservation that care must be taken to ensure that the gate–source, or gate–drain breakdown voltage is never exceeded, however briefly, or the device will be destroyed.

Power MOSFETs

The high speed and relatively high linearity of the MOSFET transistor encouraged the development of these devices in both higher voltage and higher power forms. The main difficulty in a simple development of the lower power versions arose because in the small-signal designs the device is formed in a way which makes the channel parallel to the surface of the chip. This kind of device would be described as a lateral FET, because the current flow will be from side to side, in contrast to the flow of current through a bipolar junction transistor, which would be vertical, through differently doped semiconductor layers formed one on top of another.

For higher drain currents, the impedance of the channel must be made as low as possible, which entails, among other things, making the channel length as short as possible. In normal lateral FETs the closeness of the two ends of the channel to one another is determined by the skill of the manufacturer in forming and applying photo-resist masks, through which diffusion and etching processes can be carried out. With the technology available at the time, it was difficult to produce very short channel lengths by this means, so the makers adopted 'V' and 'U' designs, in which the device was manufactured, like a bipolar transistor, with appropriately doped semiconductor layers formed one on top of another, as in a junction transistor, and then 'V' or 'U' shaped channels were etched into the device, as shown in Figures 6.53a and 6.53b. The exposed surface was then oxidized to form an insulating layer, so that the gate metallization could be applied across the now vertical channel region. In these cases, for manufacturing convenience, the channel would be formed in a very low conductivity layer which had a very small residual P type impurity level, described as P–, giving the equivalent of an NPN junction transistor in which the emitter and base are connected together.

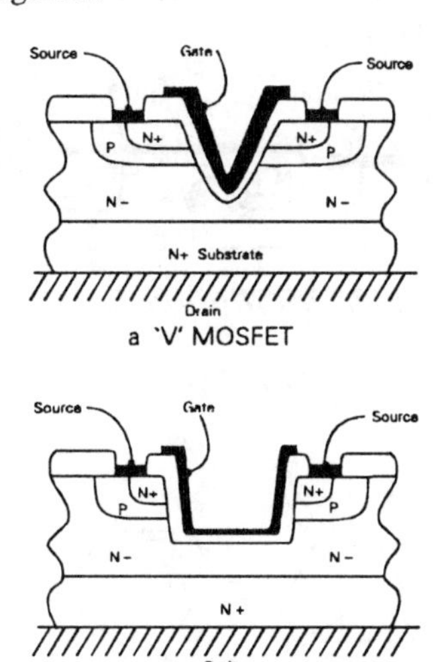

Figure 6.53

The 'U' MOSFET was somewhat better for high voltage use because the rounded bottom of the groove led to lower levels of unwanted electrostatic stress across the electrically fragile gate insulating layer. Neither of these types are particularly suited to the manufacture of P-channel MOSFETs, for which the 'T' MOSFET construction is now preferred.

The construction of this kind of device is shown in Figure 6.54, and employs the lateral diffusion of the N+ source regions into preformed P– zones to reduce the resulting effective length of the channel below that which the surface masking alone could easily provide. As in all power MOSFETs, the possible channel current flow is increased by constructing a large number of channels, connected in parallel. An elegant design of this type is that introduced by International Rectifier Corp., of the USA, in their 'HEXFET' devices, in which the the parallel MOSFET regions are fabricated as a series of hexagonal pits, as in a honeycomb, and illustrated in Figure 6.55. This paralleling of the channels also increases the mutual conductance (g_m) of the device, so the g_m is related, as a function of manufacturing technique, to the total drain current, for any given gate voltage. A typical power MOSFET gate voltage vs drain current relationship is illustrated in Figure 6.56. As with other MOSFETs, the power MOSFET has a negative temperature coefficient of drain current vs gate voltage above a certain drain current value.

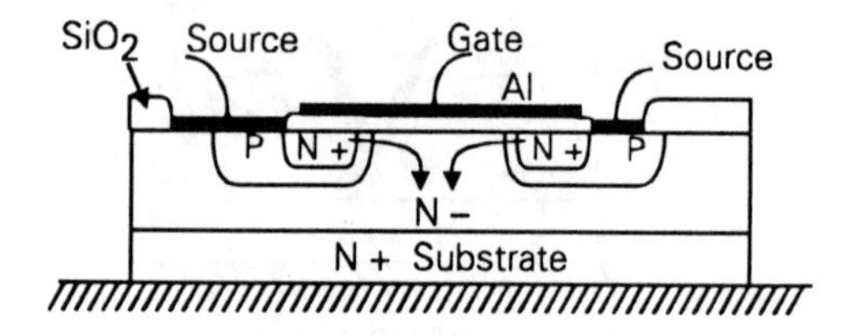

Figure 6.54 *'T' MOSFET*

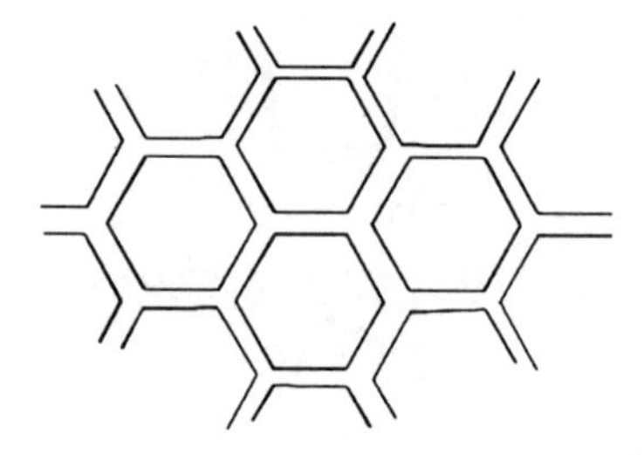

Figure 6.55 *Honeycomb construction of parallel connected MOSFET cells*

For N-channel devices g_m values in the range 0.5–1.5 S/ampere are usual, with P-channel devices falling mainly towards the lower end of this range. They will not, therefore, lend themselves to high gain stages, but they are very linear. Like junction FETs, they have an extremely flat I_d/V_d curve, shown in Figure 6.57, so that they can, within their permissible operating voltage range, make excellent constant current sources, and high dynamic impedance loads.

Two points which must be remembered in relation to power MOSFETs are that, because of their excellent

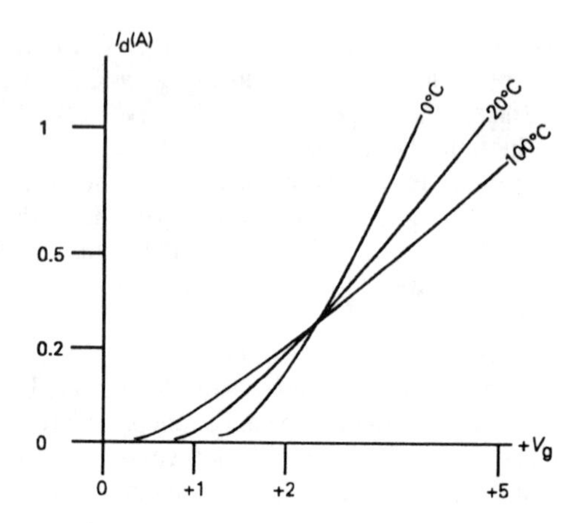

Figure 6.56

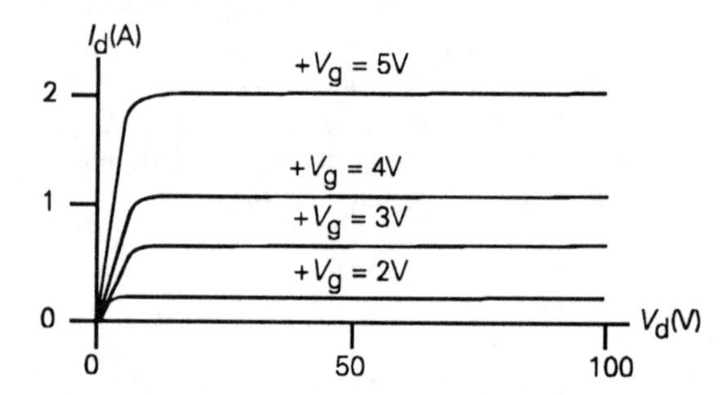

Figure 6.57

high frequency characteristics, they can readily break into unwanted – and possibly destructive – oscillation as a result of the internal capacitances of the device, and the unavoidable inductance of the gate and source leads. This tendency to parasitic oscillation is best avoided by ensuring that the gate and source leads are as short as possible, and that they do not run parallel to one another, and by including a 'gate stopper' resistor, of 50–330 ohms value, as shown in Figure 6.58. The second point is that, as a function of their method of manufacture, as with Darlington type power transistors, all power MOSFETs have an internal diode, also shown in Figure 6.58, which has identical reverse breakdown voltage and forward current carrying characteristics to the MOSFET itself.

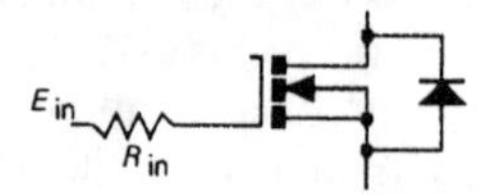

Figure 6.58 *Gate stopper resistor connection*

As with the small-signal versions of this device, the V_g/I_d characteristics are very linear, once within its conducting region, and this encourages the use of such devices as the output emitter followers (actually source followers) in the output stages of high quality audio amplifiers. Provided that their gate-channel breakdown voltage limitations are respected, and they are operated within the boundaries of the safe working area domain, shown in Figure 6.59, MOSFETs, as a whole, can be used in most of the circuit layouts shown above for bipolar transistors, allowing for their somewhat different operating characteristics, and, as an example of the simultaneous use of bipolar transistors, junction FETs and MOSFETs, a 40–50 watt, very high quality, audio amplifier is illustrated in Figure 6.60.

In this circuit, internal loop negative feedback is employed (see Chapter 7) to stabilize the performance of the circuit, and to reduce the extent of residual waveform distortions to allow, for example, a total harmonic distortion (THD) of less than 0.01%, at full output power, over the frequency range 200Hz – 10kHz.

However, in order to avoid HF instability, some loop stabilization components such as C_a, the 'Zobel network', C_b/R_b, and the output inductor, L_c, must be included; particularly in order to avoid the possibility of malfunction when used with a reactive load, such as a loudspeaker unit.

An application for which both junction FETs and MOSFETs are well suited, but especially MOSFETs, is analogue signal switching, and this is covered in a later chapter.

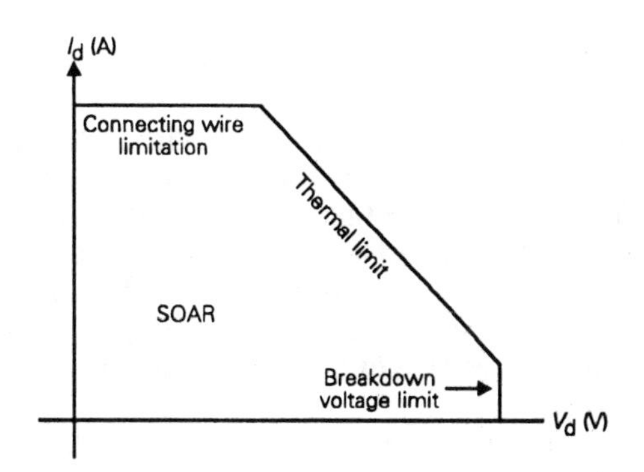

Figure 6.59 *Safe operating area for typical power MOSFET*

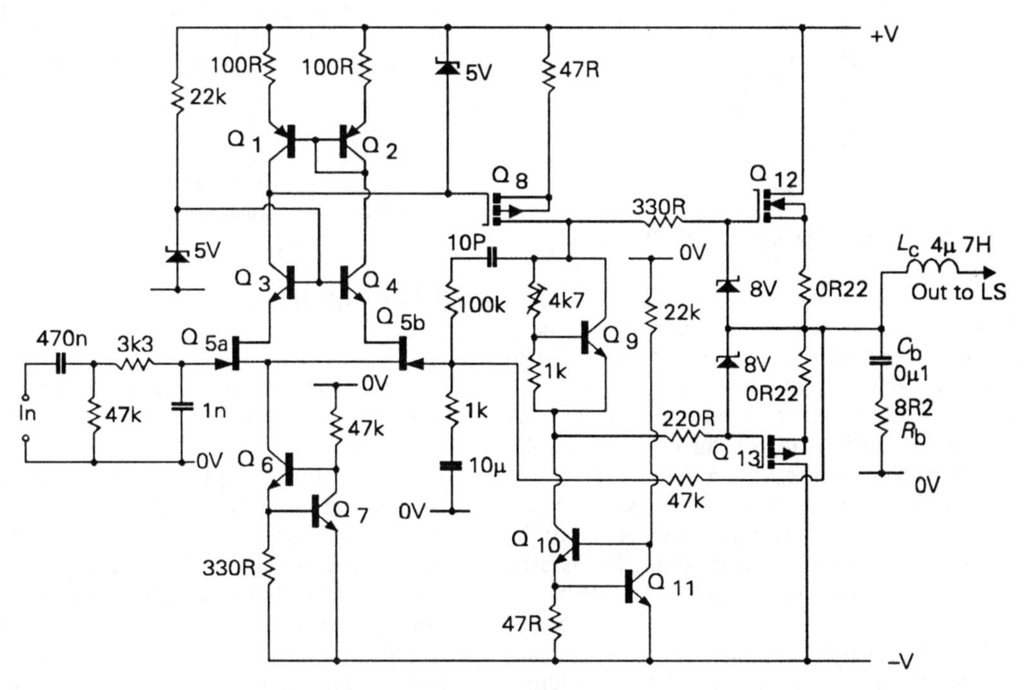

Figure 6.60 *High quality MOSFET output audio power amplifier*

Protection from gate breakdown

Although power MOSFETs do not suffer from the type of secondary breakdown which can damage power transistors due to the funneling of the emitter/collector current through small regions of the base–emitter junction, and the excessive localized heating which this will cause, they can fail catastrophically if the breakdown voltage of the gate insulating layer is exceeded.

Because power MOSFETs have very high gate-channel capacitances, up to 3nF in some designs, they are much less susceptible to inadvertent damage due to induced electrostatic charges during handling, than the small signal devices, whose gate-channel capacitance may only be a few pF. On the other hand, it is not practicable, in high power devices, to diffuse a pair of protective diodes, connected to the gate electrode as illustrated in Figure 6.61, into the MOSFET chip, as is customarily done with small signal devices, since this can lead to thyristor-type '4 layer' (PNPN) type breakdown, which would be destructive in a high power circuit.

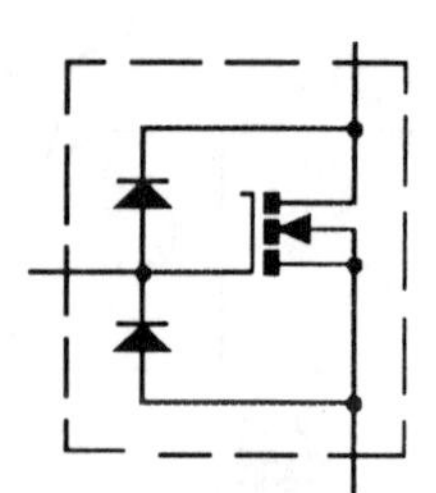

Figure 6.61 *Gate protection for small signal MOSFET*

There is, however, no difficulty in using externally connected zener diodes, as shown in Figure 6.60, to prevent some safe level of gate-channel voltage from being exceeded, and this is a sensible precaution. A somewhat less obvious reason for destructive gate-channel breakdown can be seen from Figures 6.53 and 6.54, in that there are regions of the transistor in which the gate metallizing is in proximity to that part of the transistor which is connected to the drain. In this case, where the gate–drain voltage exceeds the gate breakdown potential, the device will only survive so long as the voltage drop across the N-path between the channel and the drain is sufficiently greater than that across that part of the channel opposed to the gate electrode for the gate–drain voltage not to be exceeded, and the relative apportionment of these voltage drops will be drain current dependent. This means that under dynamic conditions, where both high drain currents and large gate–drain voltages can coexist, even briefly, gate insulation breakdown can occur, even though the limitation imposed by short-term thermal dissipation had not been exceeded, and this is a source of possible failure which must be considered.

Even so, the reliability of power MOSFETs, for high power use under difficult load conditions, is generally very much easier to ensure than with nominally equivalent bipolar junction power transistors.

Voltage reference sources

In many linear circuit applications, such as, for example, the DC amplifier layout of Figure 6.14b, or the current source of Figure 6.16., a device giving some constant potential drop, or a local source of some reference voltage is needed. The simplest answer is to use a suitable zener or avalanche diode, but these will introduce a certain amount of wideband noise, which may be unacceptable for some circuit applications. In this case, it may be preferable to use a string of small signal diodes, connected in their 'forward', i.e. conducting, mode and to choose a sufficient number of these to build up the required voltage drop from the sum of the individual 0.57–0.6V diode voltage drops, since a junction diode operated in its forward mode is a relatively low noise device.

DC Amplifiers

Requirements for stability of output voltage level

Most of the circuit applications described above have been aimed at voltage amplification, or impedance conversion, functions in which the absolute DC value of the output voltage level was not important, provided that it stayed within some satisfactory working voltage range. However, the need arises from time to time to amplify a small DC voltage signal so that it can be used to give an instrument reading, or operate some mechanism requiring a higher power input. In these cases the absolute DC voltage level at the output will be important, and 'drift' in this value, for a constant input voltage signal, must be avoided.

As has been shown in Figures 6.12a, 6.42 and 6.57, the collector or drain currents, for a fixed base or gate potential, of both bipolar junction transistors, and junction and MOSFET type field effect devices, are strongly temperature dependent, so that the output voltage any amplifier circuit using these components will be affected by changes in the ambient temperature of the devices, and also by any heating effects brought about by internal thermal dissipation.

Figures 6.44, 6.49, 6.51 and 6.56 show that, for both junction FETs and MOSFETs, there is some value of gate voltage, depending on the device design, for which the temperature dependence of the drain current can be eliminated or minimized. This is a helpful circumstance, but, in general, the lessening or elimination of such voltage offset drifts due to thermal changes will be achieved by the use of the long-tailed pair arrangement of Figure 6.15a, or the base–emitter offset potential cancelling layout of Figure 6.62, or by using differential amplification techniques, as shown in Figure 6.63, in which the comparative rather than the absolute value of the output voltage is used, for example, to operate a voltmeter. All of these approaches should be combined with means for keeping thermal dissipation effects, within the active devices, to the lowest possible level, by using the lowest usable working voltages and currents.

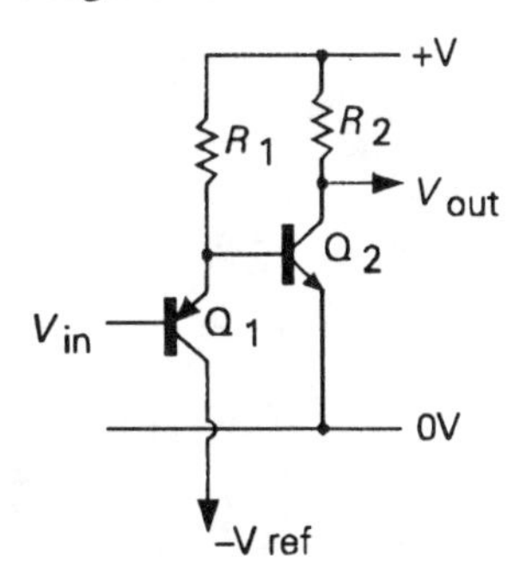

Figure 6.62 *Base–emitter offset voltage cancelling circuit*

For very high degrees of long term DC stability, the amplifier unit may be enclosed in a temperature controlled housing, or use 'chopper stabilization' of the kind shown schematically in Figure 6.64. In this, in

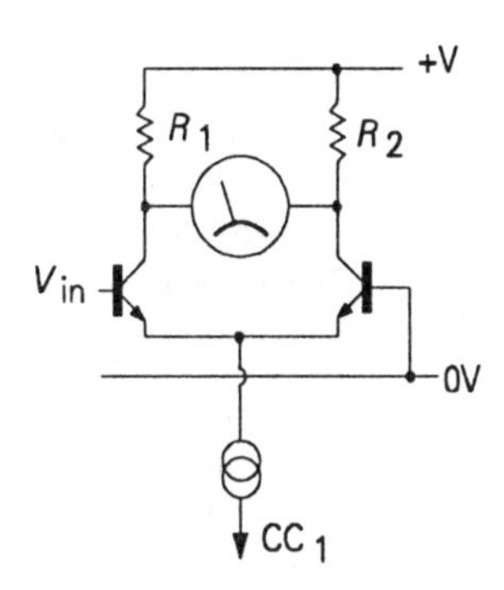

Figure 6.63

principle, the input DC signal voltage is converted into an AC voltage waveform by chopping it into a train of pulses whose peak amplitude is equivalent to the original input DC level. This pulse train can then be amplified by a conventional AC amplifier, whose AC gain is stabilized by the use of loop negative feedback (see Chapter 7). The amplified AC signal can then be converted back into a DC output voltage by means of some conventional 'rectifying' circuit. Commercially available DC amplifiers, particularly those in 'hybrid' form (i.e. encapsulated blocks containing surface mounted discrete components) may contain all these functions within a single package.

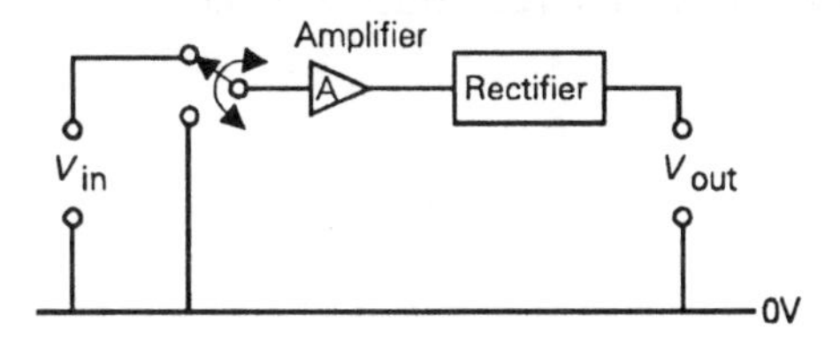

Figure 6.64 *Chopper stabilized DC amplifier*

These chopper stabilized hybrid amplifiers tend, however, to be costly, and the performance they offer is being approached by more conventional, and much less expensive, operational amplifier IC packages, and it is suggested that the use of these ICs should be considered when DC amplification is required.

These linear IC DC amplifiers either employ quite normal circuitry, with the components and layout optimized to ensure very low DC drift in the input and output offset voltages, such as the 'PMI' 'OP07' and 'NS' 'LM121' ICs, or so-called chopper systems, such as the National Semiconductors 'LMC668' (see NS *Linear Data Book*, 1988, pp 2/559–565), in which a normal amplifier stage is arranged so that its zero setting is continuously trimmed by a further amplifier within the IC package, as shown schematically in Figure 6.65. This arrangement employs two nominally identical amplifier packages, A_1 and A_2, which have external 'offset null' correction points, p and q. These amplifiers are connected to internal CMOS switches, S_1 and S_{2a}/S_{2b}, operated by an external 'clock'. At the beginning of the switch cycle, the two inputs of the 'nulling amplifier', A_2 are shorted together, and its output is connected to the two nulling points, through S_{2a} and S_{2b}. In the case of A_2, this generates a DC negative feedback loop which operates to zero the DC output voltage, and this correction, being applied to A_1 as well, has the effect of generating the necessary offset which would have been required to cancel any output errors in A_1, had its inputs been short-circuited as well. In the interval between the switching cycles, the required zero adjust voltages are stored in two (external) low leakage capacitors, C_1 and C_2.

Quoted long-term output voltage drift levels for the LMC668 are of the order of 100nV/month.

Thermal drifts in the other circuit components, of which the principal ones likely to affect the results in amplifier and other signal handling circuitry are the resistors, have now been reduced to very low levels by the choice of optimum constructional materials.

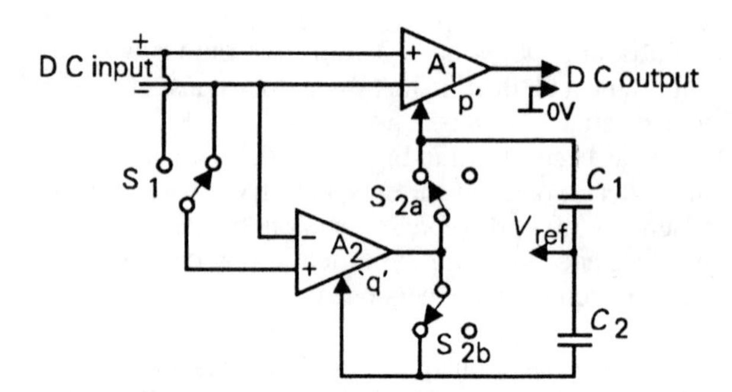

Figure 6.65 *Integrated circuit chopper stabilized operational amplifier*

The final effect which can cause DC drift, in amplifiers and other circuitry, is the change of component characteristics due to ageing, and to a lesser extent, due to thermal cycling and mechanical vibration. Once the initial burn in time has elapsed, ageing effects in semiconductors, when operated well within their maximum permissible voltage, current and working temperature ratings, are relatively small by comparison with those found in thermionic valves, where the mechanical changes due to internal outgassing and gradual loss of cathode emission are continuous throughout the lifetime of the device.

7

Feedback, negative and positive

Introduction

'Feedback' is the term which is given to the introduction of a signal voltage, derived from the output of what is usually an amplifying stage, into the input of that stage. This feedback can be present as a deliberate feature of the circuit design, but may also occur inadvertently, either because of shortcomings in the amplifying devices or other circuit components which are used, or because of oversights in the design of the circuitry in which they are employed. Whatever the cause, the presence of feedback can significantly alter the performance of the circuit, so, although this can be a rather complex aspect of circuit theory, it is necessary for its effects to be broadly understood by any engineer seeking to try his hand at circuit design.

The basic effects of feedback

If the polarity of the signal which is fed back is of the same sign as that input signal which generated it, then the feedback signal is said to be 'positive'. If its polarity is opposite to that of the input signal, it is said to be 'negative'. (Because negative feedback is such a common feature of circuit design, it is often just referred to as NFB).

In the case of positive feedback (which I shall call PFB), because the signal which is fed back is of the same polarity as the input signal, it will act to increase the size of this signal, which will, of course, then increase the size of the output signal, which will, in turn, increase the size of the voltage which is fed back. Clearly, unless there is some loss of signal somewhere in the system, or some constraint on the possible input or output voltage swing, this process would cause the gain block to produce an infinitely large output. In a DC system, this effect will usually cause the system to 'latch up'– a condition in which the output voltage is driven fully towards one or other of the voltage limits of which the output is capable. In an AC coupled system, the result will usually be a continuous and uncontrolled oscillation. Both of these effects of PFB can be utilized in circuit design.

For example, in the DC systems shown in Figure 7.1, the circuit layout can be arranged to provide either a 'bi-stable' or a 'mono-stable' function block. In the first of these forms, shown in Figure 7.1a, the output voltage can be made to jump rapidly from one stable output voltage state to another, where it will rest until

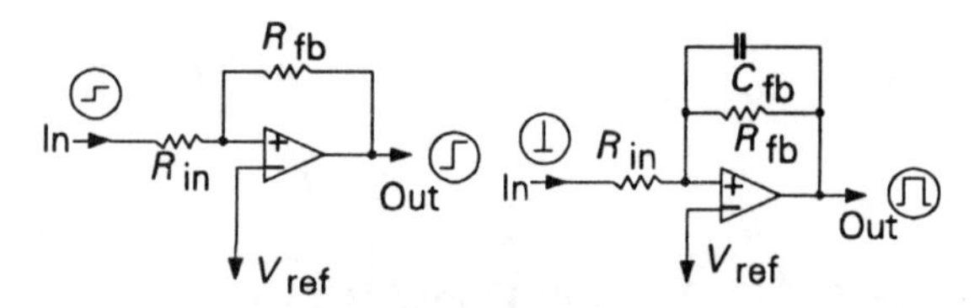

Figure 7.1 *The use of positive feedback to provide bi-stable or mono-stable logic elements*

a suitable input signal restores the original output voltage condition of the circuit. In the second case, shown in Figure 7.1b, the output voltage can be made to change from its normal rest condition to a different voltage level, where it will remain for a period of time, determined in this case by the values chosen for C_{fb} and R_{fb}, before reverting to its original state. Both of these types of function are widely used in logic elements.

Positive feedback, in an AC circuit, can be used to produce a continuous oscillatory voltage output, which could be used as a signal source, although for this purpose some means of controlling both the frequency of oscillation and the amount of feedback – which will determine the output signal voltage – is necessary. As an example of this, in the simple 'Wien bridge' oscillator circuit, shown in Figure 7.2, the system is made to oscillate continuously by the use of positive feedback, which is applied from the output to the non-inverting input of a gain block (A_1), via the *RC* Wien network – R_1C_1, R_2C_2 – (in which $R_1=R_2$, and $C_1=C_2$). This network has the interesting characteristic that, at a frequency at which the impedance of R_2+C_2 is twice that of R_1 in parallel with C_1 (for which the shorthand notation $R_1 \| C_1$ is often used), the output, at point 'A', is in phase with the input, at point 'B', and if the gain of the amplifier is equal to, or slightly exceeds 3, the circuit will give a sinusoidal output signal at a frequency defined by the equation

$$F = 1/(2\pi CR).$$

In practical oscillator circuit designs of this type, negative feedback, by way of a signal applied to the inverting input of the gain block, is then used to control the gain of the amplifier, so that is high enough to cause the circuit to oscillate, but not so high that the output sinewave signal is distorted by 'clipping', due to the limits on the output voltage swing imposed by the supply voltage rails. This can be done quite conveniently, by the use of an output voltage-dependent component, such as the low-power thermistor (TH1), shown in Figure 7.2, whose resistance value decreases as the output voltage increases, and which will thereby control the amount of the negative feedback signal applied to the inverting input of the gain block. If the value of R_{fb} is chosen correctly, this NFB input will reduce the gain so that it is exactly equal to 3 at the desired AC output voltage swing.

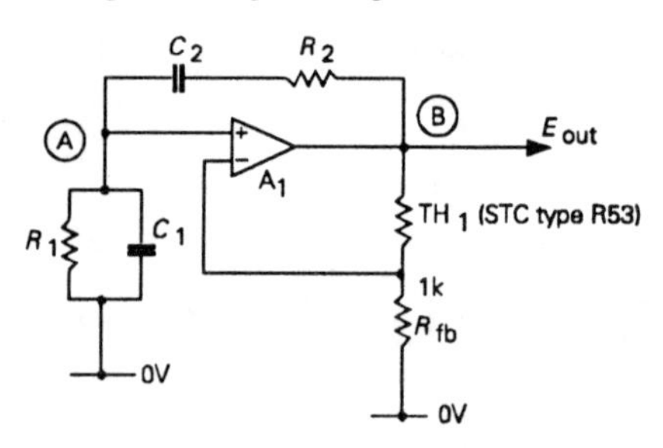

Figure 7.2 *A 'Wien bridge' oscillator using both +ve and –ve feedback*

It is convenient, for the purposes of explanation of circuit behaviour, to postulate an ideal amplifying gain block, having inverting (–in) and non-inverting (+in) inputs, an infinitely high value of internal gain, and a zero impedance output point, a concept which is called an 'operational amplifier'.

Devices such as the integrated circuit wide band gain blocks, such as the '741' type device, and particularly the later and improved FET input versions of this kind of circuit, such as an 'LF351' or the 'TL051' come very close to satisfying the operational amplifier specification, in that they have a very high gain (100,000 or greater), a low output impedance (less than 200 ohms), and, especially in the case of the FET-input versions, very high input impedance (100MΩ +), and are indeed described as 'op. amps.' in the makers' catalogues.

If an IC of this kind is used as the gain block, shown symbolically as A_1 in Figure 7.2, operated from +/– 15V supply rails, the output voltage swing is set, by the choice of R_{fb}, to lie between 1V and 3V RMS, the total harmonic distortion of the circuit would be of the order of 0.01–0.03%, at 1kHz.

In the notation used in Figure 7.3 (and which I have also used in Figures 7.1 and 7.2), the symbols '+' and '–' within the triangular op. amp. diagram, are taken to represent the 'non-inverting' and 'inverting' inputs of the amplifier. The + and – symbols outside the triangle are taken to represent connections to some external DC power supply (typically +/–15V), though these connections are often omitted from the circuit diagram unless their connections are, in some way, significant in respect of the operation of the circuit.

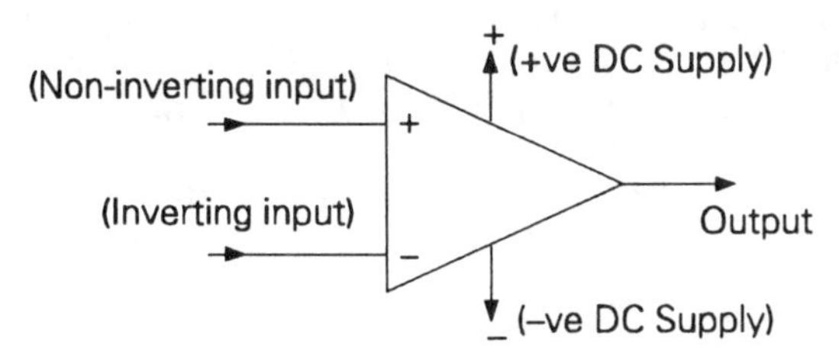

Figure 7.3 *Conventional notation for 'operational amplifier' type of gain block*

Positive feedback can also be used, with care, to increase the AC voltage gain, or output swing obtainable from an amplifier circuit. A very common example of this is the use of a 'bootstrapped' load resistor, as in the circuit of of Figure 7.4 (the term is said to derive from the fanciful idea of 'lifting oneself up by ones own bootstraps'), which is commonly used to increase the effectiveness of the driver stage in simple audio amplifier designs. In this circuit, a voltage signal derived from the outputs of the two push–pull emitter-followers (Q2/Q3) (point R), is coupled back to the +ve end of $R4$ (point S), by way of $C1$, so that if the output voltage at R, swings up or down, the potential at point S will follow this voltage excursion. This greatly increases the stage gain due to Q1, by virtue of the positive feedback signal, because it makes the dynamic impedance of $R4$ much higher than its static value. However, except in circuit layouts which are deliberately designed to function as 'free running' oscillators – a topic which is explored at greater length in Chapters 10 and 13 – positive feedback is generally avoided, and precautions are frequently taken to avoid its inadvertent occurrence, since it can lead to undesirable effects on the circuit performance.

The effect of feedback on gain

The basic characteristics of a simple series feedback circuit (i.e. one in which the feedback signal is applied

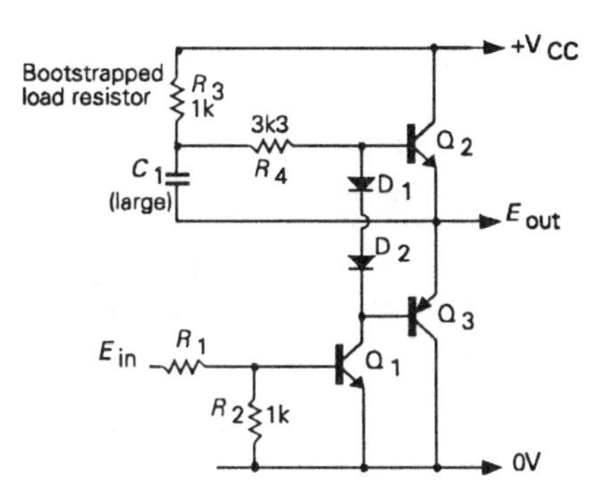

Figure 7.4 *The use of a 'bootstrapped' load resistor (R3) to increase available output voltage swing from Q1*

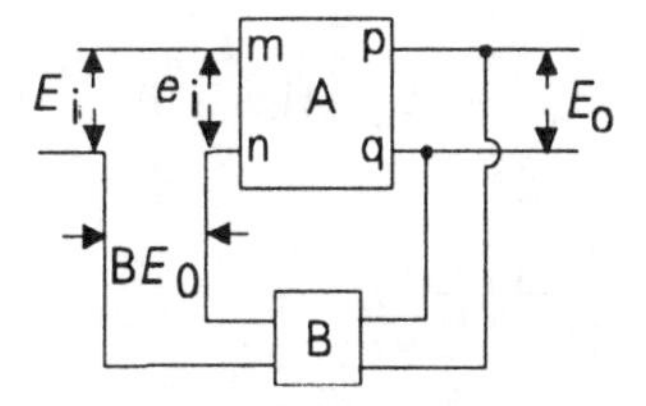

Figure 7.5 *Schematic gain block with feedback loop*

in series with the input signal) can be explained by reference to the block diagram shown in Figure 7.5, in which the square symbol is used to represent a circuit element with a gain of A, and which has two independent and isolated input and output circuits, m – n and p – q, arranged so that an input signal voltage e, applied between m and n will generate an output voltage of Ae between p and q, without sign inversion. If we define the signal actually applied between the input points m and n as e_i, and the signal appearing at the output points p and q as E_o, then the stage gain of the amplifying block, A, is

$$A = \frac{E_o}{e_i} \tag{1}$$

which is known as the 'open-loop' gain – i.e. the gain before any feedback is applied.

However, if there is an electrical network between the output and the input of this gain block which causes a proportion β of the output signal to be returned to the input circuit, in series with the applied input signal E_i, then the actual input signal fed to the gain block will be altered so that e_i will actually be $E_i + \beta E_o$, or

$$E_i = e_i - \beta E_o \qquad (2)$$

From these equations, we can determine the effect of feedback on the stage gain, bearing in mind that, in the absence of a feedback signal, $E_i = e_i$. The closed-loop stage gain, i.e. that when feedback is applied, I have called A', and is defined by the equation

$$A' = \frac{E_o}{E_i} \qquad (3)$$

Combining equations (2) and (3)

$$A' = \frac{E_o}{e_i - \beta E_o} \quad \text{but} \quad e_i = \frac{E_o}{A}$$

therefore

$$A' = \frac{E_o}{E_o/A - \beta E_o} = \frac{1}{(1/A - \beta)} \text{ or } \frac{A}{1 - \beta A}$$

from which

$$\frac{A'}{A} = \frac{1}{(1 - \beta A)} \qquad (4)$$

There are several conventions for the names used in feedback systems, but the one I prefer is that in which the term 'βA', the product of the 'open-loop' gain and the proportion of the signal reintroduced to the input by the feedback network, is referred to as the 'feedback factor'.

It must be remembered that, in this equation, the sign of β may be either positive or negative depending on the configuration of the feedback network. In the case of a NFB system, β is negative, so that the gain becomes $A/(1 + \beta A)$

[A confusing usage of the term feedback factor, which I think should be avoided, is when it is employed, in NFB systems, to describe the *amount* of negative feedback used in the design. This is usually expressed in dBs, and is normally defined as the extent to which the open-loop gain is reduced when negative feedback is applied {i.e. A'/A or $1/(1 + \beta A)$}. For example, if the closed-loop gain of an amplifier, with NFB, was reduced to one hundredth of its open-loop value, (–40dB), this would be described as 40dB of negative feedback.]

Considering the relationships shown in equation (4), in the particular case where the network is non-inverting, and the feedback signal is returned to the input in the same phase (positive feedback), so that $\{A'/A = 1/(1 - \beta A)\}$, the situation arises in which the larger the feedback factor $A\beta$, becomes, the smaller the denominator in this equation will be, and the higher the effective gain of the system – up to the point where $A\beta = 1$, when the denominator in the equation would become zero, and the closed-loop gain, A, would become infinite. This is the condition which will lead to 'latch-up' in a DC coupled circuit, or continuous oscillation in an AC coupled one, as, for example, in the Wien network oscillator described above, which will oscillate when $E_o = 3E_i$, and $\beta = 1/3$, which leads to the condition that $\beta E_o = E_i$.

If the feedback network is a phase inverting one, so that β is negative, the gain equation will become

$$A' = \frac{A}{1 + \beta A} \qquad (5)$$

In this case, the closed-loop gain will decrease as the amount of feedback is increased, up to the point at which βA is so much greater than 1 that the equation approximates to

$$A' = \frac{A}{\beta A} \quad \text{or} \quad A' = \frac{1}{\beta} \qquad (6)$$

This is particularly appropriate to those cases in which the open-loop gain (i.e. the gain before the application of feedback) is very high; as would nearly always be the case with operational amplifier integrated circuits at low to medium signal frequencies. In these circumstances, the gain is determined almost entirely by the attenuation ratio of the feedback network, so that if β is 1/10, then the gain will be 10, and if β is 1/100, then the circuit gain will be 100.

This is a very useful effect, in that it makes the gain of the circuit arrangement almost completely independent of the gain of the amplifier block itself, so long as this is high in relation to β. A specific advantage which arises in this case is that, if the feedback network has an attenuation ratio which is constant, and independent of frequency, then, so long as the gain of the amplifier block remains high in relation to β, then the system will have a stage gain of $1/\beta$ and this will also be independent of operating frequency.

The use of NFB to reduce distortion

In addition to the improvement which NFB can make to the flatness of the frequency response of an amplifier, it will also, within limits, reduce harmonic and waveform distortion, as well as any 'hum' and noise generated within the amplifying circuit, which can indeed be considered as just another aspect of signal distortion.

If the instantaneous relationship, between the input signal applied to a circuit block and the output voltage from that block, is considered to be its amplification factor, then any 'distortion' in the output signal can be regarded simply as an indication that the effective amplification factor of the system has changed during the transmission of the signal. This change might be the result of a variation in output voltage with time, within the duration of a nominally constant signal, perhaps because of the intrusion of mains hum or noise. Alternatively, it could be due to the characteristics of the system itself – as shown in the case of the square-wave signal illustrated in Figure 7.6. Again, it could be an unwanted change in the signal waveform, in the case of a continuously varying input signal voltage, because the gain of the system is influenced by the instantaneous magnitude of the input signal, due to nonlinearity in the input/output transfer characteristic, shown in Figure 7.7.

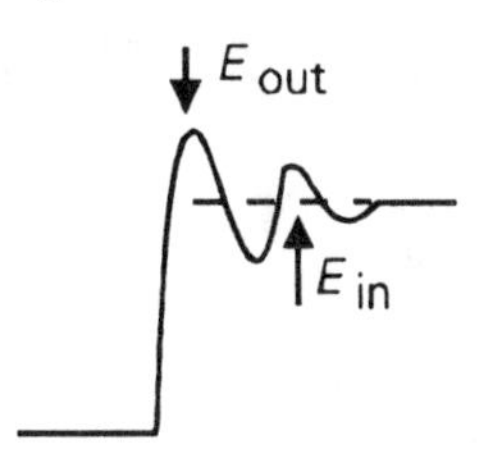

Figure 7.6 *Distortion of step-waveform due to circuit characteristics*

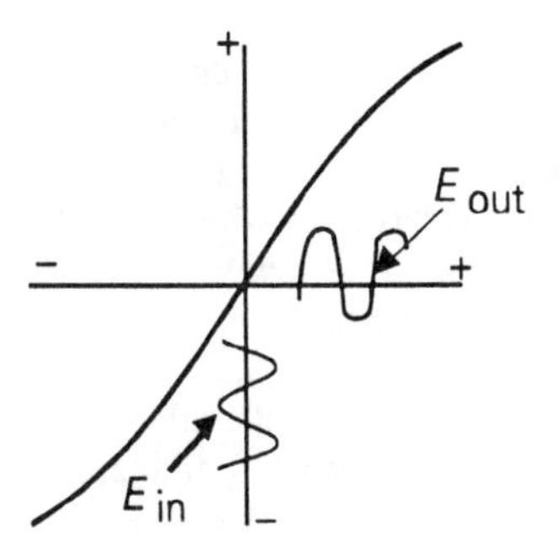

Figure 7.7 *Distortion due to nonlinear amplifier transfer function*

However, if the amplifier gain, A, is sufficiently high that the closed-loop gain (A'), becomes simply $1/\beta$, and if the factors which determine the value of β are themselves noise and hum free and distortionless, then the output signal must also be distortionless, noise and hum free, simply because variations in the open-loop amplifier gain A, as a function of input voltage or time, are no longer factors which influence the output voltage. This is only strictly true for infinitely high values of A, but even for lower values of A the distortion will still be reduced by the feedback factor $(1 + \beta A)$, as shown by the following analysis, due to Langford-Smith (*Radio Designers Handbook*, 4th Edition, Chapter 7, pp. 308–310).

Let a sinusoidal input voltage, E_i, of an amplifier employing negative feedback be defined by the term $E_{im}\cos\omega T$, where $\omega = 2\pi f$. Then, if the amplifier introduces harmonic distortion, the output will be

$$E_{om}\cos\omega T + E_{2m}\cos 2\omega T + E_{3m}\cos 3\omega T \ldots \quad (6)$$

where E_{om}, E_{2m} and E_{3m} are the peak values, respectively, of the fundamental, second and third harmonic frequencies, etc. The feedback voltage (βE_o) will therefore be

$$\beta(E_{om}\cos\omega T + E_{2m}\cos 2\omega T + E_{3m}\cos 3\omega T)$$

The voltage applied to the input of the amplifier will be

$$e_i = E_i + \beta E_o$$

$$= E_{im}\cos\omega T + \beta(E_{om}\cos\omega T + E_{2m}\cos 2\omega T + E_{3m}\cos 3\omega T)$$

$$= (E_{im} + \beta E_{om})\cos\omega T + \beta E_{2m}\cos 2\omega T + \beta_{3m}\cos 3\omega T$$

Since $E_o = Ae_i$, the output voltage will therefore be Ae_i + harmonic distortion.

$$= A[(E_{im} + \beta E_{om})\cos\omega T + \beta E_{2m}\cos 2\omega T + \beta E_{3m}\cos 3\omega T]$$

$$= A(E_{im} + \beta E_{om})(H_2\cos 2\omega T + H_3\cos 3\omega T) \quad (7)$$

where H_2 and H_3 are the ratios of the second and third harmonic voltages to the voltage of the fundamental in the amplifier without feedback.

But the output voltage has already been defined by equation (6), so, by equating the fundamental components of equations (6) and (7)

$$E_{om} = A(e_{im} + \beta E_{om}) \text{ and } e_{im} = (1 + \beta A)E_{om}/A \qquad (8)$$

Equating the second harmonic components in equations (6) and (7)

$$e_{2m} = A\beta e_{2m} + A(e_{im} + \beta E_{om})H_2$$

Inserting the value of e_{im} from (8)

$$e_{2m} = A\beta e_{2m} + [(1 + \beta A)E_{om} + A\beta E_{om}]\, H_2$$

Therefore, $e_{2m}(1 + \beta A) = E_{om}H_2$ and

$$e_{2m}/E_{om} = H_2/(1 + \beta A) \qquad (9)$$

and, by the same argument, $e_{3m}/E_{om} = H_3/(1 + \beta A)$, and so on, for any higher order harmonics. That is to say that, so long as the feedback is negative, so that the term in the denominator is $1 + \beta A$, the magnitudes of all the harmonics, and the consequent intermodulation products, caused by the nonlinearities in the amplifier, will be reduced by the same proportion by which the gain is reduced (see equation (3)). However, this conclusion is only strictly accurate if the feedback is truly negative (i.e. 180° out of phase with the input signal), and if the gain of the amplifier is the same for the harmonic frequencies as it is for the fundamental, and these conditions are unlikely to be met, precisely, in any real-life amplifying circuit. Nevertheless, this conclusion is broadly true.

A similar mathematical process can be used to show that hum and noise will also be reduced as the gain of the amplifying circuit is reduced, but it is simpler, I think, to resort to the logical argument that, if the input signal itself is hum and noise free, then if the gain of the amplifier is reduced by feedback, and the magnitude of the input signal is raised so that the output remains constant, then the proportion of hum and noise in the output signal which was introduced by the amplifier will be reduced by the same amount.

Negative feedback will reduce the output impedance of an amplifier, when measured at the point from which the feedback path returns to the amplifier input, by $(1 + \beta A)$, and, in the case of a 'series feedback' circuit, it will also increase its input impedance, in comparison with the same amplifier without NFB, by the same factor. However, this effect on input and output impedances is also only true while the open-loop gain of the amplifier remains high, and while the feedback remains negative.

At frequencies where the gain of the amplifier is low, and its internal phase errors cannot be ignored, the effects of feedback on its input and output impedances are much more difficult to predict.

Problems with negative feedback

As mentioned above, it is unlikely that NFB in any real-life circuit will ever be truly 'negative'. This difficulty arises principally because the response of any amplifying – or other – circuit to a change in the input signal level can never be truly instantaneous. In the case of a simple DC circuit, this means that if it is presented with a sudden change in its input signal level, as shown in Figure 7.8a, there will be some small, but inevitable, time delay before there is a corresponding change in the output signal level, as shown by the solid line in the diagram.

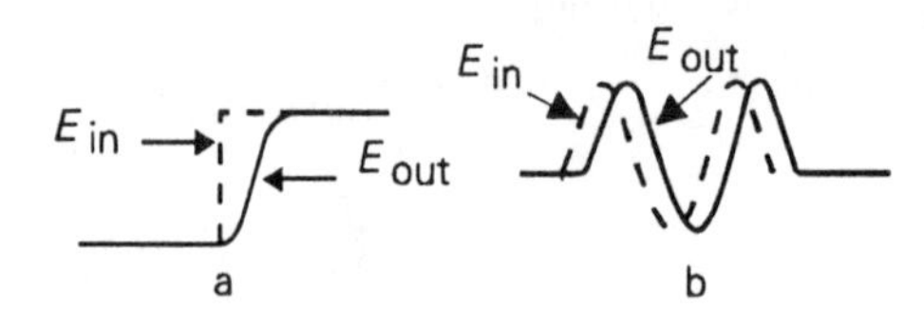

Figure 7.8 *The effect of circuit time delays on the output signal waveform*

In an AC circuit, where the input potential is continuously varying, as shown in Figure 7.8b, this same time delay within the system will result in an output signal which is shifted in phase – a 'phase lag' of the output waveform – in relation to the input one. Clearly, the higher the frequency of the input signal, the greater the effective phase lag which will be caused by the same time delay – or delays, where there are several contributing elements – and the more complex the effect of the feedback signal will be. Depending on the nature of the circuit, this could well lead to the situation where a feedback voltage, derived from the output signal of an amplifying stage, was, in reality, positive (gain increasing and stability diminishing), whereas the designer of the circuit had intended it to be negative. If the feedback factor βA, at the input frequency where this circumstance arose, became equal to unity, the circuit would become unstable and oscillate.

This incipient problem with amplifying circuits employing NFB was analysed by H. Nyquist (Regeneration Theory, *Bell System Technical Journal*, Jan. 1932, p. 126), and H. W. Bode (*Network Analysis and Feedback Amplifier Design*, published by Van Nostrand, 1945), who both proposed graphical representations of the (closed-loop) gain and phase characteristics of the circuit.

In the case of the Nyquist plot, both of these effects are shown in a single diagram, by using the length and direction of a vector to indicate, respectively, the gain and phase shift of the system on a single line graph, as shown in Figure 7.9, where the angular orientation of the vector – of which only its terminating point is usually shown in the drawing – is used to indicate the measured phase shift at each value of operating frequency, and its length is used to define the closed-loop gain. If this vector graph encloses the point –1 and 180°, the system will be unstable.

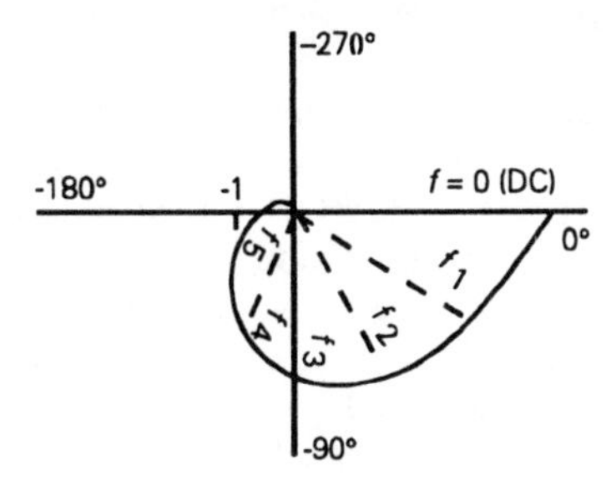

Figure 7.9 *The 'Nyquist' gain/frequency diagram*

In the case of the Bode diagram, shown in Figure 7.10, the open-loop gain of the system (*A*), and the phase shift of the feedback signal, are shown as two separate lines on the same diagram. In this case, however, an assessment as to whether the system meets the basic requirement for stability (that is to say that the open-loop gain shall be less than unity at the frequency at which the phase shift is 180°) and the determination of the system 'gain' and 'phase' margins, requires consideration of both the 'gain' and 'phase' graphs. If the gain is less than unity at the 180° phase shift frequency, but, at some other point, where the phase shift is greater than 180°, the gain exceeds unity, the circuit may not oscillate, but will be what is described as 'conditionally stable'. This is seldom a condition which is sought, although it can arise as a result of oversights or design shortcomings, and carries the risk that the system may subsequently become unstable if component characteristics change through ageing.

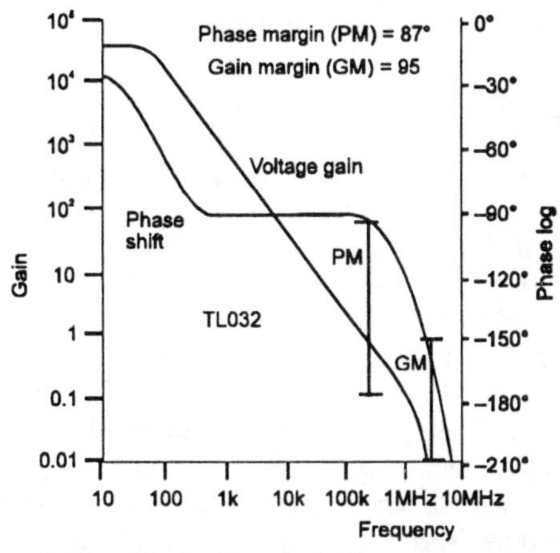

Figure 7.10 *'Bode diagram' illustrating gain and phase characteristics of commercial 'operational amplifier' IC*

'Stability margins' refer either to the extent to which, in a stable system, the gain would require to be increased at the 180° phase shift frequency for it to reach unity, a factor known as the gain margin, or to the extent to which the loop phase shift would need to be increased at the unity gain frequency, for it to reach 180°, which is known as the phase margin. Obviously, the higher the values of gain and phase margins which can be contrived within a closed-loop system the less likely it will be that changes in operating conditions or load characteristics will impair the stability of the circuit. This aspect of feedback design is particularly important in systems, such as audio amplifiers, where the characteristics of the load, such as, for example, a loudspeaker, will change greatly as a result of changes in operating frequency or drive level.

Unfortunately, in practical circuit design, it is seldom possible to arrange for the provision of high levels of gain and phase margins without incurring some other design penalty, such as, perhaps, a higher order of residual harmonic distortion than could have been achieved with a less comprehensively stable system, so it is useful for the would-be designer to understand the origins of phase shifts within circuit structures, and also the techniques by which these can be manipulated.

The relationship between frequency response and phase shift

As a general rule, any circuit arrangement which gives rise to a change in gain from input to output as a function of frequency, will introduce a shift in phase, also as a function of frequency. Also, as a general rule, the greater the change in gain with frequency, the greater the effect it will have upon the phase shift, so that, for example, some circuit arrangement which causes a change in gain of 6dB/octave will have an associated ultimate phase shift of 90°, one causing a 12dB/octave change in gain will lead to a 180° phase shift, and so on.

Taking the case of the simple *RC* lag network, shown in Figure 7.11a, and its associated Bode diagram, shown in Figure 7.12, it will be seen that the phase lag, of voltage output with relation to the voltage input, begins at $f_t/10$, where f_t is the turn-over frequency – defined as the frequency, $f_t = 1/(2\pi RC)$, at which the gain had decreased by a factor of 0.708 (–3dB).

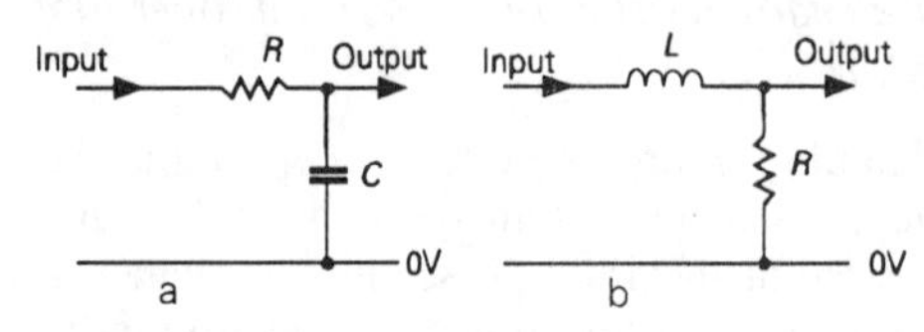

Figure 7.11 *Simple 'low-pass' networks*

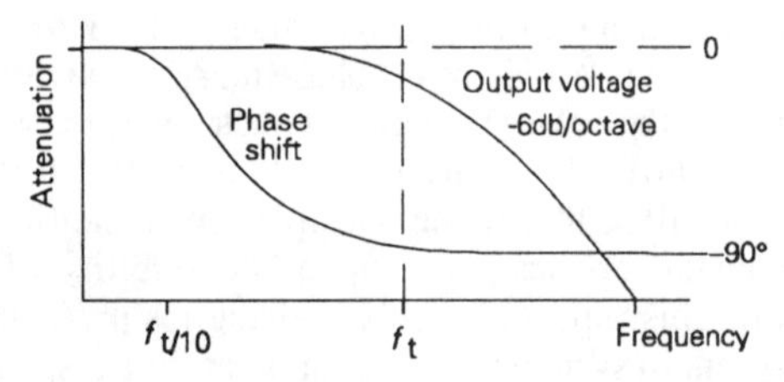

Figure 7.12 *Basic relationship between phaseshift and frequency response*

An exactly similar pair of gain/phase curves would be given by the simple *LR* network shown in Figure 7.11b, or by any other arrangement which had the same effect.

Conversely, if the characteristics of the circuit are such that it has a transmission which decreases as the frequency is reduced, then there will be a phase shift due to this which will begin to appear at $10f_t$, where, as before, f_t is the frequency at which the attenuation is –3dB. However, in this case, there will be a phase 'lead' rather than a phase lag, as shown in the gain/phase diagrams of Figure 7.13, which would apply to either the simple *RC* network of Figure 7.14a, or the *LR* network of Figure 7.14b.

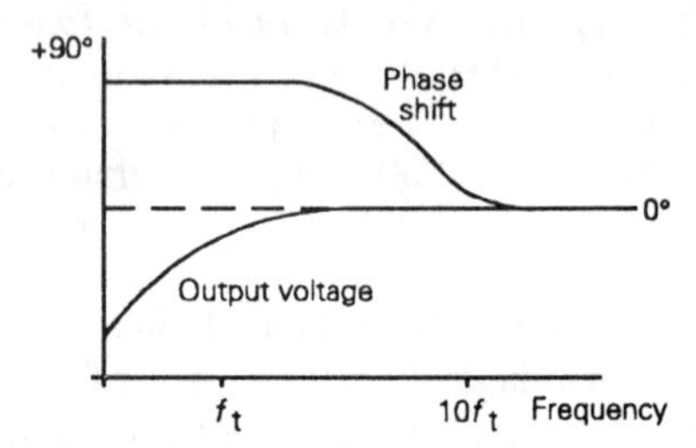

Figure 7.13 *Phase characteristics of high-pass network*

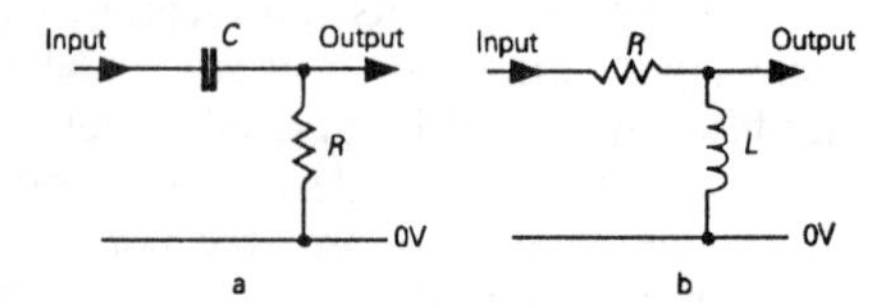

Figure 7.14 *Simple high-pass networks*

An inevitable consequence of the relationship between changes in transmission with frequency and associated phase shifts is that, for example, a transistor amplifying stage, of the kind shown schematically in Figure 7.15, whose gain would inevitably decrease with increasing frequency beyond some HF turn-over point, would also introduce a phase shift, at $f_t/10$, if f_t was the –3dB point in its HF gain curve.

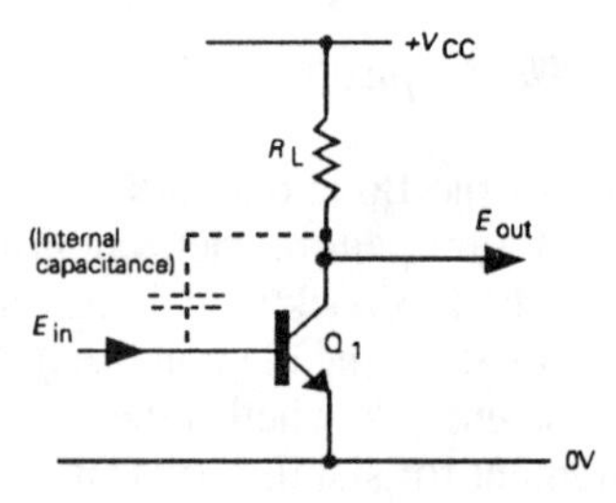

Figure 7.15 *Simple transistor amplifying stage*

In practice, the gain/frequency behaviour of even a simple transistor amplifying circuit would be much more complex than this, because, in addition to the time delays due to the charge carrier transit time within the device, there would be straightforward *RC* lag effects due to stray circuit and inter-electrode capacitances. This has the effect that if a string of, say, three transistors is connected as an amplifying circuit using negative feedback; as shown in Figure 7.16, where, with ideal devices, the output, at HF, would be exactly in antiphase with the input; it is improbable that it would be stable unless the feedback factor is very low, due to the accumulated incidental phase shifts within the circuit, mainly due to the transistors.

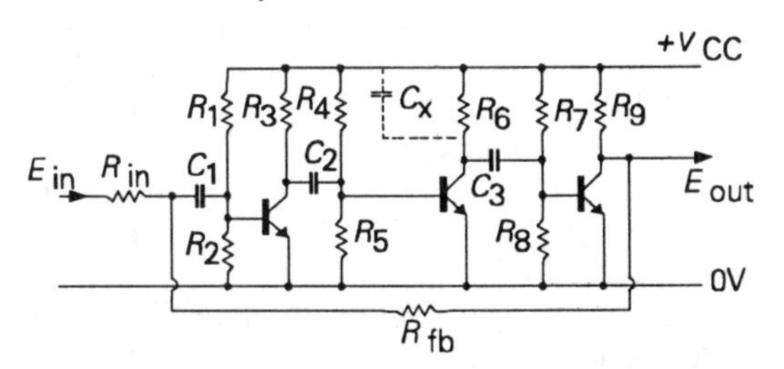

Figure 7.16 *Three-stage transistor amplifier*

A possible, though crude, method for making such a circuit stable, at HF, would be to connect a relatively large value of capacitor, C_X, across the collector load resistor of one of the transistors in the circuit, so that the gain had decreased to a sufficiently low value before the frequency was reached at which the incidental circuit and device phase shifts had become significant in their effects. This added capacitor would usually need to be large enough to reduce the HF turn-over point of the circuit to a frequency which would be, at the most, one tenth of that due to the fall-off in HF response due to the transistor circuit on its own.

Similarly, if phase shifts, due to coupling components, such as C_1, C_2 and C_3, occurred at the low frequency end of the circuit pass band, the application of closed-loop negative feedback would also lead to instability, and any simple form of LF phase compensation, to prevent continuous LF oscillation, would require a very severe curtailment of the LF response. Again, this would not indicate good circuit design.

There are several design techniques which can be used to lessen the phase error in such a circuit, of which one of the simplest is the use of a 'step' network, of the kind shown in Figure 7.17a, for which the Bode diagram is shown in Figure 7.18. Here, the phase lag, due to the reduction in gain, with increasing frequency, caused by R_1C_1, is removed as the gain/frequency graph levels off, which gives rise to a phase lead (in reality, a decreasing phase lag) at a frequency determined by R_2/C_1.

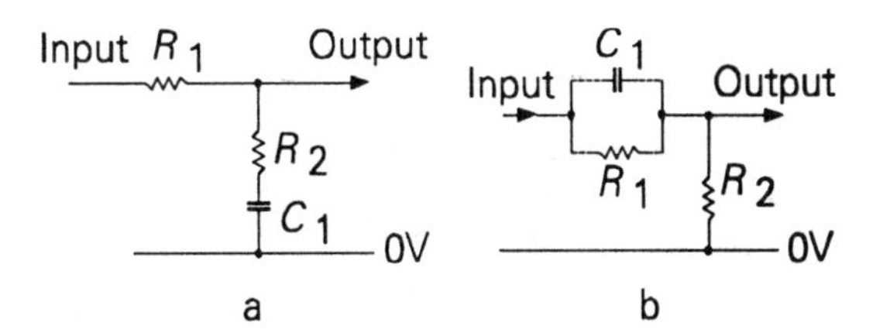

Figure 7.17 *RC step networks for HF and LF phase error correction*

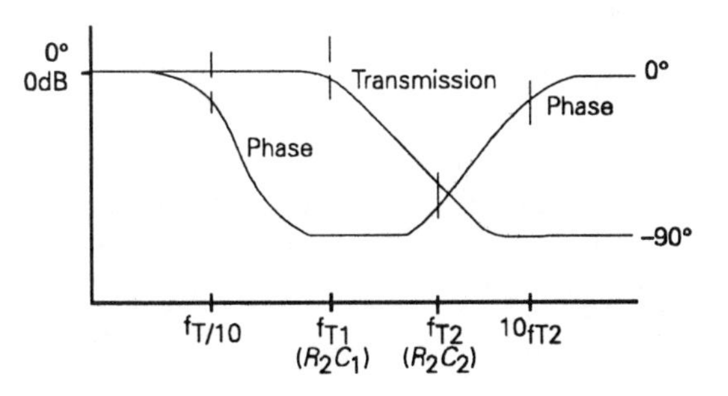

Figure 7.18 *Gain and phase characteristics of circuit of Figure 17a*

A similar step network, of the kind shown in Figure 7.17b, can, of course, be used as a phase correction method at the LF end of the spectrum. In practice, with solid-state circuit design, it is usually possible to arrange that the inter-stage DC potentials are such that it is possible to omit inter-stage coupling capacitors, so that LF closed-loop instability does not occur. Unfortunately, though, the intrinsic characteristics of all amplifying devices lead to inconvenient phase shifts at the HF end of the system pass-band, and some form of HF compensation is usually essential to avoid closed-loop instability.

As an example of this, an HF step-network could be used to stabilize a feedback amplifier– whose gain and phase characteristics, shown in the curves a and b in Figure 7.19, indicate that the amplifier would be unstable at unity gain with closed-loop feedback – by changing these to the curves a′ and b′, which would indicate stable closed-loop operation.

A typical transistor operated gain block is shown in Figure 7.20, in which transistors Q1 and Q3 are connected as an input long-tailed pair, Q2 and Q4 are constant current sources, and Q5 is the second stage amplifying transistor feeding the output emitter followers Q6 and Q7, for which a suitable forward voltage bias is provided by D3 and R_5. If loop negative

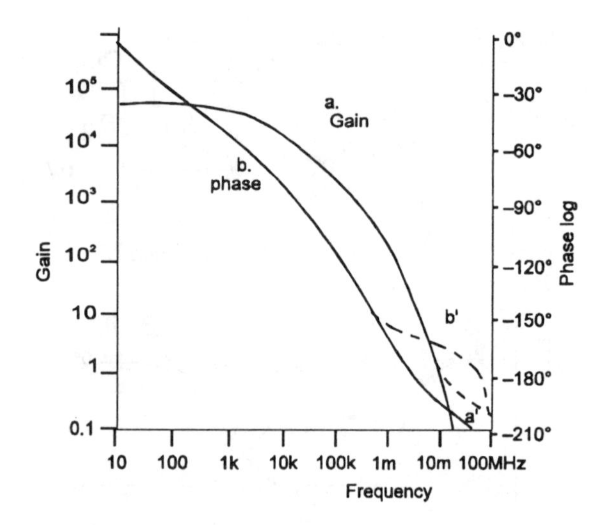

Figure 7.19

feedback is applied, by means of the external network R_a, R_b and C_b, the circuit will become unstable, and will oscillate at some relatively high frequency. It is conventional practice to stabilize the circuit by the 'compensation' capacitor, C_x, connected between the collector and base of Q5, which imposes a –6dB/octave fall in the amplifier frequency response, above some suitably chosen 'turn-over' frequency. The effect of this capacitor, in this position, is greatly increased by the voltage amplification of Q5, a phenomenon known as the Miller Effect. The reason why this happens is that if the input signal is such as to reduce the current through R_1, which would cause the base voltage of Q5 to fall, the consequent rise in the collector voltage of Q5 will cause C_x to draw increased current through R_1. This has the effect that the capacitance of C_x is multiplied by $(1 + A)$, where A is the stage gain of Q5 in that particular circuit layout.

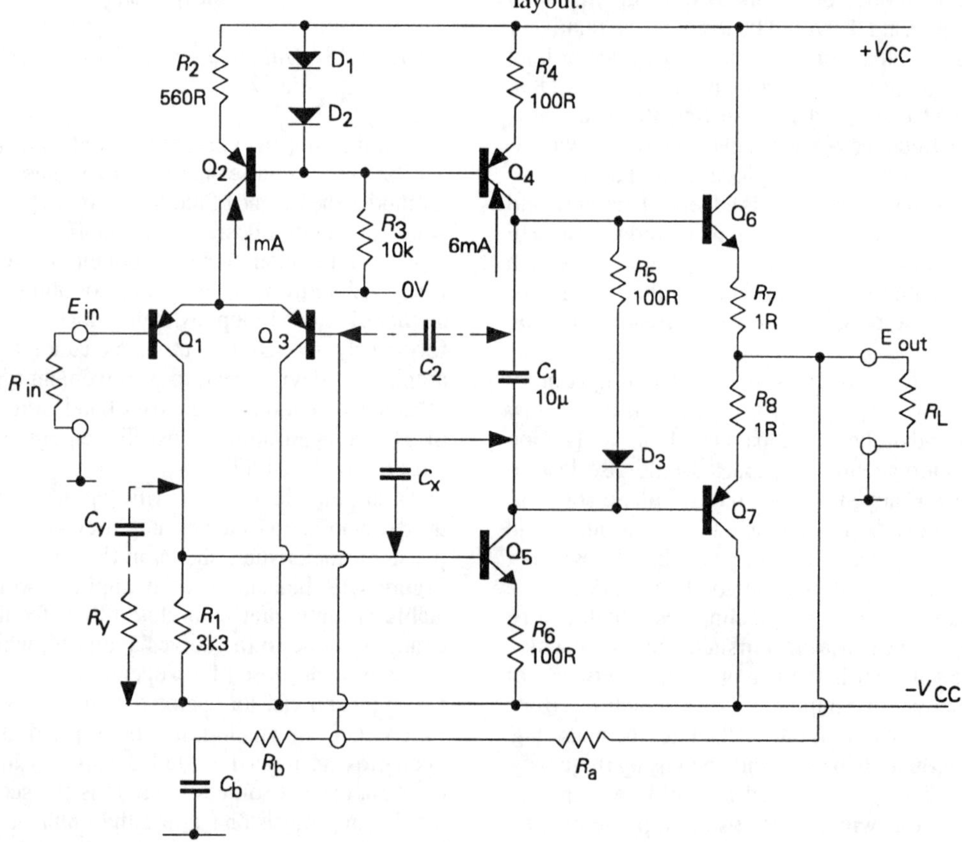

Figure 7.20 *Typical transistor gain block*

A practical snag inherent in this particular brute force method of closed-loop stabilization is that any sudden change of input signal voltage is likely to cause a temporary paralysis of the amplifier, while the capacitor C_X charges or discharges through Q_1 or R_1, a situation described as 'slew-rate' limiting. This causes the kind of response to a sudden voltage change which is shown in Figures 7.21a and b. If some other signal is present at the same time as this sudden change of input voltage, it will be obliterated during the slewing period. This type of operational defect has been called 'transient intermodulation distortion'. This problem can be minimized if an input RC network is connected to the input of the amplifier, as shown in Figure 7.22, to limit the possible rate-of-change of the input signal voltage.

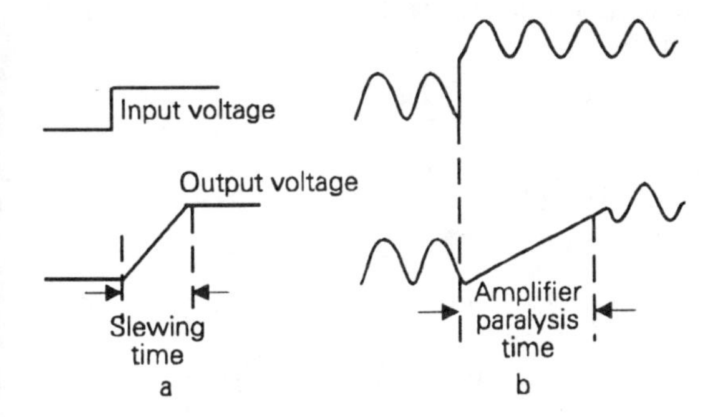

Figure 7.21 *Effect of compensation capacitor C_X on slewing rate and transient intermodulation distortion*

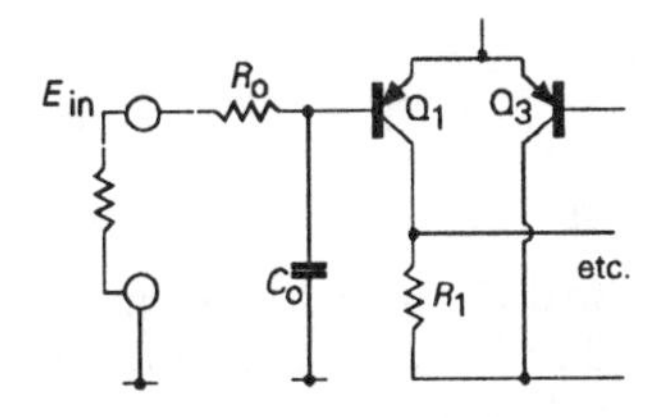

Figure 7.22 *Use of CR network to limit susceptibility of feedback amplifier to slew rate limiting on input voltage step*

A better approach to the HF stabilization of the kind of gain block shown in Figure 7.20 would be to connect a 'step' network, C_Y/R_Y, across R_1, in the collector circuit of Q_1. Alternatively, one could take advantage of the fact that the internal phase lag in the amplifier, between the base of Q_3 and the collector of Q_5, is not sufficient to cause instability if a capacitor, C_Z, is connected directly between these points. Such a capacitor can also be used to impose a –6dB fall in the HF response of the gain block, and ensure that the overall gain and phase margins are adequate to provide stable closed-loop operation, but without incurring the penalty of slew-rate limitation.

Operational amplifier systems

Operational amplifiers, devices which are now readily available as integrated circuit functional blocks, are basically amplifier units in which the internal circuit design and 'HF compensation' arrangements are such that they will operate stably with 100% NFB – a condition in which the output pin is connected directly to the inverting input pin, as shown in Figure 7.23. – an arrangement which, incidentally, provides a wide bandwidth, unity gain, 'buffer' block, having a high input impedance and a low output impedance.

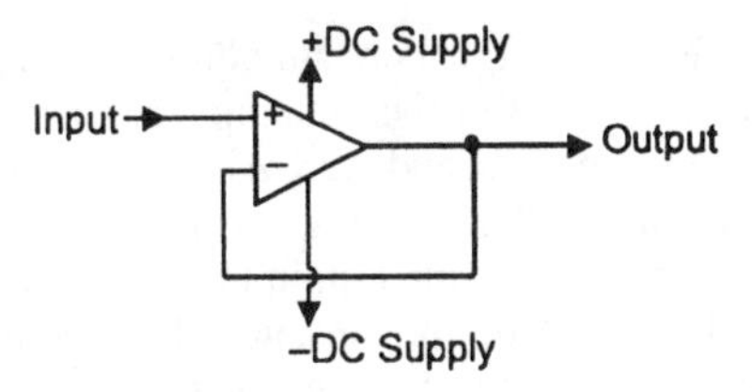

Figure 7.23 *Operational amplifier connected as unity gain buffer stage*

Commercial op. amp. ICs are normally operated from a pair of +ve and –ve supply lines, typically over the range +/–3V to +/–18V, depending somewhat on device type. (+/–15V is the most common maximum supply voltage value for normal commercial components, although there are some devices specially designed to operate from much higher or lower supply line potentials.) It is also normally practicable, by the use of a suitable circuit layout, to operate these devices between a single supply line and the 0V rail, and some op. amp. types are designed specifically for this type of application, with the 0V level being part of their input signal range.

Negative feedback can be applied either as series feedback – the method considered above, in which the feedback signal is applied in series with the input signal – or as shunt feedback, in which the input signal

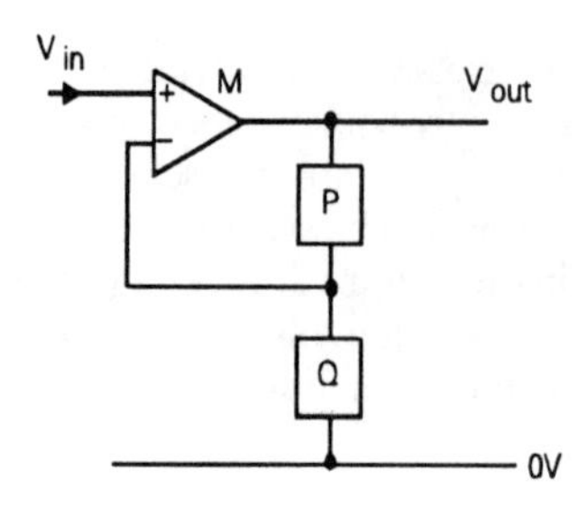

Figure 7.24 *Connection of op. amp. as series feedback system*

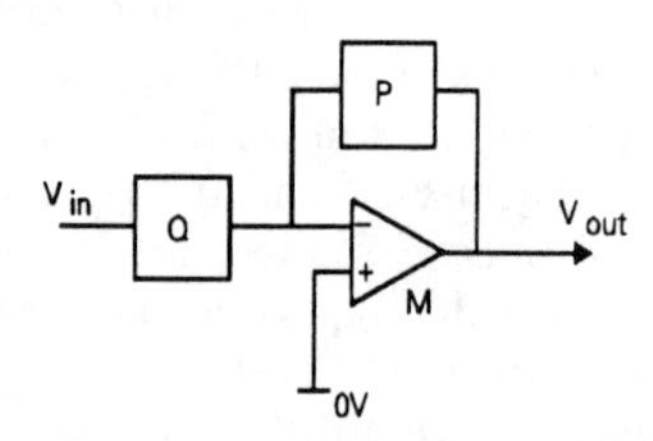

Figure 7.25 *Connection of op. amp. in shunt feedback system*

and the feedback signal are simultaneously applied to the inverting input of the op. amp., as shown respectively in Figures 7.24 and 7.25. Although both of these methods of feedback connection have a virtually identical effect upon the output impedance and bandwidth of the system, they are not identical in respect of gain and internal distortion, and have very different effects upon the input impedance of the circuit.

In the case of the series feedback system shown in Figure 7.24, in which the amplifier, M, has an open-loop gain of A.

$$V_{in} = \beta V_{out} + V_1 \quad \text{and} \quad V_{out} = AV_1$$

from which

$$V_{in} = A\beta V_1 + V_1 = V_1(1 + A\beta)$$

where β is the proportion of the output signal fed back by the feedback network [Q/(P + Q)], V_1 is the actual voltage appearing across the input of the amplifier block, and R_a is the input resistance of the amplifying block. The input resistance of the circuit

$$R_{in} = V_{in}/I_{in} \quad \text{and} \quad I_{in} = V_1/R_a$$

so, since $V_{in} = V_1(1 + A\beta)$ then

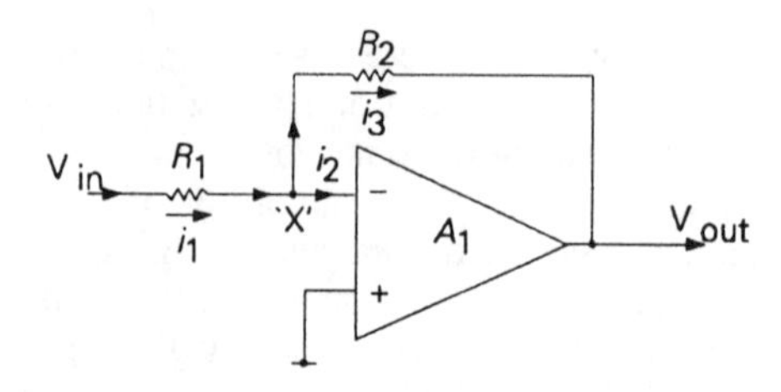

Figure 7.26 *Basic 'shunt feedback' layout*

$$R_{in} = \frac{V_1(1 + AB)}{V_1/R_a} = (1 + AB)R_a \qquad (10)$$

In the case of the shunt feedback layout shown in Figure 7.26, if we define the circuit input current through R_1 as i_1, the amplifier input current as i_2, and the current through the feedback resistor, R_2, as i_3, then:

$$i_1 = i_2 + i_3 \qquad (11)$$

and, if the current amplification of the amplifier is A, then

$$i_3 = Ai_2 \quad \text{or} \quad i_2 = i_3/A \qquad (12)$$

Combining these two equations, we get

$$i_1 = i_3/A + i_3 = i_3(1 + 1/A) \text{ from which}$$

$$i_1 = i_3\frac{(A+1)}{A}$$

so, by inversion,

$$i_3 = i_1\frac{A}{(A+1)} \qquad (13)$$

If $V_{in} = i_1R_1$ and $V_{out} = -i_3R_2$, because M is a phase inverting amplifier, we can determine the circuit gain with feedback, A', which is

$$A' = \frac{V_{out}}{V_{in}} = -\frac{R_2}{R_1}\frac{A}{(A+1)} \qquad (14)$$

when, if A is high enough $A' = -R_2/R_1$

The impedance at the input point, X, of an amplifier using shunt feedback can be determined, approximately, by the following argument:

Let us postulate an ideal amplifier, in which neither of the inputs will draw current, so that $i_1 = i_3$. However, because M is an inverting amplifier, the voltage across R_2 will be $(A + 1)V_{in}$, and $i_1 = \{(A + 1)V_{in}\}/R_2$. So

$$R_{in} = \frac{V_{in}}{I_{in}} = \frac{V_{in}}{[(A + 1)V_{in}]/R_2}$$

from which

$$R_{in} = \frac{R_2}{(A + 1)} \quad (15)$$

Since this can be very low indeed if A is high enough – much lower in practice than the input impedance of any likely amplifier circuit – neglecting the amplifier input current, i_2, does not significantly affect the result.

The very low impedance at this point gives rise to the common term applied to the input of an inverting amplifier with feedback as a 'virtual earth', and allows the circuit arrangement shown in Figure 7.27 to be used as a summing amplifier, or signal mixer, with very little interaction between the separate inputs, a, b and c, etc.

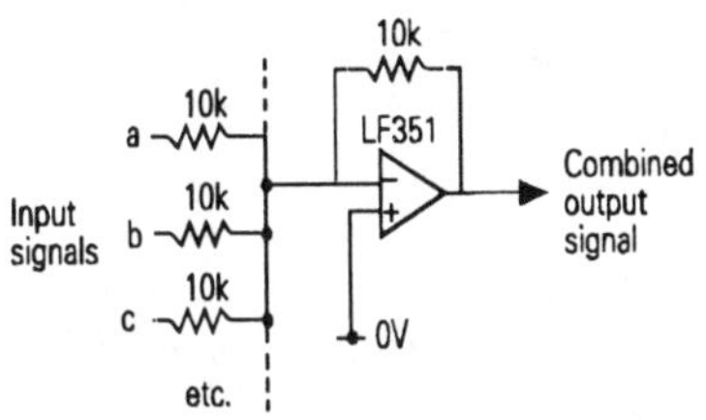

Figure 7.27 *Application of shunt feedback system as input mixer stage*

Another important difference in the way of operation of the series and shunt feedback layouts shown in Figures 7.24 and 7.25, is that the gain of the shunt feedback layout can be made to fall to zero, if that part of the feedback network, P, is reduced to zero, whereas the gain for any series feedback system can only be reduced to unity if the limb P is made to have zero impedance. This difference in characteristics can be important when the impedances of limbs P and Q are made frequency dependent, as in frequency response shaping applications.

In both the shunt and series feedback circuits, the application of negative feedback will lower the output impedance of the amplifying block, at the point from which the feedback path is taken. The magnitude of this effect can be calculated by considering the case of an amplifier, which has a gain A, and an effective output resistance, without feedback, R_o, which has its non-inverting input grounded, and which has a feedback signal returned to its inverting input through the attenuator network B, as shown in Figure 7.28. If a voltage V is applied to its output, this will produce a feedback signal at the inverting input of the amplifier having a magnitude BV. Since the input is an inverting one, this will give rise to an output from the amplifier, $V_{out} = -ABV$. The voltage across R_o (V_o), will be given by

$$V_o = V - V_{out} = [V - (-ABV)] = V(1 + AB)$$

and the current through R_o (I_o), will be

$$\frac{V_o}{R_o} = \frac{V(1 + AB)}{R_o}$$

so the output impedance, with feedback, R'_o, which is V/I_o, will be:

$$R'_o = \frac{V}{V(1 + AB)/R_o}$$

$$= \frac{R_o}{(1 + AB)} \quad (16)$$

This effect is the same for both series and shunt feedback systems.

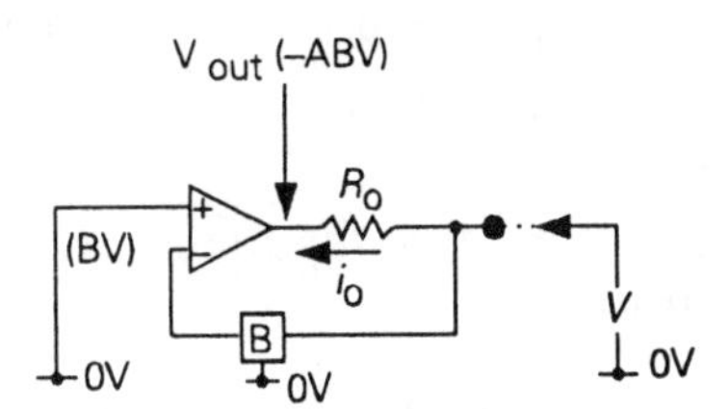

Figure 7.28

The principal use of the series FB system, shown in Figure 7.24, is to isolate the input signal from the feedback network, and allow the input impedance to be determined independently of other considerations, as shown in the gramophone input RIAA equalization stage of Figure 7.29, where the required input load is commonly 47kΩ

As mentioned above, a common application of such feedback systems is in the field of frequency response shaping circuitry, in which the networks P and Q, in Figures 7.24 and 7.25, may contain combinations of capacitive, inductive and resistive elements, so that the

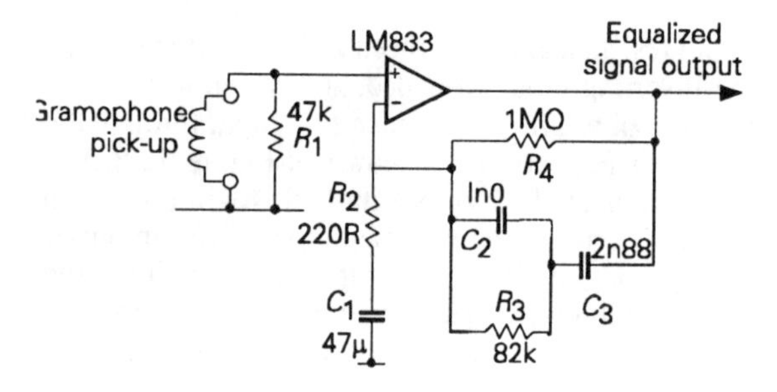

Figure 7.29 *Gramophone pick-up input amplifying stage with R. I. A. A. type frequency response equalization*

impedance of the feedback limbs will depend on frequency, which will alter the the circuit gain, as a function of frequency.

In all cases, where negative feedback is applied around an op. amp., care should be taken to make sure that any feedback network connected between the output and the input of the the operational amplifier does not cause it to operate beyond the maker's design limits for closed-loop stability, a consideration which is explored later on in this chapter. The design of these ICs will normally be chosen so that there is an adequate safety margin for all normal uses, but the possibility that the external feedback networks may influence the op. amp. performance is a point which should be borne in mind.

Effects of negative feedback on distortion

It is normally assumed that steady-state nonlinearities in any amplifying circuit will be reduced in direct proportion to the extent by which the gain of the system is reduced by negative feedback. It has been noted earlier that this assumption will only be strictly correct if the feedback is truly negative, and if the gain of the system is the same for all of the frequencies associated with the harmonic components introduced by the distortion of the signal as it is for the fundamental frequency of the waveform. However, there is also the fact that if the amplifier introduces, say, 2nd and 3rd order harmonics, these components will be present in the feedback signal, and will, in turn, be distorted in their passage through the amplifier. So, while the antiphase 2nd and 3rd harmonic feedback components will tend to cancel those introduced by the amplifier – and thereby reduce the magnitude of these unwanted signals present in the output – the nonlinearities, and the intermodulation effects of the amplifier, acting on these reintroduced signals will generate higher order, and probably less desirable harmonic components.

A graphical illustration of this phenomenon, due to P. J. Baxandall (*Wireless World*, December 1978, pp. 53–56), showing the way in which the magnitudes of the harmonic components alter as the amount of negative feedback is increased, based on measurements made on a simple FET amplifier stage, is shown in Figure 7.30.

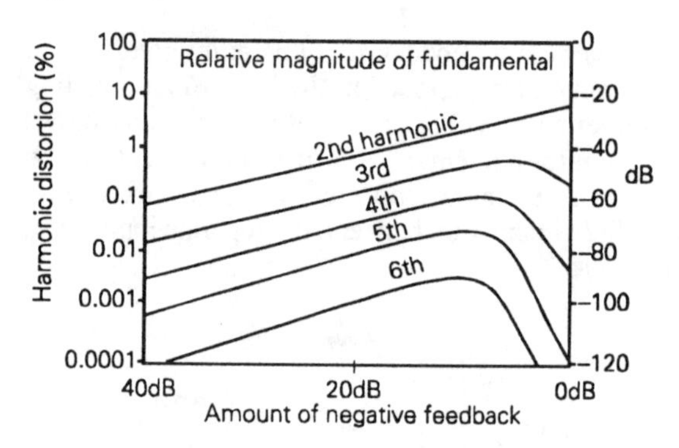

Figure 7.30 *Effect of increasing amount of NFB on distribution of harmonics in simple FET amplifier stage*

A further point related to the distortion of a feedback amplifier stage is that less distortion, other things being equal, will always be given by a shunt feedback circuit, in comparison with a series feedback layout. The reason for this is shown, using the example of a junction transistor input stage, in the circuit diagrams of Figures 7.31 and 7.32.

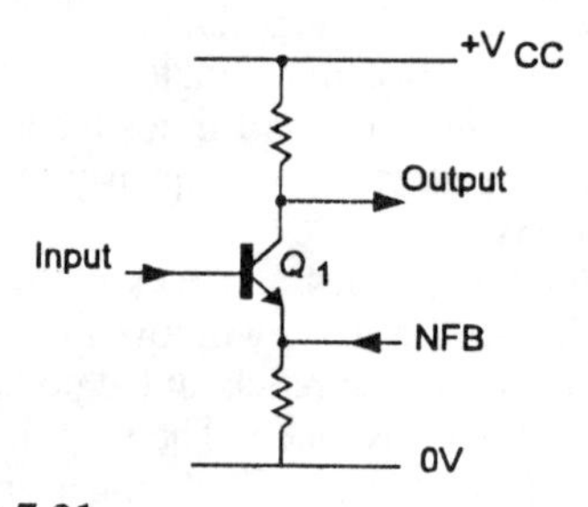

Figure 7.31

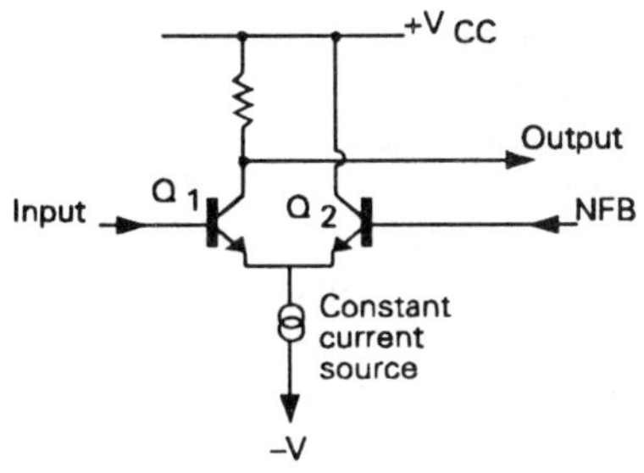

Figure 7.32

In the first of these, where the NFB is applied to the emitter circuit of the input transistor, while the input signal is applied to its base, the assumption is made that the base–emitter potential of the input transistor will remain constant, throughout the excursions of the input signal. Unfortunately, this is not a valid assumption, since the base–emitter potential will vary as the emitter current changes during the transmission of a signal. This will add a voltage component (related to the input transistor signal) to the feedback voltage, and impair its effectiveness in reducing the amplifier distortions. This leads to a worsening of the distortion as the magnitude of the input signal is increased.

In the long-tailed pair input circuit of Figure 7.32, most commonly used in operational amplifiers and other good quality amplifier systems, the problem of collector current modulation of the input base–emitter voltage in Q_1 is somewhat lessened because of the compensating effect of the balancing transistor Q_2. The way this happens is that if a positive-going signal voltage is applied to Q_1, which will increase its collector current and its base–emitter offset voltage, then this will cause a reduction in the collector current of Q_2, since both transistors are fed from a constant current source. This will then reduce the base–emitter offset voltage of Q_2, which will consequently increase the size of the NFB voltage applied to Q_1 emitter by an amount which is closely equivalent to the increase in the base–emitter offset voltage of Q_1. The result of this is that the sum of the base–emitter potentials of Q_1 and Q_2 tends to remain constant, so that there is more effective coupling of the feedback voltage into the feedback loop. Nevertheless, even in well designed amplifiers, the input signal voltage will have an effect on amplifier distortion, especially on those circuits which have a 'single-ended' input, such as that in Figure 7.31.

This effect is well illustrated in the manufacturer's application note for the 'single-ended' Harris A-5033 buffer amplifier IC, for which the internal circuit and the relationship between input signal voltage and harmonic distortion is shown in Figures 7.33 and 7.34. However, even in circuits with an input long-tailed pair arrangement, of the type shown in Figure 7.32, there is a residual difference in closed-loop distortion, as shown in the data sheet for the Texas Instruments LT1028, and shown in Figure 7.35, for various values of closed-loop gain (A_V). The extent to which such an amplifier is immune to the effect of a simultaneous application of input and feedback signals is known as the 'common-mode rejection ratio', and, for a competently designed op. amp., will normally be of the order of 100dB or more, although, like many other quoted parameters, it is likely to be quoted for the most favourable circumstances, and like other qualities it will get worse with increasing input signal frequency. In a shunt feedback arrangement, there is no common-mode signal, so the rejection ratio will always be infinite!

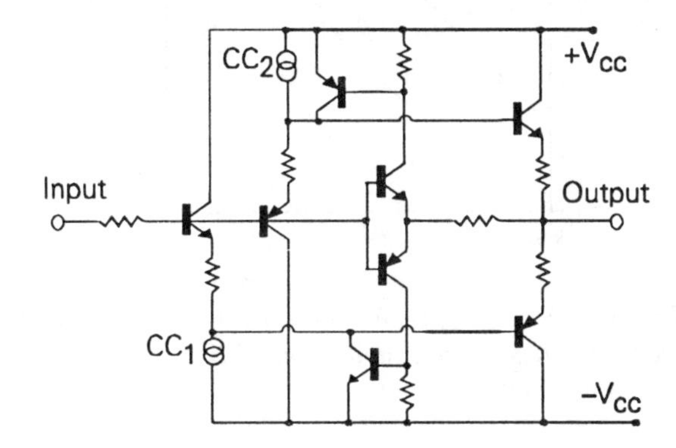

Figure 7.33 *Slightly simplified circuit layout of Harris A-5033 buffer amplification*

Appendix

Operational amplifier parameters

There are many other aspects of op. amp. performance characteristics which one would normally expect to be specified in the manufacturers' data sheets and these will relate to both the steady-state and the dynamic operating conditions of the device.

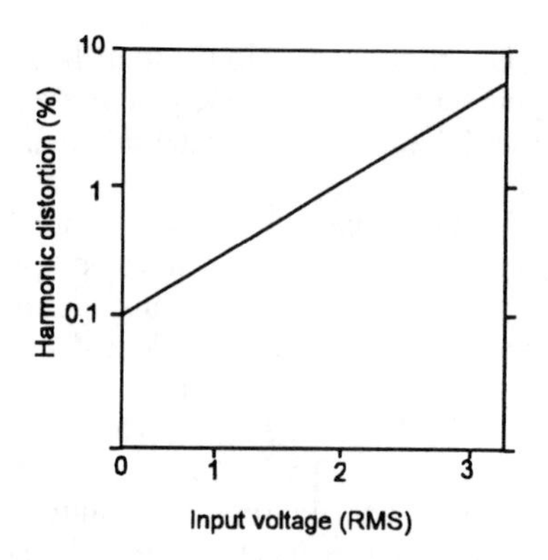

Figure 7.34 *Input signal voltage/distortion characteristics of A–5033*

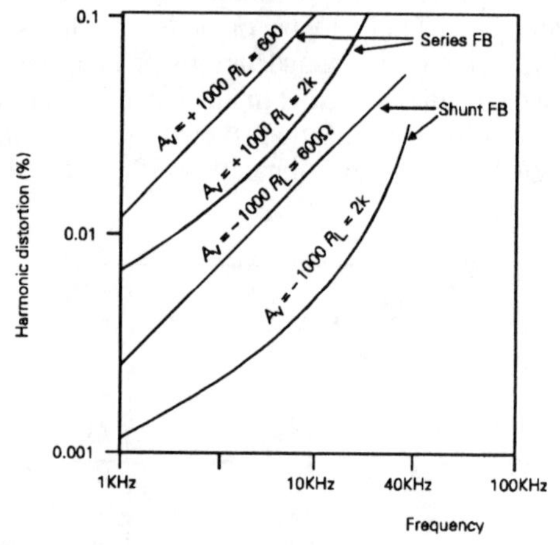

Figure 7.35 *Comparison of distortion characteristics of series and shunt NFB in high quality op. amp.*

Slew rate

Among the latter, an important characteristic in relation to the HF performance of the amplifier is the slew rate – the maximum rate at which the output voltage can change when it is presented with an instantaneous change in its input voltage. This has a rigid limit – for reasons explained above in discussing the HF stabilization of the amplifier circuit shown in Figure 7.20 – because the design choice for this purpose is almost invariably the use of a 'dominant lag' capacitor, in the circuit position C_X in the analysis of this circuit. This parameter will be defined in terms of a slope, in volts/microsecond, and will usually be somewhat different for positive- and negative-going waveform edges. If loss of signal during this slewing process cannot be tolerated, then either the size of the input signal must be restricted so that the required output voltage change does not exceed the slewing capability of the amplifier, or the high frequency components of the input signal must be curtailed by a simple input *RC* network, of the kind shown in Figure 7.11.

Gain bandwidth product (GBP)

This characteristic, like many other specifications offered by manufacturers, is misleading, in that it appears, at first sight, to indicate the usable bandwidth of the device. However, this is, in reality, very much less than the quoted value, since the gain bandwidth product defines the point at which the op.amp. gain falls to unity, and an amplifier with an open-loop gain of unity is unlikely to be of much use.

So, if the minimum open-loop gain which will be of use is, say, 10, then the maximum HF bandwidth will be the GBP value divided by ten, and so on.

Power supply rejection ratio (PSRR)

It will be obvious that the performance of any electronic circuit could be impaired if there is unwanted and unexpected intrusions of noise, hum or ripple, from the supply lines into the signal channel, and most data sheets will quote values for the ability of the circuit to reject such breakthrough.

However, although a value of, say, 110dB (300,000:1) may be quoted, which would appear to be excellent, it will probably be worse for the negative supply line than for the positive one, and, in both cases, will deteriorate with increasing frequency, so that it may only be 20dB (10:1) for a 300kHz supply line signal, as is the case for the 'LM312' IC.

The values quoted should always be for open-loop operation, since the application of overall NFB will improve this figure, in any case, except at higher frequencies where the internal phase shift of the device is significant.

Output voltage swing

This is influenced both by the supply line voltage and also by signal frequency, since it will decrease beyond some point at which slew rate limiting begins to occur.

The makers graphs of this parameter, as a function of frequency, are a much better guide to the HF performance of the device than either the GBP or the quoted slew-rate figure.

Other characteristics

Many other operating characteristics will also usually be quoted, which may allow the optimum choice of a particular type of op. amp. for a particular application. With high impedance circuitry, a low value of input bias current – the input base currents required by the input transistors – is desirable, though, for very high impedance circuitry, where very low amplifier noise is not also an essential quality, a JFET- or MOSFET-input op. amp. may be the preferrred choice.

In amplifiers designed for low noise operation, one would also expect to find details of the equivalent input noise voltage, in nanovolts/√Hz. Other details, such as output DC offset voltage, supply current, voltage gain, and the permissible or practicable input or output voltage swing, will also be given, together with information on the gain and phase margins of the device, to provide some guidance on the likely stability of the amplifier when negative feedback is applied. This is likely to be affected by output load characteristics, especially if these are inductive or capacitative.

It is becoming increasingly common for op. amp. ICs to employ internal HF compensation so that they can be used (but usually only on a resistive output load) with 100% NFB (as a unity gain buffer) without danger of HF instability, even though this will lead to a rather poorer HF response than would have been possible if the device had only been compensated for the closed-loop gain level appropriate to its intended use. For this reason, there are still a few externally compensated op. amps. available, for which the manufacturers will supply details of the necessary components which will be required to be connected to the device to ensure stability at a given closed-loop gain level. The data sheets may also, if one looks for it, give information on permissible load characteritics, where these affect closed-loop NFB stability.

8

Frequency response modifying circuits and filters

Introduction

Many applications of electronic circuits will require some adjustment to the frequency response of the system to produce an increase or a decrease in its gain at some part of its pass band in respect to another. This could entail either a gradual change in gain, as a function of input frequency, to lessen the magnitude of some signals with respect to others, or a more abrupt alteration in gain, perhaps to select some particular signal, or to cause the complete exclusion of other signals occurring at different parts of the input frequency spectrum.

Where the change in gain exceeds 6dB/octave (in which the gain is either halved or doubled by a halving or doubling of the input frequency), the generic term filter is usually employed, and such filters can be designed to operate either in low-pass, high-pass, notch, tuned response or bandpass arrangements.These filter systems can be built either in passive forms, using capacitors (impedance decreases as a function of frequency) or inductors (impedance increases as a function of frequency) in combination with resistors, or with each other, or as active layouts where the characteristics of the circuit are dependent on the effect of some form of feedback, either positive or negative, within the system. The term active is used to distinguish these circuit types from those where any gain blocks simply provide signal amplification, without altering the effect of any *LR*, *CR* or *LCR* networks which are present in the signal path.

Active circuit designs are preferable, generally, for high attenuation rate filters in low frequency applications, where large values of inductance would be needed to achieve the same result by passive *LC* systems. Such large inductors could also pose problems due to hum pick-up. For RF applications, above, say, 100kHz, passive *LCR* networks, often quite complex in design, are almost always employed. (See Chapter 11.)

For convenience, I have divided the text into passive and active portions, in order to examine the various functions possible with both of these systems.

Passive circuit layouts

Low-pass circuitry

The simplest circuit arrangements used for this purpose are just RC or LR combinations of the kind shown in Figures 8.1a, or 8.1b, which will give gain vs. frequency response curves of the kind shown in Figure 8.2a, with a turn-over frequency, f_t (that frequency at which the gain has fallen by –3dB with respect to its value at frequencies well below the turn-over point), given by the equation

$$f_t = 1/(2\pi R_a C_a) \quad \text{or} \quad f_t = R_a/(2\pi L_a)$$

and an ultimate attenuation slope of –6dB/octave. The LCR circuits shown in Figures 8.1c and 8.1d, give a hump in the frequency response curve at the turn-over frequency, but because they include two reactive components, will give an ultimate attenuation rate of –12dB/octave, as shown in Figure 8.2b. The turn-over frequency is given approximately by the equation

$$f_t = 1/(2\pi\sqrt{C_a L_a})$$

though the actual value of f_t is modified a little by the presence of R_c.

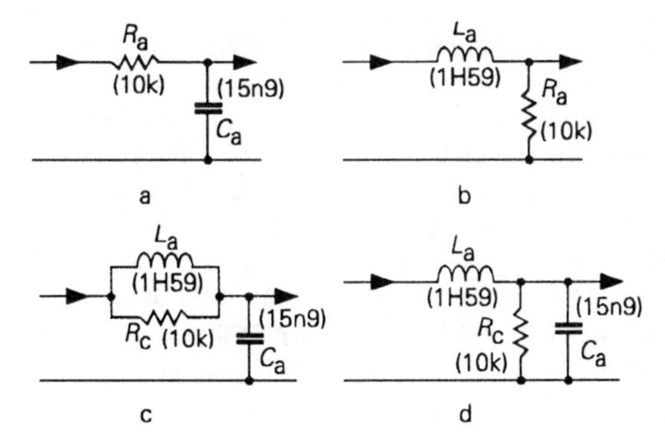

Figure 8.1 *RC, LR and LCR low-pass filter layouts (f_t = 1000 Hz)*

In most practical applications, it is necessary for L_a or C_a to be shunted by a resistance, R_c, having a value similar to that of the impedance, at f_t, of either the inductor or the capacitor, in order to prevent the circuit presenting the signal source with a near-zero load impedance at the series resonant frequency of the LC

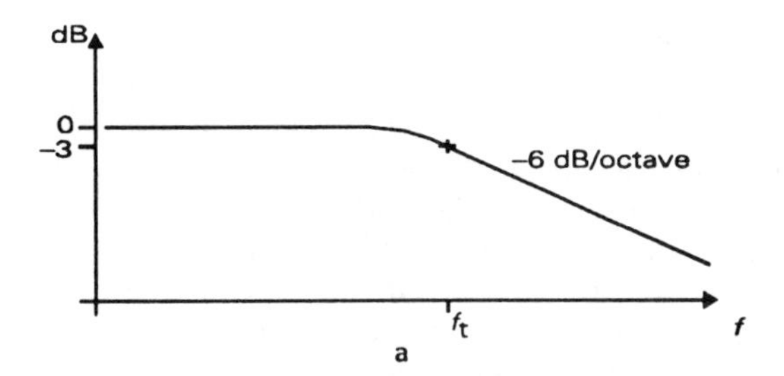

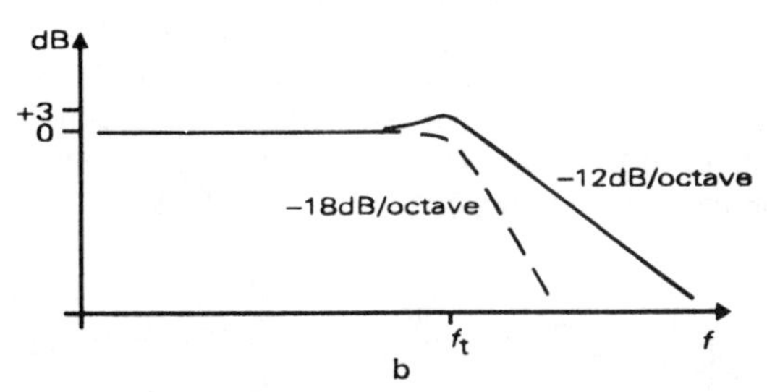

Figure 8.2 *Frequency response curves for circuits of Figure 8.1*

circuit, and to limit the size of the hump in the frequency response. However, the presence of this hump, even when damped by R_c, does allow the interesting possibility that the circuit of Figure 8.1c could be followed by a simple RC attenuator circuit, such as that of Figure 8.1a, to give the layout shown in Figure 8.3, which will, if the component values are chosen correctly, both level out the hump and give an attenuation slope of –18dB/octave, as shown by the dashed line in Figure 8.2b.

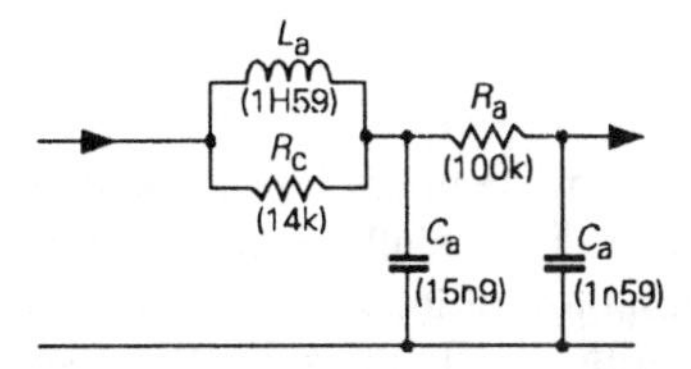

Figure 8.3 *Filter circuit for –18 dB/octave slope*

In deciding what component values should be employed for these types of circuit, it is sensible to consider at what overall impedance level one wishes to operate, and then choose values of R, and C or L, whose impedance at f_t is near to the required circuit impedance level.

It is customary, in considering the performance of all the frequency response modifying systems described in this chapter, to assume that they are driven by a signal source having a very low output impedance, and that they will feed a load which has a very high input impedance. If this is not the case, simple op. amp. buffer stages, of the kind shown in Figure 8.4a, could be interposed between the filter circuit and its source or load to satisfy this requirement. Similar frequency response characteristics are given by the shunt and series feedback layouts shown in Figures 8.4b and 8.4c, for similar values of R_a and C_a, though, in this case, there will be some gain in signal level determined by the ratios of R_a to R_b. In both of these circuits, R_a, or some equivalent DC path, is essential in order to provide a DC negative feedback path to stabilize the working point of the gain block.

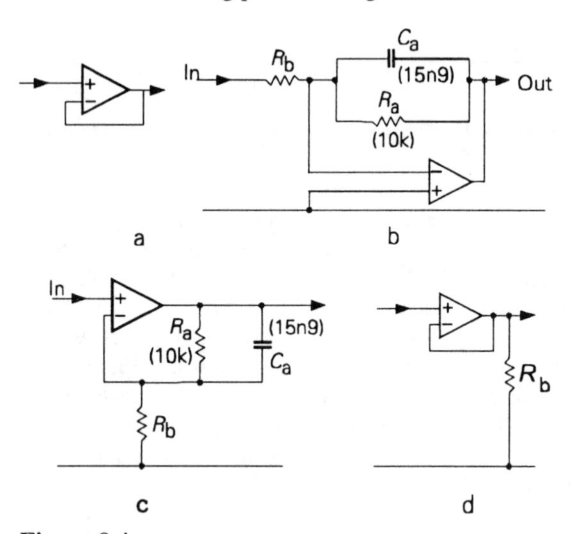

Figure 8.4

It should be remembered that, as mentioned in Chapter 7, there is a small residual difference in the frequency response of these two circuits, since, apart from the additional circuit gain, the frequency response curves of the circuits of Figures 8.1a and 8.4b are substantially identical, giving a stage gain which will fall to zero at very high frequencies.

By contrast with this, the circuit of Figure 8.4c gives a gain which decreases only to unity, because when the frequency is high enough for the impedance of C_a to be negligibly small, the circuit effectively becomes equivalent to that of Figure 8.4d, which is a simple unity gain buffer stage. This gives a frequency response of the kind shown in Figure 8.5, which is

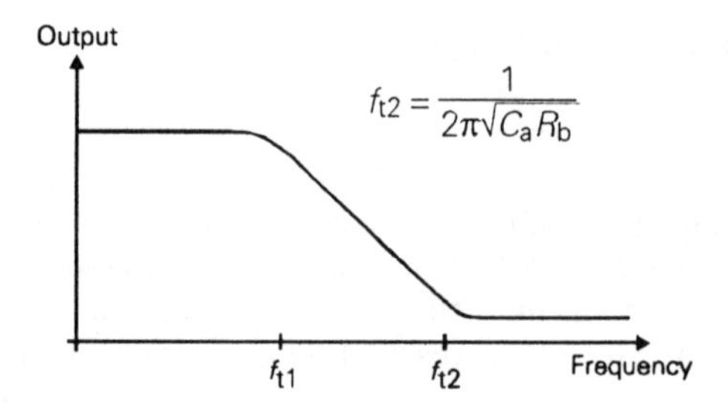

Figure 8.5 *Frequency response of circuit of Figure 8.4c*

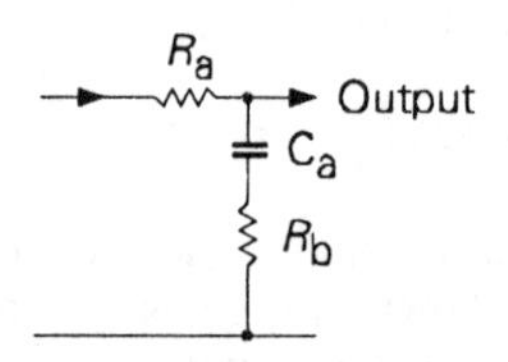

Figure 8.6

equivalent to that of the passive circuit of Figure 8.6. The series feedback circuit layout of Figure 8.4c has the significant advantage that it offers a very high input impedance, which means that the source impedance does not need to be low, and it is therefore used more often in practical circuit layouts.

Although simple, single CR, networks are often used, for example in the two audio tone control networks shown in Figures 8.7a and 8.7b, which give the type of frequency response curves shown in Figure 8.7c, more complex combinations of capacitors and resistors are often used, either as a simple network, or in the negative feedback path of a gain block, as shown in Figures 8.8 and 8.9, which are typical RIAA equalized gramophone record replay, and 'NAB' equalized cassette tape head replay amplifier arrangements, for which the frequency response curves are shown in Figures 8.10a and 8.10b.

High pass circuitry

Circuits of this type are simple rearrangements of Figures 8.1– 8.3, with the positions of some capacitors, inductors and resistors interchanged, as shown in Figures 8.11–8.12. The frequency response characteristics of the circuits of Figures 8.11 are shown in Figures 8.13a and 8.13b/c, in which the turn-over frequency is given by the same equations quoted above. These circuits can also be connected around op.

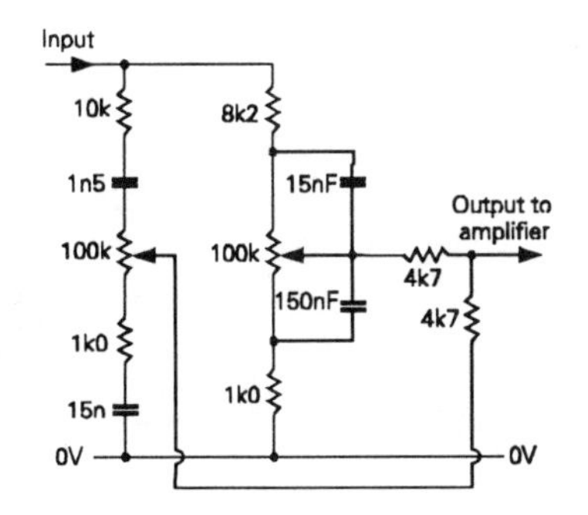

Figure 8.7a *Typical passive tone control circuit*

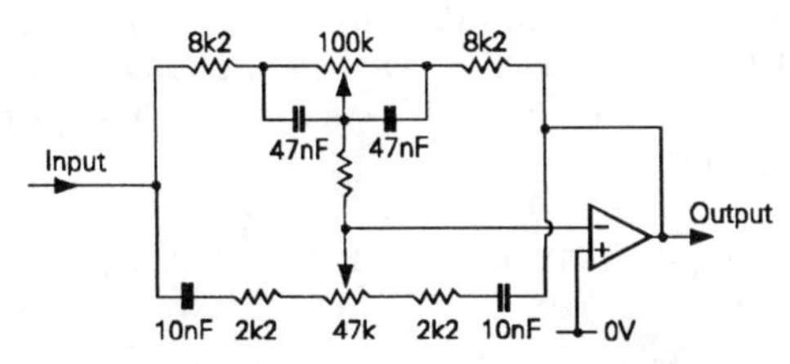

Figure 8.7b *Baxandall type negative feedback tone control circuit*

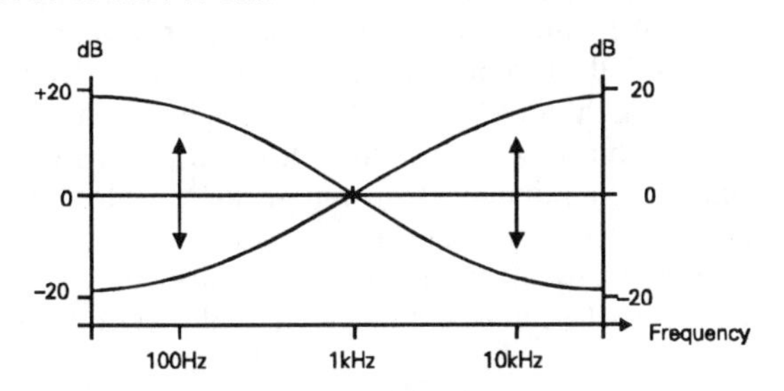

Figure 8.7c *Frequency response curves of tone control circuits*

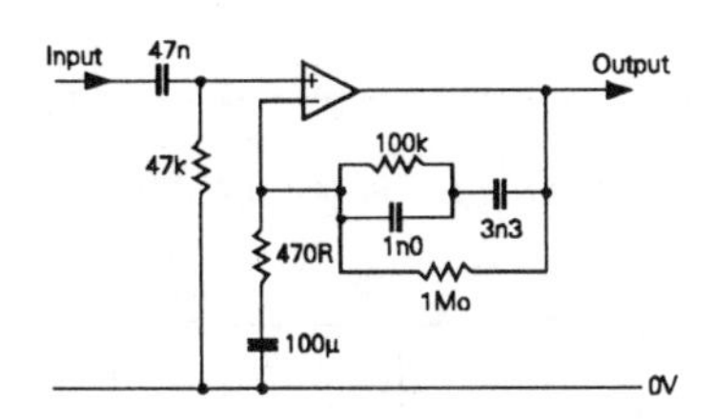

Figure 8.8 *Typical R!AA compensation pick-up input stage*

amp. gain blocks as shown in Figures 8.14 and 8.15, to give output characteristics which will rise with frequency, as shown in Figure 8.16. In the absence of R_c, the gain would increase continuously, as the fre-

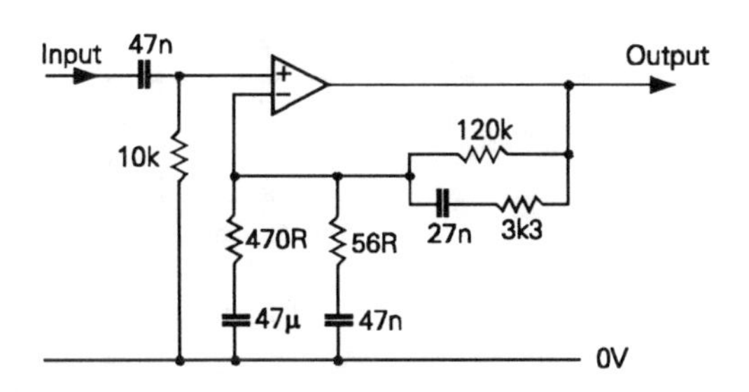

Figure 8.9 *Tape recorder (3.5 cm/sec) replay equalization circuit*

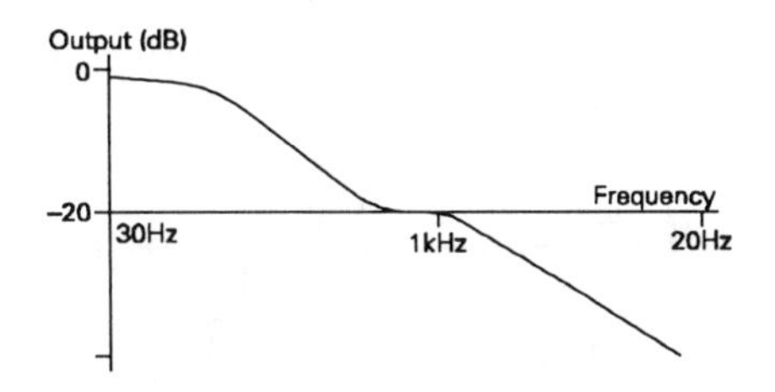

a RIAA equalized phono preamp frequency response

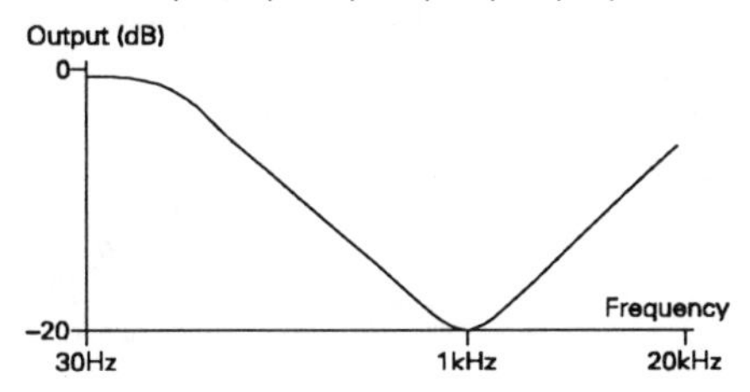

b NAB equalized cassette player frequency response

Figure 8.10

quency increased, up to the open-loop gain of the amplifying stage. With R_c, there will be two turn-over points, an LF one whose +3dB point will be given by

$$f_{t1} = 1/(2\pi R_a C_a)$$

and a higher frequency one, at which the gain will again level off, whose turn-over point is given by

$$f_{t2} = 1/(2\pi R_c C_a)$$

Notch filters

This kind of response characteristic, which is used to remove a specific single frequency from the pass-band of the circuit, can be provided by a variety of circuits,

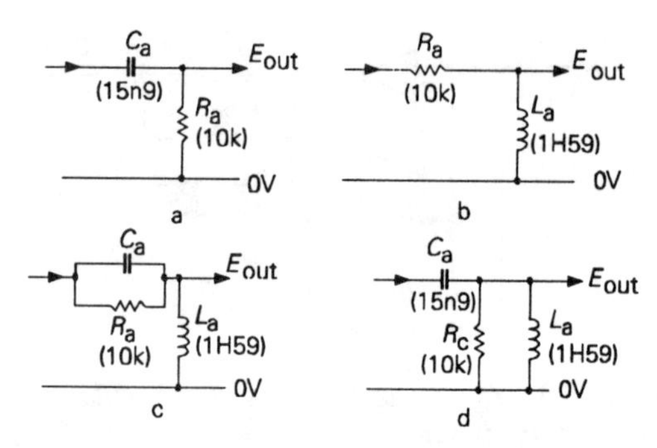

Figure 8.11 *Rearrangements of circuits of Figure 8.1 as high-pass layouts (f_t = 1000Hz for values shown)*

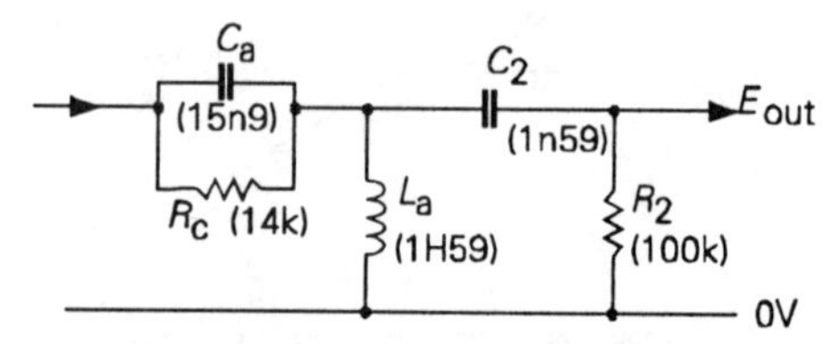

Figure 8.12 *Third-order (–18dB/octave) high-pass LCR filter*

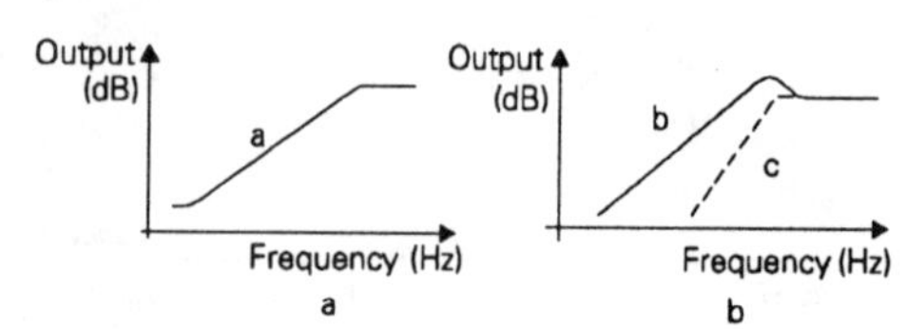

Figure 8.13

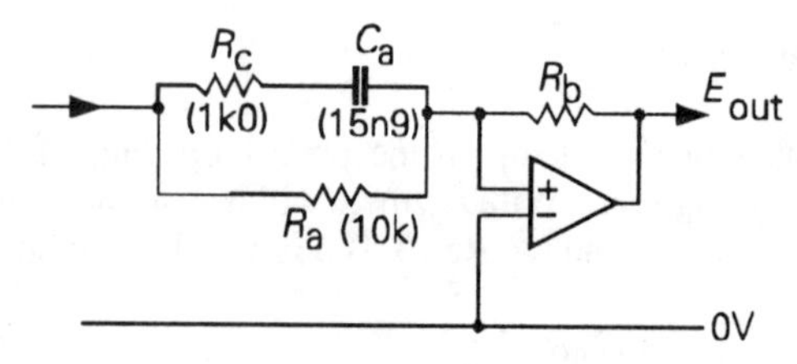

Figure 8.14

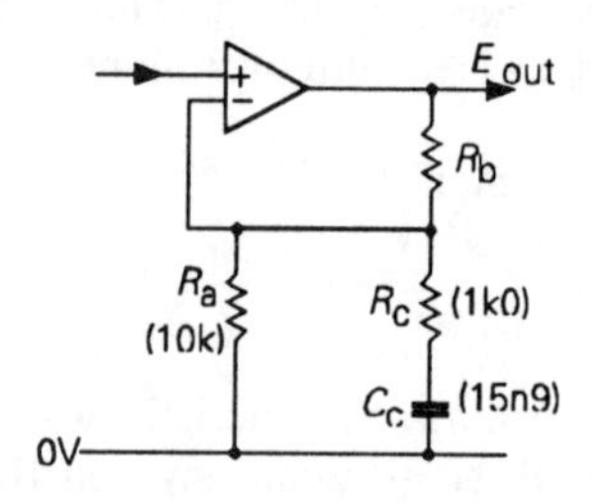

Figure 8.15

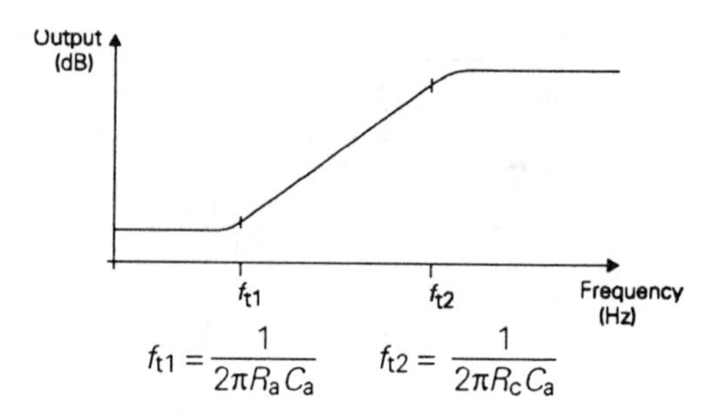

Figure 8.16

of which the two most common are the parallel resonant *LC* filter, and the 'parallel-T' circuits shown in Figures 8.17 and 8.18, for which the notch frequencies are given by

$$f_t = 1/(2\pi\sqrt{LC}) \quad \text{and} \quad f_t = 1/(2\pi CR)$$

respectively.

If low loss capacitors and high 'Q' coils (i.e. those with low internal resistive losses), are used in the circuit of Figure 8.17, and if the source resistance is low, and if the load resistance, R_L, is chosen so that its value is high in relation to the AC impedance of the inductor, but low in relation to leakage impedances, quite a good ultimate attenuation factor, coupled with sharp notch can be obtained.

The notch frequency attenuation given by the circuit of Figure 8.18 can be extremely high, if precise value

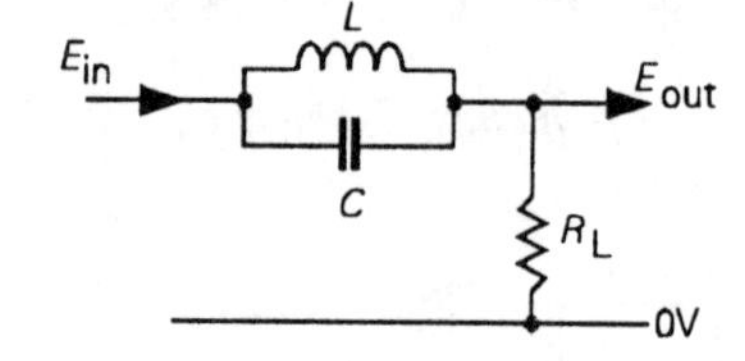

Figure 8.17 *Simple LC notch filter*

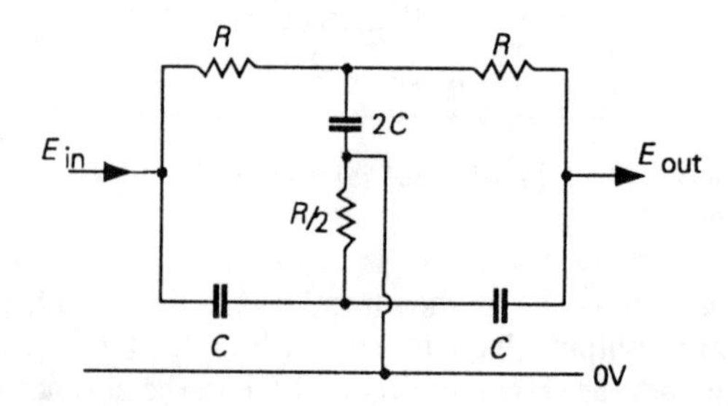

Figure 8.18

components are used, though the steepness of the cut-off slope, on either side of the notch, is not very good, as illustrated by the solid line curve in Figure 8.19. Negative feedback can be used to sharpen up the notch, with the circuit arrangement shown in Figure 8.20, which gives the type of response shown in the dashed curve in Figure 8.19. The sharpness of the notch, in this case, can be adjusted by the setting of RV_1.

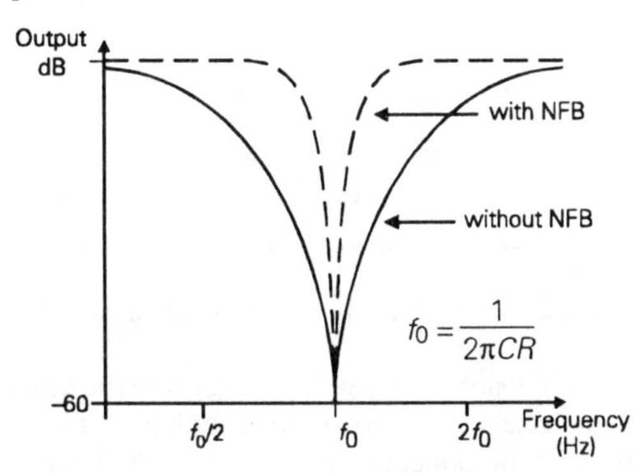

Figure 8.19 *Frequency response of parallel-T notch filter*

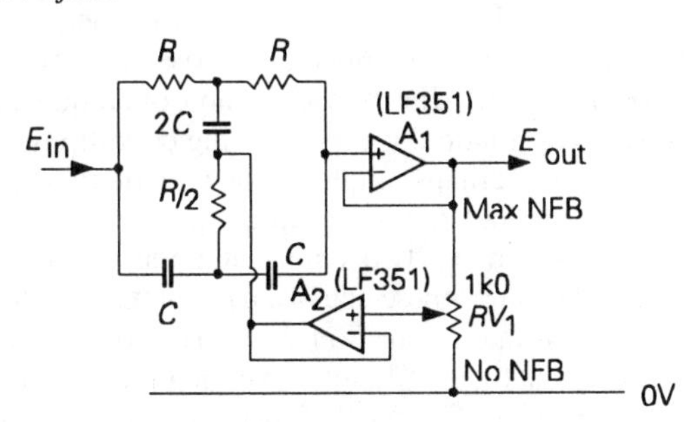

Figure 8.20 *Way of applying negative feedback around parallel-T filter circuit*

Tuned response

This kind of circuit is the opposite of the notch filter, in that it can be used to select one particular frequency from an input spectrum of signals, with the kind of frequency response shown in Figure 8.21. The LC circuits shown in Figures 8.22a and 8.22b can do this, provided that the values of R_{in} and the load resistor, R_L are chosen appropriately, with the limitations noted earlier about the bulk – and likely low Q – of large values of inductor necessary for use at low frequencies, together with their susceptibility to hum pick-up.

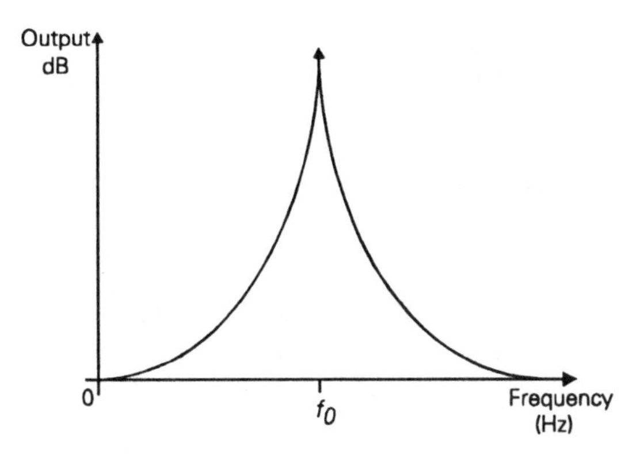

Figure 8.21 *Frequency characteristics of LC tuned response filter*

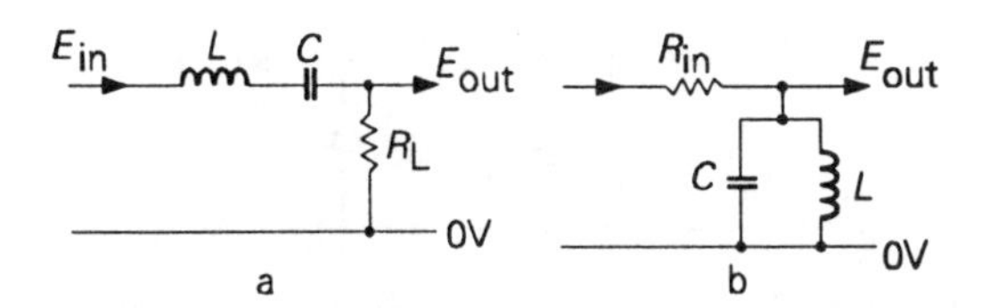

Figure 8.22 *LC tuned response filter circuits*

An alternative approach is to insert a parallel-T notch filter in the feedback path of a gain block, as shown in Figure 8.23, but this must be by-passed by a gain limiting resistor, R_s, or the circuit will oscillate continuously.

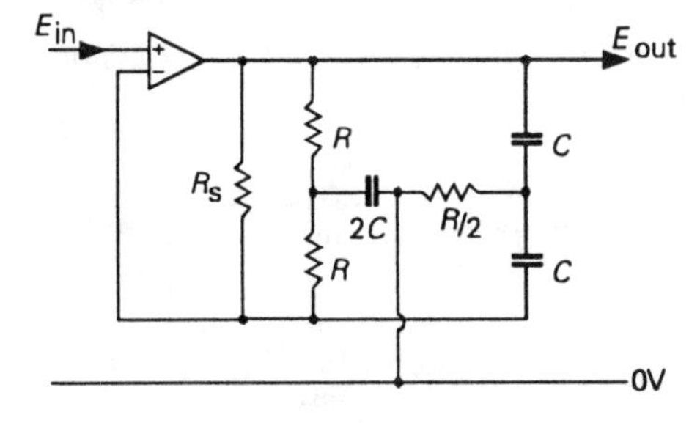

Figure 8.23 *Tuned response filter based on bridged parallel-T circuit*

Bandpass circuitry

In principle, this is merely a high-pass circuit, followed by a low-pass one, such as shown in Figures 8.24, which will give the type of response shown in Figure 8.25, where the separation between the two turn-over frequencies, f_{t1} and f_{t2}, can be adjusted to give the appropriate pass-band. However, in practice, R_2 and C_2 will need to be high in value compared with R_1 and C_1 to get a reasonable degree of independence in the operation of the two halves of the circuit. This layout, and the frequently recommended op. amp. system shown in Figure 8.26, of which the performance is shown in Figure 8.27, do not really give a high enough frequency discrimination to be of much use in practice, and active circuit layouts giving higher 'skirt' attenuation rates are generally preferred. For RF applications, this need can be met very satisfactorily by the coupled tuned circuit layout shown in Figure 8.28. The operation of this kind of circuit is examined in greater detail in Chapter 14.

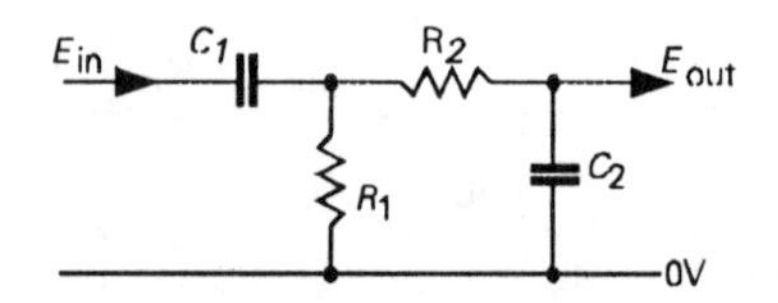

Figure 8.24 *Simple cascaded RC bandpass filter*

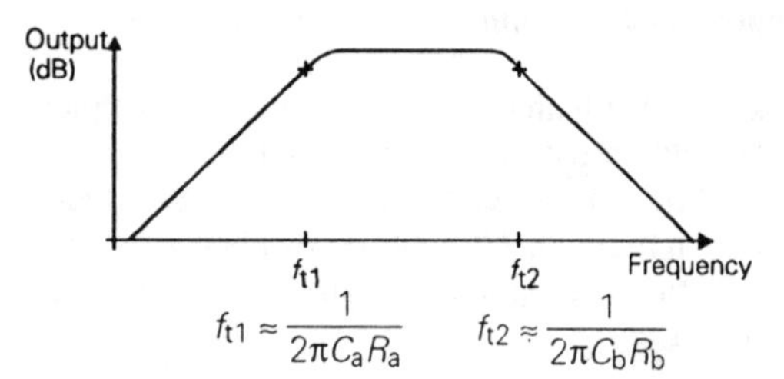

Figure 8.25 *Frequency response of circuit of Figure 8.24*

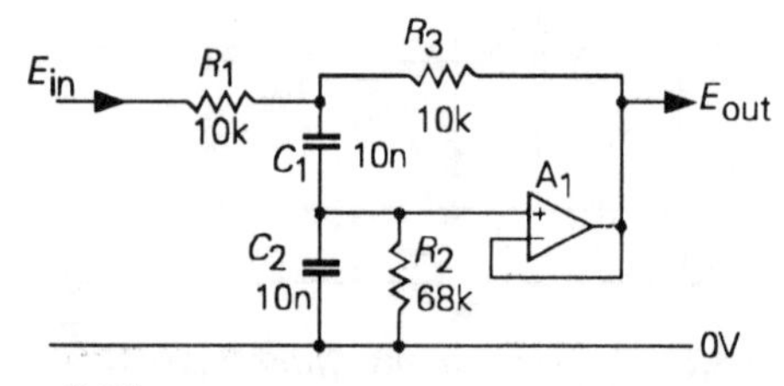

Figure 8.26

Active filter circuitry

Introduction

As a general rule, this term is used to describe systems which contain amplifying or impedance converting

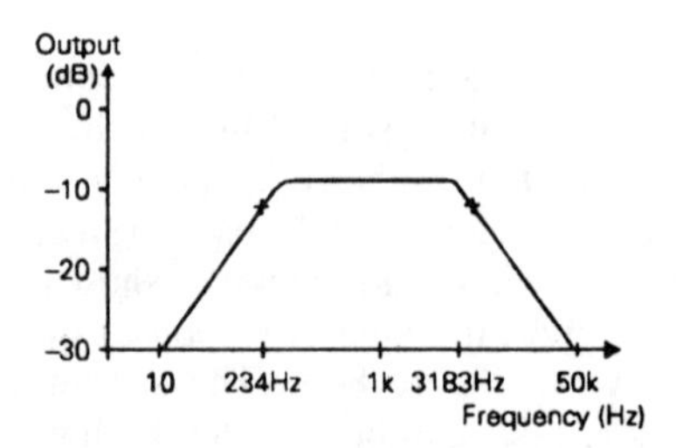

Figure 8.27 *Frequency response of circuit of Figure 8.26*

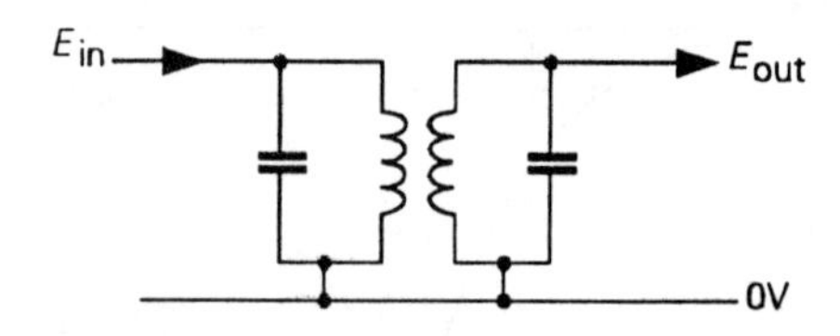

Figure 8.28 *Bandpass-coupled tuned circuits*

elements, capable of injecting energy into the system, and which can be used to modify the characteristics of the passive components, *Cs*, *Rs* and *Ls*, which are used in the circuitry.

A point which should be remembered in connection with such circuits is that, because of the energy injected by the active component, it is possible for them cross over the borderline between stable operation and continuous oscillation, if the operating conditions are incorrect – as, perhaps, might be the case of an active filter whose input becomes open-circuited.

I have lumped the two main categories of active filter together since these can often be interchanged in their mode of operation, from high-pass to low-pass and vice versa, by simply interchanging the positions of the appropriate *Cs* and *Rs*. There are a few exceptions, which will mainly concern those circuits where a simple transposition would make the DC bias paths awkward to implement, or where the intrinsic HF gain or phase characteristics of the gain block will influence the final result.

High-pass (HP) and low-pass (LP) filters

A simple example of the way a gain block can be made to increase the rate of attenuation in the gain/frequency characteristics of a circuit, above or below some specific turn-over frequencies – by the use of negative feedback, applied around the amplifying block – is shown in Figure 8.29. Without NFB, the gain of the system, as a function of frequency, shown in the dashed line of Figure 8.30, is very non-uniform. If, however, NFB is applied, the response of the system is modified, as shown by the solid line curve in Figure 8.30. This shows not only that the gain is more uniform, as a function of frequency, though at a lower gain level, but also that the LF and HF frequency response is extended beyond the open-loop turn-over points. Moreover, beyond these points, and this is the feature of particular interest, it can be seen that the application of NFB causes the response to fall at a much steeper rate than would have been the case without it. This suggests an approach by which a simple low-pass active filter could be made. If an amplifying block, such as the op. amp. shown in Figure 8.31a, is preceded by an input RC circuit, R_2/C_1, so that, left to itself, the system would have a falling gain with increasing frequency, and negative feedback is then applied, via R_3, to the junction of R_1 and R_2, then the circuit will give the type of frequency response shown in Figure 8.32, as the NFB operates to maintain a constant gain for all input frequencies.

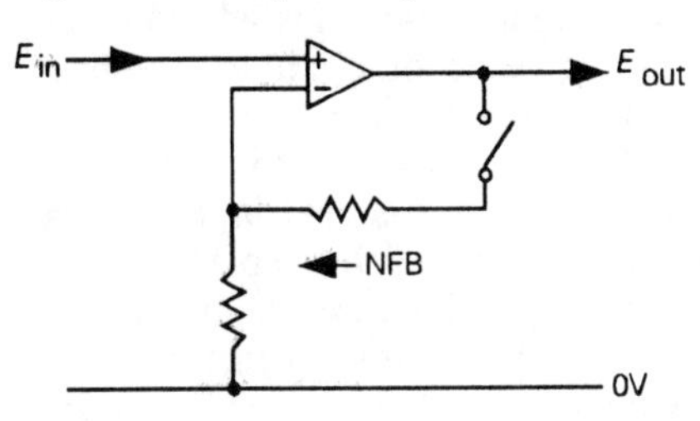

Figure 8.29

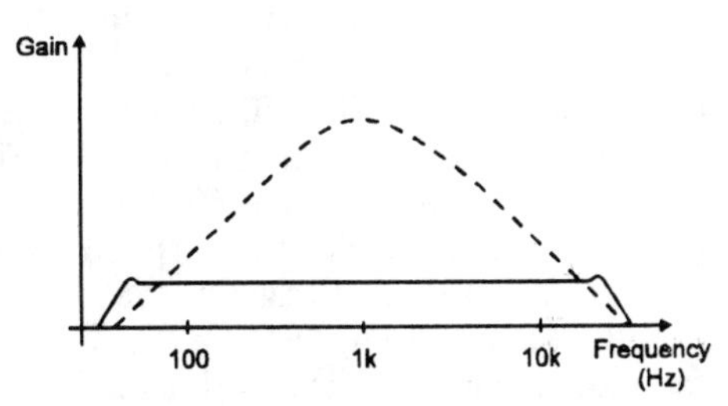

Figure 8.30 *Effect of NFB on frequency response of gain block*

The circuit layout shown in Figure 8.31b would be inappropriate for this purpose because, as shown in Chapter 7, the loop feedback would have the effect of reducing the output impedance of the op. amp. (which, in effect, will be largely that due to R_2,) by the factor

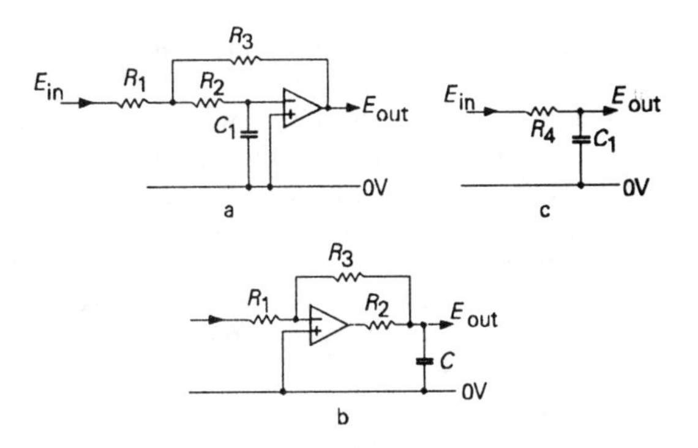

Figure 8.31 *Method of constructing an active low-pass filter*

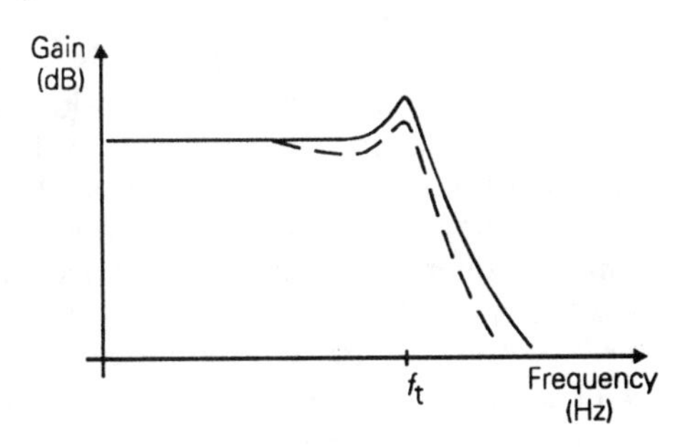

Figure 8.32

$(1 + AB)$, where A is the open-loop amplifier gain, and B is the feedback factor, giving an effective output impedance value of $R_2/(1 + AB)$. C_1 would therefore need to be very much larger in value than a simple f_t calculation based on R_1/C_1 would suggest, and it would also mean that, while the HF gain would indeed fall quite steeply beyond some turn-over point, this would mainly be due to the limited ability of the op. amp. at higher frequencies, to drive current into C_1, which would result in fairly severe distortion of the output waveform.

In both of the circuits of Figure 8.31, there would be a prominent peak in the frequency response at the point preceding the fall in gain. This is because, in all normal op. amps., as a result of their internal HF compensation components, there is an internal input/output phase shift of 90°, which, when added to the phase shift due to R_2/C_1, will cause the overall feedback to cease to be negative in character, and approach the condition of positive feedback. This will always increase the stage gain, and can, in some circumstances, lead to continuous oscillation.

If an additional RC HF attenuation network is added to the circuit, as shown in Figure 8.31c, the size of this hump in the frequency response can be reduced, and this RC network will also cause the rate of HF attenuation to be increased, as shown in the dashed line in Figure 8.32. The penalty is that, in order for the added RC network to do much about removing the hump in the curve, it will also cause some loss of gain at frequencies below the turn-over point, which will usually be undesirable. This brings up the point of the flatness of the frequency response of a filter, in relation to the ultimate rate of attenuation of the system, between which there must usually be some compromise, since the circuit considerations which improve the one will usually impair the other.

A further consideration in all filter types is the time delay which they will introduce: a factor which also tends to impose constraints on the filter type chosen. This implies that some choice must be made, in most cases, between the various options, and has led to the specification of filter characteristics as 'Butterworth', in which flatness of frequency response within the pass-band (i.e. below or above the turn-over point, in the case of LP and HP filter types) is optimized at the expense of ultimate attenuation rate.

The second option, that in which the attenuation rate is maximized at the expense of pass-band flatness, is known as the 'Chebyshev' type of response, and a variation of this, in which flatness of frequency response, even outside the pass-band, is sacrificed for steepness of cut-off characteristics, is known as the 'Cauer' response. The final common filter form is the 'Bessel', which is the type of design in which the layout is chosen to minimize time delay. The general transmission characteristics of these filter types, in comparison with a simple RC network, are shown in Figure 8.33.

So, with regard to the specification of filters, in addition to whether an HP or an LP type is required, and the desired turn-over frequency and attenuation rate, it is also necessary, in the case of the Chebyshev or Cauer forms, to stipulate the extent of 'ripple' – both within and beyond the pass-band – which is acceptable. In the case of multiple pole LCR filters used in RF circuitry, filter design has become a highly specialized field, which is largely outside the scope of this book.

Although the simple circuit shown in Figure 8.31a does work, and would come within the broad category

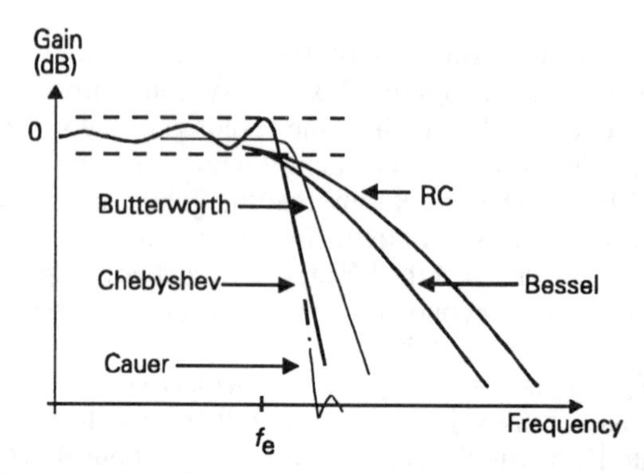

Figure 8.33 *General form of filter characteristics (illustrated for low-pass systems)*

of active filters, it is a rather crude approach to the design of such systems, of which a number of quite elegant circuits have been evolved by circuit engineers. As an example of such a design, an immediate improvement can be made to the performance of the circuit of Figure 8.31a, by adding a further capacitor, C_3, between the 0V line and the junction of R_1 and R_2, as shown in Figure 8.34a. This substitutes an external phase shifting capacitor, C_3, in the feedback path, in place of the incidental phase lag due to the amplifying block, and converts the circuit into a simple 'second order' filter system (i.e. one in which the characteristics are determined by two phase shifting elements, such as, in this case, the two RC networks), which has a Butterworth characteristic and a –12 dB/octave attenuation rate beyond the turn-over frequency, as shown in the graph of Figure 8.34c. The layout shown in Figure 8.34b is a simple modification of this circuit, which also gives excellent results, in which C_1 is introduced in the NFB limb. Its value will require to be reduced, in relation to that used in Figure 8.34a, by the amount $A'+1$, where A' is the gain of the amplifier after feedback has been applied. Both of these layouts are useful in that the filter will also provide gain, determined by the ratio R_3:R_1.

A common application for a high-pass filter is to eliminate turn-table 'rumble' in Hi-Fi audio equipment. This facility was often provided in earlier equipment designs, at a time when neither the turn-tables in domestic use, nor the record manufacturers' cutting lathes were as well engineered as they are now, and LF noise intrusion due to inadequate turntable bearings could be audibly objectionable. Such a rumble filter will usually require to have a ruler flat frequency

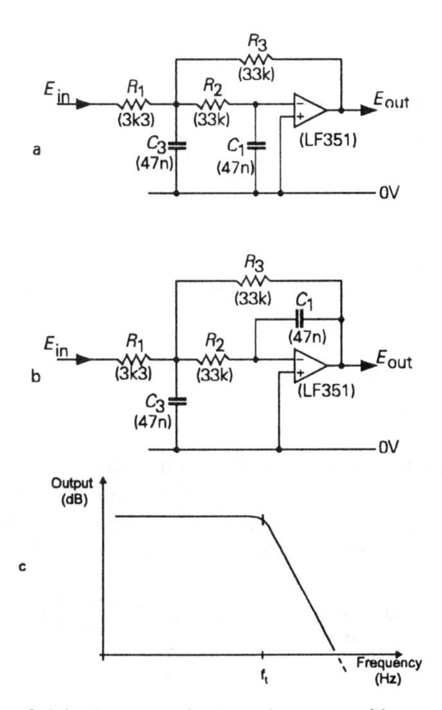

Figure 8.34 *Improved active low-pass filter circuits (component values quoted are for f_t = 1 kHz)*

response down to, say, 32Hz, and as rapid a rate of attenuation as is practicable (usually at least –20dB/octave), below this frequency, in order to exclude rumble components in the 2–10Hz frequency range without significantly reducing the magnitude of the wanted VLF components of the signal.

A typical circuit layout for this purpose is shown in Figure 8.35. This incorporates a high-pass filter network within the negative feedback path of a gain stage. The frequency response generated by this circuit is shown in Figure 8.36. A particular advantage of this kind of circuit is that the gramophone record replay frequency response correction network (the RIAA network), can also be connected between the output and inverting inputs of the amplifier, in parallel with the rumble filter components, without impairing the operation of this filter. This circuit is actually a filter of the general type classed as a bridged T design, which are usually configured as 'third-order' (–18 dB/octave) designs, with some external passive element, in this case the input capacitor/resistor network, C_1R_1, used to remove the characteristic hump in the

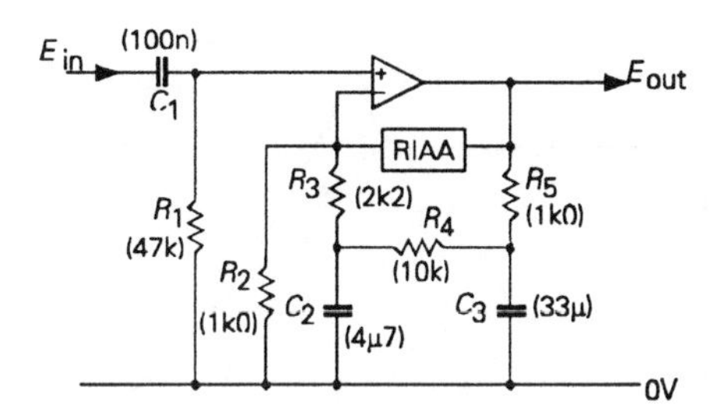

Figure 8.35 *Hi-Fi rumble filter circuit (f_t = 32Hz)*

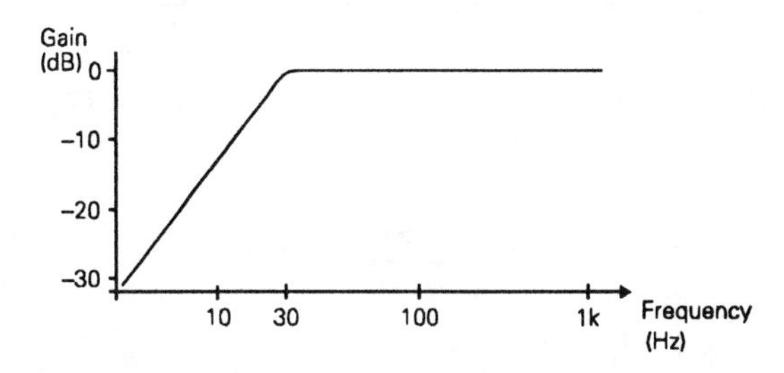

Figure 8.36 *Frequency response of circuit of Figure 8.35*

response curve at the turn-over frequency.

The somewhat simpler circuit layout of the bridged T, when used as a low-pass filter as shown in Figure 8.37a, can be used to illustrate this point. With the component values shown it will give the response curve shown in Figure 8.38, with a prominent hump at 1kHz. If an additional *RC* low-pass network, R_4C_3, is added to the circuit, as shown in Figure 8.37b, the hump can be flattened out to give a Chebyshev type of response, with a +/–3dB ripple, as shown in Figure 8.39.

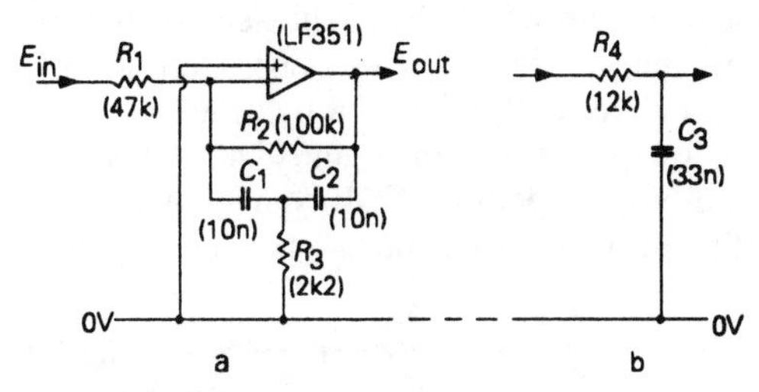

Figure 8.37 *Bridged-T filter (f_t = 1kHz)*

Two important categories of filter are the 'Sallen and Key' (Sallen, R. P., and Key, E. L., *IRE Trans. Circuit Theory*, March 1955, pp. 40–42), and the 'bootstrap' (JLH, *Electronic Engineering*, July 1976, pp. 55–58)

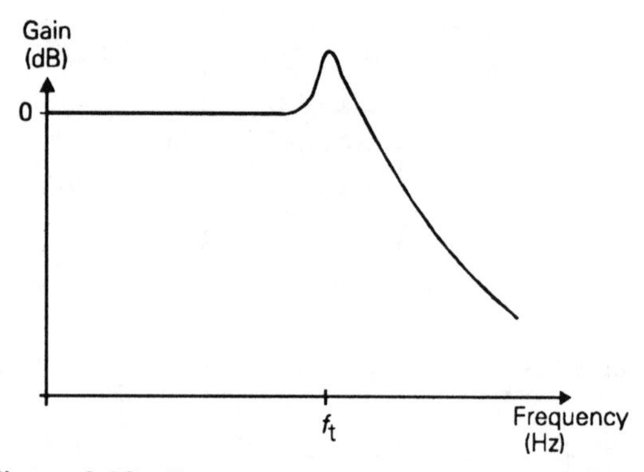

Figure 8.38 *Frequency response of Figure 8.37a*

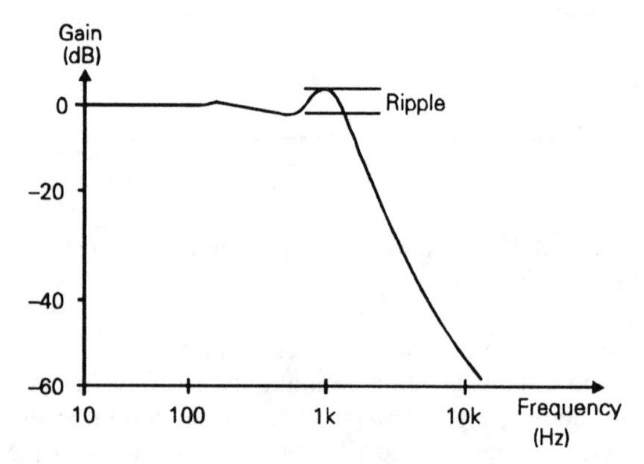

Figure 8.39 *Frequency response of circuit of Figure 8.37 with addition of R_4/C_3 network*

filter types, which are built around unity gain impedance converter blocks – though, with some adjustments to the circuit, they can also be employed with stages giving gain. The big advantage of these layouts is the ease with which the circuits can be transposed to give either LP or HP versions. Considering the Sallen and Key design first, this uses the layout shown, for a low-pass filter, in Figure 8.40. This gives a second-order response (–12dB/octave), with a Butterworth type of characteristic, as shown in Figure 8.41.

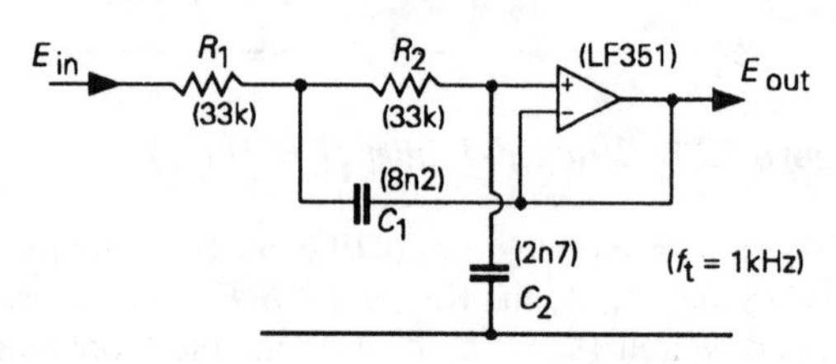

Figure 8.40 *Low-pass Sallen and Key filter circuit*

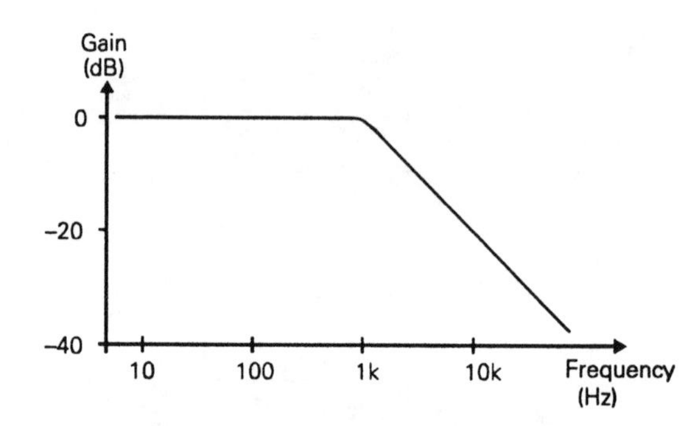

Figure 8.41 *Frequency response of circuit of Figure 8.40*

A high-pass version of the same kind of circuit, in which the position of the resistors and capacitors has been interchanged, is shown in Figure 8.42, with the frequency response illustrated in Figure 8.43.

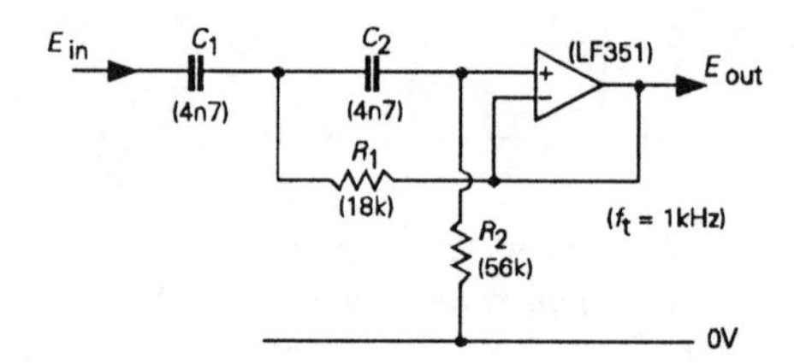

Figure 8.42 *High-pass Sallen and Key filter circuit*

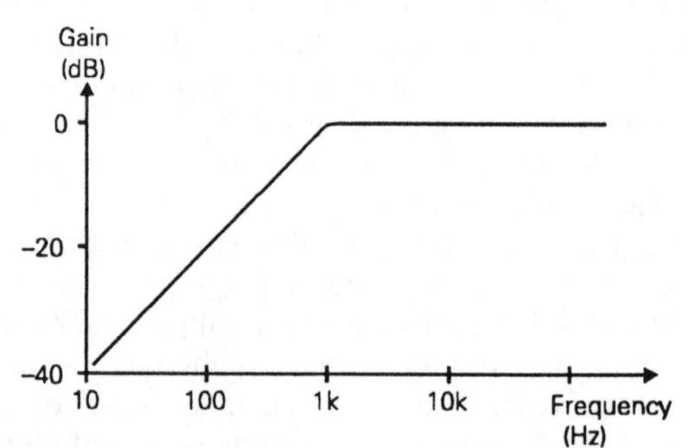

Figure 8.43 *Frequency response of circuit of Figure 8.42*

A circuit curiosity is that this kind of filter can be further elaborated as shown in Figure 8.44, to employ four pairs of *RC* networks. Sadly, this layout does not give the expected –24dB/octave attenuation rate, but only some –20dB/octave, as shown in Figure 8.45. In the application in which it was shown (Philips *Components and Applications*, August 1980, pp. 215–218),

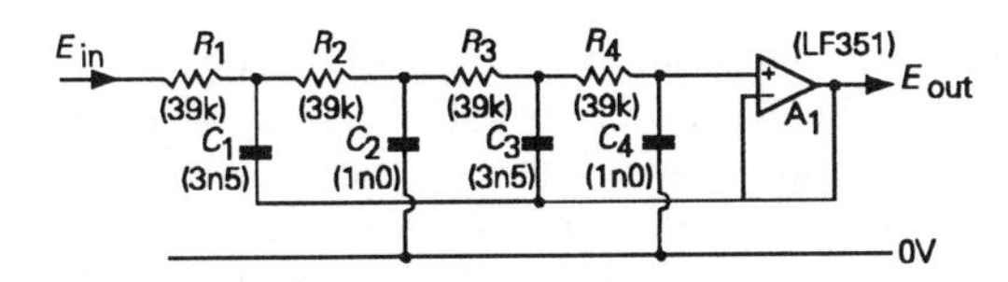

Figure 8.44 *Four element Sallen and Key filter*

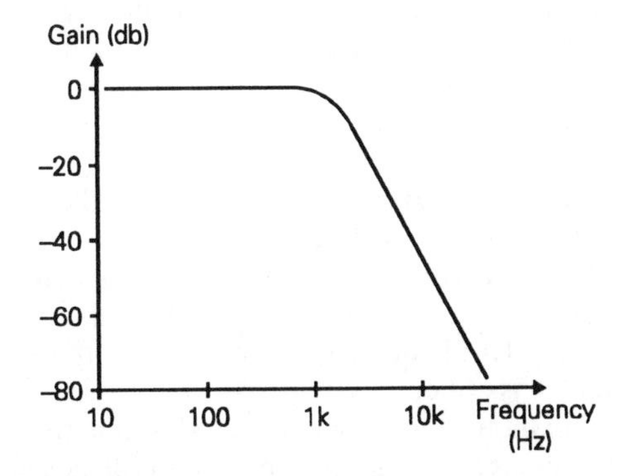

Figure 8.45 *Frequency response of circuit of Figure 8.44*

it was used simply as a means of generating a time delay in the signal path.

As noted above, Sallen and Key filters of this design will usually be configured to have a Butterworth characteristic. By contrast, the bootstrap designs are usually arranged as third-order Chebyshev response systems, with a typical pass-band ripple of +/–1dB and an initial attenuation rate of –20dB/octave. A suitable circuit design for a low-pass filter of this type is shown in Figure 8.46, with the frequency response as shown in Figure 8.47.

The equivalent high-pass bootstrap design is as shown in Figure 8.48, for which the frequency response is shown in Figure 8.49. The method by which the bootstrap filter operates can be visualized by considering the effect of R_2C_2, or C_2R_2, in the feedback path. At frequencies within the pass-band of the filter, the impedance of the feedback limb is low in relation to that of the lower input limb, which has the effect, by bootstrap action, of making the impedance of the upper half of the input limb (R_1 or C_1), appear to be high, so that there is little input signal attenuation. However, as the turn-over frequency is approached, the change in relative impedances of the feedback and lower input limbs causes the effectiveness of the bootstrap action to diminish, which causes the input circuit attenuation to increase, and so on.

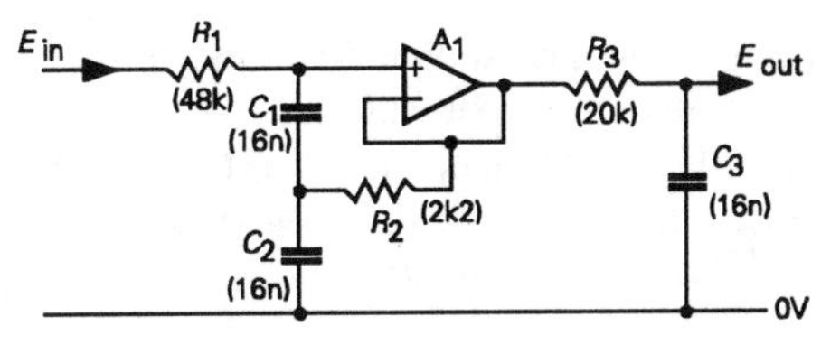

Figure 8.46 *Low-pass bootstrap filter (f_t = 1kHz)*

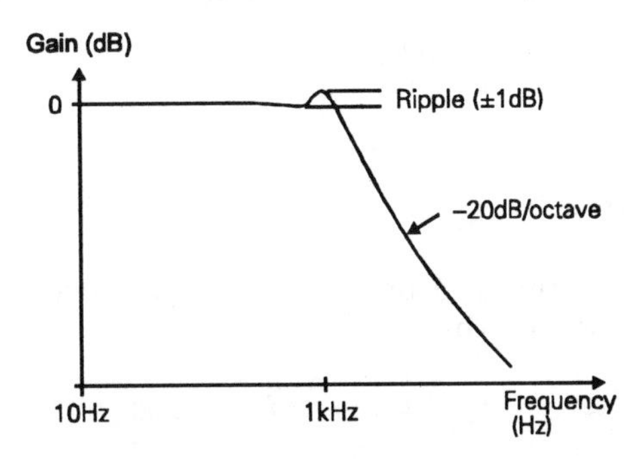

Figure 8.47 *Frequency response of circuit of Figure 8.46*

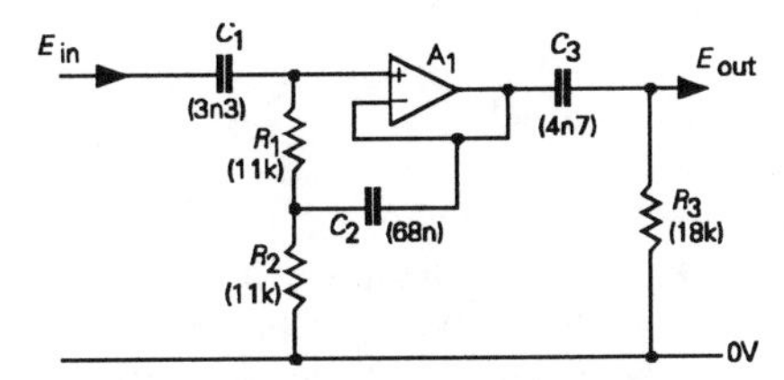

Figure 8.48 *High-pass bootstrap filter (f_t = 1kHz)*

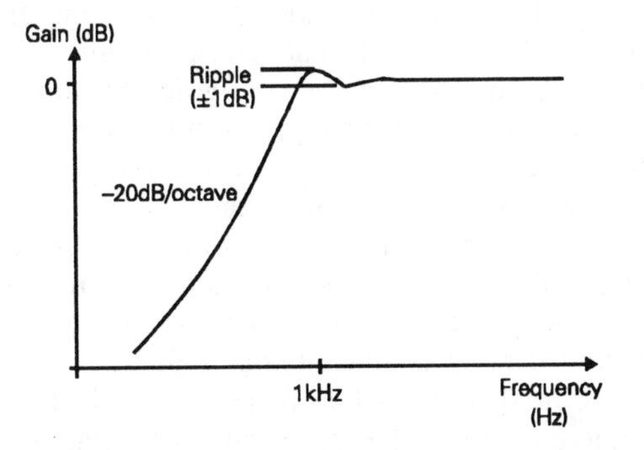

Figure 8.49 *Frequency response of circuit of Figure 8.48*

The turn-over frequencies of both the Sallen and Key and Bootstrap designs are given by the equation

$$f_t = 1/(2\pi\sqrt{R_1R_2C_1C_2})$$

though due allowance should be made for possible inaccuracies in component values.

The specific component values for the bootstrap filter can be calculated, in the case of a third-order low-pass filter, from the relationships:

$$Q = \sqrt{xy}/(1+y) \quad \text{where } y = C_2/C_1 \text{ and } x = R_1/R_2$$

$$C_1\ (\mathrm{F}) = 1/(2\pi f_t R_2 Q(1+y))$$

$$R_1 = R_2 Q^2 (1+y)^2/y$$

$$C_3\ (\mathrm{F}) = Q/(2\pi f_t R_3)$$

In the case of a third-order high-pass filter, the values are given by

$$Q = \sqrt{xy}/(1+y) \quad \text{where } y = R_1/R_2 \text{ and } x = C_2/C_1$$

$$C_1\ (\mathrm{F}) = 1/[2\pi f_t R_2 Q(1+y)]$$

$$C_2 = C_1 Q^2 (1+y)^2/y$$

$$C_3 = 1/(2\pi f_t R_3)Q$$

In practice, the calculations can be simplified by making the ratio y equal to unity, which is satisfactory for general use. For a typical Chebyshev response, with a ripple between +/–1dB and +/–3dB, Q values between 1.8 and 2.2 should be chosen.

Although I have shown most of the preceding filter designs with component values chosen to give a 1kHz turn-over point, they can be configured for different frequency values by simple proportional adjustments to the values of the resistors and capacitors.

The filter designs shown, so far, have allowed the generation of cut-off slopes up to some –20dB/octave. In applications where a steeper attenuation rate is needed the simplest solution is merely to connect two or more filter blocks, tuned to the same frequency, in series. Care should be taken, depending on the type of filter, to ensure that the pass-band ripple is not made unacceptably high by this process – perhaps by combining circuits of Butterworth and Chebyshev characteristics – and that the output and input impedances of the succeeding stages are compatible.

A further possibility which can be of interest in sharpening up the cut-off slope of a simple filter system is by combining it with a notch filter – especially if the whole arrangement is included within the negative feedback loop to flatten out the frequency response of the system within the required pass-band. The cut-off response curve given by this type of arrangement will generally be of the Cauer type, with a pronounced ripple in the transmission characteristic beyond the cut-off point. This type of filter system is examined below in the section covering notch filters.

Variable slope filters

The cut-off slopes given by the simple *CR* filter types of Figures 8.1c and 8.1d can be lessened by reducing the value of the resistor, R_c, across the capacitor or inductor, and this approach has been used in some commercial Hi-Fi circuitry where the network is used as a treble hiss filter.

A rather more elegant option is as shown in the case of the low-pass and high-pass bootstrap filters shown in Figures 8.50 and 8.51, where if a variable resistor, RV_1, is inserted in the return paths of C_3 or R_3, as shown, the slope and characteristics of the circuits can be adjusted to provide frequency response curves which approximate to either Chebyshev, Butterworth or Bessel, as needed. This can be advantageous in audio circuitry since all filters, having a cut-off slope beyond some –6dB/octave, will introduce some degree of tonal coloration, particularly noticeable with low-pass filter systems in the mid- to treble band, where the peak in the voltage response as a function of time – illustrated in Figure 8.52 for a square-wave input signal – characteristic of all steep filter cut-off rates, will tend to concentrate wide-band signal energy (e.g. that due to broad-band signal or noise components) in the region of the cut-off frequency, thereby giving it a degree of tonality.

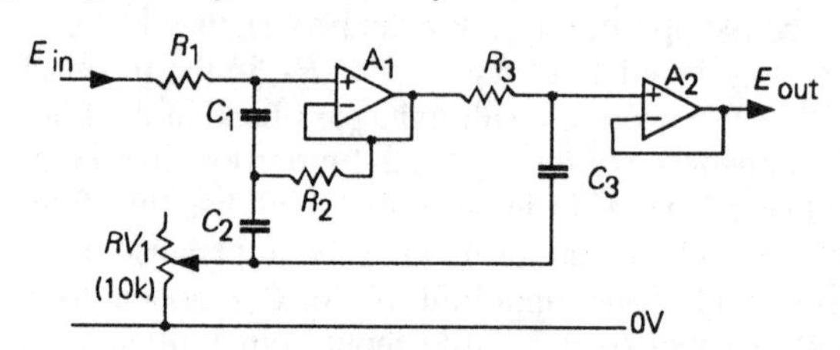

Figure 8.50 *Variable cut-off slope LP filter*

Notch filters

Two types of *CR* network are commonly used to

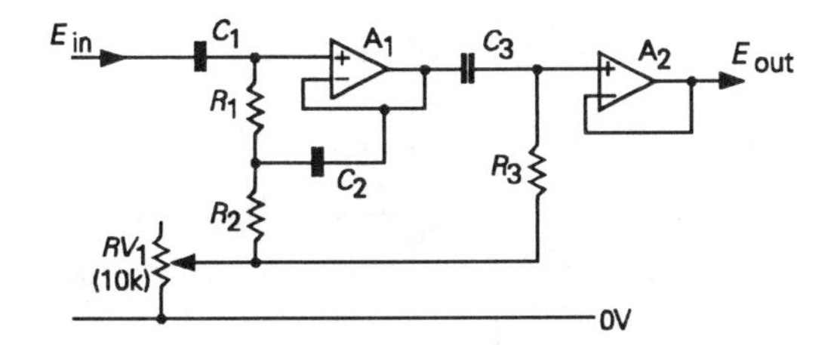

Figure 8.51 *Variable cut-off slope HP filter*

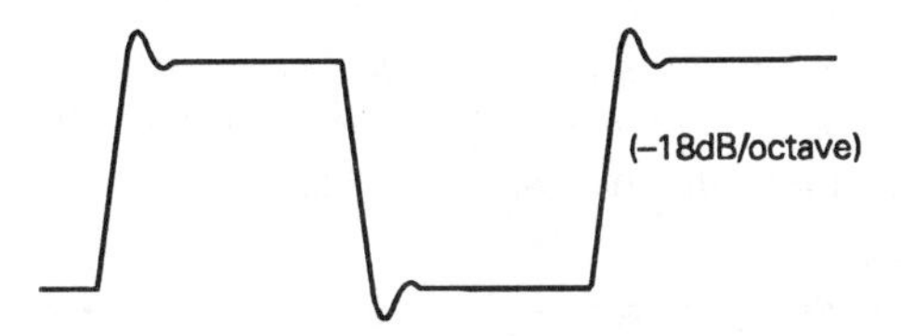

Figure 8.52 *Influence of filter slope on square-wave waveform*

generate a null output at some chosen frequency: the parallel-T, also known as the twin-T, and shown in Figure 8.19, and the Wien network, often called the Wien Bridge, shown in Figure 8.53.

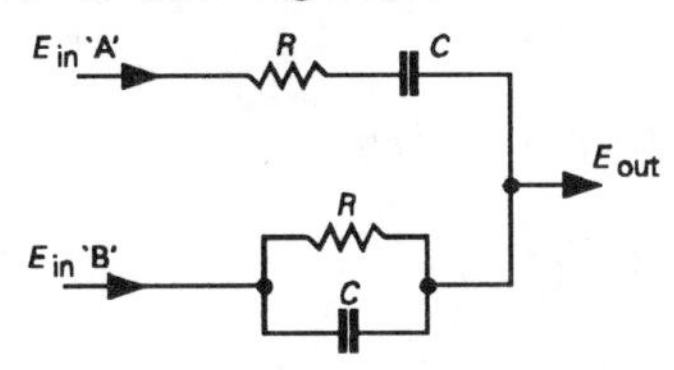

Figure 8.53 *Wien network*

Considering the Wien circuit first, the impedance of the upper limb will be nearly infinite at some very low frequency, but will decrease with increasing frequency as the impedance of C decreases, until, at some very high frequency, the impedance of this limb will approach the value of R. However, the lower limb has an impedance at VLF equal to that of R, and it will continuously decrease with increasing frequency as the impedance of C falls, asymptoting to zero. At some intermediate frequency, at which the impedance of C is equal to R, the impedance of the upper limb will be twice that of the lower, but the phase shift in each limb will be the same. This means that if this network is fed with signal inputs at A and B, and if these two inputs are in antiphase, and it is arranged that the input at A is exactly twice the magnitude of that at B, then the two signal currents at X will exactly cancel at that frequency at which the impedance of C is equal to R, or, $f_o = 1/(2\pi CR)$.

If an inverting amplifier having a gain of two is interposed, as shown in Figure 8.54, to provide the required antiphase upper limb input, the attenuation characteristics of the circuit, as a function of frequency, is as shown in Figure 8.55. With the component values indicated, f_o is approximately 1kHz (actually 994.7Hz).

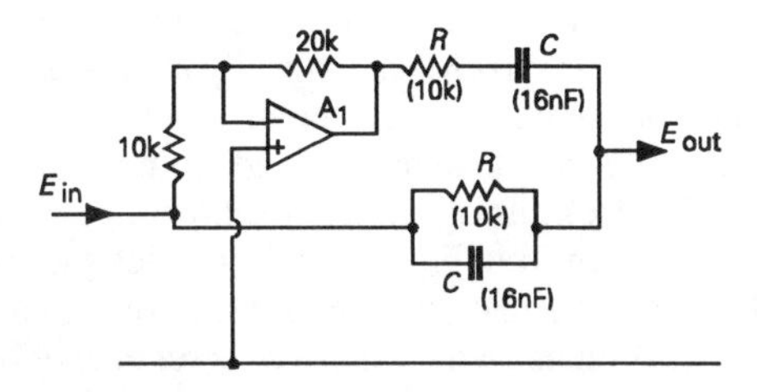

Figure 8.54 *Null network using Wien bridge*

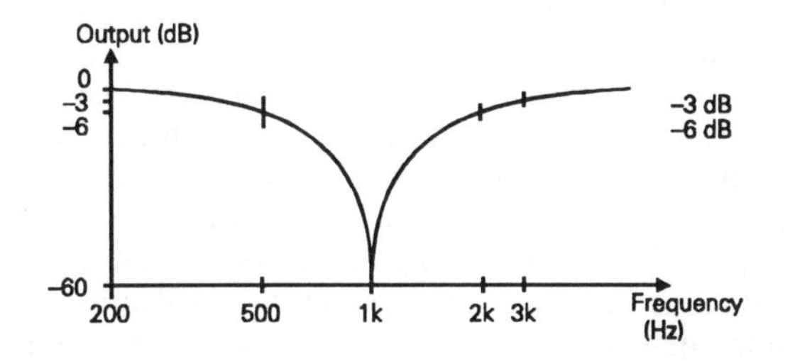

Figure 8.55 *Performance of circuit of Figure 8.54*

In the case of the parallel-T circuit, its method of operation can best be explained by considering the attenuation and phase shift given by each of the two limbs. In the *RCR* limb shown in Figure 8.56, the phase shift of the output voltage in relation to that of the input is –63.44° at the frequency at which the phase shift of the *CRC* limb, shown in Figure 8.57, is +63.44°. Since both networks will give the same attenuation at this frequency (994.7Hz for the values suggested), their combined outputs will cancel – which means that the circuit will give zero output at this frequency.

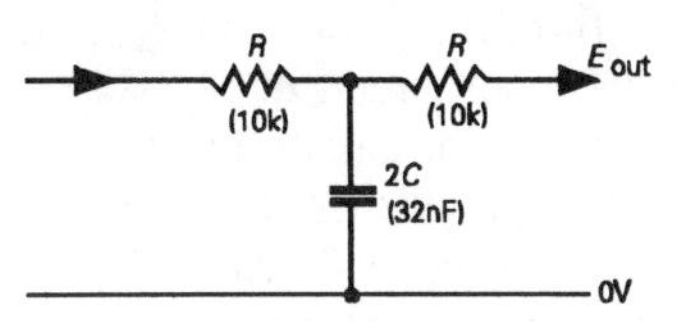

Figure 8.56

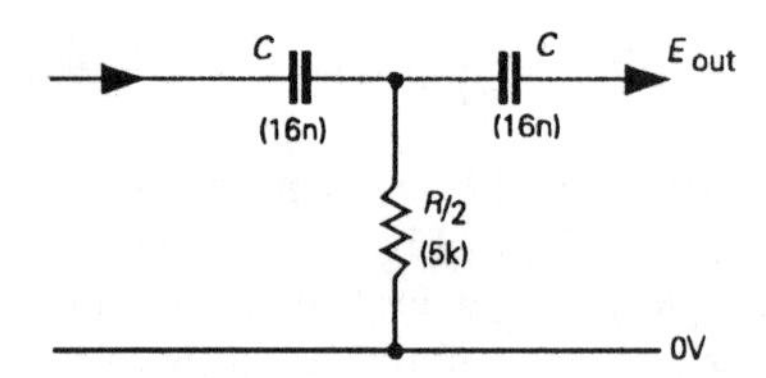

Figure 8.57

The use of a gain block to sharpen up the notch given by a parallel-T filter was mentioned earlier, and the circuit layout was shown schematically in Figure 8.19. The effect on the notch characteristics of the network of a voltage feedback signal derived from the output of the gain block, using the circuit layout of Figure 8.20, can be seen from Figure 8.19, in which the solid line shows the attenuation characteristics of the parallel-T network on its own – when RV_1 is set to give zero feedback – and the dashed line shows the effect with RV_1 set to give maximum feedback. Intermediate settings of RV_1 will give intermediate degrees of notch sharpness.

However, the same type of approach can also be used to sharpen up the characteristics of other circuits of this type, such as the Wien network shown in Figure 8.58, as R_a,C_a/R_b,C_b – where NFB applied around the loop can be used to modify the response curve from that shown in the solid line of Figure 8.59 to that shown in the dashed line. This kind of layout, because the actual notch frequency can be tuned by adjustment of C_a and C_b, which may, conveniently, be a pair of air-spaced dual gang capacitors, provides a useful basis for a 'total harmonic distortion' type of distortion meter, in which the residues of a notionally pure sinewave signal, after the fundamental frequency has been notched out, are presumed to be the harmonic distortion components together with any circuit noise and hum.

In the circuit of Figure 8.58, the input buffer amplifier, A_1, is used to provide a high input impedance, a low drive impedance for the Wien network R_5/C_1 and R_6/C_2, and also to provide a means by which NFB can be used to sharpen up the notch – by way of the feedback path from A_3 output, via R_7, to the resistor in the inverting input circuit of A_1. A_2 provides the required phase inversion and the two times increase in the signal voltage applied to R_5/C_1, with respect to that applied to R_6/C_2, needed to cause the network to provide zero transmission at the frequency determined by

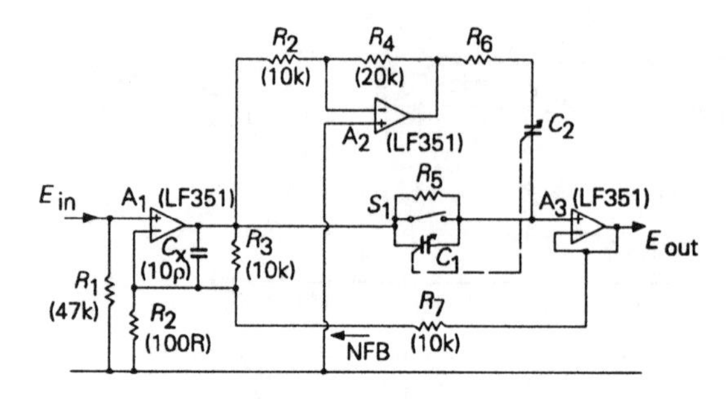

Figure 8.58 *Total harmonic distortion meter circuit based on Wien bridge network using NFB to sharpen up notch response of circuit*

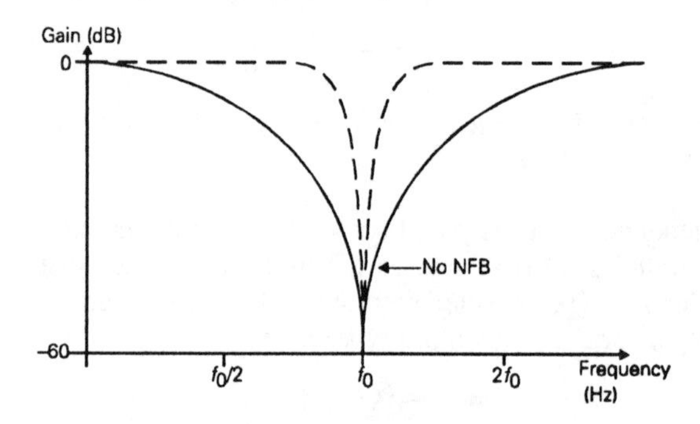

Figure 8.59 *Effect of NFB*

$$f_o = 1/(2\pi CR)$$

where the values of R and C are R_5=R_6 and C_1=C_2. The need to sharpen up the notch arises from the need to ensure that the null network will pass the second and higher harmonics (2kHz, 3kHz, 4kHz etc.) at the same magnitude as the un-notched 1kHz fundamental, of which the value is determined by closing the switch S_1.

Clearly the simple circuit of Figure 8.54; which gives the solid line response curve shown in Figure 8.59; would not satisfy this requirement. On the other hand, too tight a notch would be awkward in use, because of the great delicacy which would be needed to tune the circuit precisely to the null frequency. In the circuit example shown the amount of NFB is determined by the values of R_3 and R_7 in relation to R_2.

Examples of published total harmonic distortion meter circuits are given in Chapter 17.

A notch filter can often be added to an existing active low-pass or high-pass filter circuit, and if the notch frequency is placed appropriately in frequency, and NFB is applied over the whole loop, the superimposition of the two curves can give a very high initial cut-off rate. An example of a circuit in which a parallel T notch filter is added to a Sallen and Key low-pass filter, is shown in Figure 8.60, for which the overall frequency response is shown – once again normalized to 1kHz – in Figure 8.61.

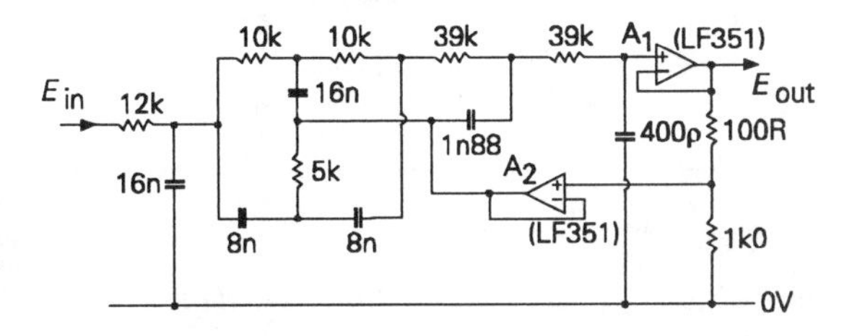

Figure 8.60 *Combination of Sallen and Key and parallel-T notch filter*

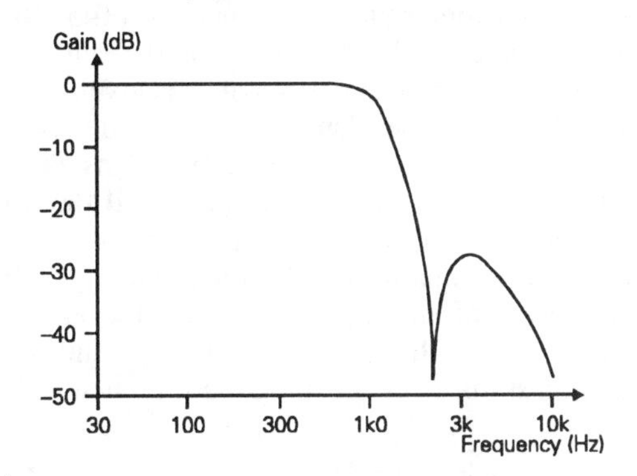

Figure 8.61 *Frequency response of circuit of Figure 8.60*

A similar layout to provide a steep cut-off high-pass filter is shown in Figure 8.62, for which the frequency response is shown in Figure 8.63.

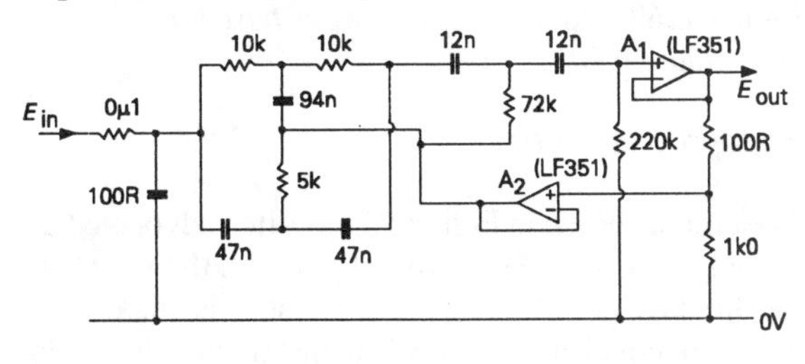

Figure 8.62 *High-pass filter using combination of Sallen and Key and parallel-T filter*

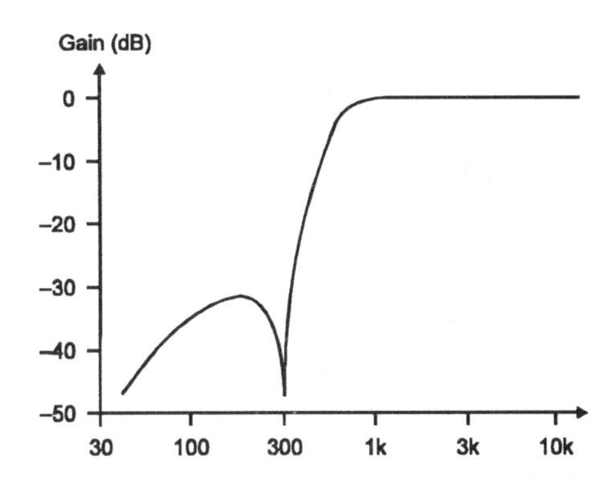

Figure 8.63 *Frequency response of circuit of Figure 8.62*

The same kind of combination can be used with a bootstrap filter to give very high initial attenuation rates – approaching –40dB/octave – and with a skirt response which does not rise above –40dB at frequencies beyond the notch. Two circuit examples are shown, with component values chosen for f_t values of 1kHz, in Figures 8.64 and 8.65. The transmission graphs shown in Figures 8.66 and 8.67, show the performance of the low-pass and high-pass designs respectively. The layout of Figure 8.65 would make a splendid Hi-Fi rumble filter using the amended capacitor values shown in brackets, as appropriate for a 30Hz turn-over frequency, with an output below –40dB for signals of 14Hz and below.

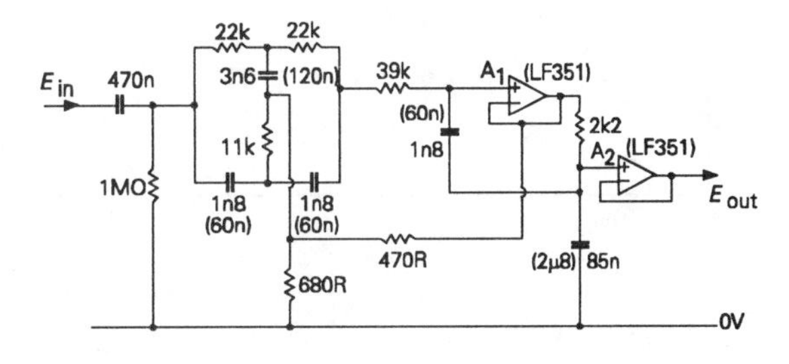

Figure 8.64 *Bootstrap and notch filter combination high-pass. Values in brackets for f_t = 30 Hz*

Tuned response filters

The use of a parallel-T notch filter in the negative feedback path of an amplifier has been shown earlier, in Figure 8.23, as a means of generating an increase in

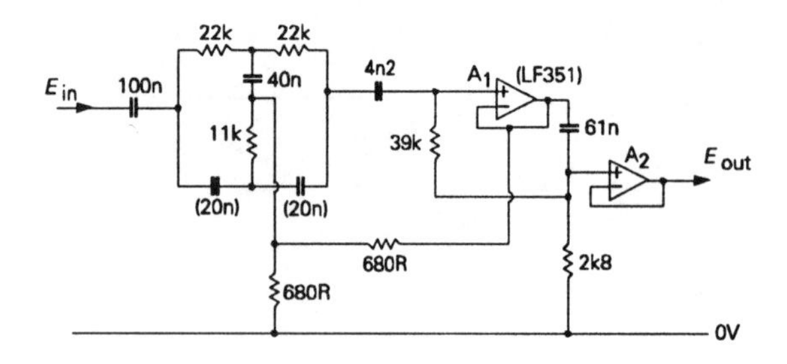

Figure 8.65 *Bootstrap and notch filter combination (f_t = 1 kHz)*

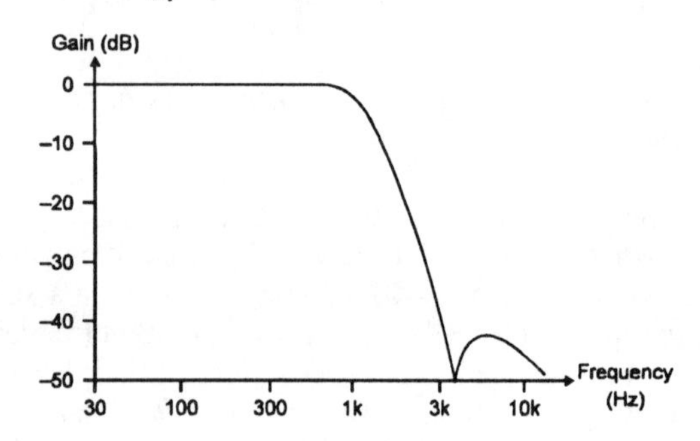

Figure 8.66

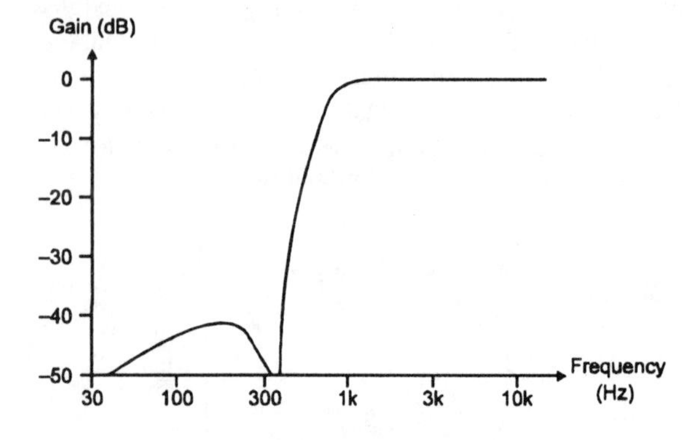

Figure 8.67

gain at some specific operating frequency. The same effect can be achieved by the use of a Wien network if it is connected around a gain block in the same manner as in a Wien bridge oscillator, but with the amount of NFB adjusted so that the circuit just fails to oscillate. A suitable circuit arrangement is shown in Figure 8.68, in which C_1R_1/C_2R_2 are the frequency determining components and RV_1 is used to set the circuit to just below the onset of continuous oscillation. The input signal can be injected into the inverting input of the op. amp. type gain block, using the virtual earth characteristics of this point to isolate the input signal from the feedback voltage.

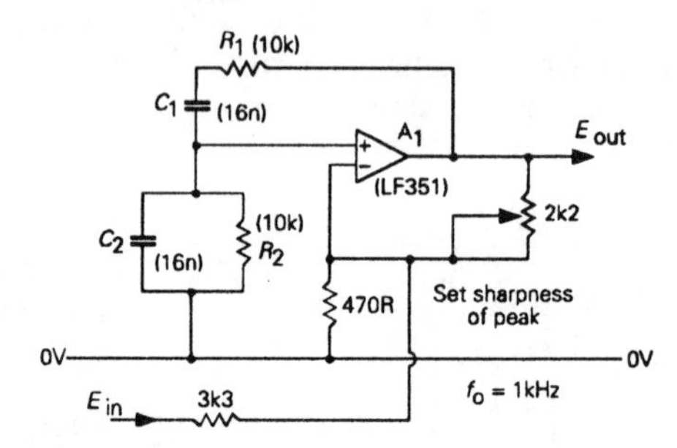

Figure 8.68 *Tuned response filter based on Wien bridge*

While this layout can give a very sharply tuned response, it suffers from the drawback that the actual circuit gain is critically dependent on the setting of RV_1. A more predictable performance is given by the use of a pair of under-damped second-order active filters in cascade, as shown in Figure 8.69, as a practical implementation of this arrangement. If the Q of each of these is set to give a response peak of, say, +6db, and the component values are chosen so that both filters operate at the same frequency, the outcome will be a circuit with a gain of +12dB, at f_o, and steep attenuation skirts on either side of this frequency.

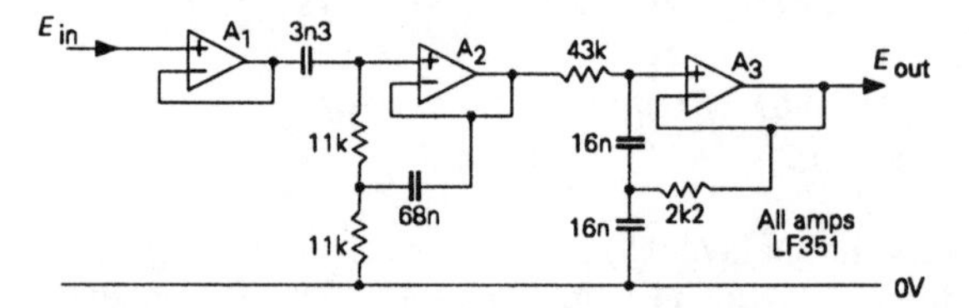

Figure 8.69 *Tuned response filter based on cascaded bootstrap filters*

Bandpass circuitry

The circuit shown in Figure 8.26 is often advocated as a bandpass layout. However, in practice, the very low (6dB/octave) skirt slopes mean that it is seldom of much practical use. A much better approach would seem to be to arrange a high-pass and a low-pass pair of suitable active filters in series, with op. amp. buffering in between them if necessary, to isolate their functions, as shown schematically in Figure 8.70.

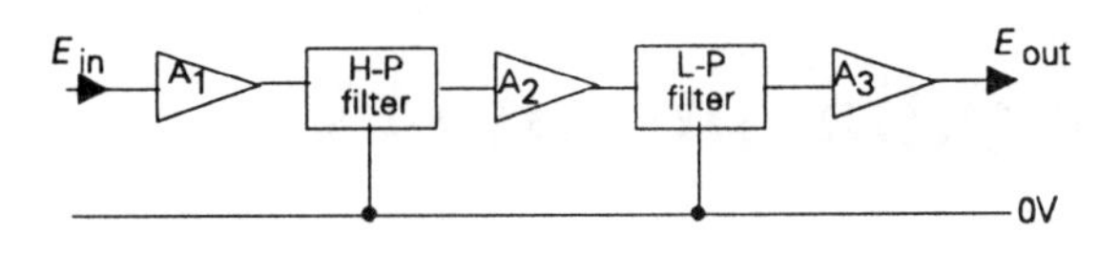

Figure 8.70

Effect of filter slope on waveform shape

One of the characteristics of any filter which modifies the frequency response of a system is that it will cause a change in the shape of a step function, or square wave input waveform, and if the slope of a HP or LP filter is greater than –6dB/octave, this effect will be particularly marked. This is an effect which is particularly noticeable in the case of low-pass filters, where the result will be a pronounced ringing on the leading and trailing edges of the waveform, as illustrated in Figure 8.52. This waveform distortion results from a combination of effects – of which the alteration in the relative phase of the individual signal components which add together to make the square waveform; the removal of the higher frequency components; and the exaggeration of those frequency components lying close to the turn-over point due to any peaks in the transmission at this frequency, are the most important.

With a wide bandwidth input signal, for example white noise, the effect of a steep filter attenuation rate can be to modify the energy distribution of the input signal, and cause an increase in relative signal energy at or near the turn-over frequency. An additional point to remember is that alterations in the frequency response of any system will cause an associated change in the relative phase of the components of that signal within one decade of the turn-over frequency, and the steeper the filter slope, the greater this effect will be.

9

Audio amplifiers

The evolution of power amplifier design

Basic problems

The design of electronic equipment for use in audio applications presents some unique problems to the circuit engineer, not least because the final result is intended to appeal to the human ear. For this reason, the relevance of specifications based purely on engineering considerations will always be a matter for some debate, and although some technical standards have been adopted because they were supported by the results of deliberate listening trials, held with audience panels composed of engineers or musicians, they exist mainly either because they are generally in accord with experience, or just because they seem to be sensible.

The contemporary preoccupation with the pursuit of perfection in both the electronics circuitry and the other hardware used in audio systems is a relatively recent phenomenon, and has come about because of the development of very high quality sound storage and reproduction systems, which are themselves mainly of fairly recent origin. In earlier times other, not usually electronic, considerations were thought to be of greater importance, if only because the possible quality of the results given by the other items in the recording and reproduction chain was not very high anyway.

Obviously, since the domestic audio equipment market has always been a very competitive one, the expectations of the user, and the apparent value for money offered by the products, have been very important considerations in the eyes of the manufacturers. However, the dominant design factor in audio amplifiers, is, and has always been, the nature of the hardware which was available at the time. For this reason, the main flow of audio design has nearly always followed the evolution of electronic circuit components and the electromechanical transducers, such as loudspeakers or gramophone pick-ups, which are used in association with it. Even the nature of the circuitry employed has usually been dictated by the advantages and limitations of the components, and the needs, in respect of input signal voltage or output power demands, of the input/output transducers.

Performance specifications

Output power and bandwidth

The levels of output power which will be needed, from any audio amplifier system, depend so heavily on the use to which it is to be put that it is difficult to suggest more than tentative figures for this part of the specification. For equipment which is to be used only in a domestic environment, with reasonably efficient loudspeaker units – with, say, input/output efficiencies above 90dB, s.p.l. (sound pressure level, per watt, at one metre distance, axially, from the speaker) – and with typical 'easy listening' or light classical music, it is unlikely that more power than 3 watts per stereo channel will ever be required, but if LS units having similar efficiencies are to be used, on the same type of music, for lecture purposes, even in only a medium sized hall, then up to 30 watts of power will probably be needed.

For the 'hard rock' or 'heavy metal' music enthusiast, with less efficient loudspeakers – say, less than 85 dB, s.p.l. per watt/metre – and very tolerant neighbours, then something over 100 watts per channel will certainly be necessary. Still larger powers would be needed to reproduce, in a large room, a realistic sound level equivalent to that of a grand piano played with panache, even with reasonably efficient (85–90dB s.p.l.) LS units.

It is useful, also, to remember that, under lecture conditions, because of the sound deadening effect of furnishings, or simply because of the presence of a number of people in an audience, the effective sound level reaching the listener will be significantly reduced, by comparison with that, in the same room, when empty. So, taking these factors into account, one might divide these various uses into groups, such as 'average domestic' – 3–10 watts. 'general purpose' – 10–30 watts. 'pop music' – 60–120 watts, and 'studio/lecture' – 80–500+ watts.

It is generally assumed, as a basis for the necessary bandwidth for audio systems, that the human ear can respond to sound frequencies over the range 20Hz–20kHz. However, in reality, very few adult men, even when young, can hear much above 16kHz, and this upper limit of frequency response will decrease gradually with increasing age. Also, very few women can hear much below about 35–40Hz. Children may hear sounds at frequencies well above 20kHz, but not usually those much below 100–150Hz. Again, few LS units, and very few listening rooms, will allow sound pressure waves to be reproduced at frequencies below 35Hz.

As for the programme sources, FM radio is restricted to an upper limit of 15kHz, though frequencies down to 30Hz can be transmitted. With a few exceptions, AM radio transmitters do not, nowadays, offer signals above 5kHz, because of international frequency allocation agreements. Vinyl discs, and some cassette tape machines, can reproduce frequencies up to 20kHz, though only at low levels, and with substantial (10% or more) waveform distortion. Similar constraints exist below 35Hz, with both tape and vinyl disks. Compact disks do, however, offer a true <20Hz–20kHz low distortion programme source.

A point which must be kept in mind, in any consideration of the bandwidth specification for an audio amplifier, is that too wide an extension of the HF or LF bandwidth can allow the intrusion of acoustically unpleasant and unwanted noises into the final sound. An unexpectedly high background hiss level on an FM signal could easily be due to this cause. This problem can arise because the inevitable HF or LF nonlinearities, and other mechanical imperfections, in the LS units or headphones used to reproduce the signal, may allow audible cross- modulation effects between signals which are themselves outside the audio band, and would normally have been quite inaudible. So, for this reason, it is sensible, in amplifier design, not to seek to preserve, purely because it looks good on the specification sheet, a high frequency response which extends too far beyond the possible frequency range of the ears of any listener. Having said that, care should also be exercised in the way in which the bandwidth is curtailed, since too steep a low-pass filter characteristic can itself impair the sound quality of the system, by modifying its transient response.

Waveform distortion

Most sounds produce air pressure waveforms which are irregular, non-sinusoidal and asymmetrical in form, especially those which are caused by percussive actions. At first glance, it would seem obvious that realism, in the reproduction of these sounds, depends on the ability of the system to generate, at the ear of the listener, air pressure patterns which are closely similar to those of the original sound. However, in a normal, medium sized, listening room, having various kinds of 'cavity resonances', filled with sound-absorbing furniture, and with reflecting walls, floors and ceilings, this is an impossible task. This can easily be demonstrated, in any listening room, by connecting a microphone to an oscilloscope and then watching the large changes in the reproduced waveform which occur as a result of relatively small movements in the microphone position, even when it is exposed to exactly the same audio signal.

So, the fact that the ear can recognize, and regard as realistic, such a wide range of possible acoustic waveforms, which can change, literally from inch to inch, with movement of the microphone or the ear, is as much a tribute to the sound recognition ability of the brain as it is to the precision of the reproducing equipment in use.

However, though the ear may be very tolerant of errors in the flatness of frequency response of the system, or the time of arrival of the individual frequency components which go to make up a complex signal, there are various types of waveform distortion which can be shown to degrade the sound quality, and one of the tasks of the audio engineer is to make all of these distortions imperceptibly small.

Under steady-state conditions, the waveform or harmonic distortion of an amplifier, using a constant amplitude single frequency (sinusoidal) input signal, can be measured quite easily, and this gives a convenient and widely used method of assessing one particular aspect of amplifier quality. Unfortunately, mainly due to the claims of the sales departments of the amplifier manufacturers and their advertising agencies, who are aware that a large number of '0's behind the decimal point in the total harmonic distortion figure looks good on paper, this part of the overall specification has assumed an excessive degree of importance in the minds of many potential purchasers of audio equipment.

It is certainly true that there are some very objectionable forms of harmonic distortion, such as the dissonant high-order odd harmonics, such as those at the 7th, 9th, 11th and upwards, which can be generated by 'crossover' effects in class B push–pull amplifiers; and it is desirable that these should be kept below 0.05%, at all signal levels, if their presence is not to be detectable. However, most musical signals are rich in naturally occurring harmonics, mostly of the lower harmonic orders, such as the 2nd, 3rd, and 4th, and the presence of these harmonics adds richness to the quality of the sound. Also, in the sounds produced by many of the stringed instruments, there are a whole range of overtones, of which the higher ones are not strictly in harmonic relationship to the fundamental tone. Musicians refer to these overtones as 'partials', and the size, and frequency distribution, of these, for any given note, determines the sound quality of the instrument, and in violins, for example, would help to distinguish a Stradivarius from a Joe Bloggs.

It is not, therefore, the presence of low order harmonic distortion which is undesirable in an audio amplifier – indeed, at levels below, say, 0.5% such harmonics are probably undetectable – but the intermodulation (IM) distortion which such nonlinearities in the transfer characteristics will cause. This kind of distortion leads to the generation of spurious composite signals, as a result of the mixing of the various components of any input signal, as ilustrated in Figure 9.1, and causes confusion and a lack of clarity or 'transparency' in the final sound.

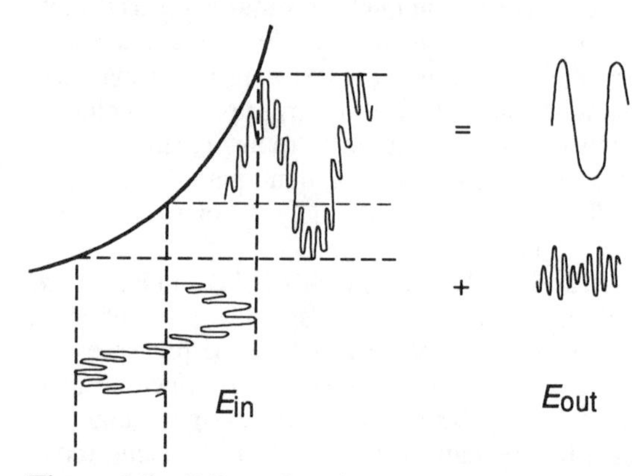

Figure 9.1 *Effect of nonlinear transfer characteristics in generating intermodulation distortion*

There is not a direct and simple relationship between harmonic and intermodulation distortions, but, as a general rule, the lower the harmonic distortion, the lower the IM distortion will be, and the more transparent the sound quality will appear. Once again, where the waveform distortion (THD) figure is below 0.05%, IM distortion will usually be negligible, and no further improvement in THD is likely to give audibly detectable benefits.

Transient distortions

No international standard has yet been adopted for the measurement, or the specification of the accuracy, of the transient response of an amplifier system, apart from the 'settling time' measurement which is sometimes quoted in high quality operational amplifiers, which is illustrated in Figure 9.2. Because no specification exists, it often seems that little attention is paid in commercial audio amplifier design to the nature of the response of the circuit to an input step-function or square-wave signal, except, perhaps, with a purely resistive output load which is quite unrepresentative of real-life operating conditions. This is a matter for regret since, now that circuit technology allows the design of amplifier circuits having a very high degree of steady-state linearity, it is probable that the residual errors in the transient response of the equipment – which can be quite large – are responsible for most of the audible differences between one otherwise high quality unit and another.

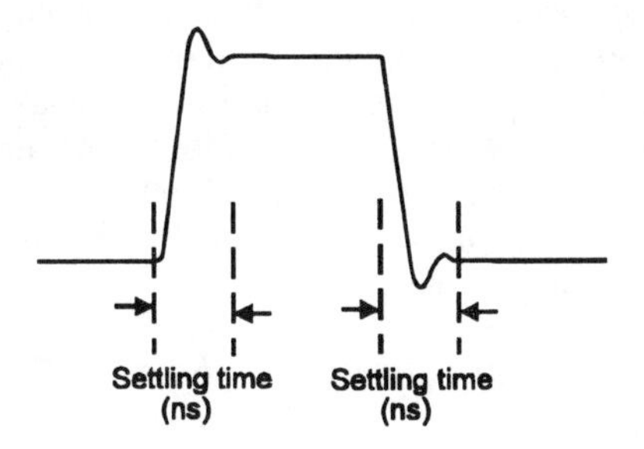

Figure 9.2 *Illustration of the settling time specification in operational amplifiers*

There are several mechanisms by which transient response errors can modify the sound of an audio system. Of these, the most common is the effect due to the presence of an 'overshoot' following the leading edge of a step-function, as shown in Figure 9.3a. This will lead to a frequency response of the type shown in Figure 9.3b, when the system is fed with a swept frequency square-wave input signal, or any other signal source of wideband characteristics. One effect of this is to redistribute the spectral energy of such a wideband signal, to give a higher output at the point of the upward kink in the frequency response, which appears to over-emphasize high frequency components in the signal and lead to a 'hard' or 'bright' quality to the sound.

A further cause of poor sound quality, as a result of transient malfunction, is that where the poor transient response is due to inadequate HF stability margins in the amplifier feedback loop, especially under highly reactive LS load conditions. This can also modify the frequency distribution of a wideband signal, to cause energy peaks at those frequencies associated with poor loop stability.

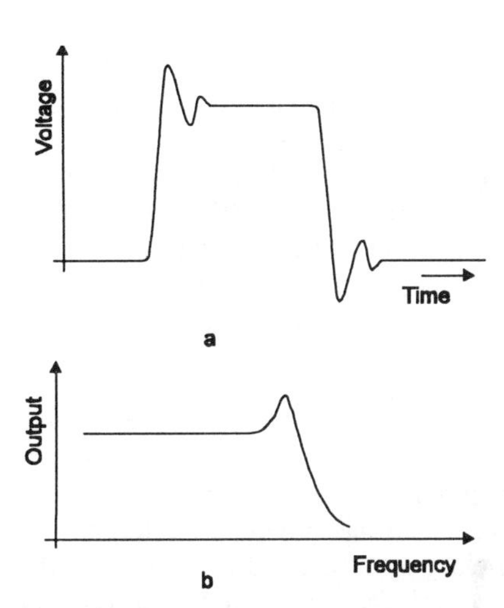

Figuer 9.3 *Relationship between transient behaviour and output voltage on wideband signal*

Many of these problems in transient response have arisen because designers have endeavoured to produce electronic power amplifier circuits with unnecessarily low levels of harmonic distortion, in order to satisfy a presumed customer demand, promoted by advertising hyperbole, and have used over complex, or inadequately stable, feedback loops in the pursuit of this end.

Load tolerance

Most audio amplifier systems will be required to operate into a loudspeaker load, and this can often present problems in matching. This difficulty arises partly because, in order to achieve the flattest practicable frequency response, the loudspeaker designers will usually employ multiple driver units, with complex inter-unit crossover networks – which can offer very low impedances at certain parts of the audio frequency pass-band – and partly because the LS unit has dynamic characteristics of its own. Because of this, in addition to the inductive or capacitative nature of the load which it presents, the LS load can provoke unexpected responses in the amplifier driving it. Unless the designer of the amplifier can be very certain of the nature of the load with which his design will be used, it is prudent for him to assume worst-case conditions, since an unsatisfactory combination of amplifier output and load characteristics may significantly worsen the performance that the amplifier might otherwise be expected to offer.

In spite of these residual problems, it is undoubtedly true that, with a few exceptions, most of the amplifier circuits currently offered as Hi-Fi units, in the mid- to upper price range, are capable of a vastly better performance, both in respect of technical specification and in terms of audible results, than all but the very best of their predecessors of some decades ago, if only because these earlier designs were aimed to satisfy much more modest specifications, were based on a more limited range of circuit components, and were evolved at a time of a more limited understanding of the field. Since there is a tendency for nostalgic feelings towards early audio amplifying equipment, which is supposed to have been capable of a degree of sonic perfection lost in the progress of circuit design, it may be instructive to look at the actual types of circuit used in the earlier days of 'audio' and at the technical standards which were attained. It should be remembered that when electronic devices were first employed to amplify audio signals it was almost sufficient that the equipment worked at all, and relatively little attention was directed towards the actual quality of its performance, so long as it was not so bad as to be actually objectionable to the user.

Early amplifier designs

Simple valve designs

In the early years of 'radio' most amplifier systems were operated from primary cells (dry batteries) as the principal source of power, at least so far as their high tension (HT) DC supplies were concerned, and since these were an expensive source of energy, economy in the use of power was a major consideration. Indeed, Briggs, in his classic book on loudspeakers (Briggs, G. A. and Cooke, R. E., *Loudspeakers*, 5th Edition (1958). pp. 11–13. (Wharfedale Wireless Works, Bradford, Yorks)), says that the success of one of his early Wharfedale LS driver designs was almost entirely due to its high sensitivity.

Similarly, the most common input transducer element, the gramophone record 'pick up', was almost invariably of piezo-electric type, in which a sandwich of foil electrodes and a piezo-electric crystal (normally Rochelle salt) was arranged so that the lateral displacement of the stylus, as it followed the undulations of the gramophone record groove, would cause a twisting action of the crystal sandwich, which would produce an audio voltage output. Because of the high intrinsic capacitance of the crystal transducer element, and the high load resistance which would normally be used with it, the high frequency response of this type of pick-up was very poor, and because of its stiffness its waveform distortion was very high.

However, in the same way that the principal quality sought, at that time, in the LS unit was that it should produce a high sound level for a modest electrical input, with wide bandwidth or flatness of frequency response being minor considerations, so, in the case of the gramophone pick-up, a high electrical output was one of the major requirements, in order to reduce the number of stages of aplification which would be required to achieve an adequate output signal level. By comparison, the relatively poor high frequency response or linearity of such crystal pick-ups was of secondary importance.

A typical battery operated audio amplifier

A typical battery operated audio amplifier for gramophone record reproduction, of this period, would be of the form shown in Figure 9.4. The anticipated output of the gramophone pick-up would be of the order of 0.5–1V RMS at 1kHz, with a 78 RPM shellac record of average 'loudness'. This would be large enough not to require a high degree of subsequent amplification, though somewhat too small to drive the output stage (V_2) directly. An input stage, V_1, would therefore be added to provide some additional amplification of the signal.

It is necessary, in almost all thermionic valves, to apply some negative bias voltage to the control grid to set the anode current to the required level, but since battery operated valves use a directly heated, barium oxide coated, filament as the electron source, and all of these filaments are connected in parallel, it is not convenient to develop the required negative grid bias voltage by just putting a resistor in the cathode circuit, as is done with valves having indirectly heated cathodes.

For the first stage, which had a high value of input resistor (R_1), the residual grid current flow through this resistor would usually provide an adequate negative

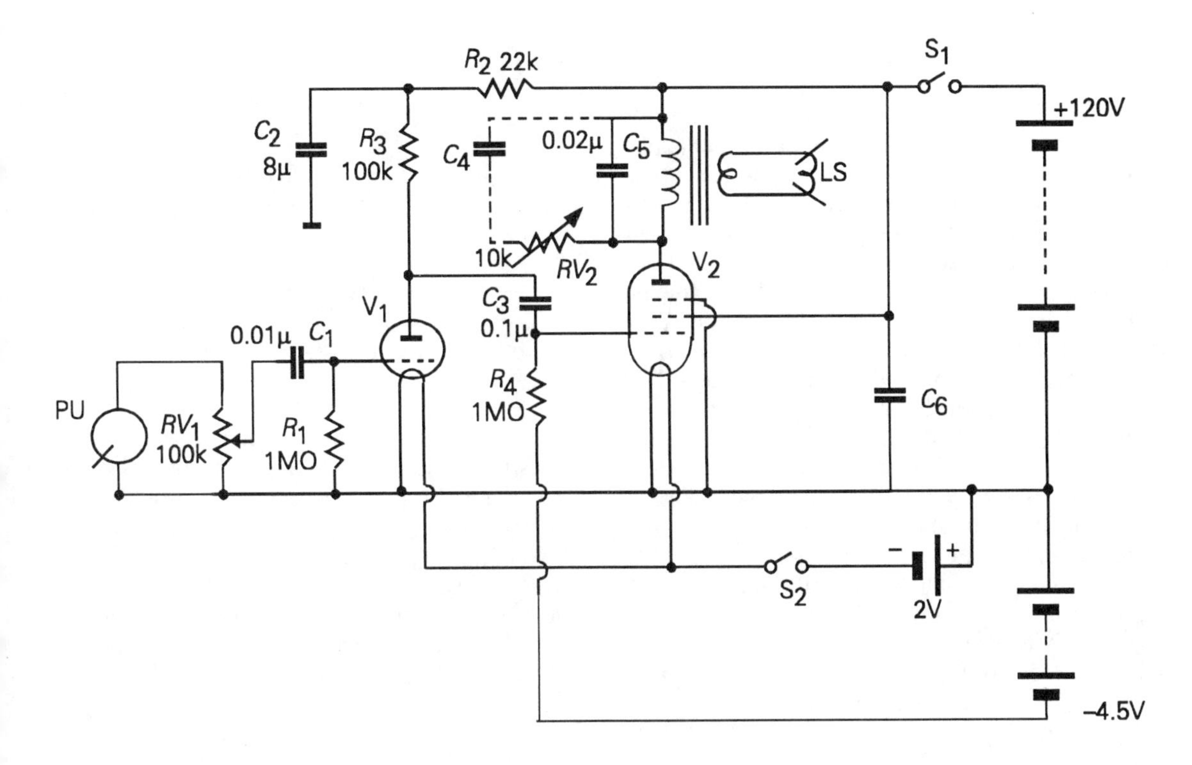

Figure 9.4 *Battery operated two-valve gramophone amplifier*

bias voltage to ensure that V_1 would operate in a reasonably linear part of its I_a/V_g characteristic. This technique is known as grid current biasing. Ideally, the grid bias voltage should be chosen to ensure that the anode potential was about half the available supply voltage, which is the preferred operating condition. However, since the available HT voltage will be between 90V and 120V, and the required output voltage swing from this stage would only be some 5V RMS, the precise anode voltage level would be far from critical. Although this type of biasing technique is simple to use, it will cause the larger positive-going peaks of the input signal to clip, through the diode action of the grid–cathode circuit, and this will lead to some worsening of the circuit distortion characteristics.

For the output stage, a pentode valve would normally be used in the interests of efficiency, and the anode current of this would normally be controlled by an externally applied negative bias voltage, typically of the order of 3V–4.5V, derived from a separate grid bias battery. These batteries would offer a range of output voltages between 1.5 and 9V so that the user could make his own choice between economy and harmonic distortion.

Although the use of a pentode output valve was preferred because of its higher efficiency, such valves also introduced a large measure of third-harmonic distortion, which gave a somewhat shrill quality to the sound of the amplifier. In order to make the sound somewhat more bland it was customary to connect a small capacitor (C_5) across the primary winding of the LS transformer.

Sometimes a simple 'tone control' circuit, consisting of a variable resistor and a rather larger capacitor (C_4/RV_2) would also be provided to allow the user, by adjusting the resistor value, to opt for a progressively more 'mellow' sound characteristic. Such an amplifier, for a consumption of some 10–15mA from the HT source, would give an output of 200–300mW, with an overall THD figure of about 10–20%, mainly composed of third harmonics, and with a bandwidth, at –3dB points, from gramophone record input to LS output, of some 150–2500Hz.

Mains operated circuitry

The desire for greater amplifier output powers, and for freedom from the cost and performance limitations imposed by battery sourced power supplies, led to the growing adoption of valves with indirectly heated cathodes. These were of a type in which the heated filament was electrically isolated from the cathode, as described in Chapter 3, and which could therefore be heated by a low voltage AC supply, derived from a low voltage winding on the main power supply transformer, without problems due to the introduction of mains frequency hum. The HT supplies could also be derived from a rectified AC voltage provided by the same transformer, fed directly from the household electric mains supply, and it became practicable to design amplifier systems in which economy in the use of input power was of minor importance, and more attention could be paid to the linearity and bandwidth of the circuit.

Two fairly typical circuits of the period are shown in Figures 9.5 and 9.6. Both of these used push–pull output stages, because these were more efficient, and the symmetrical push–pull layout of the output valves reduced the extent of even harmonic distortion for a given output power level.

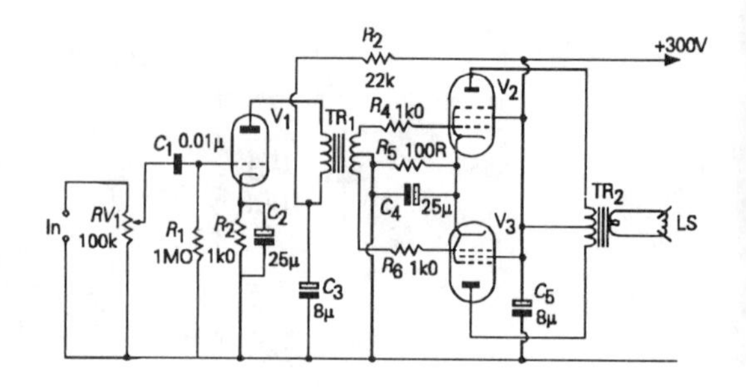

Figure 9.5 *Simple push–pull mains powered valve amplifier*

In the circuit of Figure 9.5, an inter-stage coupling transformer, TR_1, is used to provide the symmetrical anti-phase drive to the control grids of the output valves, whereas in the more elaborate circuit of Figure 9.6, a 'floating paraphase' inverter stage (V_2, with R_5/R_6 and C_4), is used to generate the required phase-inverted signal input for V_3. Both of these amplifiers would give a very acceptable quality of the output sound, and although the layout of Figure 9.6 is theoretically superior, there were some very high quality inter-stage coupling transformers, made by companies such as Varley or Ferranti, which offered an excellent transmission bandwidth, with very low distortion in

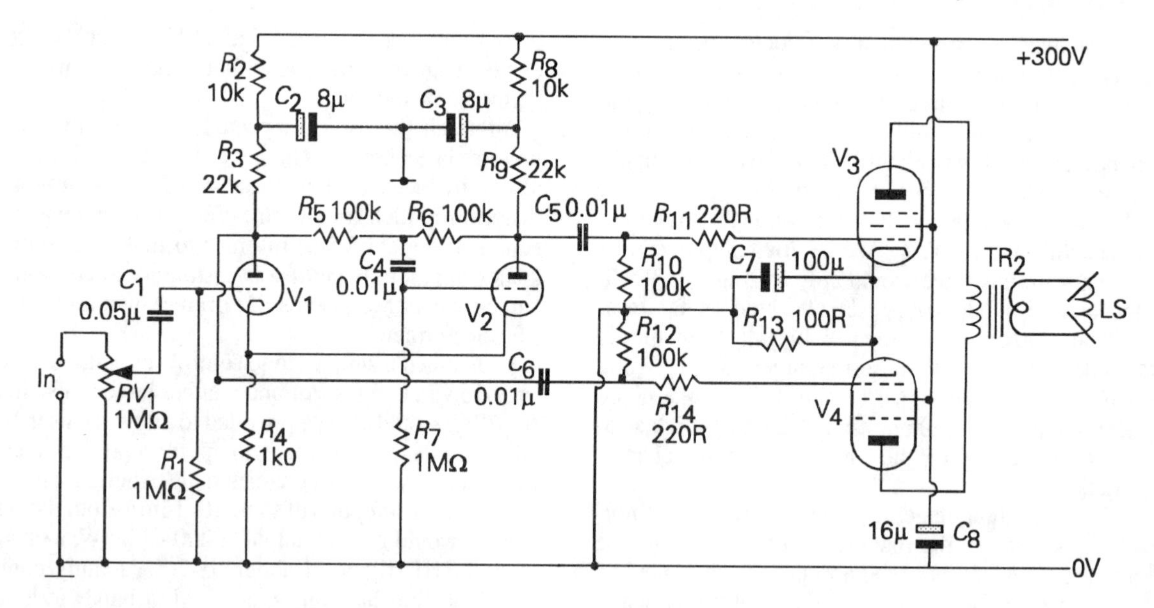

Figure 9.6 *RC-coupled push-pull audio amplifier*

the output signal, so, from the point of view of the user, both designs would be comparably good.

A point to note in all of these designs is the extensive use of supply line decoupling circuits, such as R_2/C_2 in Figures 9.4–9.6, as well as R_8/C_3, and C_8 in Figure 9.6. These were essential, since, with the typical power supply circuitry of the period, it was inevitable that the fluctuating anode current demand of the output stages would cause voltage ripple on the HT line, and this could intrude into the signal path via the inter-stage coupling capacitors, to cause a characteristic form of low frequency instability, known as motor boating from the sound it would produce in the loudspeakers.

Mains powered circuits of the type of Figures 9.5 and 9.6, would offer 6–8 watts of audio signal into a typical 3 ohm moving coil LS load, at perhaps 5% harmonic distortion – ignoring the contribution of the pick-up element – with an amplifier bandwidth of 100Hz–10kHz. These, together with the 3–4 watt output mains valve operated versions of the circuit of Figure 9.4; commonly referred to as a 'single ended' design, to distinguish it from circuits with push–pull output stages; remained typical of the type of output stages used in the bulk of audio equipment up to the late 1940s, and in many medium price radios and 'radiograms' for at least a decade beyond this period.

High quality valve operated amplifiers

The demands of radar and airborne navigation equipment had led to an enormous development in electronic circuit technology during the period 1940–45, and this led circuit designers to realize that the use of negative feedback (NFB) would allow a considerable improvement in audio system performance, both in respect of bandwidth and in respect of harmonic and intermodulation distortions.

Although many comparable audio amplifier designs were offered at this time, based on the use of various combinations of negative feedback within the circuit, the one which had the greatest impact on performance expectations in the UK and other English language countries was that published shortly after the war by D. T. N. Williamson (*Wireless World*, Aug. 1949, pp. 282–284, Oct. 1949, p. 365, Nov. 1949, p. 423). The author of this circuit worked, at the time, for the Marconi-Osram valve company, and his design exploited the qualities of the recently introduced 'KT66' output beam-tetrodes, which he used in combination with a carefully designed loudspeaker output transformer.

The Williamson amplifier

The principal stimulus to this design, and to other similar designs of the same period, was the realization that greater benefits would arise from the use of a NFB loop which enclosed the whole amplifier, including the output transformer, rather than the use of a number of separate feedback loops around individual areas of the circuit, as had been used in the past, although the adoption of such a design approach would require much greater care in the choice of the circuit and component characteristics. In particular, this use of overall NFB required that the total phase shift within the circuit must be maintained at less than 180° within the frequency band in which the loop gain was greater than unity, otherwise the circuit would be unstable, and the amplifier would break into oscillation. Moreover, even when this initial condition was satisfied, it was also essential that there should be an adequate margin of stability within the loop, otherwise bursts of oscillation could be provoked when input signals, especially those of a transient nature, occurred at parts of the frequency spectrum where the stability margin was low. This type of malfunction would greatly impair the quality of the reproduced sound.

This requirement for strictly controlled phase characteristics within the loop discouraged the use of inter-stage coupling transformers, since these would inevitably suffer from substantial phase shifts at the extremes of their frequency pass–bands. These single-loop feedback designs therefore employed other circuit arrangements for generating the required symmetrical pair of drive waveforms for the output valves.

In the Williamson design, this was done by the use of a split-load amplifier stage, of the design shown in Figure 9.7. Since this type of stage would provide only a relatively low undistorted output voltage swing, Williamson followed it by a pair of further triode valve amplifiers to drive the push-pull output valves, giving the complete design shown in Figure 9.8.

Because the anode circuit impedance of both triode and triode-connected beam-tetrode valves is high, typically in the range 500–10,000 ohms, it is necessary, in order to get the best efficiency, to use an output transformer to match the relatively low load imped-

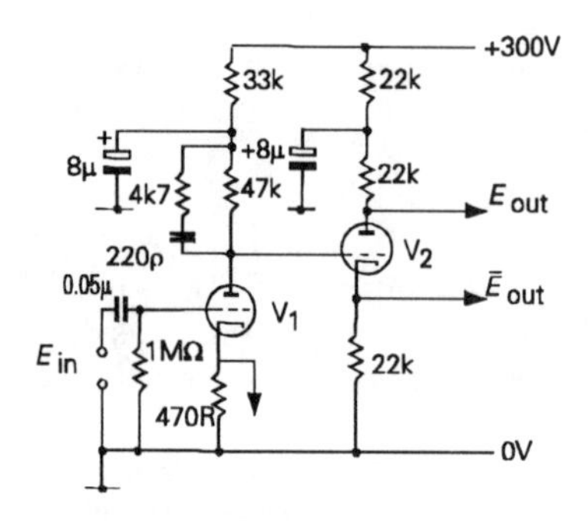

Figure 9.7 *Phase-splitting circuit used in Williamson amplifier*

ance of the loudspeaker to that of the output valves, and it is easier to minimize phase errors due to this transformer if the transformation ratio is not too high.

Although loudspeaker manufacturers responded to the evolution of these higher quality audio amplifiers by making loudspeakers with 15 ohm, rather than the previously standard 3 ohm speech coil impedance, to facilitate output transformer design, nevertheless an output coupling transformer remained essential.

From time to time, circuits did appear in which the layout was arranged to allow the loudspeaker to be directly coupled to the output valves, to eliminate the need for an output coupling transformer altogether, but the inconvenience, and low overall efficiency, of these designs prevented their becoming popular.

The performance of any amplifier using overall negative feedback derived from the output transformer secondary winding is crucially dependent on the characteristics of this transformer, for which Williamson's specification is given below. This stipulated a much higher quality component for this position than had been used previously, and this, in turn, stimulated a number of transformer manufacturers to provide improved quality components of their own designs.

The Williamson output transformer specification

Primary load impedance	10,000 ohms
Primary inductance	100 henrys, min
Series leakage inductance (whole primary to whole secondary)	30 millihenries, max
Primary resistance	250 ohms, max

Triode valves have better distortion characteristics – they generate some second harmonic, but very little third or other odd-order harmonics – than either output pentodes or beam-tetrodes, when they are used in the output stages of power amplifiers, but they are relatively inefficient in use. Also, because of the difficulty of getting adequate electron emission from an indirectly heated cathode, all of the contemporary output triode valves, such as the PX4 or the PX25, had directly heated cathodes, which led to difficulties with

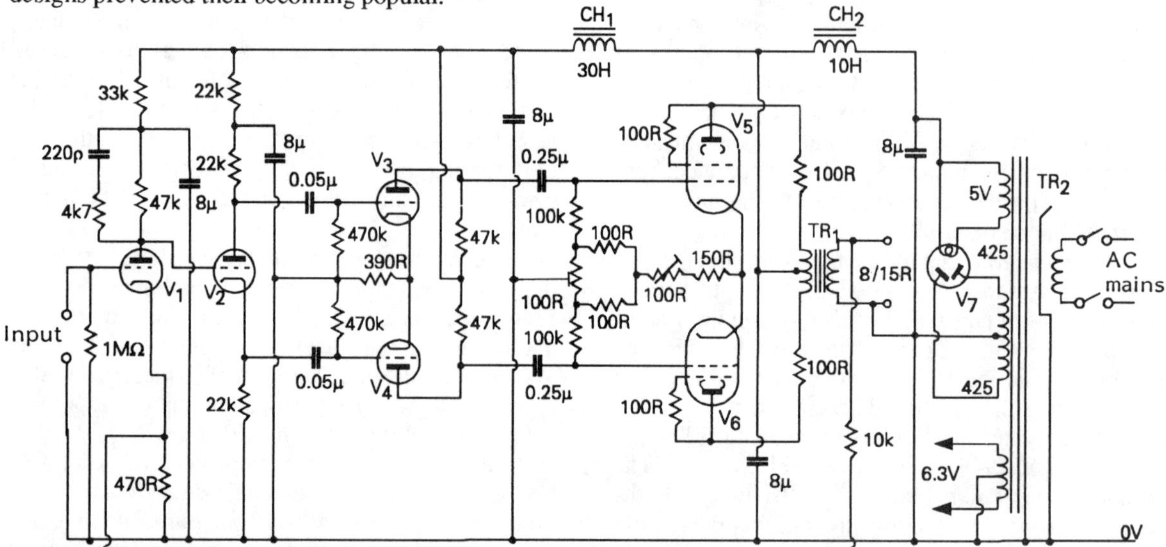

Figure 9.8 *The Williamson amplifier*

cathode bias arrangements, and a proneness to mains hum.

Beam tetrode output valves, of the KT66 type used by Williamson, generate less distortion than the output pentodes they were designed to replace, but they are still less linear than triodes. Williamson therefore used triode-connected beam tetrodes in the output stage, which avoided both the problems of pentode-type distortion characteristics and the hum problems associated with directly heated triodes, while the push–pull output arrangement tended to cancel out the remaining second-order distortion. This circuit design offered an overall performance, quoted below, which was substantially better than almost all of the contemporary audio amplifier designs.

Williamson amplifier. Performance specification

Output power, into 15 ohm load	15 W
Total Harmonic Distortion	less than 0.1% at 15W
Bandwidth	2Hz–100kHz

However, the overall efficiency of this design, at maximum output, was only of the order of 23%, and this encouraged other circuit designers to explore various alternative 'distributed load' type output stage configurations, some of which gave better efficiencies with beam-tetrodes or pentodes, without significantly worsening the output stage distortion chacteristics.

Alternative output stage connections

Of the alternatives to the simple push–pull triode-connected beam-tetrode output stage, the most popular was the so-called 'ultra-linear' or partial triode connection. In this, the screen grids of the beam-tetrodes were taken to taps on the primary of the output transformer, as shown in Figure 9.9. This allowed an increased output power for the same supply voltage and anode current.

Another arrangement, exploited in Britain by the Acoustical Manufacturing Co. (Quad), employed an output layout in which the cathodes of the output valves were taken to a second primary winding on the output transformer, as shown in Figure 9.10. A similar layout was employed in the USA by McIntosh (McIntosh, F. H., and Gow, J. G., *Audio Engineering*, Dec. 1949, p. 9), in a 50 watt design, shown in Figure 9.11. In general, such layouts did not offer a better perfor-

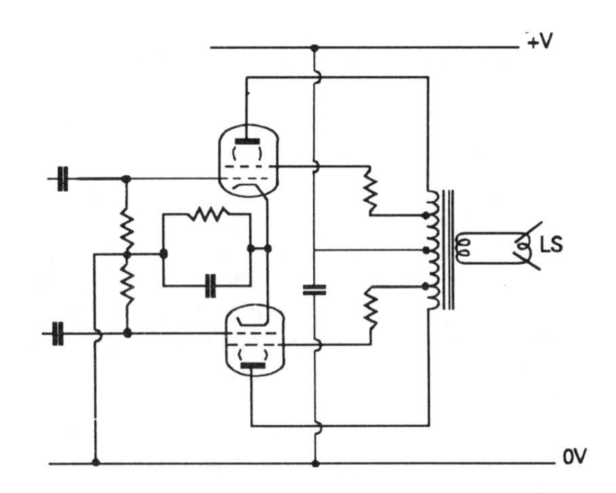

Figure 9.9 *The ultra-linear output stage connection*

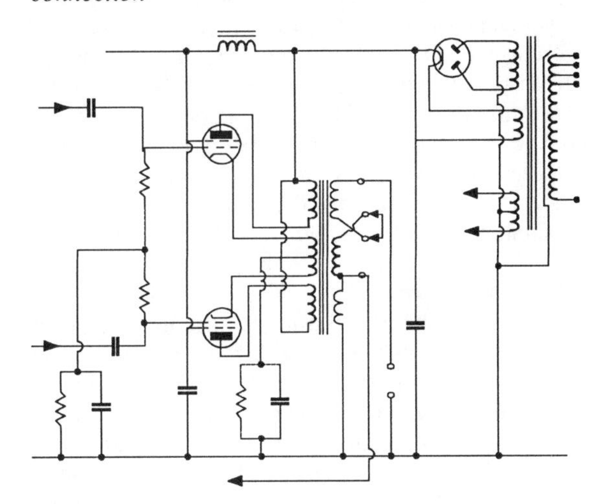

Figure 9.10 *Quad 22 power amplifier*

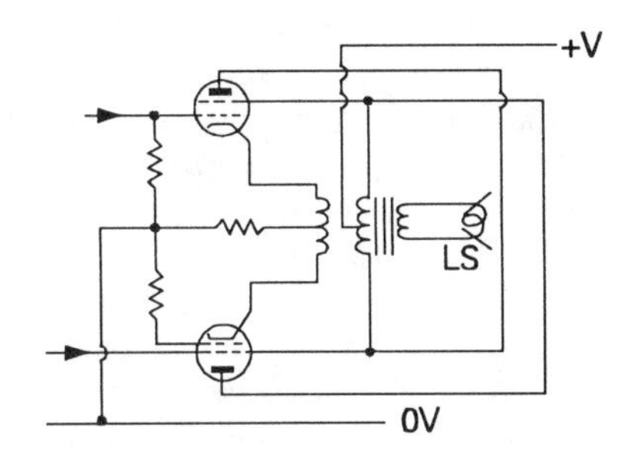

Figure 9.11 *The McIntosh cathode and anode coupled output stage*

mance than that given by the original Williamson design.

Output stage biasing

Although in audio amplifiers based on thermionic valves, the total thermal dissipation of the output stages is a matter of less consequence than in the case of transistors, which are devices having a much lower thermal inertia and generally a much lower ability to dissipate heat, nevertheless, even with valves, there is an effective upper limit on the permissible anode current and consequent heat dissipation. Beyond this level the life expectancy of the valve will be much reduced.

If the anode current under no-signal conditions could be reduced somewhat, a higher peak output power could be obtained, without increasing the average current rating. The current which flows in the valve is largely controlled by the bias voltage applied to the control grid, and the possible operating conditions are normally described as being in one or other of the various 'classes', such as 'class A', in which the bias level is chosen so that the anode current remains substantially the same, within the whole of the output power range, so long as this is below the 'clipping level'. If the control grid negative bias is increased to the level at which little or no anode current flows under quiescent conditions, then the stage is said to operate in 'class B'. Similarly, if the negative grid bias is increased still further, so that anode current only flows during the peak positive swings of the grid drive voltage, this is referred to as 'class C'. This latter condition is not used in audio amplifiers, which all operate either in class A or at some intermediate quiescent current level, some way towards the class B condition. This is loosely described as 'class AB'.

All of the valve output stages considered so far operated in class A, defined, above, as a condition in which the mean anode current is the same in the quiescent state as it is at full power. This has the great benefit that the harmonic distortion due to the output stage will generally decrease as the output power is reduced, whereas in class B, or other conditions in which the zero signal anode current is less than that at full power, not only will this reduction in distortion not occur, but the actual output circuit contribution to the THD may actually increase as the signal level is reduced. The failure of circuit designers to appreciate this point led to the unsatisfactory performance of almost all of the early transistor operated audio amplifiers – all of which operated either in class B or in some class AB operating condition which was close to this – and led, in due course, to the emergence of a band of Hi-Fi devotees who claimed that the only true high quality audio reproduction was that obtained from valve operated amplifiers.

Modern audio amplifier circuit design

Early transistor operated audio power amplifiers

Although transistors had been available since the early 1950s, it was not until the end of that decade that circuit engineers felt adequately confident of their design skills to offer audio amplifiers, based on these devices, which they felt could compare in performance or output power with the very highly developed thermionic valve operated audio power amplifiers which were then commonplace. This was partly because, during the 1950s, the only mass produced transistors were germanium junction types, which were generally only available in PNP forms. Although a few NPN germanium and silicon transistor types were also marketed, these generally had a low performance and were only offered in very low power versions.

A truly complementary output stage layout, using symmetrical PNP and NPN output transistors, of the type now typical of almost all contemporary audio power amplifiers, was therefore impracticable and most of the transistor designs used inter-stage transformer coupling of the types shown in Figures 9.12 and 9.13. The latter design is the output stage of a 15 watt audio power amplifier, using germanium transistors, due to Mullard (Mullard Ltd., *Reference Manual of Transistor Circuits* 2nd Edition (1961), pp. 178–180).

As with the previously described valve amplifier designs, the use of overall loop negative feedback (NFB) was desirable in order to improve the linearity and flatness of the frequency response of the system. The need for inter-stage and output coupling transformers greatly limited the extent to which NFB could be employed without sacrifice of loop stability. Moreover, as seen in Chapter 7, unless an adequate amount

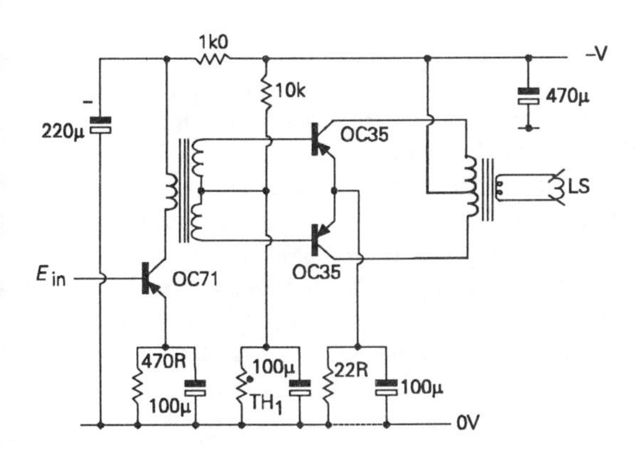

Figure 9.12 *Typical transformer-coupled germanium transistor power amplifier*

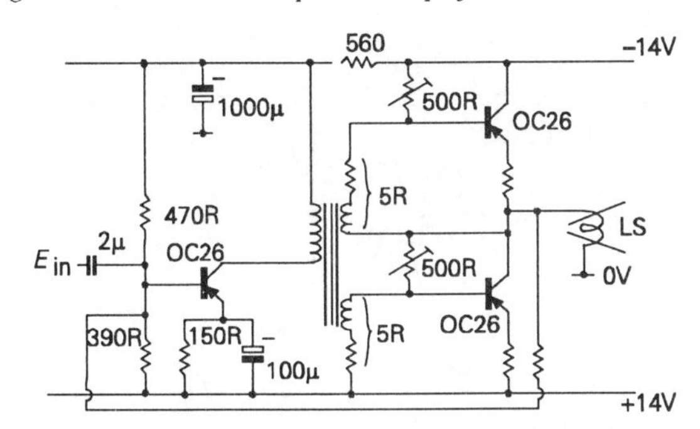

Figure 9.13 *Mullard 15 watt power amplifier (c.1960)*

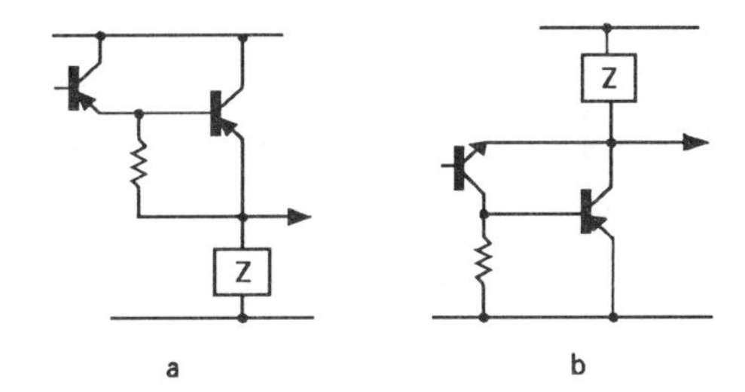

Figure 9.14 *Impedance conversion stages*

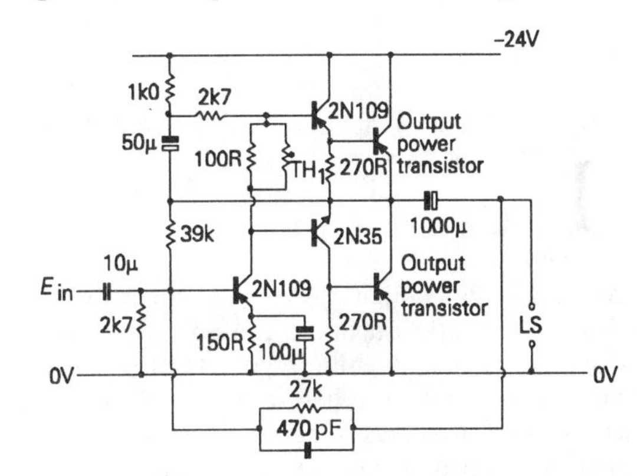

Figure 9.15 *The LIN quasi-complementary output-stage configuration*

of NFB could be employed it might impair rather than improve the performance of the system.

The Lin quasi-complementary circuit layout

The major design breakthrough in this field came with the introduction of the so-called quasi-complementary output stage, due to Lin (Lin, H. C., *Electronics*, Sept. 1956, pp. 173–175), which largely set the type of all subsequent transistor audio amplifier designs. In this, a small-signal voltage amplifier stage, operating in class A, was used to drive a combination of Darlington and compound emitter follower layouts, of the forms shown in Figures 9.14a and 9.14b, which acted as an impedance converter between the output of the small-signal amplifier and the loudspeaker, and allowed the construction of a push–pull output stage, without the need for either input or output transformers, as shown in Figure 9.15. This layout appeared to offer the design solution for which the circuit engineers had been waiting, in that it allowed the construction of a high quality transformerless audio amplifier, which could be built with identical types of power output transistor. However, herein lay a hidden source of trouble.

At the time (1956) that Lin proposed his output stage layout, the only commercially available transistors were of germanium type. In these, at room temperatures and above, the relationship between forward base voltage and collector current, shown in Figure 9.16a, had a much more gradual turn-on characteristic at the origin of the graph than is the case with silicon devices, whose base-voltage/collector-current relationship is as shown in Figure 9.16b. This meant that a push–pull arrangement using germanium transistors had a much less abrupt discontinuity at the crossover point – from one output device conducting to the other – than would be the case if silicon transistors were used. Unfortunately, by the time this circuit came into popular use among the more adventurous of the Hi-Fi amplifier manufacturers, germanium transistors had largely been replaced by silicon types, which led to the prob-

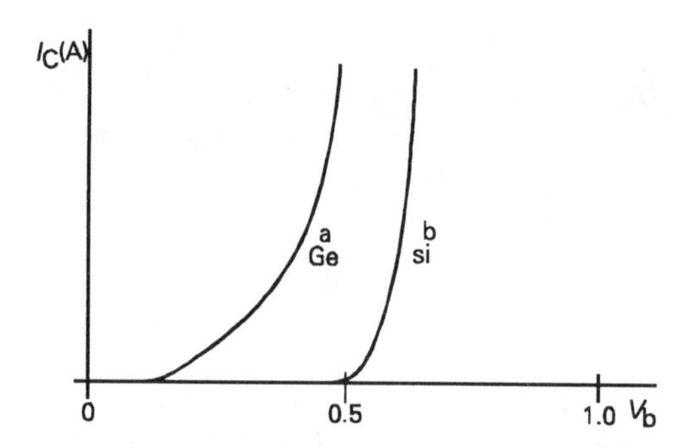

Figure 9.16

lem of inadequately specified distortion characteristics noted above.

Crossover distortion

In any push–pull output system, in which the two halves of the output stage are caused to handle the positive- and negative-going segments of the output waveform, sequentially, there is likely to be some discontinuity at the crossover point unless both halves operate at a linear region of their transfer characteristics, as would invariably be the case with the previous, class A biased, valve operated power amplifiers. However, with transistors, it was customary to choose a level of bias at which they would operate in class AB or class B, with little or no quiescent collector current. This was done in order to lessen the thermal dissipation of the output stages, which would give a longer life expectancy, and allow a larger output power for any given heat-sink size. This was necessary, at the time, because the manufacturing techniques employed, both in respect of the integrity of the solder joint between the silicon chip and the metal 'header' – the mounting plate of the device – and the stability of the thickness of the diffused junction regions were inadequate to prevent deterioration of the devices during use. This was a major problem with germanium power transistors, but was still a likely failure mechanism with silicon transistors, even though these had markedly better thermal characteristics than was the case with germanium devices. Since these problems were worsened at higher junction temperatures, it was very desirable to keep the output devices as cool as possible.

Operation in class B or AB inevitably leads to the generation of high order types of harmonic distortion (7th, 9th, 11th, etc), due to the presence of a kink in the transfer characteristics of the push–pull stage at the point of the crossover from one half of the output pair to the other. This kind of distortion would be inevitable, even if the two halves of the output pair are identical, but in the case of a quasi-complementary layout using silicon transistors this problem is made much worse because the input/output transfer slopes of the two quasi-complementary halves are markedly different, as shown in Figures 9.17a and 9.17b.

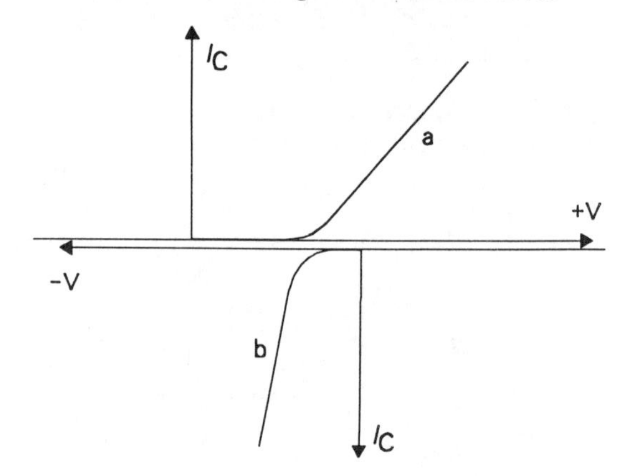

Figure 9.17 *Asymmetry in upper and lower halves of quasi-complementary pair*

Accurate choice of the quiescent forward bias applied to the output transistors could lessen the extent of the distortion due to this cause, but this would be difficult to guarantee in a commercial product intended for use in a domestic setting because the precise bias necessary would depend on the junction temperature of the output transistors and would also change with time. If too high a quiescent current setting was chosen, thermal runaway might occur, where an increase in junction temperature would lead to an increase in quiescent current which would, in turn, lead to a further increase in junction temperature. This would cause an uncontrolled increase in operating current and could lead to the failure of the output transistors. Such an event would reflect badly on the manufacturers of the equipment.

For this reason, commercial designs tended to favour the use of little or no forward bias on the output transistors, and rely on the use of high amounts of negative feedback to linearize the output characteristics. This would readily allow the achievement of

THD figures, at full output power, which were at least as good as those given by the traditional valve amplifier designs – in which the possible amount of NFB was limited by the phase errors introduced by the loudspeaker coupling transformer.

Amplifier sound and listener fatigue

After the initial enthusiasm generated by the novelty of this new breed of amplifier design, which offered higher powers in a much more compact housing, and which dissipated so little heat that it could be housed almost anywhere, without a specific need to allow for ventilation, some dissatisfaction began to be expressed by a growing number of users in respect of the sound quality these amplifiers gave. This dissatisfaction gave rise to the descriptions of 'amplifier sound' or 'listener fatigue', and related mainly to the general thinness and lack of body or 'warmth' in the sound quality of these new types of amplifier, coupled with various other, difficult to define, aspects of their performance, which made the user reluctant to spend much time listening to them.

Although there was a fairly lively debate at the time, in the technical press, about just what it was that was responsible for this reaction on the part of the user, several points soon emerged. Of these, the first was that – although the use of NFB could lessen the extent of the measured nonlinearity of the amplifier at high output powers – because the gain of the system fell to near zero at the crossover point, even apparently very high amounts of NFB could not eliminate the high-order harmonic distortion due to the kink in the crossover characteristic, and this kind of distortion would become progressively more conspicuous at low output-power levels. This meant that a graph of total harmonic distortion (THD), versus output power would show the characteristic illustrated in Figure 9.18a, with an increasing distortion level as the output power was reduced, rather than that of Figure 9.18b, which was typical of most earlier valve operated amplifiers. Not only was this high level of small-signal distortion unsatisfactory for the majority of listeners, who would do the bulk of their listening at low power levels, but this type of defect would lead to the kind of signal input vs. signal output transfer characteristic shown in Figure 9.19. where the relative gain of the amplifier was greatly reduced for input signal levels around the zero signal datum line. This had the predictable result that all small signals, such as the dying-away sounds of instrument tones, would be suppressed, along with a lot of other low signal level sounds, and amply explained the complaints of 'thinness' of tone.

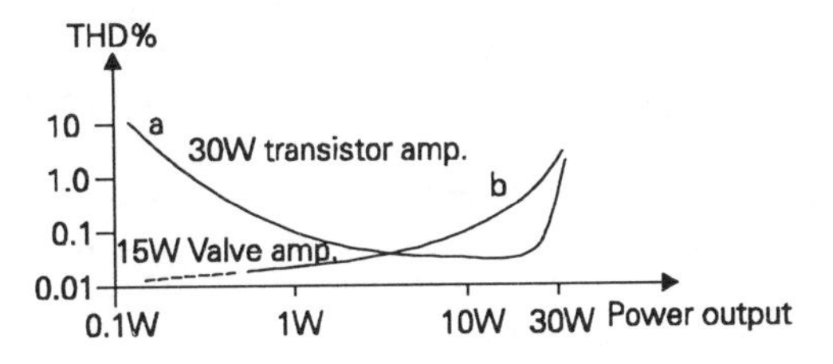

Figure 9.18 *Comparison between distortion characteristics of valve vs. quasi-complementary transistor output stages*

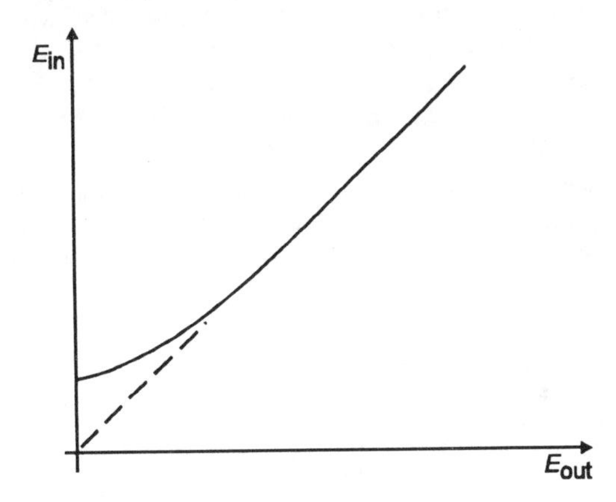

Figure 9.19 *Input/output transfer characteristics of incorrectly biased push–pull stage*

Listener fatigue was probably a combination of the ear's objection to exposure to alien high-order harmonic distortions, due to crossover effects, and of the bad transient performance of the amplifiers, especially on 'awkward' types of LS load, where the injudicious use of too high a level of NFB, in the interests of a good apparent full output power THD figure, had eroded the stability margins of the amplifier and might have caused sporadic instability following sudden changes in signal level.

Improved transistor amplifier designs

Output circuit alternatives

With the realization that much greater care must be exercised in the design of transistor amplifier output stages, if a performance was to be obtained which would compare with that given by earlier valve operated designs – particularly in respect of low signal level distortion characteristics – came a great amount of design activity aimed at remedying crossover errors.

Fully complementary output stages

This took two routes; of these the first was that of improving the symmetry of the output stage by the use of more truly complementary output devices, which were beginning to become available from the semiconductor manufacturers in the mid- to late 1960s, of which a typical example is that due to Bailey (Bailey, A. R., *Wireless World*, May 1968, pp. 94–98), whose circuit is shown in Figure 9.20. This approach was popular among designers in the USA where fully complementary power output transistors had first become readily available.

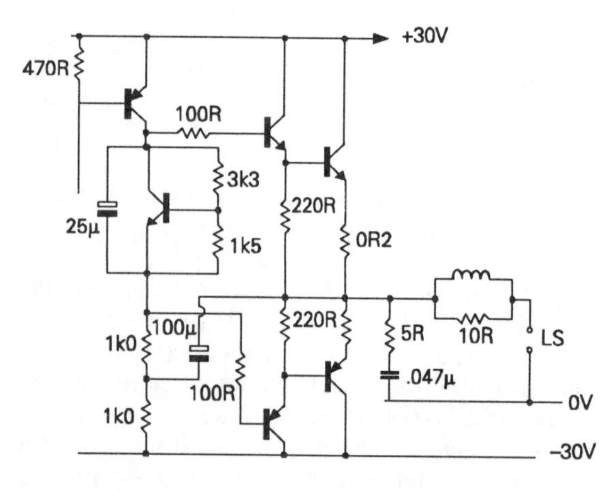

Figure 9.20 *Fully complementary output stage of Bailey 30W amplifier*

The Shaw improved quasi-complementary layout

Alternatively, steps could be taken to improve the symmetry of the Lin quasi-complementary layout, and a circuit for this purpose was disclosed by Shaw (Shaw, I. M., *Wireless World*, June 1969, pp. 148–153), and is shown in Figure 9.21. Two later improvements to Shaw's circuit, of which the first was proposed by Baxendall (Baxandall, P. J., *Wireless World*, Sept. 1969, pp. 416–417), and the second by the present author (Linsley Hood, J. L., *Hi-Fi News and Record Review*, Nov. 1972, pp. 2120–2123), are shown in Figures 9.22a and 9.22b. Versions of these latter layouts are still in commercial use, in well thought-of amplifier designs.

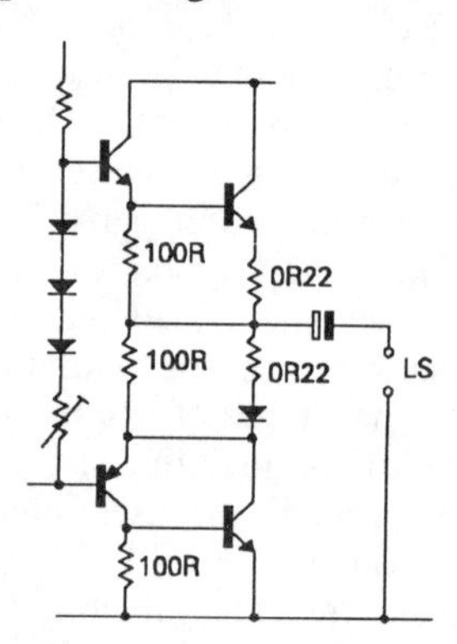

Figure 9.21 *Shaw's improved quasi-complementary output stage*

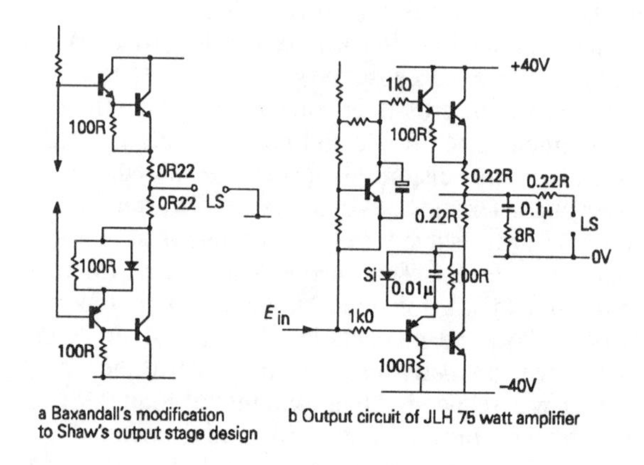

a Baxandall's modification to Shaw's output stage design

b Output circuit of JLH 75 watt amplifier

Figure 9.22

Operating current stabilization

All of these improved circuits were intended to operate in class AB, that is to say that it was necessary that, even at zero signal level, there would always be some output stage operating current, usually in the range 30–150mA. This would be chosen, on test, to give the least distortion at the crossover point, and it was essential to ensure that this chosen value was maintained during use, preferably without the need for further adjustment, if the desired low extent of residual distortion was to be sustained.

The amplified diode

Although various combinations of diodes and negative coefficient thermistors have been proposed, undoubtedly the most popular technique remains the use of an 'amplified diode', using the circuit layout shown in Figure 9.23. In this, a small signal transistor, Q_1, is inserted in the load current path of the output stage driver transistor, Q_2, between the connections to the bases of Q_3 and Q_4. A voltage drop will then appear across Q_1 of sufficient magnitude that the proportion of it which is developed across R_1 will cause Q_1 to conduct. This allows the total voltage across Q_1 to be adjusted by alteration of RV_1. Since the forward diode potential of the base–emitter junction of Q_1 will change with temperature, some measure of compensation for the temperature dependence of the output stage quiescent current can be obtained by mounting Q_1 in thermal contact with the output transistors themselves. This allows the mean value of quiescent current to be maintained at a fairly stable value over a moderately wide range of ambient temperatures, which would be far from true if no such temperature compensation was used.

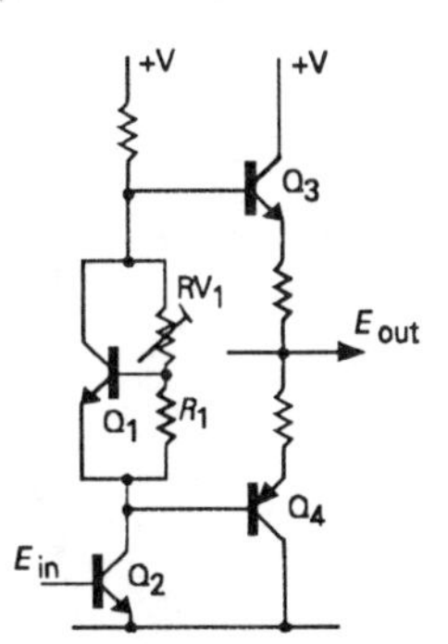

Figure 9.23 *Amplified diode circuit for providing forward bias on output transistors*

The problem with this system is that, in operation, there will always be a time lag between any rise or fall in the junction temperature of the output devices, and a compensating change in the temperature of Q_1. This could mean that on a sudden high power demand, occasioned perhaps by a sudden loud passage in the music being reproduced, the output transistor junction temperature and quiescent current would both increase, because the transistor, Q_1, in the compensating circuit, is still at the temperature it was prior to the event, so that the bias voltage which it causes to be applied to the output transistors would be too high. Then, as the increased temperature of the output heat sink was communicated to the compensating circuit transistor, the voltage across this, and the forward bias on the output devices would be reduced. However, perhaps by this time the loud passage in the music would have passed, and the output transistor junctions will have cooled down, leaving the quiescent current setting temporarily too low. So, although very commonly used, this particular technique is still somewhat less than perfect.

Other techniques

The need to ensure that the output stage quiescent current remains precisely set, at some optimum value, has encouraged some manufacturers to develop quite elaborate systems for this purpose, as, for example, the Japanese manufacturer Pioneer, who employ a purpose built IC specifically for this purpose, to upgrade the performance of an otherwise relatively straightforward amplifier design.

Class A designs

The other route to the avoidance of crossover distortion was to return to the use of class A operation, and to employ adequate heat sinks for the output transistors to allow the output transistors to remain at an acceptable temperature. Since this will be considerably above ambient, and will not increase with output power, the problem of thermal compensation for the output devices will be relatively easy to solve.

The adoption of class A operation for the output devices offered many advantages. Of these, the principal one was that because of the high intrinsic system linearity which would be possible, much less negative feedback would be required to achieve the same level of output distortion than would be the case with a comparable class AB layout, and this would greatly reduce the possibility of unexpected instability in use. It would also allow a greatly improved transient response, especially since the output impedance of the transistors would be sufficiently low that no output transformer would be required to match the amplifier to the LS load. A 10 watt class A transistor power amplifier design of my own (Linsley Hood, J. L., *Wireless World*, April 1969, pp.148–153), is shown in Figure 9.24.

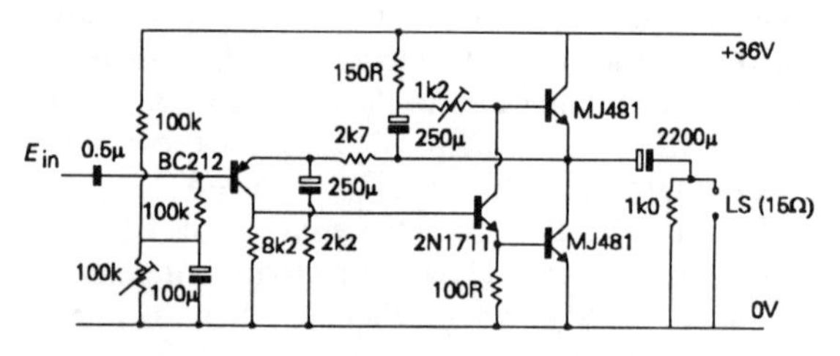

Figure 9.24 *10 watt class-A transistor power amplifier*

At the time of this design I had an audio system based on a Williamson 15 watt valve amplifier. However, this was only a 'mono' system, and I was keen to take advantage of the growing number of 'stereophonic' gramophone records which were becoming available, and which offered a much greater degree of realism in the reproduced sound. I had used the Williamson amplifier for many years, and did not welcome the difficulty and expense of building another power amplifier of this type. On the other hand, side by side comparisons between the Williamson and all of the transistor power amplifier designs which had been offered up to that date (1967) showed that these amplifiers were much inferior to the Williamson in tonal quality.

The class A design of Figure 9.24 had a very similar performance specification to the Williamson in respect of bandwidth and harmonic distortion, over a wide range of output power levels and frequencies, and had an identical tonal quality, so far as I was able to tell. However, all class A systems run hot in use, because the zero signal operating current will be the same as that at full power. An unfortunate corollary to this is that the permitted quiescent thermal dissipation will then set a limit to the maximum power output. This constraint is a matter of little importance in the case of valve amplifiers, since these will run hot anyway, and their bulk ensures that there is adequate scope for normal air convection cooling, but goes against the whole recent concept of transistor audio amplifiers, which are seen as offering high output powers from physically small, cool-running packages. All recent design work has therefore been aimed at achieving an improved performance from transistor power output stages operated in either class AB or in class B, and a varity of ingenious designs have been proposed.

Floating class A bias systems

The idea of arranging that the quiescent current of the output stage is increased automatically as the signal level increases – so that the amplifier can, effectively, operate in class A all the time – though superficially attractive, and proposed by several designers in the 1960s, has proved difficult to implement in any satisfactory manner, and has therefore been abandoned, in favour of improvements in the operation of the basic class AB layout.

The Blomley design

A very ingenious approach to this problem was suggested by Blomley (Blomley, P., *Wireless World*, Feb. 1971, pp. 57–61, and March, 1971, pp. 127–131), who proposed that the output devices should be caused to operate at some fixed, optimum, non-zero, level of quiescent current, during the whole of the output voltage swing, with the input signal being divided into two, cleanly separated, halves by a preceding small-signal switching stage. Blomley's circuit is shown in somewhat simplified form in Figure 9.25.

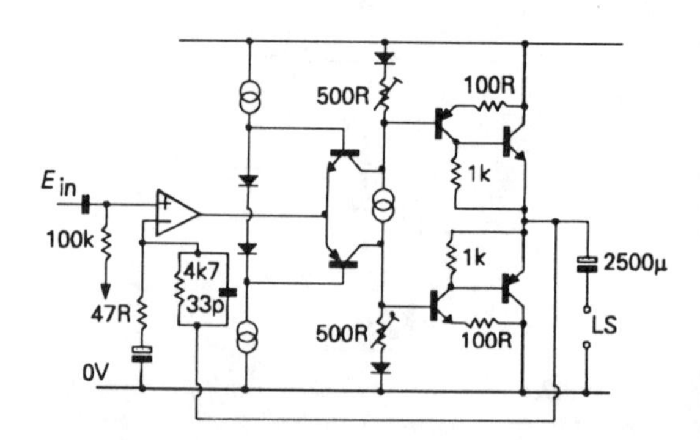

Figure 9.25 *The Blomley amplifier*

This type of operating system has not been widely

adopted, presumably because it simply refers the requirement for accurate biasing of the push–pull stage back to the small-signal driver-transistor stage, which displaces the problem rather than solving it.

Super class A

A further scheme which has been proposed, and which has, indeed, been adopted by one Japanese audio amplifier manufacturer, employs the apparently ingenious scheme of operating the output stages of the amplifier in true class A, but from a low voltage power supply, so that the power dissipation in the output transistors is kept to a relatively low level. The DC mid-point potential of this power supply is then caused to swing, in synchronism with the input signal, by means of a high-power, class B – or AB – amplifier, so that the class A amplifier is always able to accomodate the required voltage excursions of the signal, as seen at the output load.

This layout is shown schematically in Figure 9.26, and has been described as 'Super A'. The fallacy with this arrangement, as can be seen from analysis of the operation of the circuit, is that the return path of the signal, from the output of the class A amplifier, through the loudspeaker, and the amplifier power supply, and back to the class A amplifier again, is by way of the high power class B unit, so the overall performance of the system can never be better than that of the class B amplifier.

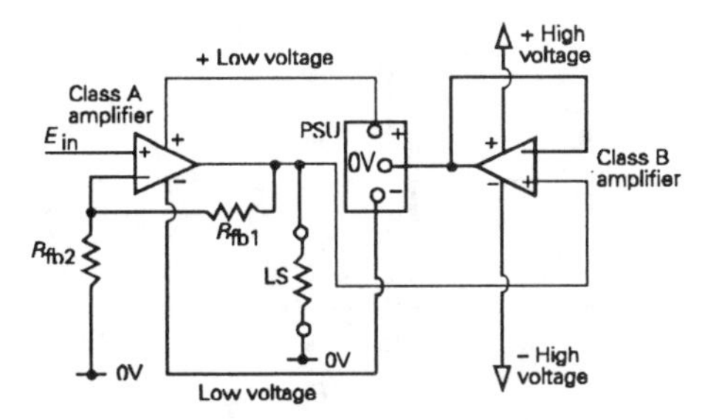

Figure 9.26 *Floating power supply system for super class A operation*

So, although the actual amplifier in question does indeed give a satisfactory performance, this circumstance cannot really be attributed to the supposed design breakthrough.

Contemporary high quality audio amplifier designs

Improved Lin type

Transistor operated audio power amplifier designs fall, in general, into three categories. Of these, the first, which represents by far the most commonly adopted type of circuit layout, is based on straightforward developments of the basic Lin type circuit, in which a linear low-power voltage amplifier stage is followed by a pair of emitter followers, connected in push-pull, to reduce the output impedance to a level suitable for driving a loudspeaker load.

In modern circuitry the output emitter followers will nearly always be devices having fully complementary symmetry (NPN/PNP, or N-channel/P-channel), using one or other of the forms shown in Figure 9.27, which are essentially similar in use to that used by Bailey (Figure 9.19).

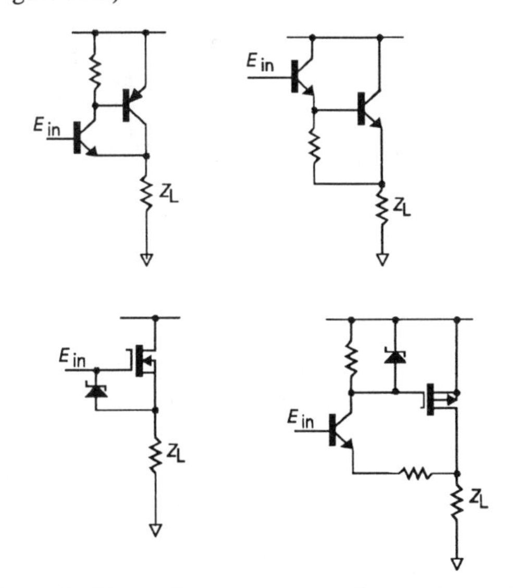

Figure 9.27 *Various output impedance conversion layouts*

It is arguable that true equivalence between such apparently complementary devices is not really possible, because of the differences in their way of operation, but, in practice, this lack of symmetry only becomes evident at operating frequencies towards the upper end of their frequency range. Also, technical developments since this type of design first appeared, in the late 1960s, have made available much better devices, such as higher power, higher voltage, higher gain, and higher transition frequency junction transistors, or very linear and robust high voltage power MOSFETs, and the availability of these components has greatly improved solid-state amplifier performance, when used as output devices in this kind of circuitry.

The overall linearity of the amplifier has also been improved by the use of more highly developed circuitry in the input voltage amplifying stages, as, for example, that used in a design due to the author (Linsley Hood, J. L., *Electronics Today International*, July 1984, pp. 44–49), for use with power MOSFET output transistors, and shown in a slightly simplified form in Figure 9.28, or in designs intended for use with conventional junction transistor outputs, in which the performance has been improved by the use of fully symmetrical driver layouts, of the kind shown, schematically, in Figure 9.29 (Borbely, E., *Wireless World*, March 1983, pp. 69–75, and also in *Audio Amateur* (USA), Feb. 1984, pp. 13–24).

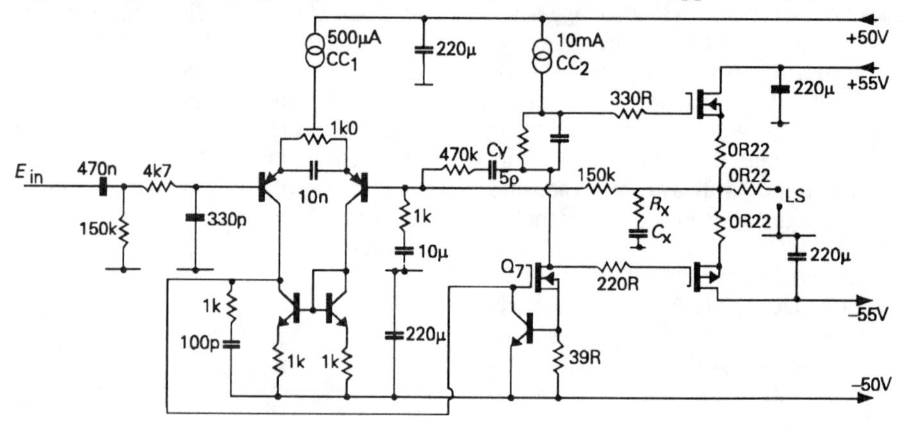

Figure 9.28 *80–100 W power amplifier based on MOSFET devices*

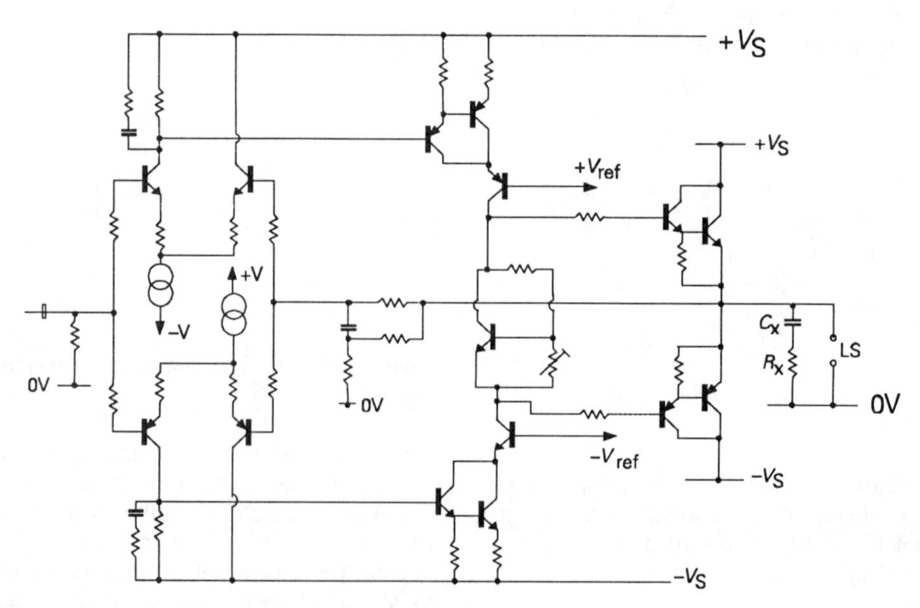

Figure 9.29 *Power amplifier stage based on fully symmetrical layout*

Unfortunately, the performance of all of these circuits will still depend on the maintenance, during use, over a wide range of ambient temperatures, of an accurately chosen level of output stage quiescent current, though the use of power MOSFETs will make the precise value required very much less critical.

Negative feedback and HF compensation

Assuming that adequate care is taken in the circuit layout to maintain the correct level of output quiescent current, and to avoid damage to the output transistors due to inadvertent misuse, the design of such a circuit is really a quite straightforward combination of a suitable voltage gain stage with an output impedance-converting emitter follower pair. Care must also be taken, in any new design, to ensure that the amplifier remains stable when overall negative feedback is applied around the system. As a general 'rule of thumb', NFB can be applied around a circuit consisting of just two stages – an amplifier followed by an emitter follower, or two successive voltage amplifying stages – without the system becoming unstable under normal resistive load conditions.

In circuits in which the NFB loop encloses three or more such stages, such as two voltage amplifier stages followed by an emitter follower, instability will almost certainly occur, usually at some fairly high frequency, for reasons examined in Chapter 6, and some arrangement to provide HF compensation will be necesssary. This will usually consist, in simple audio amplifier designs, of an output 'Zobel' network (a resistor in series with a small capacitor to ensure that the circuit sees a suitable load under open-circuit conditions), labelled R_x/C_x in Figures 9.28 and 9.30, together with some means for ensuring that the open-loop gain will decrease with increasing frequency. This will normally take the form of a simple capacitor, shown as C_y in the schematic layout of Figure 9.30, connected to impose a dominant lag on the phase-shift/output-frequency characteristics.

Although this type of HF compensation is effective in securing HF loop stability, and very widely used for this reason, it suffers from the snag that the rate at which C_y can charge or discharge is limited by the output current from Q_1, and this imposes an upper limit on the speed with which the voltage output of the gain block can slew, following a sudden change in the voltage level presented at its input. This effect is

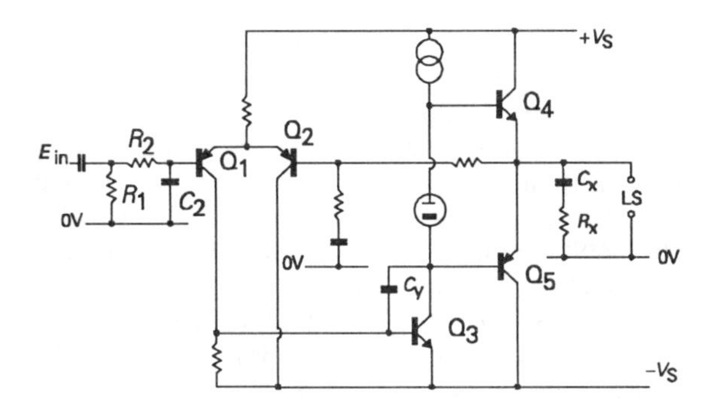

Figure 9.30 *Dominant-lag method of HF compensation for feedback amplifier*

known as slew rate limiting and gives rise to the phenomenon described by Otala (Otala, M. J., *J. Audio Eng. Soc.*, 1972, No. 6, pp. 396–399), as 'transient intermodulation distortion' (or TID), in which all signals accompanying such a sudden change in the input signal voltage level will be obliterated until the circuit returns to its steady-state condition. This effect is audibly quite unpleasant, but can be lessened, though not entirely avoided, by the use of a network, R_2/C_2, which serves to lessen the rate at which the input voltage can change.

The use of the stabilization component, C_y, in the position shown in Figure 9.28, avoids this snag, in that while it also serves to reduce the gain of the voltage amplifying stages with increasing frequency, and thereby impose a phase characteristic which assists loop stability, there is much less difficulty in providing the charging current from the drain of Q_7, and therefore much less tendency to slew-rate limiting. This HF stabilization method allows the retention of good overall phase margins within the feedback loop, which, in turn, makes for pleasant amplifier sound quality. Regrettably, the HF compensation layout shown in Figure 9.30 is normally used in commercial audio amplifier designs because it gives slightly better harmonic distortion figures at higher audio frequencies.

Additonal HF phase correction components will almost certainly need to be used with more complex circuit layouts in order to ensure adequate overall loop stability, but their position and value will need to be determined specifically for each new design.

The Quad current dumping amplifier

Because the performance of all normal class AB audio amplifiers will depend on the accuracy of setting of the quiescent current of the output stages, which will usually require some form of adjustment, with instrumental monitoring of the results, before the amplifier leaves the manufacturer's assembly line, and, because the required operating current, even when initially correctly set, may drift away from the desired value during use, the manufacturers of such units can never guarantee that the desired performance is always given.

Various attempts have therefore been made over the years to devise systems which will operate satisfactorily, with low distortion, with output stages which operate at zero quiescent current. Of these, the only design which has proved really satisfactory is that described by Walker and Albinson (Walker, P. J., *Wireless World*, Dec. 1975, pp. 560–562), and introduced by Quad (The Acoustical Manufacturing Company), in their '405' and subsequent power amplifier systems.

The idea employed is, in principle, very simple. The power amplifier consists of a basic class B system, with which is associated a low power, but very low distortion, voltage amplifier, and this low power amplifier is arranged to fill in those regions of the output waveform in which the output of the power amplifier departs from the ideal. Since these departures are likely to be small, probably less than 1% of the total, the output power requirements from the low distortion amplifier will also be small. However, although this idea is simple in concept, its implementation requires some subtlety in design, and the way in which it is done is shown shown in a simplified schematic form in Figure 9.31.

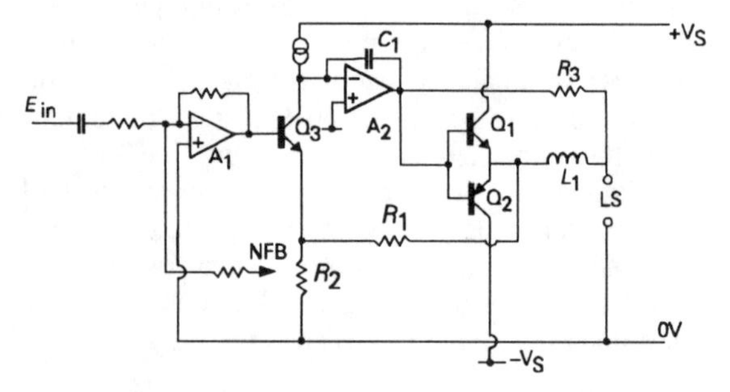

Figure 9.31 *The basic layout of the QUAD current dumping amplifier*

The operation of this circuit can be explained by consideration of a simple amplifier of the form shown in Figure 9.32a, in which a high gain linear amplifier (A_1) is arranged to drive a load through an unbiased push–pull pair of output emitter follower transistors, Q_1 and Q_2. Without any negative feedback to straighten out its transfer characteristics, the input/output transfer curve of this arrangement would have the shape shown by line 'a' in Figure 9.33. The slope of this line would be steep from D to E, when transistor Q_1 was conducting, and feeding current into the load resistor Z_L. It would, however, be much less steep from E to F, when Q_1 had ceased to conduct, and the load was only driven by A_1, by way of R_3. It would then steepen up once more, from points F to G, when transistor Q_2 began to conduct. If enough overall negative feedback was applied, by way of R_1, this could straighten out the kinks somewhat, to give the kind of characteristic shown in line 'b', but this type of performance would still be unacceptable as a Hi-Fi design. What is required is some way of reducing the gain of the circuit during the periods when Q_1 and Q_2 are conducting, so that the slope of the transfer graph is the same from D to E, and from F to G as it is from E to F. This can be done if an additional small resistor, R_4, is connected between the junction of Q_1 and Q_2, and the load resistor, Z_L, and if the negative feeback through R_1 is then taken from the junction of Q_1 and Q_2, rather than from the junction between R_4 and Z_L. When the ratios between R_2 and R_3 and R_1 and R_4 are correct, the discontinuity in the amplifier transfer characteristic disappears, in spite of the fact that Q_1 and Q_2 operate with zero forward bias.

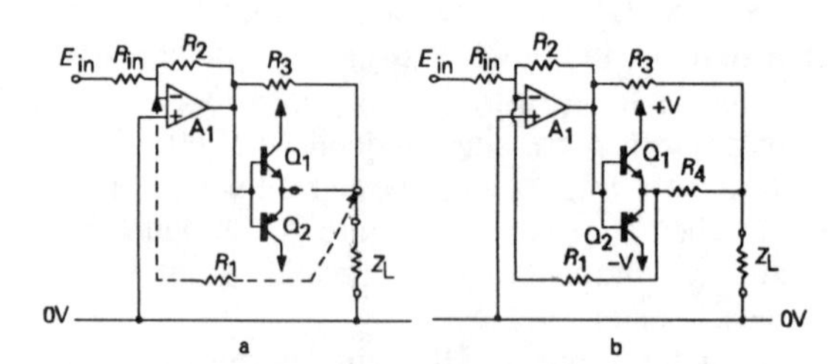

Figure 9.32 *Method of operation of current dumping power output stage*

In practice, R_4 would be wasteful of output power, so Walker and Albinson proposed the use of a small, low resistance inductor (L_1) instead, with R_2 being replaced by a suitable value of capacitor (C_1) to ensure

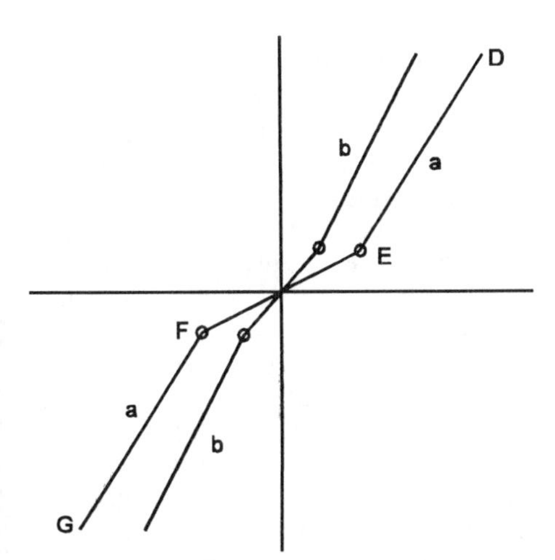

Figure 9.33 *Input – output transfer curve of circuit of Figure 9.32a*

that the ratios of the impedances, R_1/L_1 and C_1/R_3, remain the same over the audio frequency band. This modification of the circuit, while entirely effective in practice, complicates the theoretical analysis of the circuit, and gave rise to a lot of subsequent discussion and analysis, of which the most detailed was that of McLoughlin (McLoughlin, M., *Wireless World*, Sept. 1983, pp. 39–43, and Oct. 1983, pp. 35–41).

Sandman's Class S design

A further, very interesting design innovation, having a similar intention – making the bias of the output power transistors a less important factor in determining the amplifier performance – was that due to Sandman (Sandman, A. M., *Wireless World*, Sept. 1982, pp. 38–39), which is shown in schematic form in Figure 9.34. This design was the culmination of an investigation, by Sandman, of the possibility of reducing amplifier distortion by the use of 'error feed-forward' techniques, to supplement the existing negative feedback technology.

The particular circuit design employed was based on the observation that the performance of most amplifiers is better with a high-impedance than with a low-impedance load, and the circuit is arranged so that the input amplifier, A_1, sees the load as if it were a much higher impedance than it really is. This is achieved by the use of a high gain error amplifier, A_2, which is connected to sense the difference between the

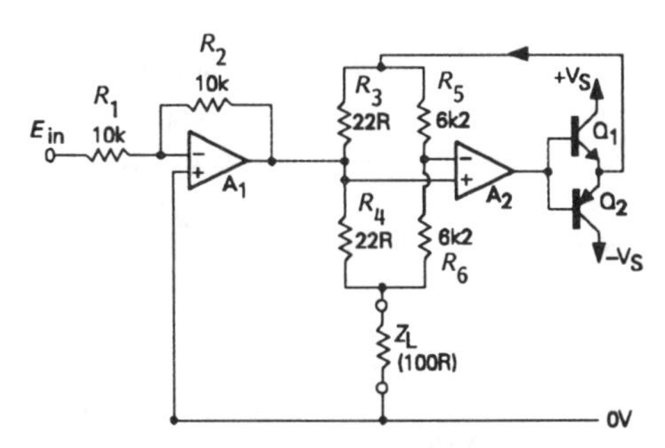

Figure 9.34 *The Sandman class S design*

output voltage of A_1 and that from an unbiased emitter-follower pair (Q_1/Q_2), which also is connected so that it feeds power into the load, and to drive Q_1/Q_2 so that this difference is made as small as possible, thereby increasing the apparent impedance of the load, as seen by A_1. Once again, as in the current dumping amplifier, the small power amplifier, A_1, is able to fill in the deficiencies of the higher power circuit, and if the ratios between R_3 and R_4, and R_5 and R_6 are correctly chosen, the crossover-type distortion, due to the unbiased power transistors (Q_1/Q_2), will disappear.

The basic philosophy of this system has been adopted in several Japanese audio amplifier designs, and a representative example of this approach, used by Technics, in their SE-A100 amplifier, is shown in Figure 9.35.

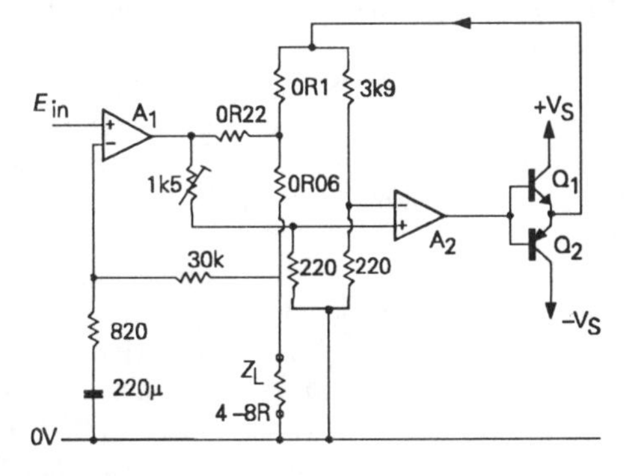

Figure 9.35 *Simplified layout of Technics SE-A100 power amplifier*

The future

Whether or not it is practicable to make a sufficient improvement to the best of the existing audio power amplifiers, that an impartial listener would be able to detect any difference in the final output sound quality, is debatable. Certainly, a number of comparative trials have been set up, over the years, with the specific aim of trying to establish whether critical listeners can distinguish, on the basis of 'sound quality', between different, high quality, amplifiers, as is commonly claimed in the Hi-Fi press, during equipment reviews. Indeed, it is this claim which provides the reason for the existence of the periodicals in question.

In order to ensure fairness, some of these trials were conducted with audience panels which were deliberately chosen to include those journalists and other critics who were convinced that all amplifiers sound different, and whose own 'in-house' tests had shown that they could hear obvious differences between known systems. They were also permitted to select the programme material and the ancillary equipment – loudspeakers, turntables, etc. – which was to be used during these tests. However, in the event, with tests based on statistically valid sampling techniques, and 'double-blind' methodology (in which neither the listening panel, nor the operators in charge of the test were aware of which amplifier was being auditioned at the time), all of these trials have failed to show that the audience was able to distinguish between one unit and another, more consistently than would have been the case for a purely random response.

For myself, I think that there are still some small remaining differences in sound quality between different power amplifier circuit designs, and that there is still, therefore, some purpose in attempting to improve on existing system performance, quite apart from the normal engineering search for ways of making such equipment simpler, more efficient, more reliable, and less costly.

The problem with any kind of multiple comparison test which can be conducted is that the sampling duration may either be individually too brief to allow the listener to pick up various small acoustic effects, which might allow a preference to be formed, or, alternatively, it may be so long that the memory of comparative sound characteristics will fade, where the differences sought are, of themselves, relatively small. On the other hand, a more leisurely side-by-side comparison between a pair of systems, conducted over a period of time, can allow the identification of particular, and individual, sound effects which could very easily be overlooked in such group tests.

An example, in point, is the difference between a good, 38cm/second analogue tape recording, and a '16 bit' encoded digital recording made of the same performance. Both of these can offer very high quality sound images, of which the principal difference lies in the nature of the background hiss. Given time, it is possible to distinguish between one and another, but in a multiple, short duration, test it may prove impossible to tell them apart.

Since, at the moment, most integrated circuit components are restricted to power supply line voltages less than +/–20V, and this is too low to allow enough output voltage swing at the LS load of a power amplifier, the great majority of audio power amplifier designs are based on discrete component circuitry, and the circuitry used is specific to each individual equipment manufacturer. This situation is bound to change with time, as IC technology is improved, to allow higher output voltages and powers, and when this happens, the choice between most audio power amplifiers will probably become simply the choice between one IC and another.

Preamplifiers

Basic requirements

The purpose of an audio power amplifier is to take a relatively low level input signal and to amplify it to an output voltage and power level which is suitable for driving loudspeakers or some other type of load. The purpose of a pre-amplifier – which may be either a separate unit, or a piece of circuitry incorporated in the same box as the power amplifier with which it will be used – is to select input signals from one or more programme sources, and to amplify and modify these, as necessary, so that they are of the size and form needed to drive the power amplifier. This will usually require some adjustment of both the signal level and the line impedance. For example, the programme sources may work best with load impedances which may be anywhere between 100 ohms and 100 kilohms, while the output signal levels from these sources may be between 50 microvolts – in the case of a low-output moving coil gramophone pick-up – and 2 volts, RMS, in the case of a typical compact disk (CD) player. On the other hand, the power amplifier may require an input signal in the range 300 millivolts up to 6 volts,

RMS, or more, to provide its full power output, and may have a fixed input impedance anywhere between 1k and 100kohms.

With any audio system it will nearly always be necessary to provide some means of adjusting the overall signal level, usually called a 'gain' or 'volume' control, and in a stereophonic system it will also be desirable to have a 'balance' control somewhere in the signal chain, to adjust the relative gain levels of the two stereo channels so that the final sound image is correctly placed. These controls will usually be placed somewhere within that part of the signal conditioning circuit called the preamplifier.

In the earlier years of High Fidelity audio amplifying equipment, it was often considered necessary to provide 'tone controls' and bandwidth limiting filters to modify the frequency response of the system, but, with improved signal sources, these are less often thought to be necessary in modern equipment.

Whereas the major problems with audio power amplifiers are concerned with producing an acceptable performance at the interface between the equipment and the listener, those of the preamplifier circuitry are more related to the purely technical aspects of the design, where there still remain a number of practical difficulties, such as that of avoiding signal voltage overload in the circuitry. This is generally referred to by the term 'headroom'.

Headroom

There is an absolute limit to the output voltage excursion which can be obtained from any simple electronic circuit – and this limit is imposed by the design of the circuit and the supply line voltages. For example, if a circuit is powered from a +/–24 volt supply rail pair, the maximum amplitude of any possible output voltage swing can not be greater than 48V. In practice, even with optimally designed circuitry, it will always be somewhat less than this, say 46V, peak to peak. This is equivalent to a symmetrical sinusoidal output of 16.26V RMS. If the stage has a gain of 100, the maximum sinusoidal input which could be handled without overload, and consequent peak clipping, will be 162.6mV. The concern of the designer is therefore to make sure that the output voltage swing, which is available from any part of the circuit, is greater than the likely input signal voltage multiplied by the stage gain.

Although it seems to be fashionable in the Hi-Fi press to portray this problem as one which requires the design of all preamplifier stages preceding the gain control so that they can handle very high signal voltages without overload, in practice most signal sources will, in any case, have a finite limit to their possible output voltage. For, example, in the case of signals derived from a vinyl gramophone record disk, the limiting factor is the ability of the pick-up (PU) stylus to track the undulations in the record groove, and the ability of the disk manufacturer to cut such lateral excursions without distortion, or breakthrough into adjacent grooves. These aspects were examined by Walton (Walton, J., *Wireless World*, Dec. 1967, pp. 581–588), who derived the relationships between groove radius, recorded signal frequency and maximum practicable stylus velocity shown in Figure 9.36. The recording levels normally found on a commercial disk are shown in curve 'd' in this drawing, and this approaches the ability of a high quality pick-up cartridge to follow the path of the groove. Allowing some margin for error, therefore, it is sensible to expect that the pick-up output will not exceed some 30dB, at 1–2kHz, with reference to a 1cm/s recording velocity. If the cartridge has an output of 3mV for a typical 5cm/s recorded velocity, which is typical of a high quality moving magnet (MM) cartridge type, the maximum output obtainable before acoustically unpleasant, and groove damaging, mistracking occurs will be 19mV.

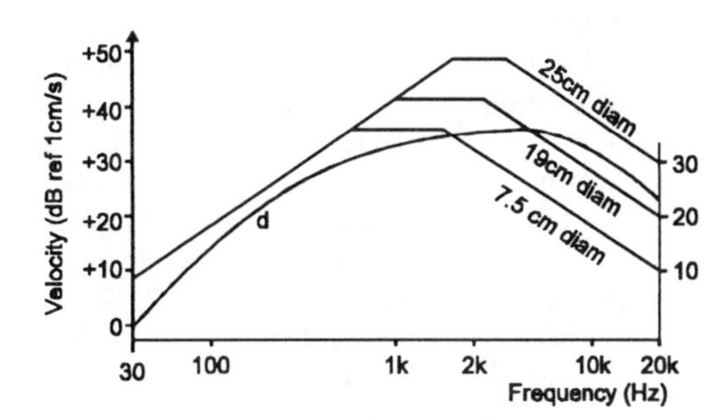

Figure 9.36 *Walton's analysis of maximum recorded levels on 12" vinyl disc records*

If a typical normal signal level to the power amplifier corresponds to a cartridge output of 2mV, then a preamplifier, without any facility of internal gain adjustment must be able to handle an overload factor of the order of 9.5 without clipping. For a typical transistorized power amplifier, requiring an input signal level of 0.77V RMS, such a preamplifier will require to be able to handle a peak output voltage of 7.3V RMS, which is equivalent to 21V p–p. A +/–15V supply line would therefore be adequate.

If the preamplifier gain control were to be positioned nearer to its input, a lesser degree of headroom would be adequate, but this would worsen the extent of the background noise level at the 'zero gain control' setting because there will be a higher overall gain, between the gain control and the amplifier output at such a 'zero level' setting, and this will amplify all the circuit noise generated by the stages following the volume control. For this reason, there is a conflict between the need to have the gain control at a circuit as near to the preamp. output as possible, to keep the background hum and hiss levels low, and the desire to offer higher levels of headroom. A customary compromise is to separate out the input circuits used for various input sources and to choose optimum, or possibly pre-set, gain levels for each of them.

Although the case of a moving magnet PU cartridge was quoted, the same arguments concerning tracking capability would apply to a low output-voltage moving-coil (MC) cartridge. The MC input channel would simply need a higher gain. However, since a low output level MC cartridge will usually also require a lower load impedance than an MM type (typically 100R, rather than 47k), most normal pre-amp. systems will provide a separate MC input channel, with appropriate gain and input impedance levels.

Similar constraints exist in tape, radio, and CD inputs. For example, if the preferred 'loud signal' recording level for a cassette tape is chosen by the user to be that shown as '0dB' by his recording level meters (where, typically, the 'off-tape' distortion might be 0.5%), then, at '+6dB', the tape third-harmonic distortion could be of the order of 3%, while at '+12dB' this distortion could have increased to some 15%. Since most listeners would regard such a distortion level as unacceptable, in practice they would choose to record at a peak signal level which was lower than this. So, in this case, a preamplifier input overload margin of some 12dB, (4×), would also be adequate.

In the case of an output from an FM tuner, the maximum signal level will correspond to a carrier deviation of +/–75kHz, equivalent to 100% modulation. (In the case of a stereo signal, the maximum modulation level is, in fact, limited to 90%). For a typical programme level of 10–15% modulation, the headroom required would be 10× (20dB) at the most. In the case of a CD player, typical maximum output levels are 2V RMS. Excursions above this level are precluded by the digitally encoded form of the input medium, where 2V RMS may represent the 'all 1s' condition of the digital signal. So, if the power amplifier has a 0.77V RMS input requirement for a full output signal, the interface circuitry between the CD payer and the power amp. input will only require to have a headroom of 2.6× (8.3dB).

All of these considerations assume that the preamplifier gain levels have been chosen to be appropriate to the input signal source being used, either by pre-set input gain settings, or, more usually, by the selection of circuit blocks with specific applications to a certain type of input.

Tone controls

The general purpose of these is to alter the relative gain of the circuit at one part of the audio spectrum in relation to another, to compensate for deficiencies in the characteristics of the input signal or in the loudspeakers and other hardware used with the system. A very wide variety of circuitry has been devised for this use, ranging from simple 'bass/treble' 'boost/cut' systems, through 'tilt' or 'slope' controls, to elaborate octave-band 'graphic equalizers'. All of these arrangements have their own advantages and shortcomings.

Bass/treble boost/cut systems

These were, historically, the first tone control systems to be used, and, typically, were expected to provide a frequency response curve which could be adjusted, as shown in Figure 9.37, to allow up to 20dB lift or cut in the 'bass' or 'treble' parts of the frequency band, in relation to that given at 1kHz, or some other mid-band frequency.

This frequency response adjustment could be provided by a simple frequency-selective attenuator circuit, of the type shown in Figure 9.38, followed by an amplifier, to restore the mid-band, 'flat response' gain to its previous level. This type of arrangement is usually described as a 'passive' tone control circuit because it relies for its operation entirely on passive components, and uses no internal gain stage.

The actual response characteristics of the circuit

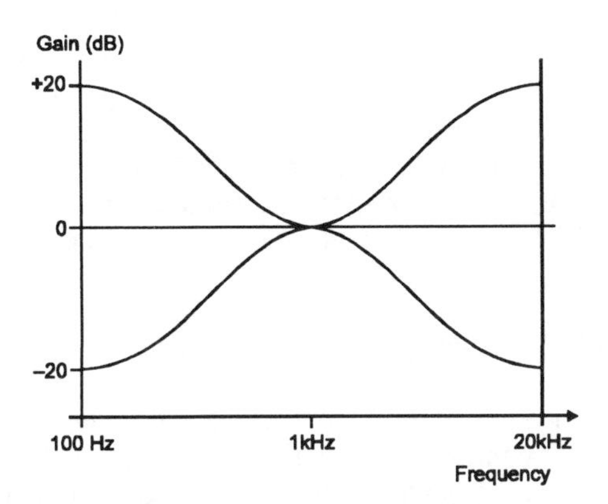

Figure 9.37 *Frequency response of feedback-type tone control*

shown in Figure 9.38, with the component values indicated, are as shown in Figure 9.39. A practical difficulty with this type of circuit is that a flat frequency response output is not provided at the mid-point resistance setting of 'linear' law potentiometers, and, even with 'logarithmic' law potentiometers, the calibration of the controls is awkward.

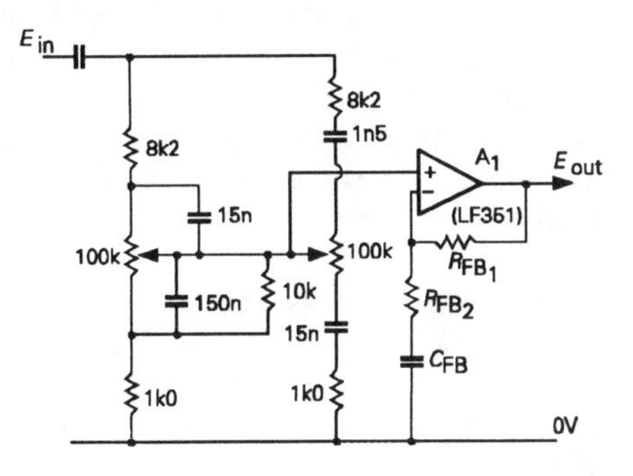

Figure 9.38 *Typical passive tone control*

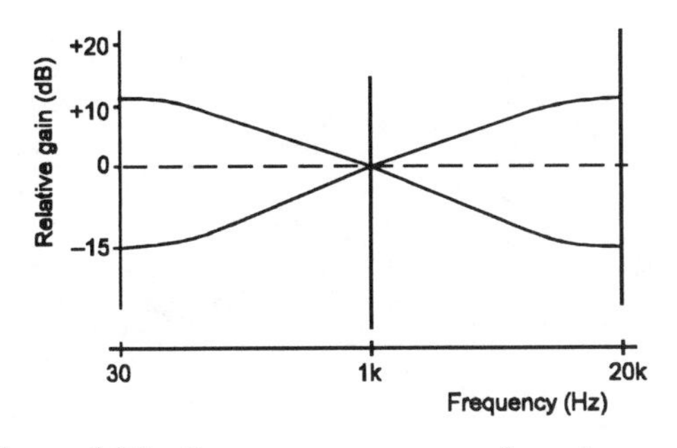

Figure 9.39 *Frequency response of passive tone control*

Feedback tone controls

This problem is avoided by the use of a circuit of the kind shown in Figure 9.40, in which the frequency selective *RC* networks are housed within the negative feedback loop of a wide band amplifier. This type of tone control circuit is often called a 'Baxandall-type' tone control, after its originator (Baxandall, P. J., *Wireless World*, Oct. 1952, pp. 402–405). The frequency response of the circuit shown in Figure 9.40 is as illustrated in Figure 9.37. With close-tolerance component values this circuit will give a flat frequency response at the mid-point settings of linear boost/cut control potentiometers.

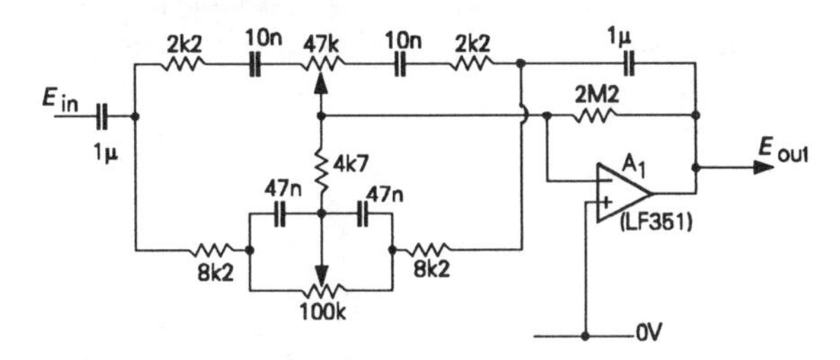

Figure 9.40 *Feedback type tone control*

A simple elaboration of this circuit (Linsley Hood., J. L., *Hi-Fi News and Record Review*, Jan. 1973, pp. 60–63), to increase its versatility, is to allow the values of the capacitors in the feedback networks to be changed by switching, to provide differing boost/cut turnover frequencies, as shown in Figure 9.41.

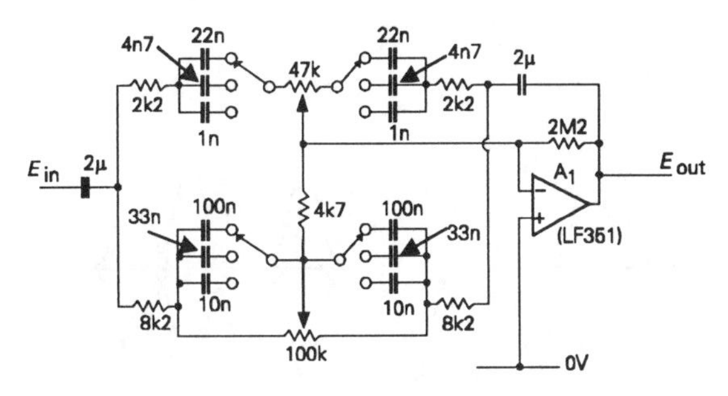

Figure 9.41 *Feedback tone control circuit with switchable lift and cut frequencies*

Graphic equalizers

Because it is obvious that no simple adjustment to the

relative gain of the bass or treble portions of the audio pass band will remedy all of the likely tonal errors in the reproduction of a musical signal – particularly where these arise through unwanted peaks and troughs in the overall frequency response of the system – various circuit layouts have been proposed which allow the gain at any part of the audio spectrum to be altered relative to the remainder.

It is customary to arrange that the parts of the spectrum on which the gain adjustments are made are divided into eight or nine octave segments, to cover the audio band from, say, 30Hz to 20kHz. A typical circuit layout, due to Williamson (Williamson, R., *Hi-Fi News and Record Review*, Aug. 1973, pp. 1484–1491), is shown in Figure 9.42. The possible response characteristics given by this circuit are shown in Figure 9.43.

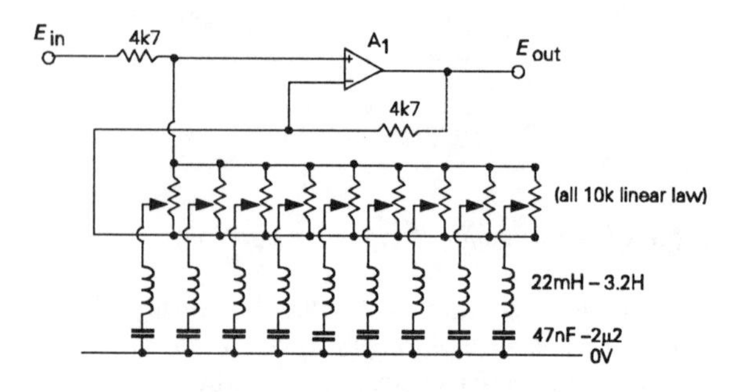

Figure 9.42 *Graphic equalizer circuit due to R. Williamson*

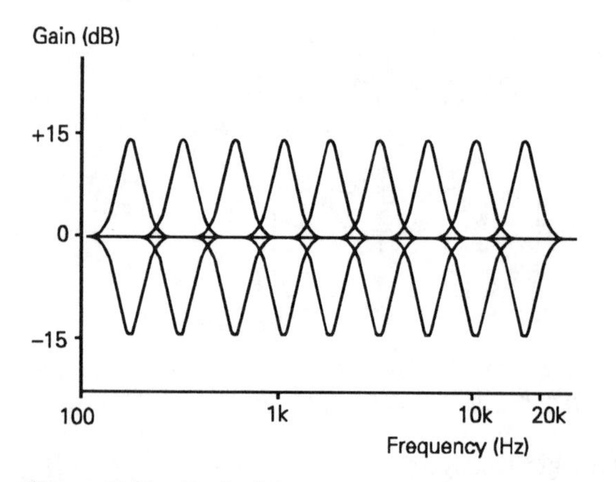

Figure 9.43 *Typical frequency response of graphic equalizer*

The difficulty with this type of system, apart from its complexity, and the tendency for the inductors to pick up hum because of their inadvertent interaction with stray mains- frequency magnetic fields, is that a ripple-free frequency response can only be obtained if all the controls are set to the mid-point position.

Parametric equalizers

A modification of this type of system, sometimes called a 'parametric equalizer' uses only one or two such narrow-passband lift or cut circuits, but with some means of altering the operating frequency, perhaps by altering the value of the capacitor in the *LC* segment of the circuit.

Tilt controls

In practice, with contemporary programme sources, and associated transducers, it is unlikely that any major errors in frequency response will be found, but, nevertheless, the user of the equipment may feel that the overall reproduction, at any given time, is too 'bright', or too 'bass heavy', and several circuits have been offered which allow some adjustment to be made to the overall slope of the frequency response across the pass band.

A circuit for this purpose was described by Bingham (Bingham, J., *Hi-Fi News and Record Review*, Dec. 1982, pp. 64–65). The layout used, and the possible frequency response adjustments allowed, are shown in Figures 9.44 and 9.45.

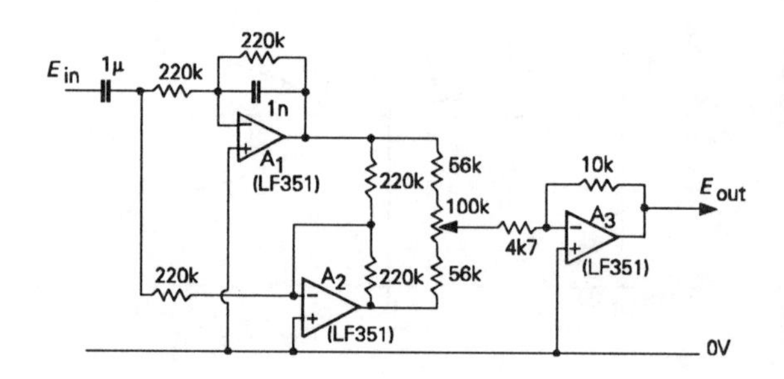

Figure 9.44 *Tilt control due to J. Bingham*

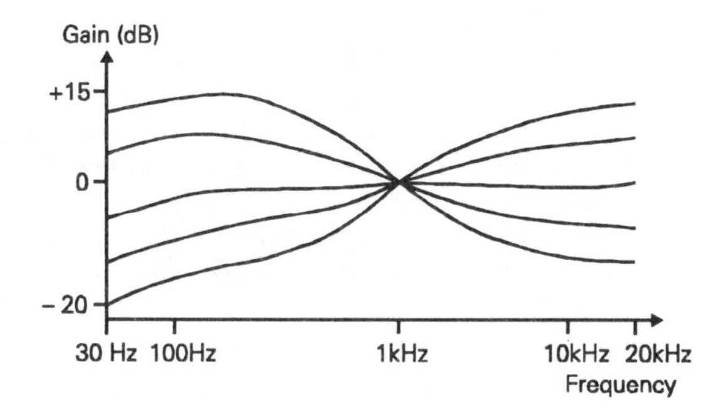

Figure 9.45 *Frequency response of Bingham's tilt control*

Filters

In the early days of audio, most programme sources suffered from unwanted additions to the signal, such as low frequency 'rumble' type noises originating from poor quality turntable bearings, 'hiss' due to the emery powder loading of 78 RPM shellac discs, or whistles due to adjacent broadcasting stations during radio reception, and various techniques were adopted to allow the unwanted part of the signal to be filtered out. Inevitably, there will be some loss of wanted programme material, so filtration is usually kept to the minimum level which is effective. Modern programme sources are generally of a higher quality, and less in need of filtering, so contemporary preamplifier units offer fewer of these facilities than would have been common in earlier Hi-Fi systems.

The circuits which can be used for this purpose are shown in Chapter 8, and the only point which needs to be made, in a specific audio context, is that the slope of the attenuation characteristic which is used needs to be chosen with care, since too steep an attenuation slope can cause audible coloration with wide-band programme material. This is not usually a problem with high-pass rumble filters, since room resonances and cabinet or driver unit resonances in the *LS* systems will mask such effects, but, at higher frequencies, low-pass treble filters having too steep an attenuation rate may give a noticeable tonality to wide-band white noise. Filters having a variable slope, as well as an adjustable turn-over frequency, provide the best approach.

Magnetic tape and gramophone record replay equalization

Magnetic tape replay equalization

In the early days of sound recording on magnetic tape, it became clear that there was an unavoidable loss of higher recorded frequencies due to the need for a finite gap in the pole pieces of the record head, shown schematically in Figure 9.46, across which any point on the magnetic tape will take a finite time to pass, and because the effects of magnetic cancellation within the magnetic dipoles in the tape coating become more evident as the recorded wavelength is shortened. Both of these effects cause a high frequency loss which increases as the tape speed is reduced, and some replay frequency compensation will be needed to offset this loss.

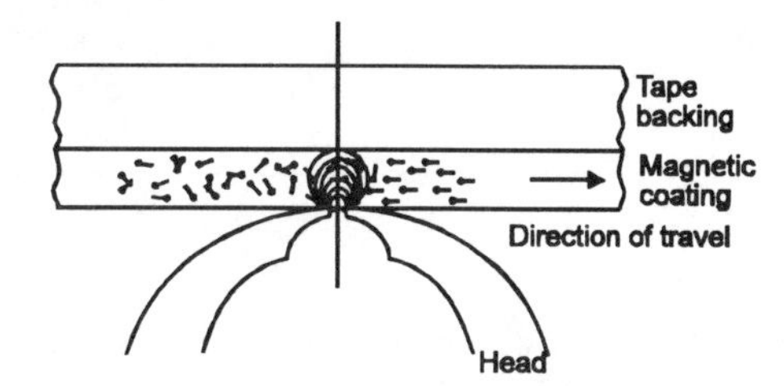

Figure 9.46 *Action of magnetic recording head*

In order to make tapes recorded on one machine compatible in replay on another, various standards have been proposed, which are listed in Table 9.1, which make the asssumption that the relative remanent magnetic flux on the tape, at various record/replay tape speeds, will be as shown in Figure 9.47. These mainly concern HF signal loss, but in the case of the 'NAB' proposals, and also in the case of all of the 4.76cm/s (cassette tape) recording systems, some degree of LF pre-emphasis is also proposed so that replay de-emphasis, either electronically introduced or inherent in the replay head characteristics, will reduce the extent of hum pick-up and improve the LF signal to noise (s/n) ratio.

The actual resistor–capacitor networks which will be necessary to provide the required record and replay frequency response will be similar to those shown in Figure 9.48, though the actual component values required will depend on the characteristics of the record and replay heads, and on the tape type and equalization

Table 9.1 *Frequency response equalization standards*

Tape speed	Standard	Time constants, μS (TC2) HF	(TC1) LF	–3dB points HF(kHz)	LF(Hz)
15"/s (38.1cm/s)	NAB	50	3180	3.18	50
	IEC/DIN	35	—	4.55	—
7.5"/s (19.05cm/s)	NAB	50	3180	3.18	50
	IEC/DIN	70	—	2.27	—
3.5"/s (9.53cm/s)	NAB/IEC /DIN	90	3180	1.77	50
1.875"/s (4.76cm/s)	DIN (Fe)	120	3180	1.33	50
	(Cr)	70	3180	2.27	50

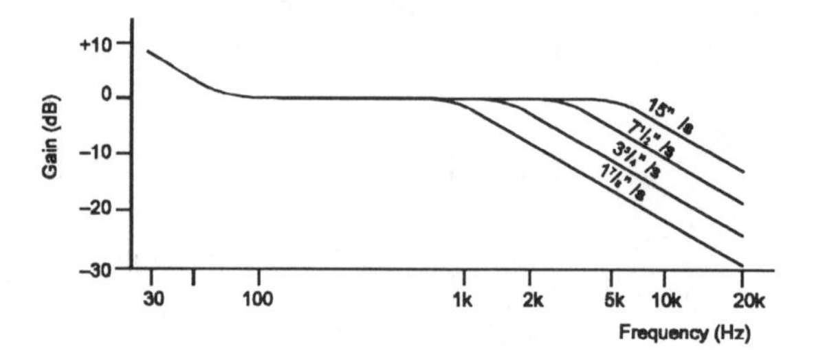

Figure 9.47 *Assumed remanent flux on recording tape equalized for various tape speeds*

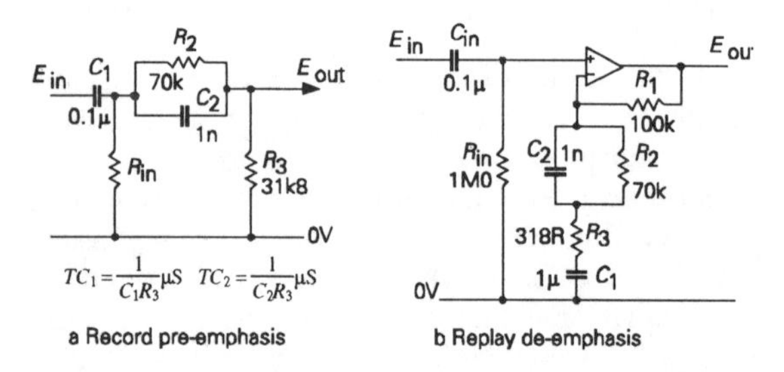

Figure 9.48 *Tape replay equalization network*

time constants chosen – bearing in mind that the design requirement is that the final record/replay frequency response of the system shall be as flat as possible.

Gramophone record replay equalization

In early shellac (78 RPM) gramophone records attempts were usually made to lessen the obtrusiveness of the replay hiss, due to the emery powder loading of the disc, by reducing the HF response of the replay amplifier. The record company RCA, in the USA, therefore proposed that some degree of pre-emphasis should be applied, at the recording stage, so that the HF response of the recording could be preserved in spite of this replay roll-off, and records having this characteristic were marketed under their 'Dynagroove' label.

With the development of vinyl long playing records in the early 1950s, this basic concept of recording pre-emphasis, to allow HF roll-off on replay to reduce surface noise, coupled with an LF de-emphasis to reduce the width of groove-to-groove spacing, was adopted by the Radio Industries Association of America (RIAA), who published a proposal, now accepted on a world-wide basis, for a recording frequency response which would require a replay characteristic – when reproduced by a velocity sensitive pick-up cartridge – of the type shown in Figure 9.49. The time- constants and 3dB points associated with this are 3180μs (50.05Hz); 318μs (500.5Hz); and 75μs (2122Hz), and one or other of the passive or active networks shown in Figure 9.50 can be used to generate this type of replay frequency response. (*Note.* It has recently been proposed that the RIAA replay response characteristic should be modified to include a further low-frequency roll-off, to reduce the effect of replay turntable rumble, having a –3dB point at 20Hz (7950μs), but this is not included in the networks shown in Figures 9.48–9.51).

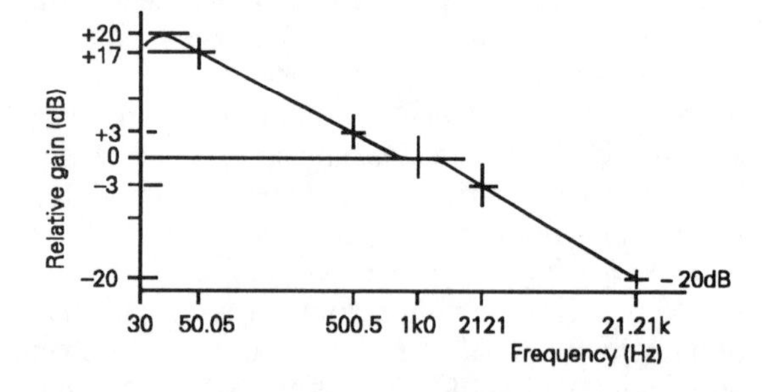

Figure 9.49 *RIAA equalized gramophone record replay response*

To achieve the required frequency response curve, in either of the arrangements of 9.50a or 9.50b, R_a/R_b = 6.88, C_aR_a = 2187μs, and C_bR_b = 109μs, while in the circuit layouts of Figures 9.50c and 9.50d, R_a/R_b = 12.40, C_aR_a = 297μs and C_bR_b = 81.21μs. In practice,

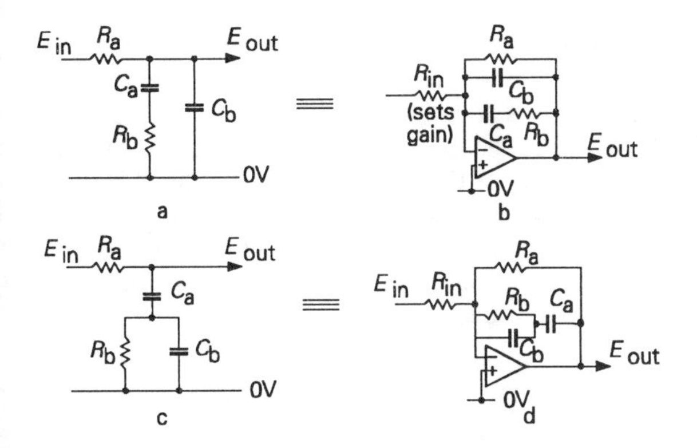

Figure 9.50 *R I A A equalization networks*

component values from the preferred numerical series may often be found which will provide close approximations to the required ratios.

It is, of course, also possible to generate the required frequency response characteristics by the use of a series of separate *CR* networks, each chosen to generate a portion of the curve, and each isolated from the others by a suitable buffer stage, as shown in Figure 9.51.

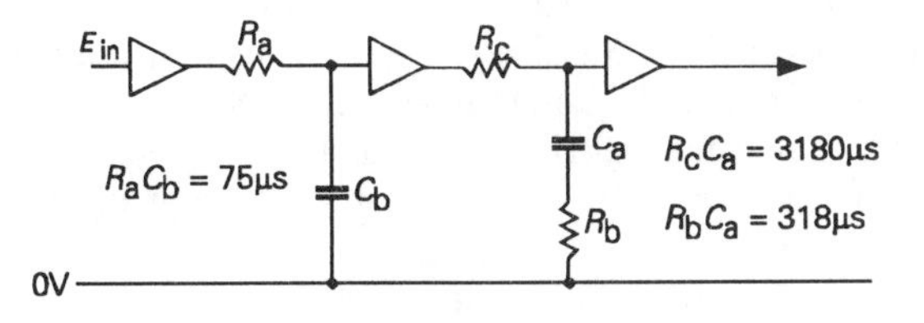

Figure 9.51 *Multiple stage RIAA equalization circuit*

Low noise circuitry

The reproduction of signals from vinyl (LP or EP) discs can provide particular problems in attaining a desirably high signal to noise (s/n), ratio since many gramophone pick-up cartridges have very low output voltage levels. A normal good quality moving magnet pick-up cartridge, such as the Shure M75E, may have an output of 5mV for a groove modulation of 5cm/s at 1kHz, but the higher quality Shure V15/5 cartridge has an output of only 3mV for the same groove modulation velocity. This input signal level may need to be amplified to 0.77V RMS in order to drive the audio power amplifier fully, and a s/n ratio of better than +65dB is typically sought. This requires that the total mains induced hum and noise, from all sources, must not exceed 1.7 microvolts, referred to the input. This requires that the screening of the input leads shall be adequate, that care will be taken to avoid hum inducing earth loops, and also that the input circuitry is chosen so that the minimum amount of circuit and component noise is introduced.

Understandably, this problem is aggravated when very low output voltage pick-up cartridge types are used, such a those using moving-coil generator systems, which can give output voltages as low as 70 microvolts/cm/s groove velocity, where a 65dB s/n ratio would require an input hum and noise level of less than some 40 nanovolts.

A typical contemporary gramophone pick-up input stage, using audio quality operational amplifier ICs, is illustrated in Figure 9.52, in which the required frequency response shaping is accomplished by R_3–R_4 and C_4–C_3 in the feedback network. This method of connection is called 'series' feedback, and is usually chosen because it allows the impedance values in both the input circuit and the feedback network to be kept low, with consequently low thermal noise values. The input impedance presented to this circuit is that due to the cartridge load resistor, R_1, and the pick-up coil connected in parallel with this – which could have an inductance of 1H and a coil winding resistance of 2k ohms. At low frequencies the input circuit impedance due to these component values will be about 1k9 ohms, but as the frequency increases, so the impedance due to the pick-up coil inductance will begin to dominate, and the input impedance and the thermal noise due to this will begin to increase.

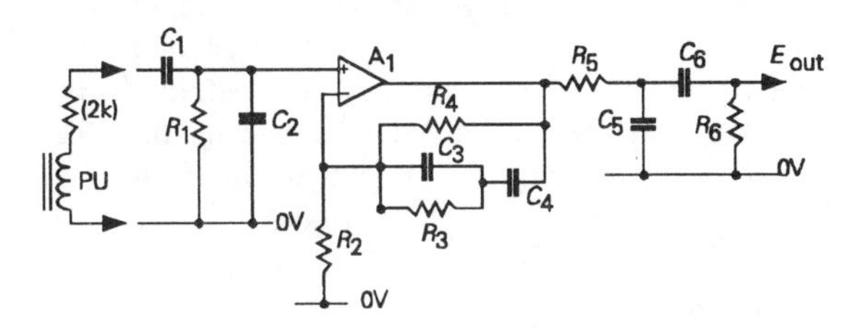

Figure 9.52 *Typical high quality input RIAA equalization circuit*

In the case of the shunt fedback connection, shown in Figure 9.53, the input impedance seen by the circuit is that due to both the pick-up coil and the load resistor,

R_1, in series with it, giving a higher impedance at low audio frequencies, and a higher thermal noise level in this part of the audio band. However, in this case, as the operating frequency is increased, so the gain of the circuit, in respect of the input noise component, will decrease. This leads to a difference in thermal noise characteristics between these two circuits, in this application, in which that of the shunt feedback circuit is more of a 'rustle', whereas that of the series arrangement is a higher pitched hiss.

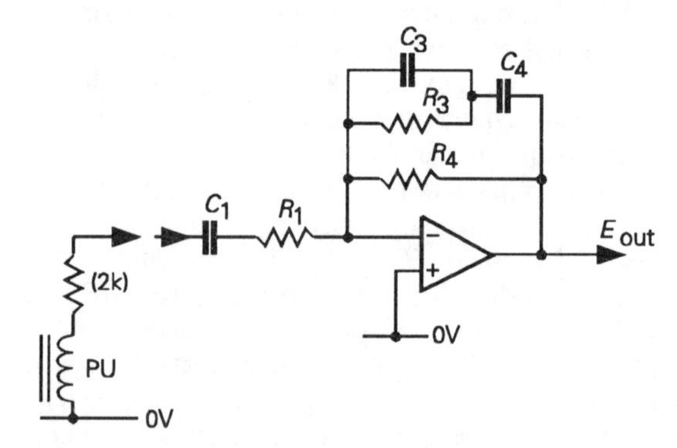

Figure 9.53 *Shunt feedback RIAA equalization stage*

Overall, the relative performance of these two circuit arrangements, in respect of thermal noise, favours the series feedback layout, below about 5kHz, and it is therefore the preferred form in commercial use.

The somewhat inaccurate adherence of this type of circuit – in that part of its frequency response above 5kHz – to the frequency response characteristics defined by the RIAA specification, shown in Figure 9.50, due to the fact that the stage gain asymptotes to unity rather than zero, at HF, can be corrected by a small additional R/C lag circuit, R_5–C_5. Alternatively, the error may be minimized by ensuring that the closed-loop gain is high, by keeping the impedance of the feedback limb high in relation to R_2. This aspect of circuit design is considered in greater detail in Chapter 6.

In general, the target of low circuit noise is achieved by keeping all of the circuit impedances as low as possible, especially at the input of the circuit; by trying to ensure that the gain of the input stage is sufficiently high that noise introduced by later stages will be small in comparison with the input noise component; and by careful choice of components.

In the case of bipolar transistors, the input noise is mainly that due to the 'base spreading resistance' – the apparent resistance between base and emitter – and this will depend on the input base current and the effective area of the base–emitter junction. This can be minimized by increasing the effective base junction area, either by the use of a transistor intended for use at higher power levels, or by connecting a number of small-signal transistors in parallel, as shown in Figure 9.54.

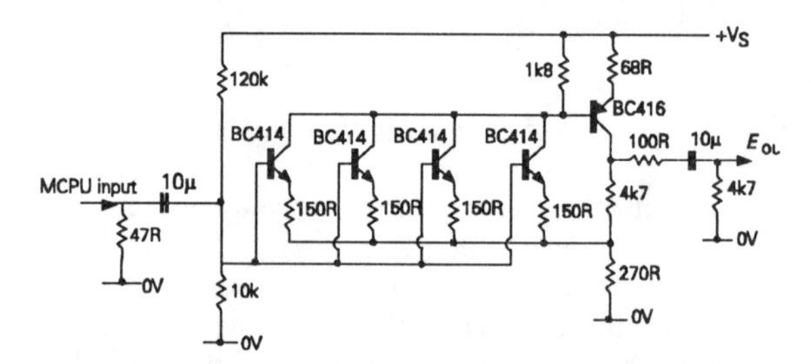

Figure 9.54 *Low-noise PU head amplifier using parallel connected input transistors*

There are a number of commercially available integrated circuits, such as the LM394, in which a closely matched pair of transistors is made up from a large number of parallel-connected small-signal bipolar devices, with the specific aim of achieving a very low input base spreading resistance, and a consequent very low input noise.

The other major noise sources are those due to 'shot' noise – effectively that due to the random arrival of electrons at the collector – which is statistically more important as the current through the circuit is reduced, and '$1/f$' or 'flicker' noise, due to the random choice of current paths through the device, and this is a mechanical feature of device construction, and will vary from one type of device to another.

So, in constructing low-noise circuitry, the actual specification of the devices to be used should be examined as well as the intended circuit layout. For example, some transistors have a better noise specification, for the same operating current value, than others, and some resistors – such as metal film – have a lower flicker noise figure than, for example, carbon composition types, because film types have a smaller cross sectional area of conducting path.

Two design examples of low noise input stages, intended for use with low impedance moving coil

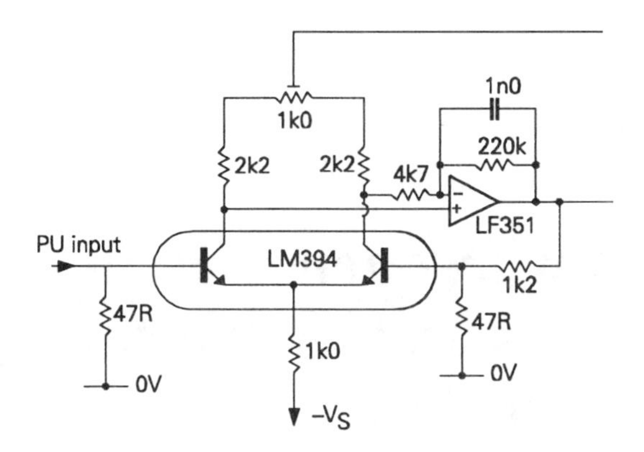

Figure 9.55 *Head amplifier using LM394 transistor array*

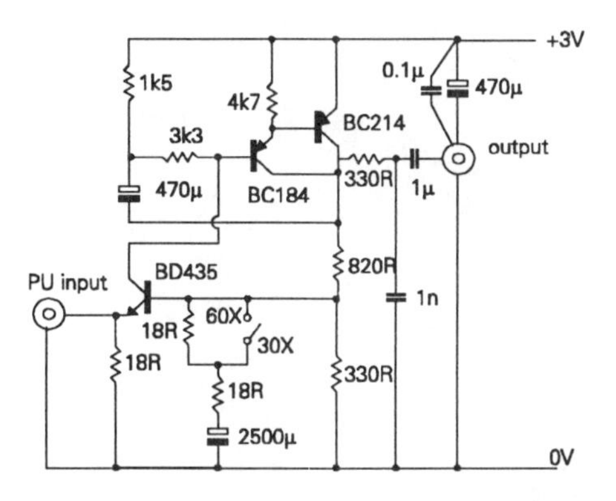

Figure 9.56 *Cascode input moving coil head amplifier*

gramophone pick-up cartridges, are shown, schematically, in Figures 9.55 and 9.56.

Similar considerations apply to the design of microphone amplifiers, and also in the design of the input stages of cassette tape recorders, which have a very low output voltage, typically in the range 500μV to 2mV RMS max., because of the very low tape speed, and the very narrow head pole-piece gaps, necessary to reproduce high audio frequencies.

A useful design feature, employed in many commercial cassette player systems, is to arrange that the amplifier circuit is muted while the tape is stationary, so that the user is never made aware of the amplifier background noise, which will be more audible in the absence of the reproduced audio signal.

Integrated circuit systems

During the past two decades, there has been a lot of activity on the part of semiconductor manufacturers aimed at the production of 'application specific' integrated circuits, intended to perform most of the functions which are required for audio preamplifier use. These have found ready application in low-cost Hi-Fi modules, but, in general, not only has the performance specification for these ICs been relatively low, but they have tended to become obsolete, and to be replaced by later, not necessarily compatible designs, after a relatively brief period of availability.

For this reason, there has been an increasing trend among audio system designers to base their circuitry upon IC operational amplifier gain blocks, of which there are an increasing number having an exceedingly high performance specification, coupled with a relatively low unit cost. This approach allows adequate flexibility in design and also allows the system to be upgraded, with great ease, as newer, or improved performance IC gain blocks become available.

10

Low frequency oscillators and waveform generators

Introduction

Most of the applications of linear electronic circuits are in the amplification of alternating electrical signals, of sinusoidal, or similar repetitive form, and it will usually have been necessary, at some earlier point in the system, to generate this signal, either as the output of some electro-mechanical transducer, or from some other electronic circuit arrangement.

Repetitive signals, of sinusoidal or other forms, may also be used as a means of testing equipment which is intended for use in other applications. In this case, the kind of signal which is used will depend on the type of test which is to be made. Nearly all of the oscillators and signal generators used in electronic circuitry are based on the use of positive feedback – see Chapter 7 – from the output of an amplifying or impedance converting stage back to its input, though there are a few exceptions to this, as in negative impedance oscillators and 'noise' sources, which will be considered later.

Sinewave oscillators

A number of techniques have been evolved for generating pure (i.e. noise, hum, and harmonic free) sinewave signals, of which the most common is to pass the feedback signal from a low distortion amplifier through a phase-shifting network chosen so that the required 360° loop phase shift is provided. These are generally classed as 'phase shift oscillators', even though, for example, in the case of the Wien Bridge and similar systems, the condition required of the *RC* network is that it should provide zero phase shift at the required operating frequency.

For a pure sinewave output, assuming a distortion free amplifier block, the gain of the system must be adjusted so that the overall feedback loop has unity gain, or either the system will not oscillate at all, or the amplifier will be driven into overload, which will cause the peaks of the output waveform to be clipped.

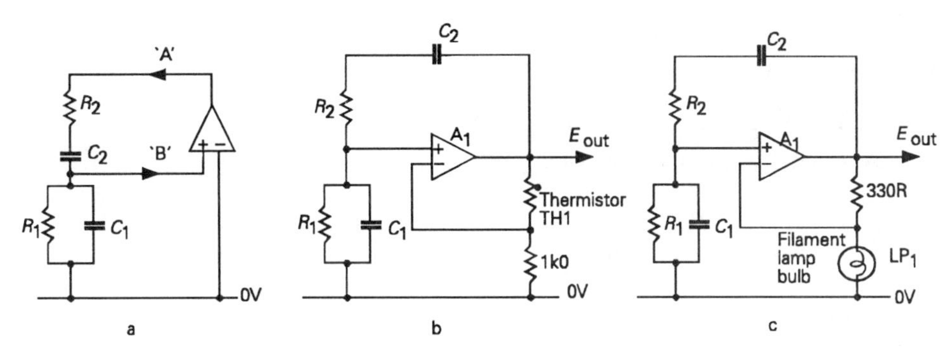

Figure 10.1 *Wien bridge oscillator layouts*

Phase-shift oscillators

Wien bridge systems

The simplest style of Wien Bridge oscillator is shown in Figure 10.1a. This design takes advantage of the fact that the Wien network, $R_1R_2C_1C_2$, shown in the practical design given in Figure 10.1b, has zero phase shift, from point 'A' to point 'B', at a frequency given by the equation

$$f_o = 1/(2\pi CR)$$

in the case where $C_1 = C_2$, and $R_1 = R_2$. At this frequency the upper limb, C_2R_2, has twice the impedance of the lower, C_1R_1, so the transmission of the network is 1/3.

If some means is used to adjust the system gain to 3, the circuit will oscillate, and, provided that the amplifier itself is linear, and the output voltage swing is not too large, a low distortion sinusoidal output signal can be obtained.

Clearly, some means for automatically adjusting the loop gain of the amplifier is desirable, and this is commonly done by the use of either a thermistor (TH_1), whose resistance will decrease, increasing the amount of NFB and reducing the loop gain, as the output AC signal increases, or a low power filament bulb (LP_1), as shown in the design given in Figure 10.1c, whose resistance will increase as the applied AC voltage is increased; which gives the same result. This type of circuit can be improved by separating out the two portions of the Wien network, as shown in Figure 10.2. In this arrangement, A_1 is used as a summing amplifier, combining the negative feedback path through R_1C_1 with the positive feedback signal through R_2C_2. The overall loop gain is stabilized, in this case, by connecting a thermistor, TH_1, as a negative feedback element across A_2.

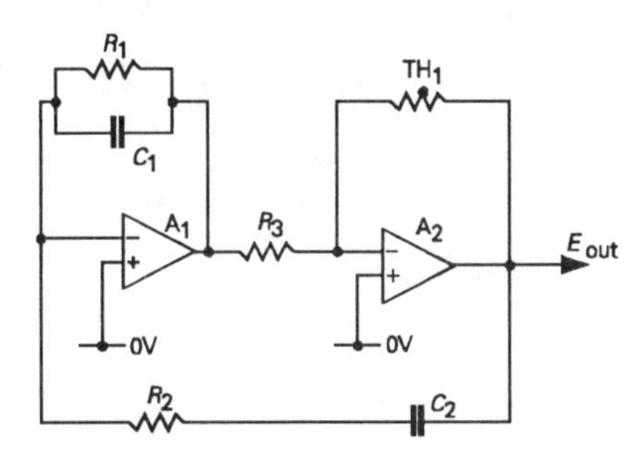

Figure 10.2 *Improved Wien bridge oscillator*

Because both of the amplifiers are operated in the phase inverting mode, so that the 'common mode input' (the signal appearing simultaneously on both the inverting and non-inverting inputs of the amplifier block) is zero, very low orders of harmonic distortion can be obtained from this layout. A typical total harmonic distortion figure for the circuit shown in Figure 10.2 is <0.002% @ 1kHz and 1V RMS output.

Lag/lead and lead/lag oscillators

There are two other layouts, shown in Figures 10.3a and 3b, which, for the same R and C values, have identical phase and transmission characteristics to the Wien network, and which can be employed in oscillator systems, as shown in the designs given in Figures 10.4a and 10.4b. Once again, these can be recast in circuit layouts free from common mode distortion effects, as shown in the circuit arrangements, due to the author, of Figures 10.5a and 5b. As in the case of the circuit of Figure 10.2, these designs are charcterized by the very low distortion of the output signal, which will consist of third harmonic residues, and be almost entirely due to the amplitude modulation effect, within the waveform, of the amplitude stabilization component.

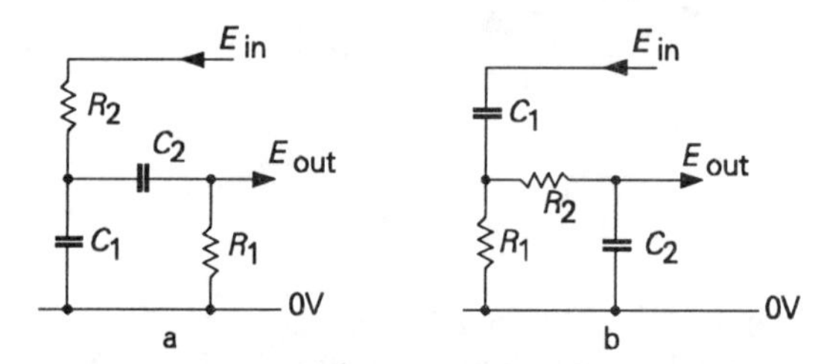

Figure 10.3 *Phase shift networks having an equivalent performance to the Wien bridge*

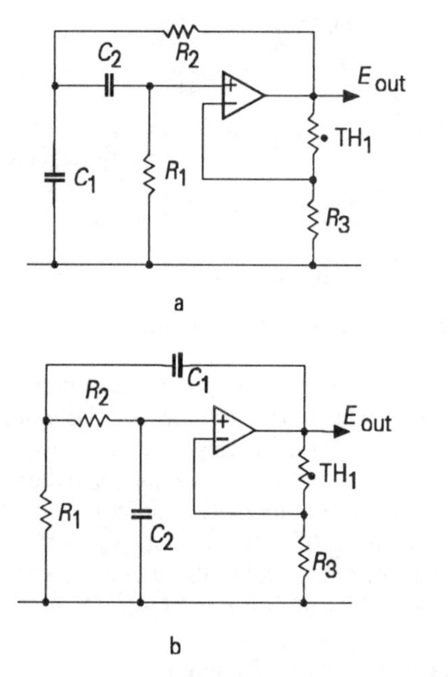

Figure 10.4 *Practical oscillator designs based on lag–lead and lead–lag networks*

The phase-shift or Dippy oscillator

This type of design, analysed fully by Gamertsfelder and Holdam (*Waveforms*, Chapter 4, MIT/McGraw-Hill, 1949), enjoyed considerable popularity some 40–50 years ago, but has now largely been forgotten, though it is capable of excellent results.

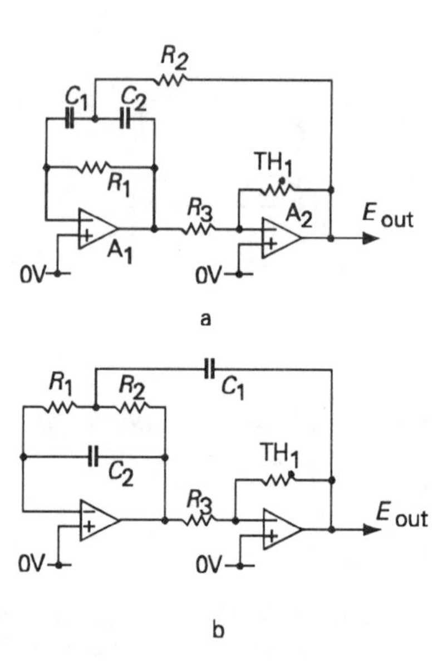

Figure 10.5 *Rearrangements of lag–lead and lead–lag oscillators to avoid common-mode type distortions*

Its method of operation is based on the fact that any simple *RC* network will give a phase shift of 60° at some input frequency, and that a group of these networks – at least three are needed – can therefore be made to give a total phase shift of 180°, to provide the necessary 360° loop phase shift, when connected between the output of an amplifying block and an inverting input. There are two ways of doing this, shown in Figures 10.6a and 10.6b. If all the values of *R* and of *C* are identical, these circuits will oscillate at frequencies given by the equations

$$f_o = \sqrt{6}/(2\pi CR) \quad \text{and} \quad 1/(2\sqrt{6}\pi CR)$$

respectively, and a loop gain of 29.25dB (29) is required.

In any practical design in which the purity of the output waveform is important, the circuit layout of Figure 10.6a is preferable – in spite of the fact that, for the same operating frequency, it will require the values of the capacitors in the phase shifting chain to be six times greater than that of the layout of Figure 10.6b – because it has a somewhat lower harmonic distortion, due to greater attenuation of high frequency harmonic components in the *RC* chain. Also, unless a bridging resistor, R_x, is included in parallel with the *CR* chain,

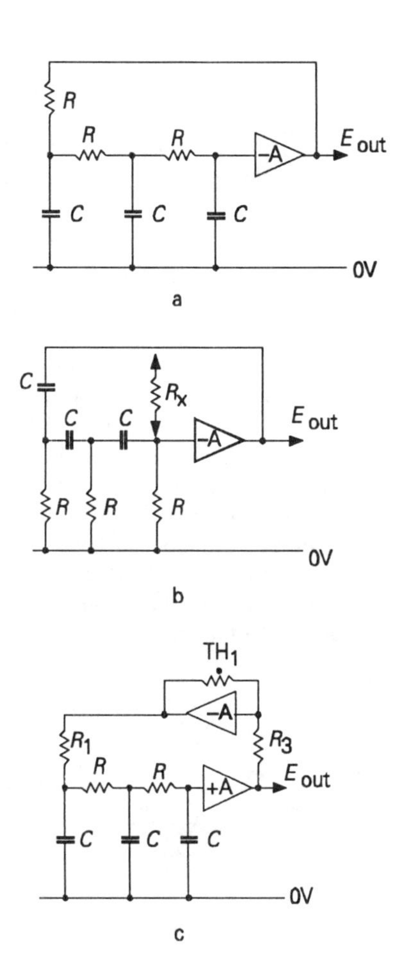

Figure 10.6 *Phase-shift or Dippy oscillator layouts*

which somewhat alters the loop gain and the operating frequency, the layout of Figure 10.6b does not provide the DC negative feedback path required to give stability of the amplifier DC level. Automatic control of the system loop gain to sustain oscillation at a suitable level, below the point of overload, can be provided by including a second amplifier, with a thermistor in its feedback path, as shown in Figure 10.6c.

A modern version of the phase-shift oscillator is made from a cascaded series of op. amps., or digital buffer amplifier ICs, each of which has a capacitor between its output and inverting input, as shown in Figure 10.7a. If a CMOS inverter gate is used, the time lags in each of these, due to propagation delays, will allow a phase-shift oscillator without any further components, as shown in Figure 10.7b. The operating frequency of such an oscillator is dependent on the supply voltage, allowing its use as a simple voltage controlled oscillator (VCO), but such an oscillator is very temperature dependent, and its long-term frequency stability is poor.

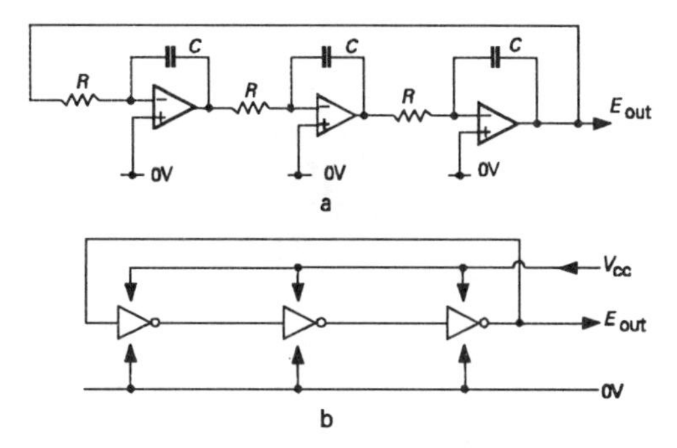

Figure 10.7 *Modern versions of phase-shift oscillator circuits*

The parallel-T oscillator

This is a very important type of oscillator, and has the general layout shown in Figure 10.8a. Its method of operation is simply that, since the parallel-T network will have zero transmission at some specific frequency, depending on the component values, so the amplifier will be free to amplify all those components of circuit noise, which occur at that frequency, with its full open-loop gain. Since all other frequencies will be subject to a large amount of negative feedback, the magnitude of these distortion components will be low.

In reality, though, the method of operation is rather more subtle. Most practical amplifier blocks will have an internal open-loop phase shift of at least –90° over a fairly wide part of their working frequency range, so since the transmission characteristics if the parallel-T shift abruptly from a phase lag of –90° to a phase lead of +90° on transition through the null transmission frequency, as shown in Figure 10.9, there will be a frequency, just below the null point, at which the overall phase shift in the feedback loop will be –180°, and the circuit will oscillate.

The circuit can be elaborated slightly, as shown in Figure 10.8b, to allow automatic stabilization of the circuit gain. Circuits of this kind can offer very low levels of harmonic distortion, and have formed the basis of commercial ultra-low distortion test oscillator designs.

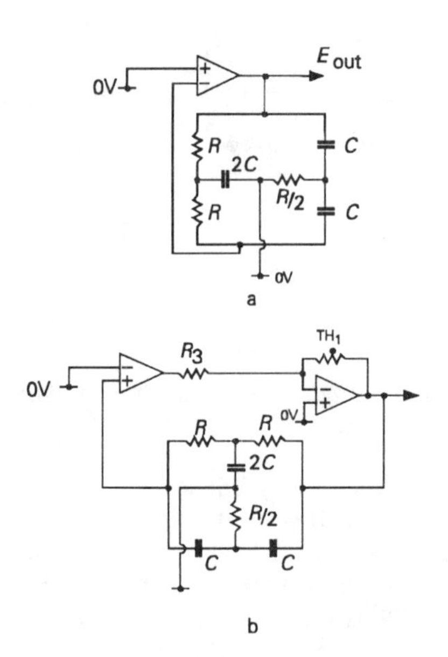

Figure 10.8 *Parallel-T oscillator layouts*

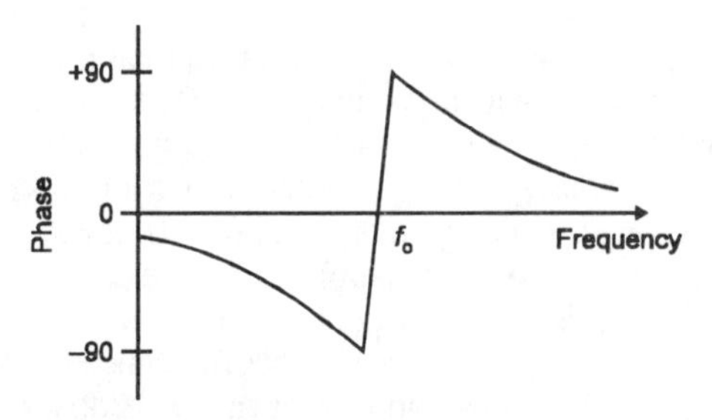

Figure 10.9 *Phase shift of parallel-T network on passing through f_o*

The Fraser oscillator

This circuit (Fraser, W., *Electronic Eng.*, May 1956), was originally offered as a thermionic valve design, but shown here with op. amp. gain blocks for simplicity, is based on the use of a pair of all-pass filter networks – see Chapter 8 – each of which will give a phase shift of 90°, interposed between amplifier blocks, as shown in Figure 10.10. A third gain control block, A_3, can be used, as shown, to stabilize the overall loop gain.

Beat-frequency oscillators (BFOs)

This type of oscillator, which is the pre-eminent choice

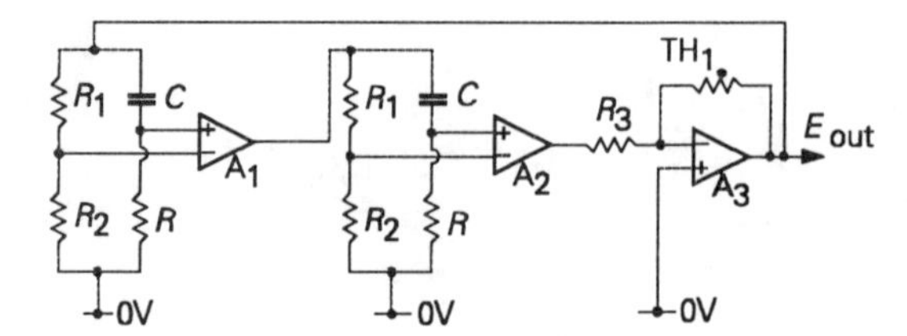

Figure 10.10 *The Fraser oscillator*

where a wide output frequency range is to be provided from a single knob frequency control, employs the type of design, shown in schematic form in Figure 10.11, in which the outputs from a pair of HF oscillators, OSC_1 and OSC_2, are combined in a 'sum and difference' frequency generator, PD_1, and the resultant output is filtered to remove the 'sum' frequency components. The output 'difference' frequency can then be made to sweep a range from zero up to as high a frequency as is required, depending on the range of the variable frequency oscillator.

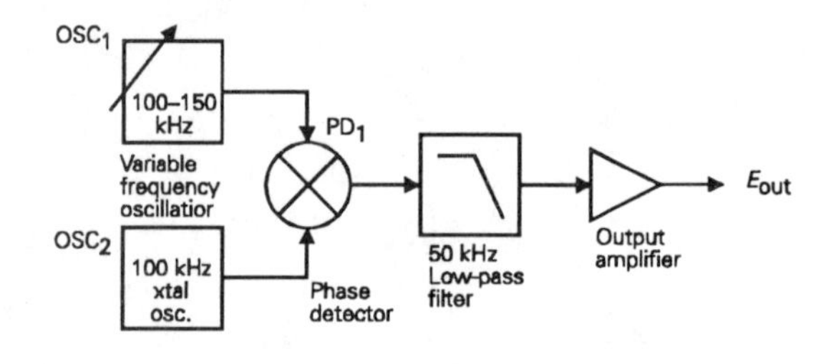

Figure 10.11 *Layout of beat frequency oscillator*

At least one of the input HF oscillators should have a low distortion output if the output signal is to be of high purity, since the difference frequency between any harmonic components present in both oscillators will generate a harmonic component in the output. If one oscillator is harmonic free, the spurious HF signals generated by harmonic components in the other can be removed by subsequent low-pass filtering. To ensure good frequency stability in the final output signal, very high stability is required from the two input oscillators. However, the fixed frequency oscillator can be crystal controlled, which will also assist in achieving a low harmonic content in the output signal.

The function of a sum and difference generator in a BFO can be performed by combining the signals at the input to almost any nonlinear circuit element, but the highest purity in the output difference waveform is best achieved by the use of a 'double balanced modulator' system, such as the Motorola MC1496 integrated circuit.

Amplitude control and amplitude bounce

All of the oscillators described above will only give a true, low-distortion, sinewave output if the amplitude of the output waveform is kept within the linear range of the amplifying devices. This requirement generally demands some method of automatic output voltage control, and I have sketched, in several of the circuits shown above, methods, using either negative temperature coefficient thermistors or positive temperature coefficient heated wire systems, by which this can be done. Unfortunately, all high purity sinewave generator circuits simulate, in their characteristics, a high Q tuned circuit (see Chapter 11), of which one of the most notable characteristics is that its bandwidth, in relation to its operating frequency, is narrow; and the higher the Q of the circuit, the narrower the bandwidth will be. This means that the system will only respond slowly to a change in the magnitude of any input sinusoidal signal applied to it. In this it resembles the operation of an *RC* integrating network, in which the time constant of integration is related to the Q of the circuit, and to the purity of the output waveform.

It is also necessary, in any amplitude control system, that there should be a long time constant, typically several seconds, in its response to a change in the signal amplitude, in order to limit the extent to which the wave shape of low frequency signals is distorted by the system gain being modulated by the instantaneous value of the waveform, in terms of its displacement from its mid-point potential. So the control network, itself, is required to simulate an *RC* integrating network. This leads to the amplitude control system having the general form shown in Figure 10.12, where C_1 and R_1 represent the effect of the narrow frequency response of the system (related to the sinewave output purity), and C_2 and R_2 represent the time constant of the thermistor or filament bulb element.

Typically, if a step-function voltage input, shown as line (a) in Figure 10.13, is applied to such a feedback system, it will lead to an output response, as shown by the line (b), and this is typical of the behaviour of high purity sinewave oscillators following initial switch-on or any subsequent disturbance – in which the output signal amplitude will exhibit a bounce as a function of time, exactly as shown in Figure 10.13b.

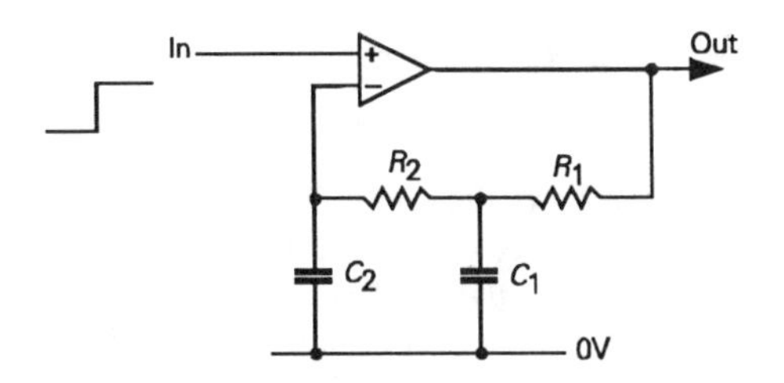

Figure 10.12 *Equivalent circuit of low distortion gain stabilized amplifier*

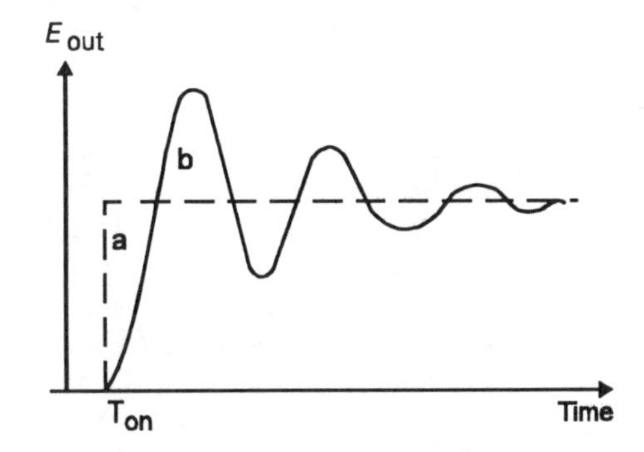

Figure 10.13 *Output amplitude bounce effect*

For any given amplitude control method, the duration of the bounce period, before the oscillator settles down, will lengthen as the purity of the waveform is improved, and as the operating frequency is lowered. Indeed, the duration of the bounce, in any simple amplitude control system, for any given operating frequency, can be taken as an indication of output waveform purity. This output signal amplitude instability can be a nuisance in practice, especially as an adjustment to the operating frequency in a variable frequency oscillator, is likely to trigger a sequence of amplitude bounces in the output signal.

Various methods have been proposed to avoid this snag, such as the use of two control systems, one to control the amplitude in the normal manner, and the second to 'clip' the output voltage swing at, say, 10% above its normal level: a system which will limit the duration of the amplitude instability.

However, the only really effective system is to use an input waveform which is already amplitude limited, such as an amplitude clipped triangular wave, and then to filter this waveform to remove unwanted harmonics, and to improve its purity.

Relaxation, or capacitor charge/discharge oscillators

This type of oscillator design, in its various versions, can generate a very wide range of output waveforms, almost all of which will be non-sinusoidal. These may be triangular, or 'sawtooth' (a triangular waveform in which the slopes of the 'rise' and 'fall' transitions are not identical), or rectangular, or square-wave (a rectangular waveform in which the 'on' duration is identical to the 'off' duration – a condition described as having an equal 'mark-to-space' ratio), or 'staircase, or any combination of these. Most of these circuits were developed for use with thermionic valve gain blocks, though for practical convenience I have shown these, modified as necessary, in their contemporary bipolar transistor, field effect device, or op. amp. forms. (A useful general survey of the more commonly found types, arranged for use with bipolar junction transistors, is given in the Ferranti 'E-line' transistor application report, 9th Edition, 1980.)

It is perhaps appropriate to note at this point that, in logic circuit nomenclature, free-running waveform generator systems are referred to by the general term 'astable'. Most of these astable designs can be configured, with some modification in circuit layout, so that they will become 'one shot' oscillators, and because of the demands of digital circuit technology there is now a wide range of such triggerable logic-function ICs.

However, in order to keep the scope of this book within reasonable limits, I propose to confine this description to continuous waveform generators, and to omit analysis of those multivibrator or similar systems which are 'mono-stable' – a term which is used to describe circuit arrangements which will always revert, following a cycle of operation, to some specific 'rest' condition, – or 'bi-stable' – a classification which denotes systems which can reside in either of two possible rest states.

The multivibrator

This is one of the longest established of all the rectangular waveform generating circuits, and is represented best, in modern circuit practice, by the junction FET layout shown in Figure 10.14. Its method of operation is that, on switch-on, the current flow through one or other of the FETs, Q_1 or Q_2, will be slightly greater than that of the other, and that, when the voltage drop which this causes, through either R_2 or R_3, is coupled to the gate electrode of the device passing the lower initial current, by way of C_1 or C_2, it will cause the current through that device to be reduced still more. This causes a further reduction in its drain current, and a consequent increase in the forward gate bias of the conducting FET, with the result that the cumulative effects of these changes causes the drain current of one FET to become completely cut off, while the other is passing a value of current which is characteristic of its zero gate bias state.

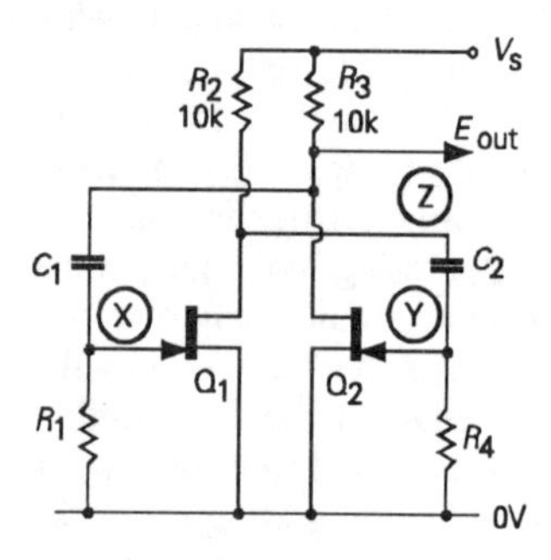

Figure 10.14 *Simple FET multivibrator*

However, the negative gate bias voltage developed across C_1 or C_2, in the circuit of Figure 10.14, will decay away with time through R_1 or R_4, with a time constant dependent on the values of R_1/C_1 or R_4/C_2, giving a gate voltage waveform of the type shown in Figures 10.15a or 10.15b. When the voltage has decayed to the extent that the device which is cut off begins to conduct once again, the fall in its drain voltage is coupled to the gate of the conducting FET, causing its drain current to become less, and causing a positive voltage to be applied to the gate of the FET which had been cut off. Once again, the cumulative effect of these voltage changes causes the gate voltage of this transistor to swing rapidly positive, up to the point at which gate current begins to flow (remember that the gate electrode of a junction FET is just a simple junction diode), which limits the forward voltage excursion to about +0.6V. This voltage will again decay back towards zero.

The circuit voltage waveforms for the points X, Y and Z are shown in Figures 10.15a, 10.15b and 10.15c.

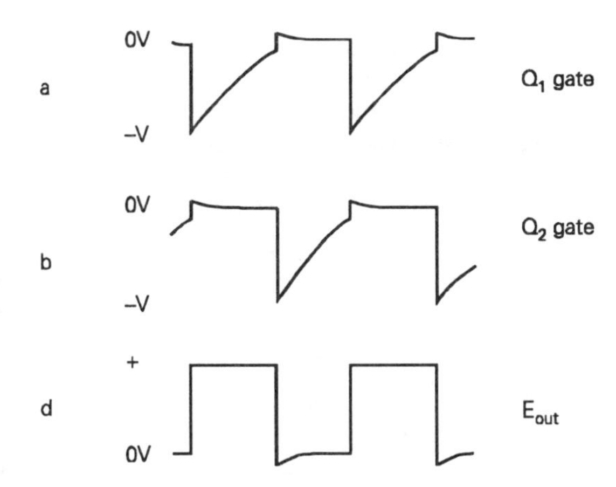

Figure 10.15 *Output waveforms of circuit of Figure 10.14*

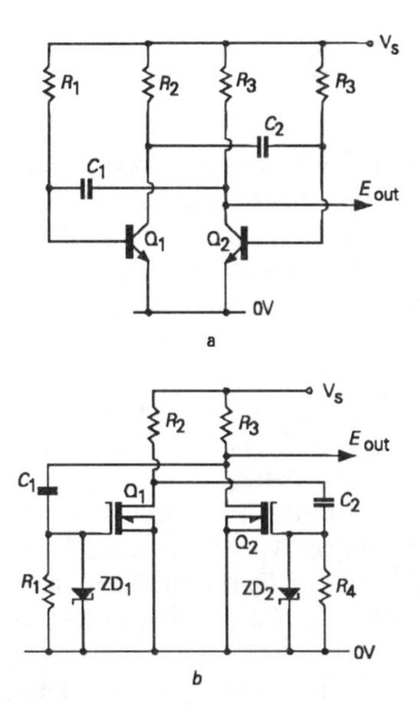

Figure 10.16 *Multivibrator layouts using bipolar and MOSFET transistors*

Because there is a small change in the forward gate bias on the conducting FET, following turn-on, the lower edge of the 'off' part of the waveform will not be quite horizontal in profile. The duration of the 'on' and 'off' segments of the output waveform, at Z (the mark-to-space ratio), is largely determined by the time constants of R_1C_1 and R_2C_4, and the conducting characteristics of Q_1 and Q_2. If all components are accurately matched, this circuit will give a good 1:1 mark-to-space ratio, though there are better ways of achieving this aim.

This circuit can – of course – be built using both bipolar and MOSFET type transistors, using the modified layouts shown in Figures 10.16a and 10.16b: a type of transposition which could equally well be done with any of the circuits described below, where I have chosen to illustrate the type of device which I think to be the most appropriate for the circuit chosen.

The source-coupled oscillator

This type of layout, shown in Figure 10.17 as an arrangement based on MOSFETs, though it could also use junction FETs or bipolar transistors, is based on the long-tailed pair layout described in Chapter 6, and achieves free-running multivibrator action with the use of only one timing *RC* network.

Its method of operation is that, if, because of an initial imbalance in the device characteristics, Q_1, say, is driven towards cut-off by a voltage applied via *C* to its gate, the loss of Q_1 source current through R_2 will cause the voltage, V_x, developed across R_2, to decrease in relation to V_{ref}, turning Q_2 more fully on, with a consequent reduction in its gate potential – which reinforces the action of switching off Q_1: a process which is very rapid in effect. When the cut-off bias on Q_1 decays away, and source current once more begins to flow through R_2, the voltage V_x will begin to increase, the current through Q_2 will begin to decrease, and its drain voltage will become more positive, turning Q_1 more fully on. Once again, this action is cumulative (sometimes called 'regenerative') in effect, so that this 'off' to 'on' transition is also very rapid.

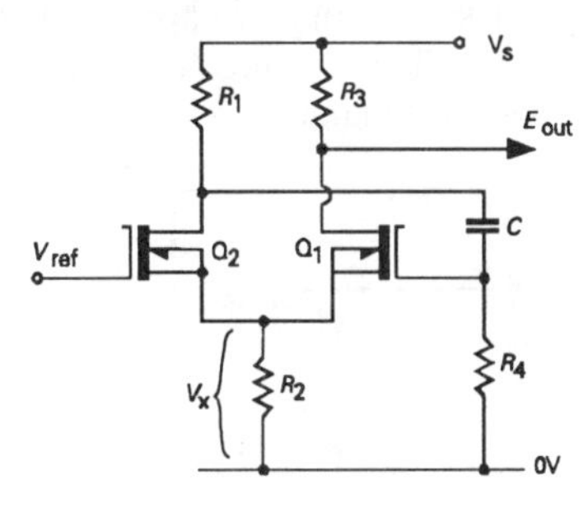

Figure 10.17 *Source-coupled oscillator*

The relaxation process is now one in which the gate and source voltages of Q_1 decay back towards 0V, until the point is reached at which Q_2 again begins to conduct, and the circuit switches back to its alternative state.

Because of its lower loop gain, the transition times for this circuit may not be quite as rapid as those of Figure 10.14, but it offers the advantage that the output signal can be derived from the drain of Q_1, an electrode which is not involved in the multivibrator action, so that the oscillatory frequency is not altered significantly by the nature of the output load impedance.

A more modern version of this kind of multivibrator is shown in Figure 10.18, based on an op. amp., A_1. In this the circuit is fundamentally unstable because of positive feedback through R_2, from the output to the non-inverting input of A_1, which means that the circuit will, on switch-on, rapidly assume a DC state in which the output is either as far positive as the circuit will allow, or as far negative. Current flow through R into C, connected to the inverting input, will then cause this point to become slightly more positive (or negative, as the case may be), which will cause the output voltage of the amplifier to switch into the opposite state, when the discharging, or charging, of C will begin again in an endlessly repetitive manner.

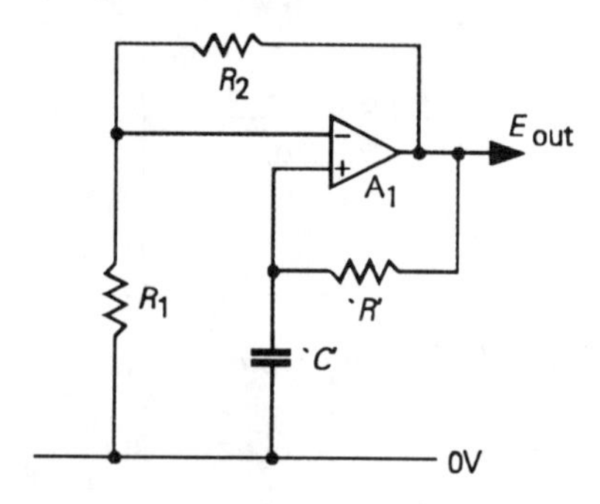

Figure 10.18 *Op. amp. astable circuit*

This circuit will give an excellent symmetry in output mark-to-space ratio, but the rapidity of transition from off to on states, and vice versa, will depend on the op. amp. HF characteristics.

The 555 circuit

In view of the proliferation of application specific integrated circuits, it is hardly surprising that ICs for use as free-running and triggered multivibrator circuits have been included in the range. Some of these circuits have survived for only a short period in the makers' catalogues before being superseded, but the '555' type device, in both bipolar and CMOS versions, has remained as a widely used component, used for both timing applications, and as a free-running oscillator, using the connection layout shown in Figure 10.19.

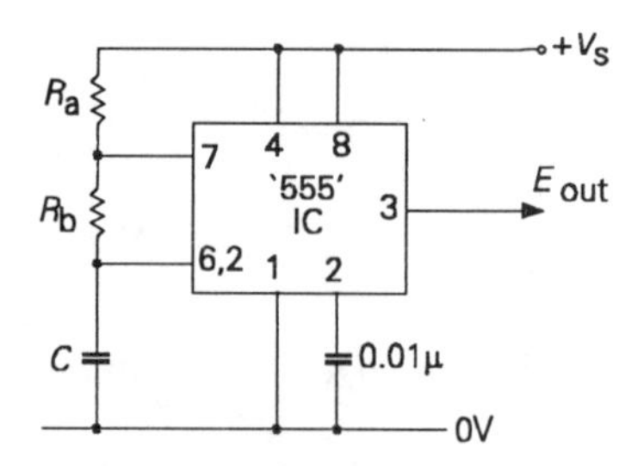

Figure 10.19 *Astable circuit based on 555 IC*

The on (1) and off (0) durations of the output waveform, in its free-running state are given approximately by

$$T_1 = 0.7(R_a + R_b)C \text{ and } T_2 = 0.7R_bC \text{ seconds}$$

The 'Bowes' multivibrator

A modification of the circuit of Figure 10.17, variously attributed to both Bowes and White, in which the timing cycle is accomplished by the emitter circuit components, is shown in Figure 10.20. (See Smith, J. H., *El. Eng.*, p.426, Aug. 1963). Because of the low circuit impedances, and the fact that neither transistor is driven into saturation, this layout offers very rapid transition times.

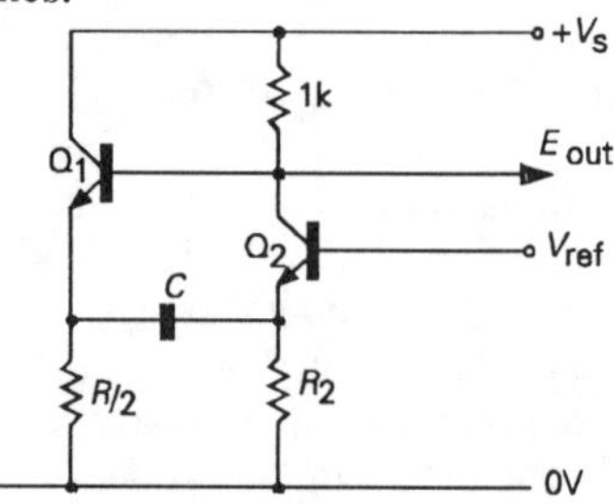

Figure 10.20 *The Bowes emitter coupled multivibrator*

The 'Spany' pulse generator

This circuit (Spany, V., *El. Eng.*, Aug. 1961), shown in Figure 10.21, employs a regenerative pair of transistors connected in series, so that either both devices or neither must conduct. This has the effect of generating a very brief output pulse, at a repetition rate determined by whichever of the two time-constant networks (C_1R_1 or C_2R_2), is the longer.

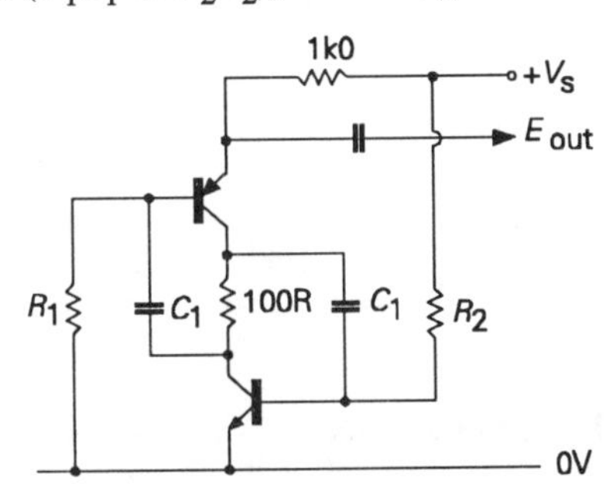

Figure 10.21 *The Spany pulse generator*

The blocking oscillator

This arrangement, shown in Figure 10.22, employs a pulse transformer (T_1), connected so that an increase in current through the primary winding (P), causes a positive potential to be developed at the end of the secondary winding (S_1), connected to the base of Q_1, which will, in turn, cause an increase in the collector current passing through the primary winding of the transformer. The positive potential applied to the base of the transistor causes current to flow through S_1 into C until the increase in base current can no longer be sustained by regenerative action. At this point, the current through P begins to decrease, and the potential at the base of Q_1 begins to swing negative. To this potential is now added the accumulated negative charge acquired by C, which causes Q_1 to be driven hard into cut-off.

The output waveforms which are given by this circuit are indicated in the diagram, and consist of a brief duration output spike from S_2, and a large, and fairly linear, sawtooth waveform across C.

This circuit, in a valve operated version, enjoyed great popularity as a sawtooth generation system for use in picture frame scanning in television equipment. In the junction transistor operated version of the circuit

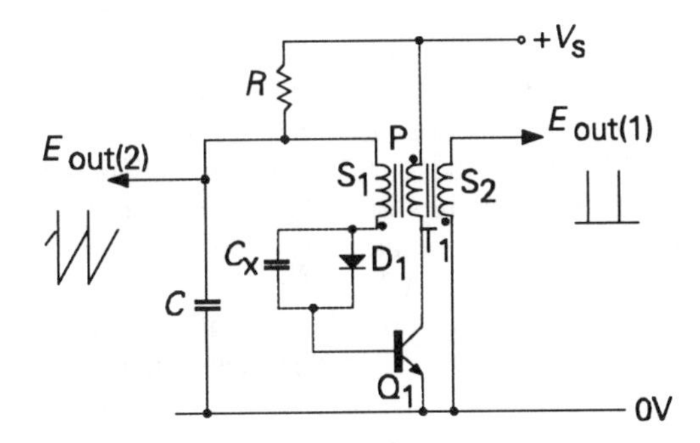

Figure 10.22 *Blocking oscillator circuit*

shown in Figure 10.22, a diode, D_1, must be included in the transistor base lead, with perhaps a 'speed-up' capacitor connected across it, to prevent zener diode action in Q_1 base–emitter circuit limiting the possible negative swing on C.

Logic element based square-wave generators

Quite apart from the various digital IC types, usually described as 'flip flops', which can be configured to operate as free-running multivibrators, there are two circuits, shown in Figures 10.23 and 10.24, which are based on CMOS inverting buffers. The method of operation of the circuit of Figure 10.23 is that if the output of IC_1 is 'high', the output of IC_2 will be 'low', and vice versa. This causes C_1 to charge or discharge through R_1 until the potential at the junction of R_1 and C_1 reaches the threshold input potential of IC_1, when the output of this amplifier will change state, and the cycle will repeat itself, giving a continuous rectangular-wave output waveform, of nearly 1:1 mark-to-space ratio, and a frequency of approximately

$$f = 1/R_1C_1$$

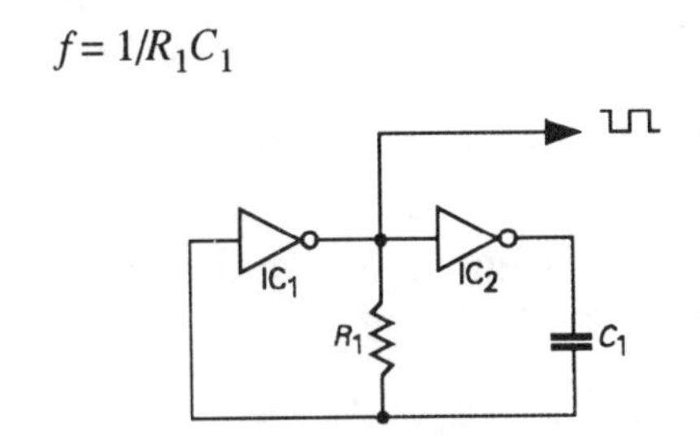

Figure 10.23 *Logic gate square-wave generator*

The behaviour of the circuit of Figure 10.24 is very similar, except that a third inverting buffer, IC_3, is connected as an active integration stage. If the output of IC_2 is high, the output of IC_3 will ramp downwards until, once again, the input threshold potential of IC_1 is crossed, when the output of IC_1 will go abruptly high, and the output of IC_2 will make a very rapid negative transition – reversing the input potential to IC_3, and causing the cycle to start over again. This circuit gives a triangular waveform, from IC_3 output, as well as a pair of complementary phase, nearly square, waveforms from IC_1 and IC_2 outputs.

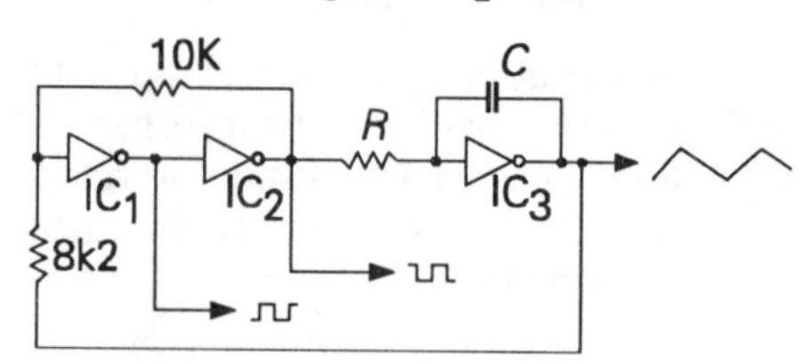

Figure 10.24 *Square and triangular wave generator*

If an input sinewave is available, CMOS inverting buffer amplifiers can be used to convert it into a nearly symmetrical square-wave, having rapid rise and fall times, by the use of the circuit shown in Figure 10.25. The resistor, R_1, across A_1 biases the amplifier into an input voltage setting which approximates to its DC mid-point, and assists in achieving the symmetry of the output waveform. Because a chain of three amplifier blocks is used, the final on/off transitions are rapid. As with the circuit of Figure 10.24, complementary phase output waveforms are available.

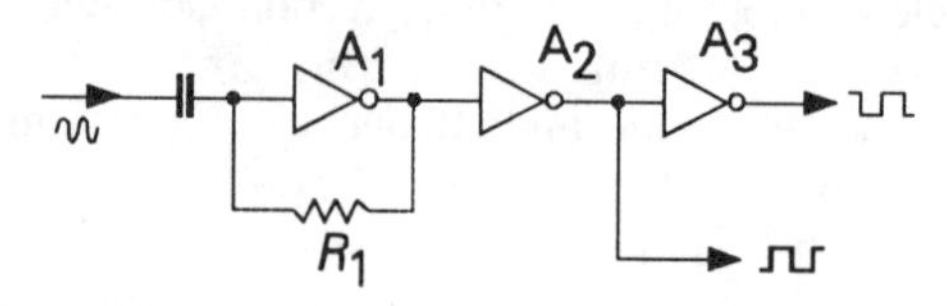

Figure 10.25 *High speed squaring circuit*

Sawtooth, triangular and staircase waveform generators

Several of the circuits shown above will generate triangular shaped output voltage waveforms which are useful for various circuit functions. However, the scanning of an oscilloscope, or TV tube, screen nearly always requires that the position of the spot shall move linearly across the screen, as a function of time, and this demands an input waveform which has a very linear relationship between output voltage and time. In addition, it is usually desirable that the voltage ramp is terminated by an abrupt flyback transition, in order that the ramp may begin again, giving a voltage/time relationship of the kind shown in Figure 10.26.

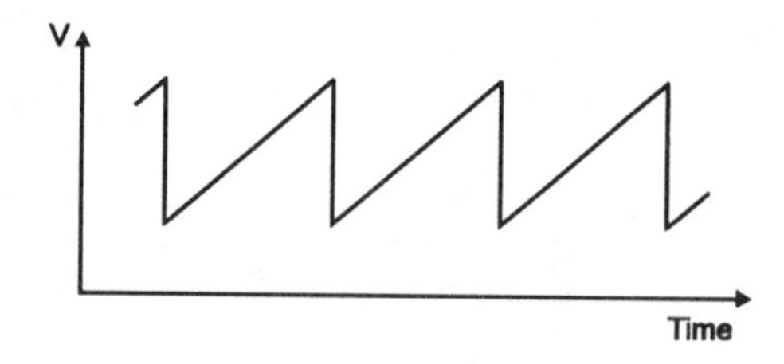

Figure 10.26 *Oscilloscope X-axis scanning waveform*

Three approaches are commonly employed for the purpose of generating a linear ramp waveform, of which the simplest is the use of a standard *CR* charging circuit, which is fed from a relatively high source voltage, as shown in Figure 10.27a, so that only the initial, and relatively linear portion of the exponential charging slope, seen as the potential across the capacitor, and shown in Figure 10.27b, need be employed.

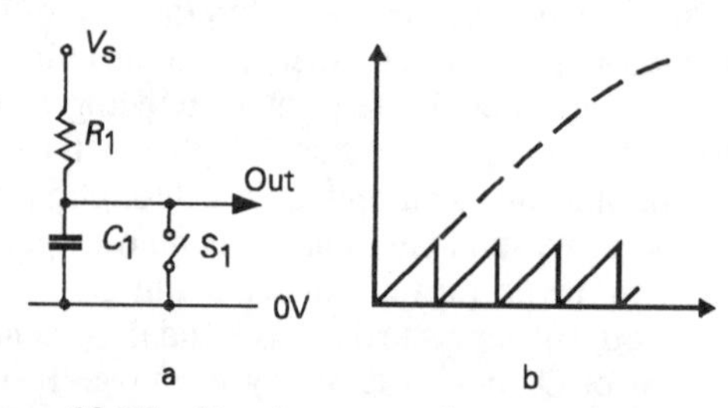

Figure 10.27 *Simple sawtooth scan waveform generator circuit*

The second technique is to cause the timing capacitor (C_1), to charge through a constant current source (CC_1), as shown in Figure 10.28a, so that the rate of change of voltage with time, shown in Figure 10.28b, is constant, within the limits imposed by the constant current source, regardless of the proportion of the output waveform sampled.

The third method used it to connect the timing

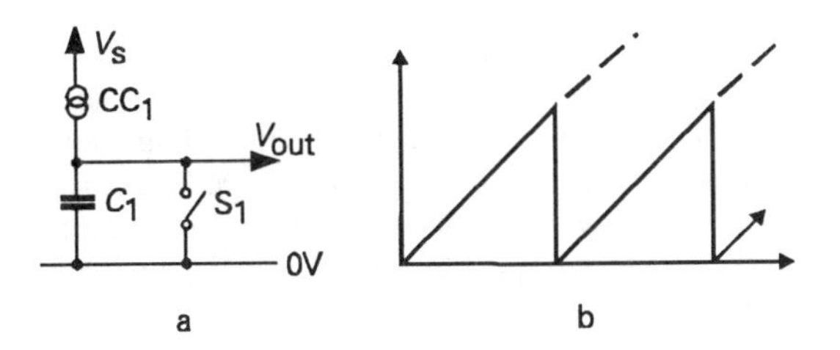

Figure 10.28 *Improved sawtooth generator system*

capacitor across a linear inverting DC amplifier, as shown in Figure 10.29. If the input current source, R_1, is connected to a stable voltage reference point, the input current will remain virtually constant, as the output voltage ramps upwards or downwards – which, in turn, will depend on the polarity of the input voltage reference – since the action of the amplifier, if its gain is high enough, will be to generate a virtual earth node at its input, which, by definition, remains at a nearly constant potential.

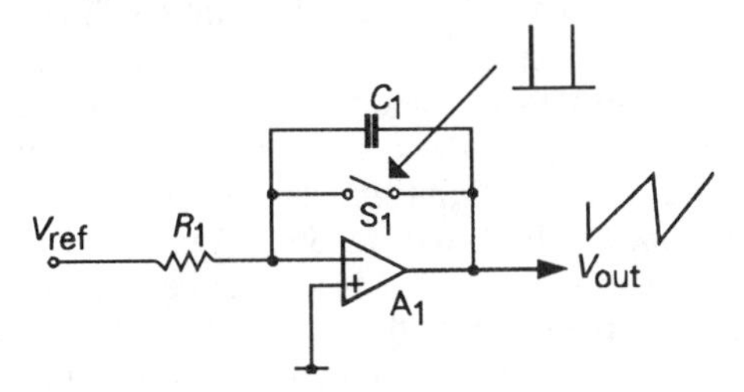

Figure 10.29 *Negative feedback sawtooth generation circuit*

In all of these examples, the circuit will be reset by discharging the timing capacitor through some form of electronic switch, triggered repetitively by some externally applied trigger waveform. In most cases, it is more useful if the reset function is triggered by the potential reached by the output voltage ramp, rather than just by some arbitrarily applied time interval.

In the case of the sawtooth circuit of Figure 10.29, the output waveform can be modified to that of triangular profile – this term is usually employed to denote a waveform in which both linear slopes are of comparable duration – by switching the input from a positive to a negative input reference potential. A practical example of such a waveform generator layout is shown in Figure 10.30.

A type of waveform which has similar characteristics to the sawtooth is the staircase, shown in

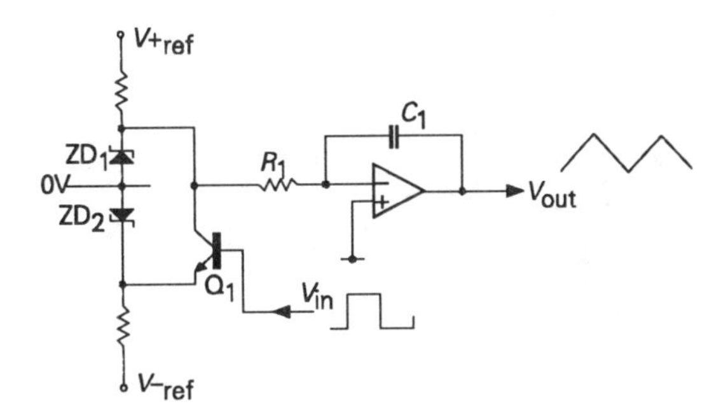

Figure 10.30 *Circuit allowing adjustable slope triangular waveform output*

Figure 10.31. This type of waveform can be generated, as for example in the circuit layout of Figure 10.32, by introducing a uniform, unidirectional, sequence of pulses of charge into the input circuit of a high impedance integrating amplifier (A_1). In the circuit shown, these pulses of input current are derived, from a charged capacitor (C_1) by way of an electronic switch (S_1) actuated by a square-wave control signal. The circuit can be reset by actuating S_2, perhaps by means of an output ramp voltage level detecting circuit (A_2). This approach is widely used, in a single or double ramp system, as a means of converting an input voltage level into a sequence of pulses, as, for example, in a pulse-counting type digital voltmeter.

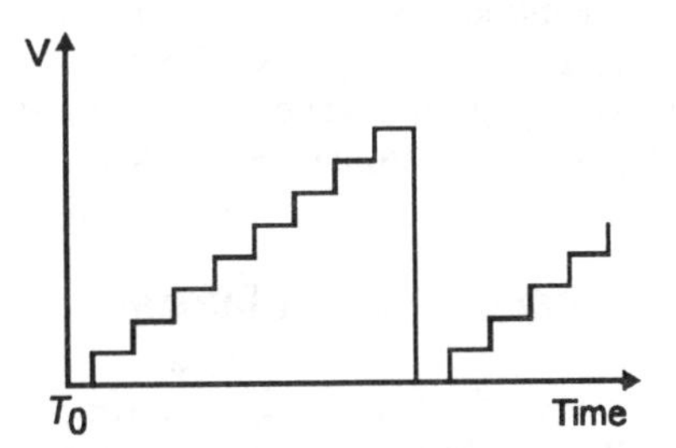

Figure 10.31 *Staircase waveform*

Mark-to-space ratio adjustment

One of the useful purposes to which a triangular or double-ramp staircase waveform can be put is in the generation of rectangular waveforms in which the mark-to-space ratio can be adjusted by some externally applied control voltage. This can be done by using a triangular waveform as the input signal to a narrow input threshold amplifier, as shown in Figure

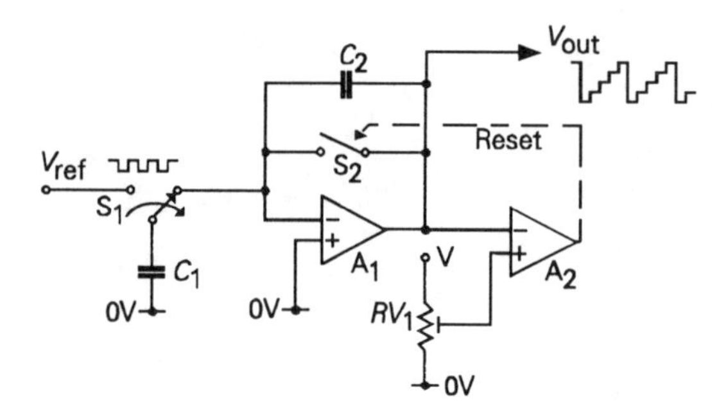

Figure 10.32 *Staircase waveform generator circuit*

10.33. If the input reference voltage is adjusted, up or down, the slicing level of the amplifier will be modified, and the relative on and off durations of the output rectangular wave will be altered.

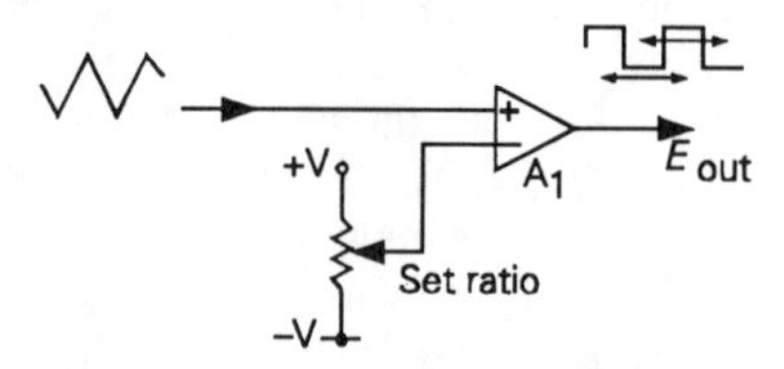

Figure 10.33 *Mark-to-space ratio adjustment*

In this application, the use of a double-ramp staircase waveform input will allow greater precision in the on and off durations, which will be precisely synchronized to the leading edges of the initial ramp generating control square-wave, but at the cost of the loss of the potentially infinite variability of this ratio.

Negative impedance oscillators

These fall into two broad classes: those which employ a simple *CR* or *LR* timing circuit as the frequency determining element, and those which use an *LC* resonant circuit for this purpose. Of the first category, the simplest arrangement is that using a gas discharge device, such as a low pressure neon tube (V_n) in the type of circuit shown in Figure 10.34. In this, when a suitable potential is applied to the free end of R_1, the voltage across C_1 will begin to climb exponentially towards the level of the applied input potential. Assuming that this is greater than the breakdown voltage of the gas discharge tube, at some point on the charging voltage ramp the tube will 'strike', and current from R_1 and C_1 will flow through the tube. At this point, the tube will exhibit a negative impedance, in that the greater the current flow, the lower the voltage across the tube – because the conducting impedance of the tube is inversely proportional to the number of current carrying ions, and this increases as the discharge current increases.

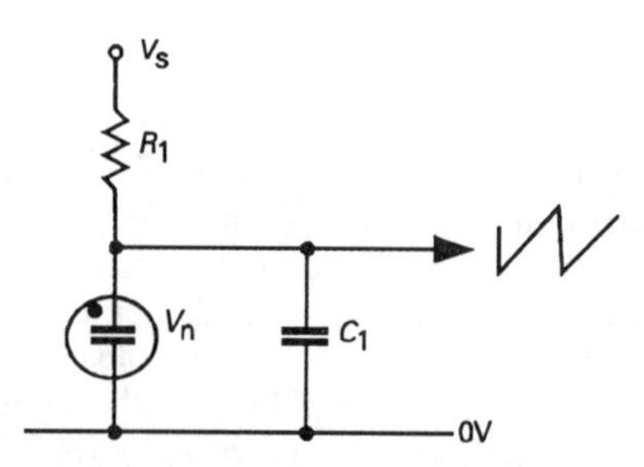

Figure 10.34 *Gas discharge tube sawtooth oscillator*

Since the voltage drop across the tube will depend on the product of current flow and impedance, this negative impedance characteristic causes the output voltage to fall as C_1 discharges through V_n, until the system reaches equilibrium, and no more current flows from C_1. If R_1 has a value which is relatively high in relation to the conductive impedance of the tube, when the current contribution from C_1 ceases, the level of internal ionisation in the tube will fall and its impedance will rise. However, because of the presence of C_1, the voltage across the tube can only rise relatively slowly, and the conditions required for ionic conduction, at that voltage, are no longer met, and so the discharge will extinguish. This allows the voltage across C_1 to rise, and the discharge sequence to repeat itself, giving a sawtooth output waveform, as shown.

A similar type of repetitive charge/discharge behaviour will be given, though at lower applied voltages, by a range of modern semiconductor devices, such as the DIAC shown in Figure 10.35, the unijunction transistor of Figure 10.36, and the silicon controlled switch shown in Figure 10.37.

A more general category of negative impedance oscillator is that in which an *LC* tuned circuit, or other resonant system, is connected to an external negative impedance generator, such as for example the two-stage feedback amplifier in the Franklin oscillator – see Chapter 12 – or where the active agent is a tunnel

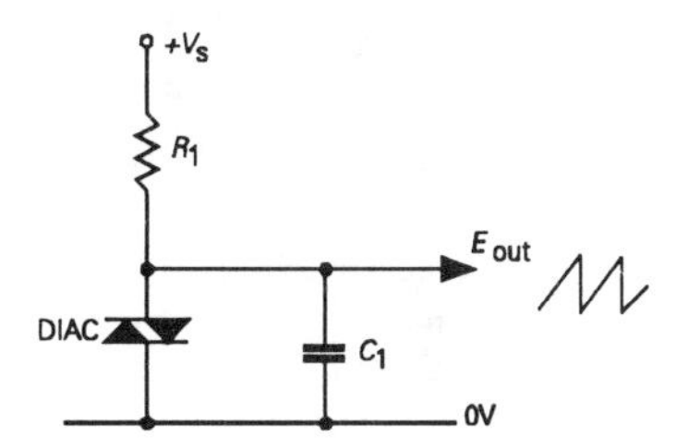

Figure 10.35 *Relaxation oscillator based on DIAC*

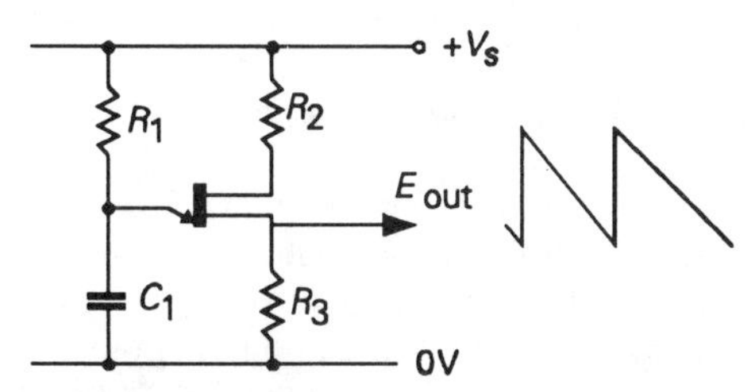

Figure 10.36 *Sawtooth oscillator based on unijunction transistor*

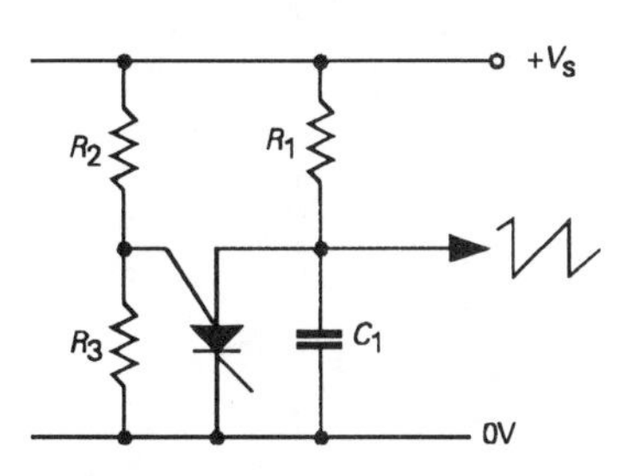

Figure 10.37 *Relaxation oscillator using silicon controlled switch*

diode, a device examined in Chapter 5, which has a voltage/current curve of the kind shown in Figure 10.38. This type of device is notable for the fact that, for a significant range of applied voltage, any increase in applied potential will result in an reduction in device current – a true negative impedance condition. When biased into this state, components such as this can be used, in principle, to make 'two-terminal' *LC* oscillator systems of the type shown in Figure 10.39. However, because of their relatively high cost, tunnel diodes are mainly used in VHF circuitry where other techniques are less conveniently applicable.

Devices such as the impatt diode, and the Gunn diode – a highly doped three layer device, usually fabricated from gallium arsenide – also exhibit negative impedance regions, parts of the voltage/current

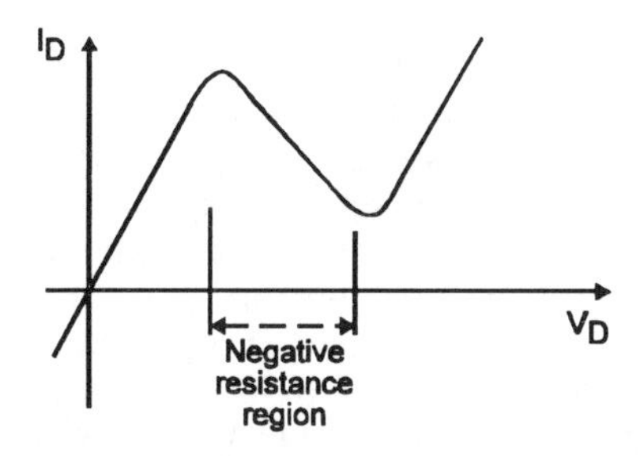

Figure 10.38 *Tunnel diode current vs. voltage characteristics*

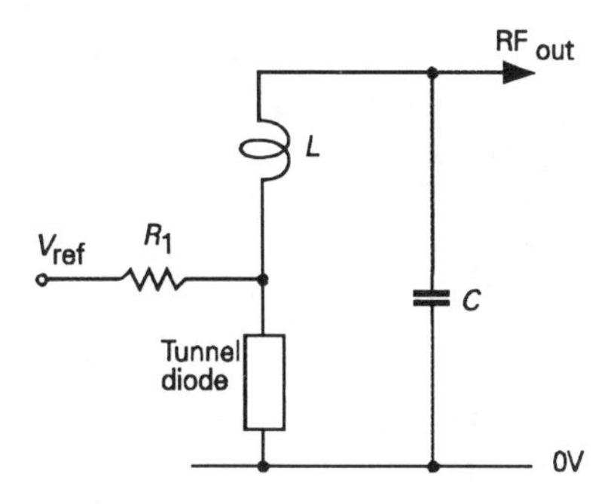

Figure 10.39 *Tunnel diode negative resistance oscillator*

characteristics in which the current flow falls as the applied voltage is increased, and these are also used in low power oscillator systems, though almost exclusively as microwave signal sources.

Noise sources

'Noise' is generally regarded as an inconvenient, but unavoidable, accompaniment to all electronic signal sources and all subsequent amplification or manipulation processes, and care is usually taken, in the circuit design, to limit the proportion of the output signal which is due to noise – as defined as a signal voltage of random occurrence and frequency distribution.

White and pink noise

For some types of test procedure it is very useful to have a calibrated noise source. This type of signal generator will generally be a source of 'white' noise –

a term which is given to a noise signal which has the characteristic that the output power is constant for any equal unit of bandwidth. If converted into an audible signal, this kind of noise would have the sound of the hissing escape of gas from a high pressure system.

The type of noise known as 'pink' noise is that which has a constant output power for each octave of bandwidth, and in audible form, would sound like the rustling of leaves, or the sound of the wind through long grasses.

Since circuit elements designed to produce noise will generally have a white noise power distribution, within their useful bandwidth, some form of output filtering will be necessary to achieve the –3dB/octave reduction in voltage output, as a function of increasing frequency, required for pink noise characteristics. This can be done by an *RC* network of the kind shown in Figure 10.40.

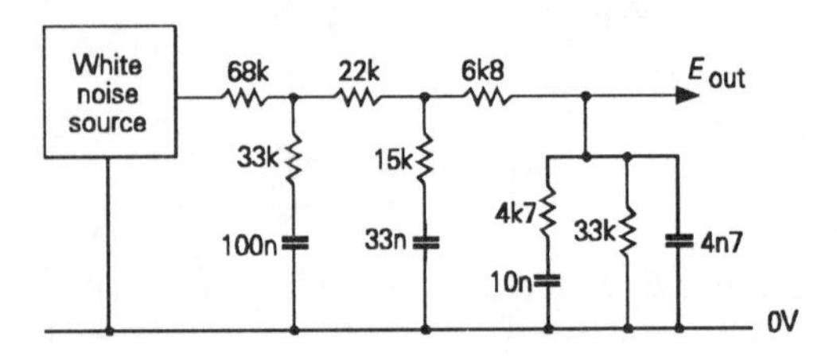

Figure 10.40 *Noise shaping circuit*

Noise source systems and bandwidths

Several circuit layouts can be employed as wide-band noise sources, of which the simplest is the Edison diode, a directly heated filament, mounted in an evacuated envelope, in proximity to a positively charged metal anode, as shown in Figure 10.41. Electrons will be emitted, and immediately collected by the metal plate, in a quite random manner, to give a noise (current) output which is directly proportional to the square root of the measurement bandwidth and the anode current.

With suitable components, the type of circuit shown in Figure 10.41 has a usable output over the frequency range 10Hz–1000MHz, a frequency at which inter-electrode transit time effects and electron 'bunching' are beginning to impair the randomness of the output signal.

A commonly used alternative to the Edison diode as a noise source is the simple layout based on a gas

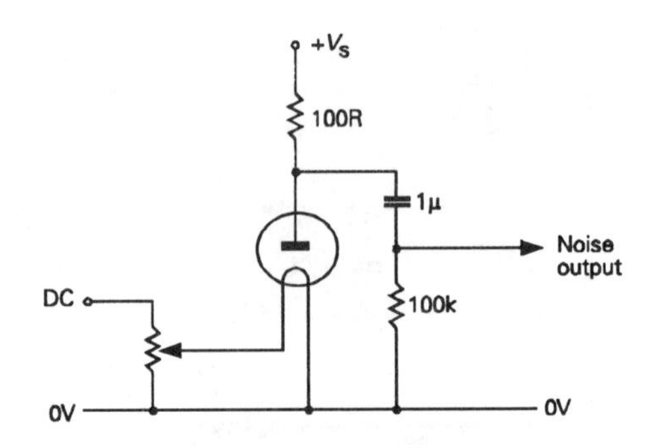

Figure 10.41 *Noise diode circuit*

discharge tube shown in Figure 10.42. In this the random collisions between the ionized atoms of gas in the space between the two electrodes give rise to a fluctuating output current which approximates to a true white noise source over the frequency range 100Hz–20kHz, which is adequate for audio purposes.

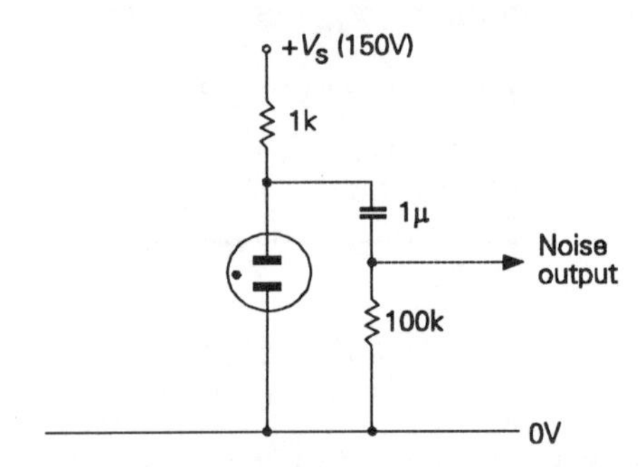

Figure 10.42 *Gas discharge tube noise generator*

A somewhat better performance is obtainable from a reverse biased avalanche diode (i.e. most zener diodes with a breakdown voltage above 5.5V), used in the circuit shown in Figure 10.43. This is usable over the range 30Hz–100kHz. With a specially designed noise diode the bandwidth can be extended upward beyond 1GHz.

Comparable performance to that from an avalanche diode is given by the use of a silicon transistor base–emitter junction which is biased into reverse breakdown, as shown in Figures 10.44 and 10.45. The largest noise output is given at a bias voltage just greater than the base–emitter breakdown voltage, but this may exhibit random peaks in output voltage at unpredictable parts of the spectrum, so it is prudent to choose an applied reverse voltage rather higher than

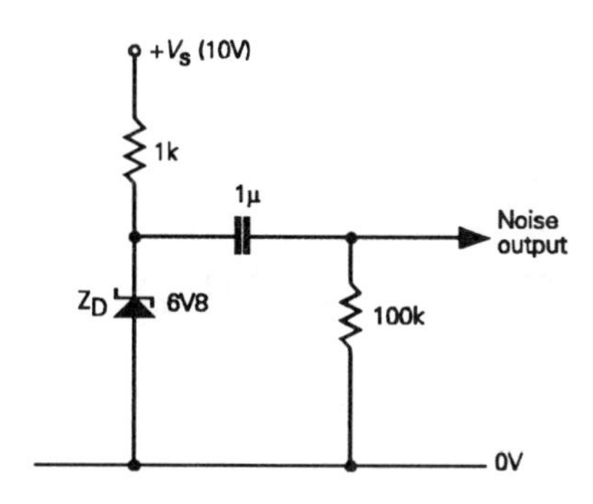

Figure 10.43 *Zener diode noise source*

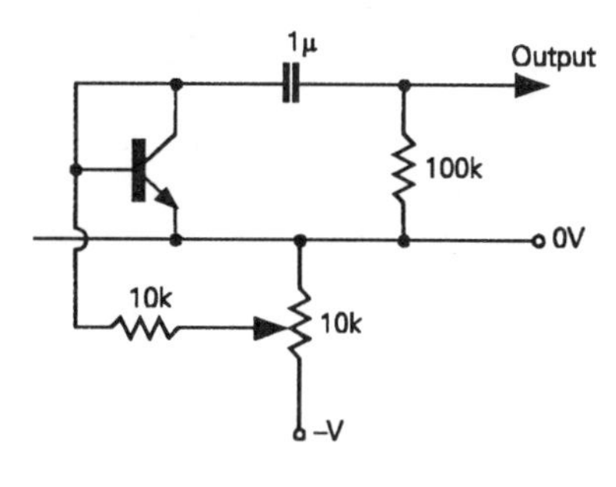

Figure 10.45 *Junction reverse breakdown noise source*

this level, say, twice the breakdown value.

A further simple circuit is that based on a MOSFET input op. amp. such as the RCA CA3140, shown in Figure 10.46, which has a usable and fairly level noise voltage output over the range 10Hz–200kHz.

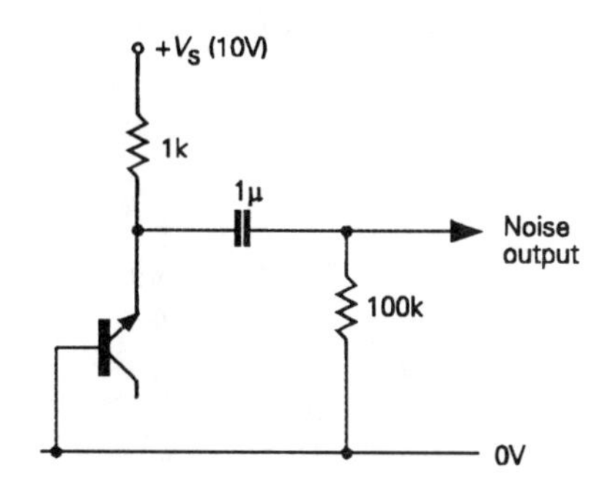

Figure 10.44 *Reverse biased emitter-base junction noise source*

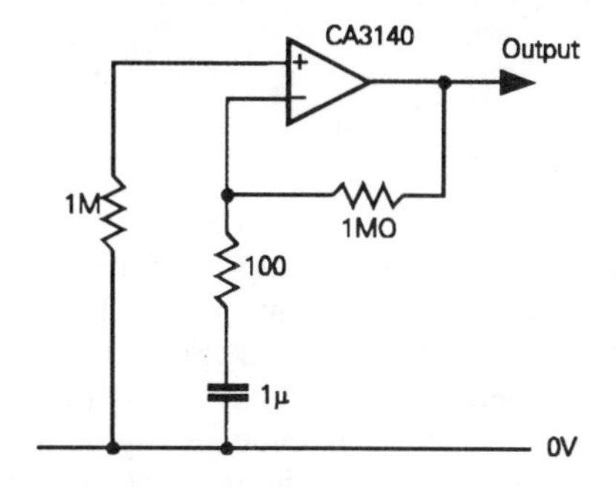

Figure 10.46 *Op. amp noise source*

11

Tuned Circuits

Introduction

There are a number of naturally occurring systems, such as a swinging pendulum, in which there is a periodic interchange between the kinetic energy and the potential energy stored in the system, which results in an oscillatory motion at some clearly defined frequency. In the case of the pendulum, the kinetic energy is the mechanical momentum of the bob as it swings past the mid-point of its travel, and the potential energy is that which is stored in the bob, when it momentarily comes to rest, at the highest point of its swing.

There is an exact electrical analogue to this action in the behaviour of a combination of capacitance and inductance in the circuit shown in Figure 11.1a, where the kinetic energy is the flow of current through the inductor, sustained by its associated magnetic field, at the point of zero oscillatory voltage, and the potential energy is that due to the electrical charge accumulated in the capacitor at the peak of the oscillatory voltage swing developed across it.

There must also be some element of energy loss present in any real system, which I have denoted as a resistor r, connected in series with the inductor, although, in reality, it will be composed of a variety of energy absorbing mechanisms, distributed through the

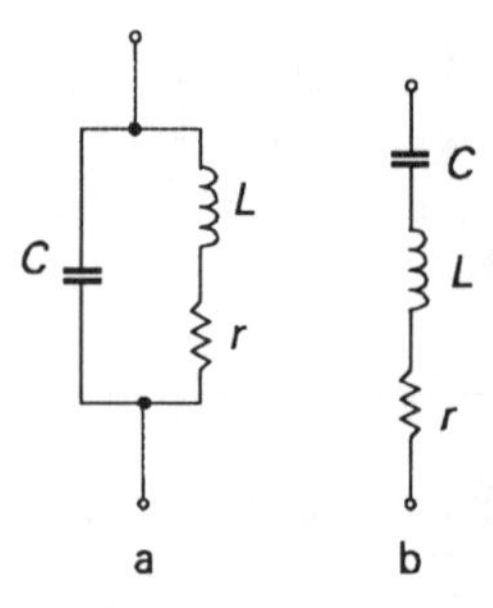

Figure 11.1 *Parallel and series resonant tuned circuits*

circuit, for which the resistor r is just a mathematically convenient equivalent. (*Note*. It is conventional to use the symbol r to represent resistance in series with the inductor, and R, to denote resistance in parallel with it. Both types of connection can be used to simulate the effect of system losses, and there is a mathematical relationship, shown below, between the resistance values needed to give the same effect in either case.)

The value of this type of combination of inductance and capacitance in electronic circuit arrangements is enormous, because of the phenomenon of electromagnetic resonance, which will occur at some specific and unique frequency. At this frequency the dynamic impedance of the parallel tuned circuit shown in Figure

11.1a, becomes extremely high, and at which the impedance of the alternative series tuned circuit layout, shown in Figure 11.1b, collapses to a value equivalent to that of its equivalent series loss resistance, r.

The term 'tuned circuit', which is applied to the circuit layouts shown in Figure 11.1, is used because if the value of the inductor, or, more conveniently, the capacitor, is made variable, the frequency of resonance can be made adjustable, and this was the original method by which radio receivers were tuned to respond to the desired incoming signal.

There is also, in the case of the circuit layouts shown in Figure 11.1, a 'circuit magnification factor', denoted sometimes by the symbol 'm', but more commonly referred to as the 'Q' of the tuned circuit. In a parallel tuned circuit, this term refers to the extent to which an externally induced current, at the resonant frequency of the circuit, will be magnified by the effect of the circuit resonance. In the case of the parallel tuned circuit, this gives the type of frequency response curve shown in Figure 11.2, when an input signal, applied to the circuit of Figure 11.1a, is swept through the frequency of resonance. This diagram also illustrates the effect on the output voltage, and the sharpness of tuning, of different values of circuit Q.

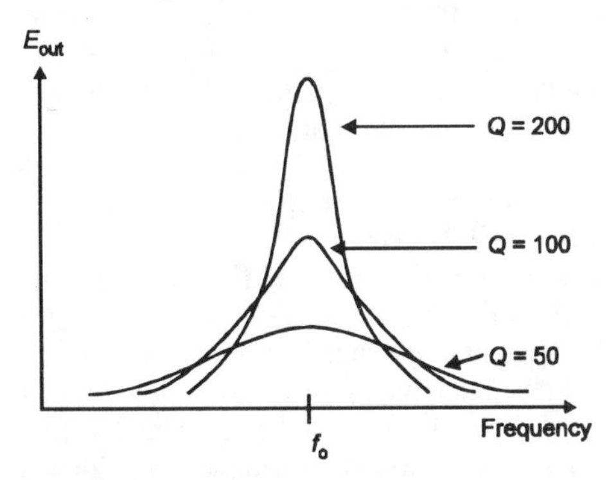

Figure 11.2 *Frequency curve of single LC tuned circuit*

The importance of this effect, in radio communication, is that it allows a degree of 'selectivity' in the choice of frequency at which a radio receiver will have its maximum sensitivity, to allow the preferential reception of one incoming radio signal rather than another. It also provides, in the design of high frequency oscillators, a convenient means for selecting the frequency at which the circuit will operate.

Calculation of resonant frequency, dynamic impedance and Q

There are a number of methods by which the electrical characteristics of a tuned circuit can be derived, but one of the simplest is to use the relationships which define the electrical impedance of an inductor

$$Z_L = j\omega L$$

and that of a capacitor

$$Z_c = 1/(j\omega C)$$

where j denotes the square root of -1, which is a convenient way of representing the 'imaginary', or phase quadrature, components of voltage, current or impedance, and ω is used as a short-hand symbol for frequency in radians/second ($= 2\pi f$): while L and C are the values of inductance and capacitance in henries and farads respectively.

Using this notation, the impedance of the series circuit of Figure 11.1b will be

$$Z_{ser} = r + j\omega L + 1/(j\omega C)$$
$$= r + j[\omega L - 1/(\omega C)]$$

At resonance, when $\omega L = 1/(\omega C)$, the terms in the square brackets become zero, the imaginary terms in this equation disappear (which implies that there are no residual quadrature currents, and the circuit looks like a pure resistance), and

$$Z_{ser} = r \quad (1)$$

Similarly, the frequency of resonance can be determined from the relationship $\omega L = 1/(\omega C)$, which can be rewritten as

$$w_0^2 = 1/(LC)$$

or

$$f_0 = 1/(2\pi\sqrt{LC}) \quad (2)$$

In the case of the parallel resonant circuit, the equation defining its impedance, at any frequency, is

$$Z_{par} = \frac{(1/j\omega C)\,(j\omega L + r)}{1/(j\omega C) + j\omega L + r}$$

multiplying through by $j\omega C$, this becomes

$$Z_{par} = \frac{r + j\omega L}{(1 - \omega^2 LC) + j\omega Cr} \quad (3)$$

this impedance is at a maximum when

$$\omega_o{}^2 LC = 1 \quad \text{or} \quad \omega_o{}^2 = 1/LC \quad (4)$$

if r can be neglected, this can be written as

$$f_o = 1/2\pi\sqrt{LC}$$

which gives the same frequency as in the series circuit.

Using the relationships which exist at resonance, as shown in equation (4), allows equation (3) to be expressed in a greatly simplified form, as

$$Z_{o\,par} = \frac{r + j\omega L}{j\omega Cr}$$

the absolute magnitude of which, at resonance, is

$$|Z_o| = \sqrt{\frac{r^2 + \omega_o^2 L^2}{\omega_o^2 C^2 r^2}} \quad (5)$$

If r is small in relation to ωL, this approximates to

$$|Z_0| = \frac{L}{Cr} \quad (6)$$

which is the dynamic impedance of a parallel tuned circuit at resonance, or using the relationships shown in equation (4), $1/C = \omega_o{}^2 L$, the equation can be redrawn as

$$|Z_o| = \frac{\omega_o^2 L^2}{r} \quad (7)$$

The presence of the series resistor somewhat alters the resonant frequency of the circuit, to give a modified value, called the 'natural resonant frequency', f_n

$$f_n \approx \frac{1}{2\pi}\sqrt{\frac{1}{LC} - \frac{r^2}{4L^2}} \quad (8)$$

from which the presence of circuit losses can be seen to lower somewhat the resonant frequency of the circuit.

However, it must be borne in mind that a simplifying assumption was made in equation (6), that r was small in relation to ωL, which implies that the subsequent relationships will only be accurate if this condition is met. In particular, equation (8), which is commonly quoted as an exact relationship, is quite obviously only valid when $r^2/4L^2$ is small in relation to $1/(LC)$.

Circuit magnification factor Q

This characteristic of a resonant tuned circuit, which is usually simply called the Q (or quality factor) of the circuit – would be, in numerical terms, infinitely high in a resonant circuit which was free from all losses, and which will become lower as the losses increase – determines, as shown above, the sharpness of tuning of the circuit.

High selectivity is advantageous in the sense that it allows greater ease in discriminating between input signals in a radio receiver – and the associated high Q value will also augment the receiver sensitivity by increasing the size of the signal voltage developed, at resonance, across the tuned circuit. However, there is a snag, in that amplitude modulated radio signals carry their modulation information as a series of sidebands, distributed as sum and difference frequencies on either side of the nominal 'carrier' frequency, so too high a Q value will restrict the amplitude of the sideband signals which will be received.

In a parallel tuned circuit, of the kind shown in Figure 11.1a, Q is defined as the ratio of the current externally applied to the tuned circuit to the internally circulating current, which for any given applied EMF, is the ratio of the dynamic impedance of the tuned circuit to the series resonant impedance: relationships which have been shown in equations (1) and (7).

From these equations

$$Q = \sqrt{\frac{\omega_o^2 L^2}{r} \times \frac{1}{r}} = \frac{\omega_o L}{r} \quad (9)$$

or, since, at resonance, $\omega_o L = 1/(\omega_o C)$

$$Q = \frac{1}{r}\sqrt{\frac{L}{C}} \qquad (10)$$

The relationship between the Q of the tuned circuit and the 'half power' (–3dB in terms of the RMS voltage developed across the circuit) points on the frequency response curve of Figure 11.2, is

$$\text{Bandwidth} = f_o/Q \quad \text{(Hz)} \qquad (11)$$

However, since the bandwidth refers to the actual pass-band between these –3dB points, the actual frequency difference between the frequency of resonance and the –3dB output voltage point on either side of resonance is only half this value, i.e. $f_o/2Q$. For an actual off-tune frequency difference of f_o/Q, the output voltage will be –6dB with respect to that at resonance.

The general equation for the selectivity of a single tuned circuit of the type shown in Figure 11.1a is given by the relationship

$$\frac{i_o}{i} = \frac{E_o}{E} = \sqrt{1 + Q^2\left\{\frac{f}{f_o} - \frac{f_o}{f}\right\}^2} \qquad (12)$$

where i_o and E_o are the circulating current within, and the voltage developed across, the tuned circuit at resonance (f_o), and i and E represent the currents and voltages at a frequency, f, somewhat away from resonance. By fitting values into this equation, the frequency response curve for a single parallel resonant tuned circuit can be constructed from the relationships:

f	E_o/E
f_o	1
f_o +/– $f_o/2Q$	–3dB
f_o +/– f_o/Q	–6dB
f_o +/– $2f_o/Q$	–12dB
f_o +/– $4f_o/Q$	–18dB
f_o +/– $8f_o/Q$	–24dB

Additionally, the relationship between the Q of a tuned circuit and the equivalent series (r) and parallel (R) loss resistance, at resonance, is given by

$$R/Q = Q/r \qquad (13)$$

The phase angle between an applied voltage and the circulating current in a parallel tuned circuit undergoes an abrupt transition as the input signal passes through resonance, and is given by the equation

$$\text{Tan}\ \varphi = Q\left[\frac{f_o}{f} - \frac{f}{f_o}\right] \qquad (14)$$

Bandpass-coupled tuned circuits

The problem associated with the attenuation of broadcast sidebands by receivers whose selectivity is determined by single tuned circuits, of the kind shown in Figure 11.1a – a problem which is increased if several such tuned circuits are connected in cascade – can be lessened either by reducing the Q of the individual circuits or, in the case of a series of such circuits, by staggering the frequencies to which the individual circuits are tuned.

However, a better approach is to combine a pair of tuned circuits as a 'bandpass-coupled' layout, of the kind shown in Figure 11.3. In this, which is the simplest of the possible circuit arrangements usable for this purpose, if the inductors of two identical tuned circuits are brought into proximity with one another, the interaction of the magnetic fields associated with the currents flowing in the inductors will cause a mutual induction of currents from one to the other.

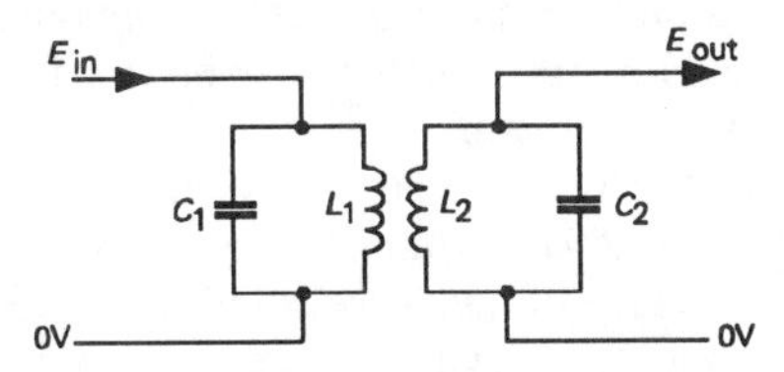

Figure 11.3 *Bandpass coupled pair of tuned circuits*

If an oscillatory input signal is swept in frequency through the point of resonance of the tuned circuits, the transfer of energy, by this means, from one tuned circuit to the other – which will be at its greatest at the frequency of resonance because the circulating current through the inductors will be greatest at this frequency – will cause a reduction in the magnitude of the output voltage peak at this frequency. The extent of the dip which this will cause in the combined frequency response curve will depend on the extent to which the circuits are coupled together, as shown for various coupling factors in Fig, 11.4. From the point of view of radio reception, the optimum condition is that in which the response has the widest flat topped region, compatible with the avoidance of a significant trough in the output, at resonance. This condition is described as critical coupling, or k_c, which has the value

$$k_c = 1/Q \tag{15}$$

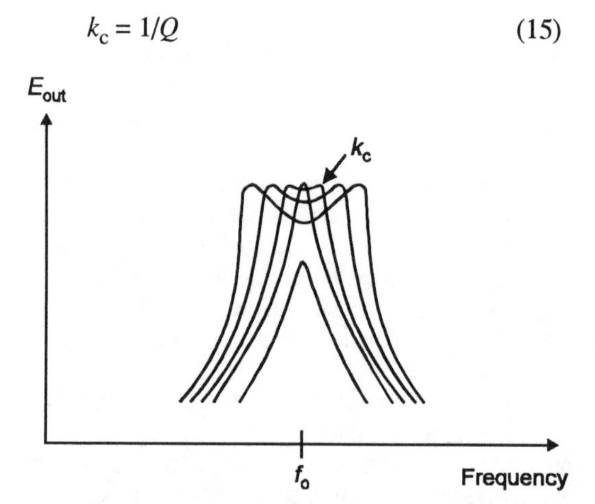

Figure 11.4 *Frequency response of bandpass-coupled pair of tuned circuits for various coupling factors*

There are a number of other circuit arrangements which will provide bandpass-coupled layouts, shown in Figure 11.5, for which the coupling factors are shown on the diagram.

For optimization of the bandwidth of a receiver using such bandpass-coupled circuits, not only must the correct value of bandpass coupling coefficient be selected, but also the value of the Q of the individual circuits must also be correctly chosen, and adjusted, if necessary, by the use of damping resistors, usually connected in parallel with each part of the tuned circuit.

For very high orders of selectivity, it is possible to connect a number of high Q bandpass-coupled tuned circuits in cascade, usually in the 'intermediate frequency' amplifier stages in a superhet radio, but in modern practice a quartz crystal filter or some form of electro-mechanical 'ladder' filter will generally be preferred.

A 'universal selectivity chart' for bandpass-coupled

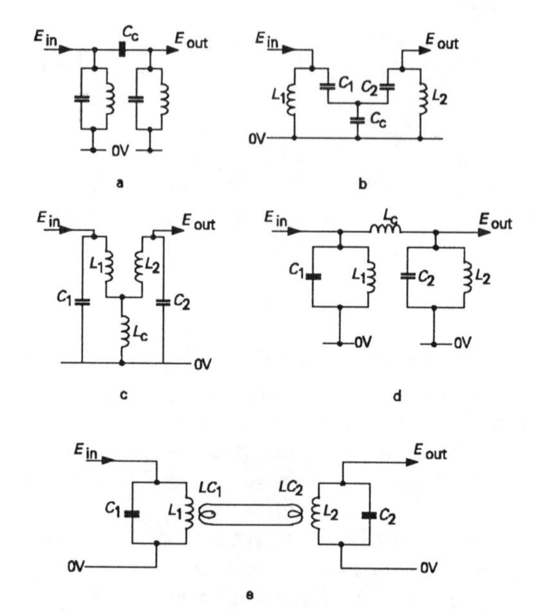

Figure 11.5 *Some common bandpass-coupled circuit layouts*

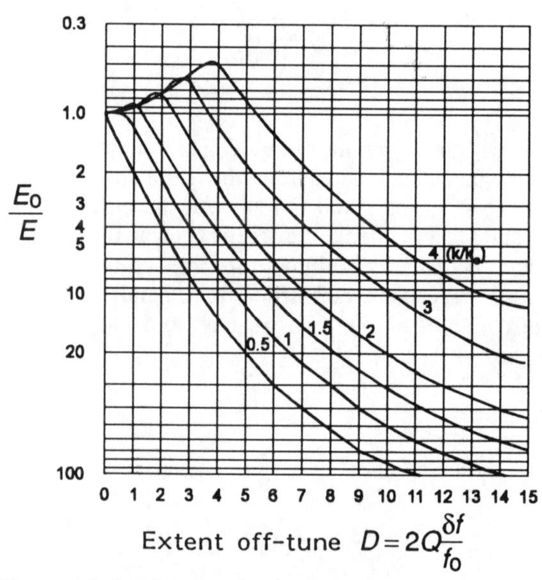

Figure 11.6 *Universal selectivity chart for bandpass-coupled tuned circuits*

tuned LC circuits is shown in Figure 11.6. This shows the relationship between the 'skirt' attenuation of a coupled pair of tuned circuits and the degree of off-tuning of the signal, expressed as a proportion (f/f_o), of the resonant frequency. It is apparent from this that the higher the operating frequency (f_0), the wider the pass-band, in kHz, for any value of Q and coupling factor.

Quartz crystals and other mechanical resonators

The phenomenon of 'piezo-electricity' was first observed by the Curie brothers in 1880, and refers to the physical effect that certain materials, usually crystalline in form, will develop an electrical potential on opposed surfaces, when subjected to a mechanical strain. Conversely, in such materials, if an electrical potential is applied to physically opposed surfaces it will result in a small mechanical distortion of the material.

Quartz crystal resonators

Crystalline quartz (SiO_2) is one of the materials which exhibits this phenomenon, and it was found that if thin slices of quartz were cut from a crystal, and held between fixed metal electrodes, several very sharply tuned mechanical resonances could be excited, corresponding to the crystal slice 'ringing' in its torsion, compression and shear modes.

The usual mode of operation for high frequency use is for the crystal to be cut and mounted in such a way as to force the excitation of the first order (fundamental frequency) shear mode type of oscillation, shown in Figure 11.7a, though all of the feasible resonance modes can be exploited by the crystal manufacturers for specific applications.

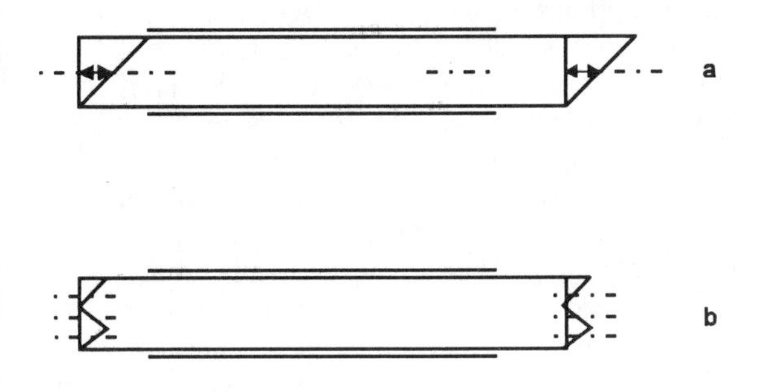

Figure 11.7 *Resonance modes in quartz crystal*

The frequency of oscillation of a quartz crystal is given by the approximate relationship

$$f(\text{MHz}) \approx \frac{1.67}{\text{thickness (mm)}} \qquad (16)$$

which would imply a thickness for the quartz slice of only 0.067mm for a 25MHz crystal – a typical upper limit for crystals operating at their fundamental resonant frequency – so for higher frequeny operation, it is customary to operate the crystal in a harmonic mode, for which the shear mode vibration pattern is shown in Figure 11.7b.

In contemporary practice, the modes in which quartz crystals are operated are as indicated in Table 11.1.

Table 11.1

Mode	Frequency range
Fundamental	Up to 25MHz
3rd overtone	20–75MHz
5th overtone	60–125MHz
7th overtone	90–175MHz

The equivalent electrical circuit of a typical quartz crystal resonator is shown in Figure 11.8, in which Co is the intrinsic capacitance of the crystal mounting electrodes and the holder, and will be typically of the order of 3–5pF. R_1, C_1 and L_1 are the so-called 'motional' parameters of the crystal, and are characteristic of the crystalline homogeneity of the material, and the precision and type of cut. This gives a relationship between impedance and frequency of the kind shown in Figure 11.9. At very low frequencies, the impedance of the crystal is very high, and is due entirely to the stray capacitance of the crystal mounting plates and holder. With increasing frequency, this impedance falls, but remains capacitative, until the frequency of series resonance is reached, when the impedance abruptly plummets to a value equivalent to the loss resistor, R_1 (see equation (1) above).

Other things being equal, the motional resistance component, R_1, decreases linearly with the area of the crystal slice, and is inversely proportional to its thickness, but a practical limit is imposed, at high frequencies, on the area of the slice by the fragility of the crystal.

As in a conventional series resonant LC circuit, the

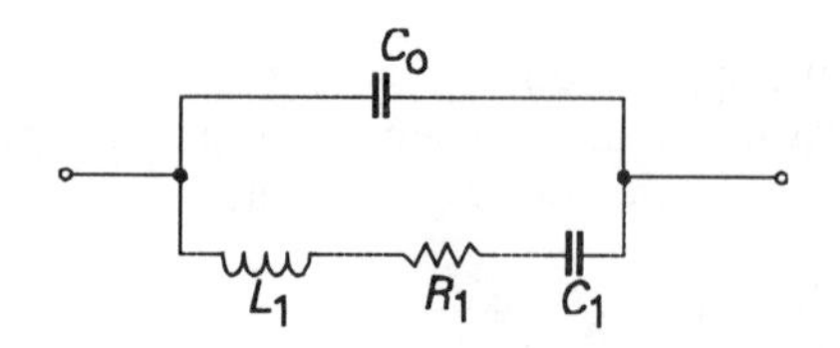

Figure 11.8 *Equivalent electrical circuit of quartz crystal resonator*

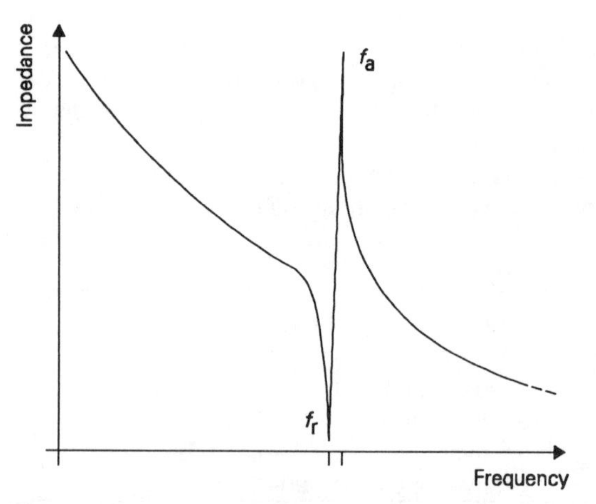

Figure 11.9 *Resonance and anti-resonance modes in quartz crystal resonator*

frequency of resonance (f_r) is defined quite normally by the equation

$$f_r = \frac{1}{2\pi\sqrt{L_1 C_1}}$$

However, at a frequency slightly above this value, the motional inductance component of the crystal will also allow the crystal to resonate with the holder capacitance, C_o, in a parallel, or 'anti-resonant', mode (f_a). This gives a high impedance peak in the impedance curve, at a frequency defined by the relationship

$$f_a = \frac{1}{2\pi\sqrt{L_1 \frac{C_1 C_o}{C_1 + C_o}}} \quad (17)$$

The width of the region between these two resonant frequencies, where the crystal presents an inductive impedance, is given by

$$f_a - f_r = f_r C_1 / 2C_o \quad (18)$$

The frequency of resonance of a quartz crystal will be influenced by the size of any external parallel or series connected capacitor, C_L, which allows some adjustment of the actual operating frequency. This modified operating frequency, of the crystal, f_L, will then be defined by

$$f_L = f_r\left(1 + \frac{C_1}{2(C_o + C_L)}\right) \quad (19)$$

and the crystal will be calibrated by its manufacturers for some specified load capacitance, usually on the assumption that the crystal will be operated in its normal series resonant mode, with an external circuit which will give a near-zero value of phase shift at the operating frequency.

Modern crystal resonators offer extremely high values of Q – in the range from 50,000 to 2,000,000 – due to the combination of relatively low levels of internal lossses in the crystal and its mounting, giving an effective value for R_1 in the range 12–200 ohms, associated with high values of L_1 – in the range 100–2000mH – and exceedingly low values for C_1 – typically 0.1 – 5fF – for crystals resonating in their fundamental mode at 10MHz.

Very high Q values are useful in quartz crystal oscillators, in maintaining a high precision in the frequency of operation, purity of waveform, and good signal/noise ratios. However, the 'activity' of a given crystal – a measure of the ease with which it will oscillate – bears an inverse relationship to Q. It may therefore prove more difficult in practice to cause circuits using very high Q crystals to oscillate, even with an appropriately chosen design, and if the external loop gain is increased to make sure of oscillation, care must then be taken to ensure that the input drive power does not exceed the makers recommended limits, to avoid damage to the crystal or excessive frequency drift due to heating. Typical maximum input drive powers recommended for quartz crystal oscillators usually range from a few milliwatts for a general purpose system down to a few microwatts where the crystal is used in a precision frequency reference.

The natural resonant frequency of all quartz crystal resonators is somewhat temperature dependent, an effect which is greatly influenced by the angle at which the slice is cut from the bulk crystal. A commonly employed type of cut, which exhibits a very low degree of thermal drift, around a median temperature of 20°C, is called the 'AT'. The relationship between cutting angle and temperature coefficient for such an AT cut crystal, is shown in Figure 11.10.

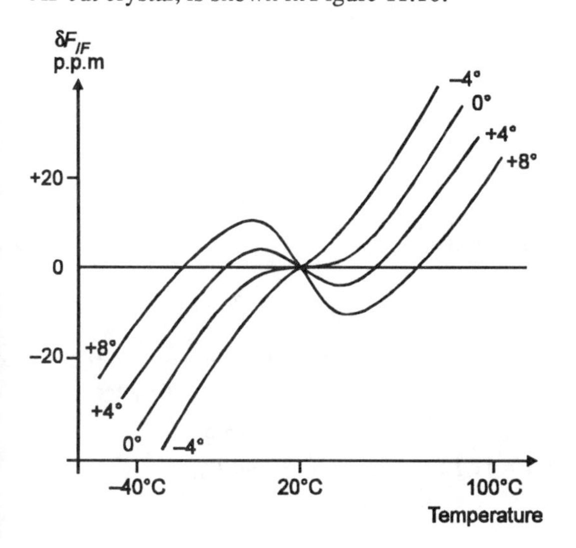

Figure 11.10 *Effect of cutting angle in an AT cut crystal on temperature coefficient*

Quite apart from their value as precise and stable frequency defining components for oscillator circuits, the sharply defined notch in the impedance curve of a quartz crystal resonator allows the design of narrow pass-band filters for use in radio receivers. The simplest circuit layout of this type is that shown in Figure 11.11a, where the effect of the shunt capacitance of the crystal, Y_1, is phased out by a variable capacitor, CV_1, in an anti-phase connected limb in the circuit, to give a resultant frequency response of the kind shown in Figure 11.11b.

A modification to this layout to give band-pass characteristics is shown in Figure 11.12a, which employs two crystals, Y_1 and Y_2, whose resonant frequencies are slightly different, to give the frequency response curve shown in Figure 11.12b.

Ladder and Surface Acoustic Wave filters

The quartz crystal type of bandpass filter is only really

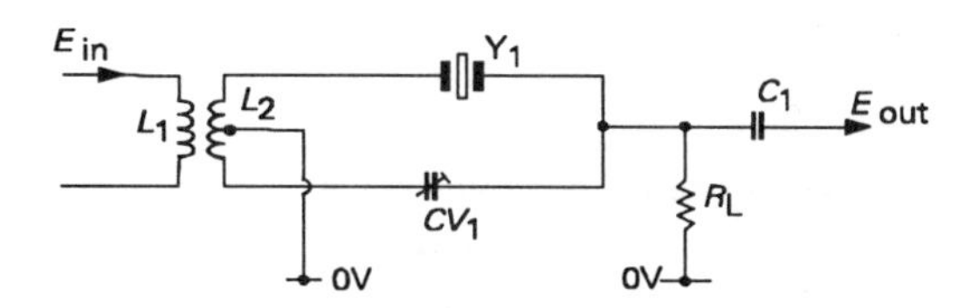

a Simple quartz crystal filter

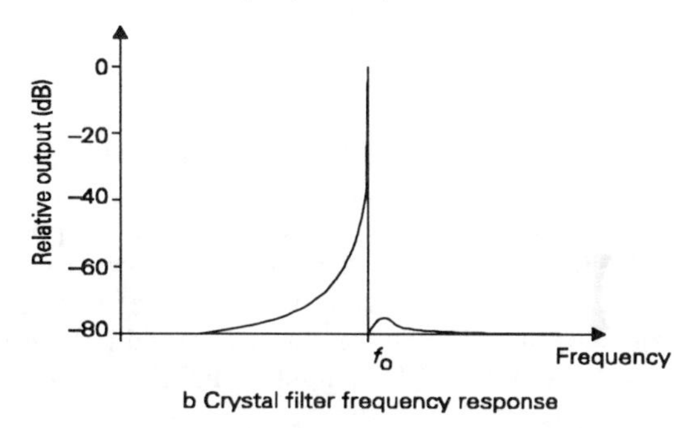

b Crystal filter frequency response

Figure 11.11

appropriate for use where relatively narrow pass-band widths are required. For wider bandwidths, other techniques are more suitable, of which the two most widely used are the series resonator ladder filter, and the surface acoustic wave (SAW) filter types. Of these, the ladder filter consists of a stack of piezo-electric ceramic disks, which are electromechanically resonant in their fundamental radial mode, and which are mounted in a suitable metal screening tube, as a string of up to 19 elements, so that they are electrically connected but mechanically and acoustically isolated. Because of their series connection, stray capacitance leakage effects are small, and stop-band rejection levels of better than –80dB are feasible in the 455kHz radio receiver IF pass-band.

In operation, this type of filter behaves like a string of bandpass coupled parallel resonant tuned circuits, whose pass-band is determined by their Q and coupling coefficient, and their skirt attenuation rate by the number of stages connected in series.

Surface acoustic wave (SAW) type filters

These devices, sometimes referred to as 'inter-digital filters' because of their method of manufacture, exist in two basic forms, the 'transversal', in which a 'travelling wave' is launched along the surface of a strip of piezo-electric material, and the 'resonant' type, in which the electrical excitation is used to produce a sharply tuned 'standing wave' pattern.

(A useful general description of these types of component is given by R. J. Murray and P. D. White (Philips Research Laboratories) in *Wireless World*, March and April, 1981.)

Transversal types

The transversal 'T-type' filter came into widespread use in the late 1960s because of the requirements of television and FM radio receivers. Both of these applications required wide, substantially flat topped selectivity curves for the receiver IF (intermediate frequency) amplifier stages – which would be up to 7MHz for a 45MHz TV IF amplifier, and up to 300kHz in a 10.7MHz FM IF stage.

Although this type of pass-band can be provided by a suitably tuned series of shunt resistor damped band-pass coupled tuned circuits, it is an expensive approach, especially in terms of the labour required to adjust the tuning of the individual circuits, and the SAW T-type filter allowed great manufacturing economies to be made in this respect.

In the case of the T-type (travelling wave) design, the piezo-electric effect is used to launch a mechanical ripple along the surface of a flat strip of suitable material. This is done by depositing an interlaced pattern of electrically conductive stripes on the surface of the strip, as shown, schematically, in Figure 11.13. If the pattern of the conductors is accurately deposited on the surface of the slice – a task which is similar to, but, in practice, rather less demanding than, the similar processes performed in the manufacture of integrated circuits – then if an electrical signal, at an appropriate frequency, is applied to the input (transmitting) group of conductors, the mechanical surface ripple which it causes will be propagated along the slice, and will, in turn, generate an electrical output when it passes under the output (receiving) group of electrodes.

The pass-band characteristics of this layout arise because the signal has effectively been made to pass through a series of delay paths, both at the transmitter and receiver ends of the strip, from which the outputs are added together. In the 'pass' band, these signals will add constructively, while in the 'stop' band they will tend to cancel. By careful choice of the inter-

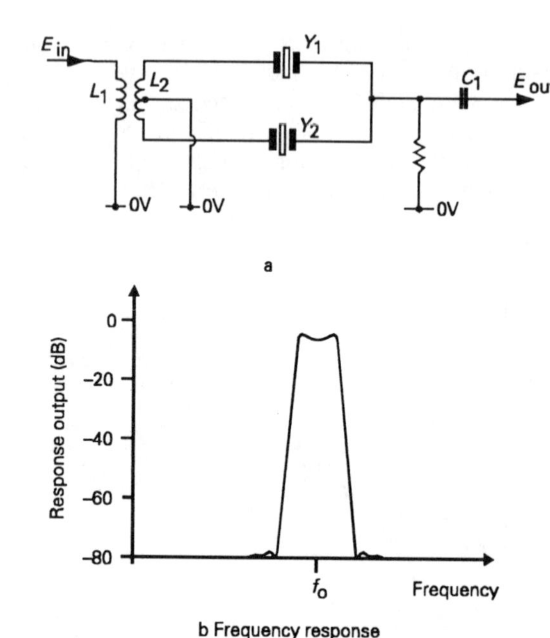

Figure 11.12 *Band-pass crystal filter*

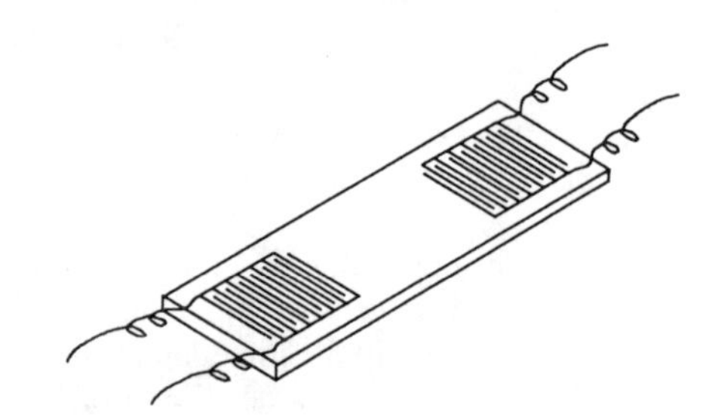

Figure 11.13 *T-type surface acoustic wave filter*

digital spacings, and the extent of the interleaving of the individual digits, it is possible to generate flat-topped transmission curves for these filters of the type shown in Figures 11.14 and 11.15. As in the case of the examples shown, these filters are widely used in TV and FM radio receivers, because they are inexpensive, and do not require frequency tuning following installation.

Some of the low cost filters of this type use a substrate of a thin flat strip of highly polished piezo-electric ceramic, such as lead zirconate-titanate (PZT), and this has given the generic name 'ceramic filter' to this kind of component.

Higher quality devices may employ similar strips, made from mono-crystalline lithium niobate

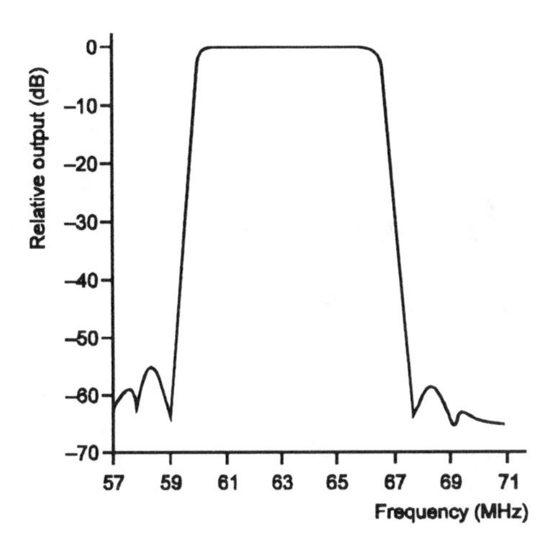

Figure 11.14 *SAW band-pass filter for TV IF*

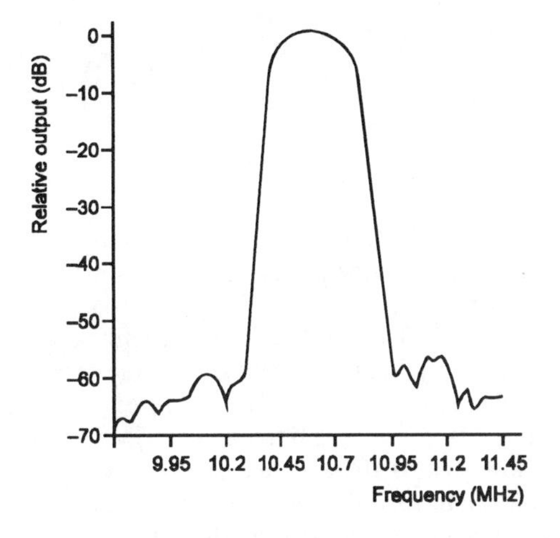

Figure 11.15 *SAW band-pass filter for 10.7 MHz FM IF stages*

($LiNbO_3$), bismuth germanium oxide ($Bi_{12}GeO_{20}$), lithium tantalate ($LiTaO_3$), or quartz. Of these, lithium niobate is preferred for wide pass-band designs, and quartz for narrow pass-band types. Unlike standard *LC* filters, it is possible, in this kind of design, to control both the amplitude and phase of the transmitted signal independently of one another. The type of performance which can be obtained is shown in Table 11.2.

Since the velocity of propagation of such a surface wave is of the order of 3000m/s, the physical dimensions of a component of this type can be quite compact. Also since 95% of the wave energy is confined within

Table 11.2

Centre frequency range	10MHz – 1.5GHz*
Minimum bandwidth	100kHz or 0.2% of operating frequency, whichever is the greater
'Skirt' bandwidth	As above
Group delay	1–5μs
Insertion loss	15–25dB (increases somewhat with bandwidth)
Pass-band ripple	Typically +/– 0.25dB
Stop-band breakthrough	Typically better than –50dB, can be as low as –80dB

* The upper frequency limit is imposed, largely, by the shortcomings of the photo-lithographic techniques used to produce the conductor pattern.

one wavelength of the surface, if the strip is more than a few wavelengths in thickness, the rear face of the resonator strip can be cemented to a rigid substrate to give a mechanically robust and vibration resistant component. To prevent spurious end-reflections, the ends of the strip are normally cut at an angle, and a layer of some acoustically absorbent material is normally applied to the regions outside the expected path of the wave pattern.

Resonant SAW systems

The other application of SAW techniques is in the production of resonant (R-type) filters, where end-reflections are deliberately exploited to assist in the formation of standing wave patterns. Such a component has much in common with a quartz crystal resonator, except that the SAW device has superior HF capabilities, covering the frequency range 100kHz–1.5GHz.

The signal/noise ratio of oscillators based on resonant SAW devices is excellent, with a quoted performance for a 400MHz system as better than –150dB at 10kHz from the resonant frequency. At these frequencies there are no comparable quartz crystal resonators. However, at lower frequencies the frequency stability of an SAW resonator, using a mono-crystalline quartz substrate, is somewhat inferior to that of a normal quartz crystal frequency standard, with temperature coefficients typically in the range 5–90ppm/°C, over a temperature range of +/–50°C.

12

High frequency amplifiers

Introduction

Some of the circuit techniques which are useful in the design of amplifiers for low frequency signals have been examined in Chapter 6, and while most of the circuit layouts which work at low frequencies will also work at higher ones, they do not generally work as well, and there is nearly always a problem in obtaining a useful stage gain from simple circuitry at high frequencies. This leads to practical difficulties, particularly with small signals, in getting a good enough overall signal to noise ratio, bearing in mind that every amplifier stage will introduce some additional circuit and device noise of its own. It is therefore essential to ensure that the individual stage gains – especially at the front end of an amplifier – are sufficiently high that the amplified noise at their inputs will always be greater than that generated by following stages.

The use of a resonant tuned circuit, of the kind analysed in Chapter 11, is a great help in this respect since the voltage magnification given by the Q of such a circuit will provide an almost noise-free element of gain, so that the 'gain' stage may only be needed to act as an impedance transforming 'buffer' to convert the high output impedance of a parallel resonant tuned circuit into a low impedance signal source which can drive following stages. However, this approach is only useful if amplification at a single frequency, or a narrow band of frequencies, is acceptable. This was almost always the case in early radio receivers, where the main task of the 'radio frequency' (RF) amplifier stage was just to increase the size of the received signal, from the aerial input, to a level where a simple rectifying device could perform the function of separating the modulation from the radio frequency 'carrier'. The RF stage might also give a useful increase in selectivity to help the user to separate a wanted signal from another on an adjacent carrier frequency.

Practical circuitry

In the early days of radio, when triode valves were the only types available, such a 'tuned' RF amplifier stage would use the type of circuit layout shown schematically in Figure 12.1, using both an input and an output tuned circuit. However, this arrangement suffers from the snag that when these two *LC* circuits are nearly in tune, the circuit will almost inevitably burst into oscillation. This problem can be mitigated if the tuned circuits are fairly heavily damped by the parallel connection of suitable value resistors, but this will then reduce the effectiveness of the tuned circuits, in respect of the gain and selectivity which they could provide. This instability occurs because the small anode/grid 'stray capacitance' (C_{fb}), will allow the unwanted feedback of energy from the amplified signal at the valve anode into the input tuned circuit: a problem which will worsen as the operating frequency is increased, because this will also reduce the RF impedance of the stray capacitance feedback path.

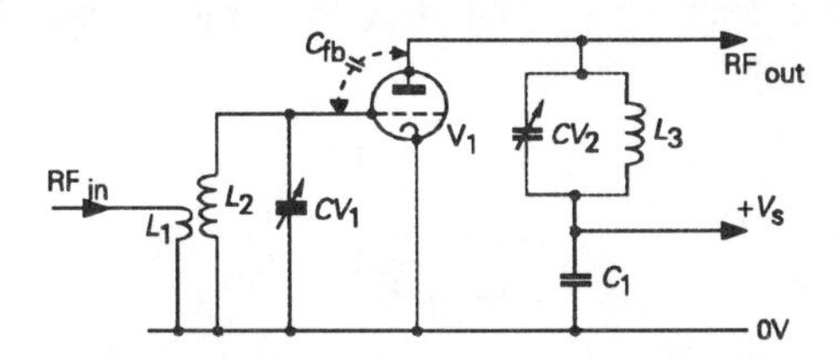

Figure 12.1 *Typical triode valve RF amplifier stage*

The first part of the requirement for continuous oscillation, that the loop gain should exceed unity, will be met when the impedance of the feedback capacitance, at the frequency in question, is less than or equal to the product of the dynamic impedance of the input tuned circuit and the voltage gain of the stage. The second part of this requirement, that the feedback voltage should be exactly in phase with the input signal at the frequency at which oscillation will occur, will be satisfied at some frequency near the frequency of resonance, because of the combined effect of the phase shift in the signal passing through the feedback capacitance and the additional, and quite abrupt, phase shift in an *LC* tuned circuit, illustrated in Figure 12.2, between its output voltage and input current, when the input signal passes through its resonant frequency.

Although it appears possible to tune the input and output circuits so that oscillation is avoided, this is, in practice, very difficult to achieve. This means that practical triode RF amplifiers must use either a gain stage of the kind shown in Figure 12.3, in which the effect of the feedback capacitance is neutralized, or a 'grounded grid' type layout, of the kind shown in Figure 12.4.

In the circuit of Figure 12.3, the mid-point of the output tuned circuit is earthed at RF, so that a signal

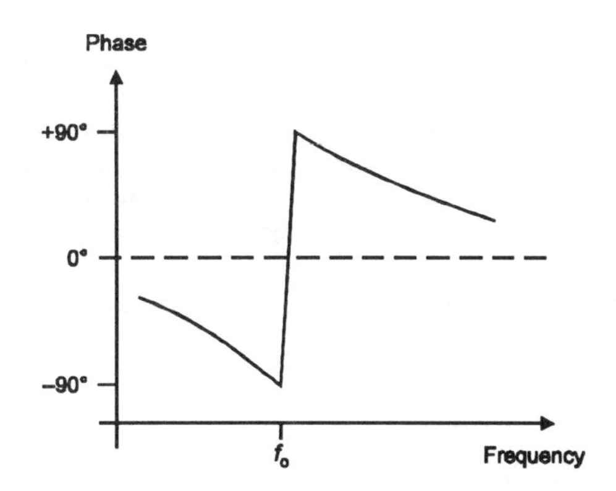

Figure 12.2 *Phase angle of tuned circuit as a function of applied signal frequency*

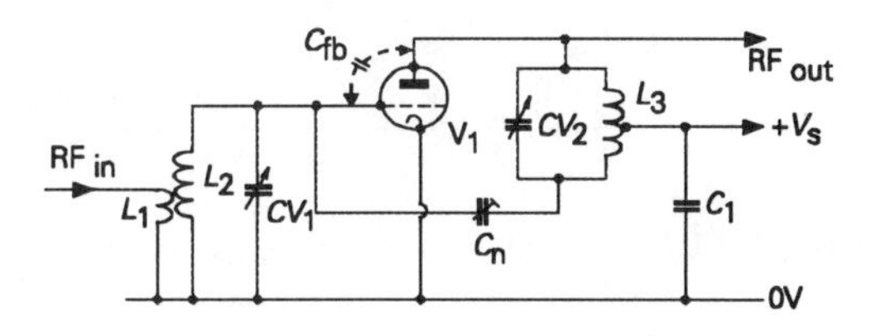

Figure 12.3 *Neutralized triode valve RF amplifier stage*

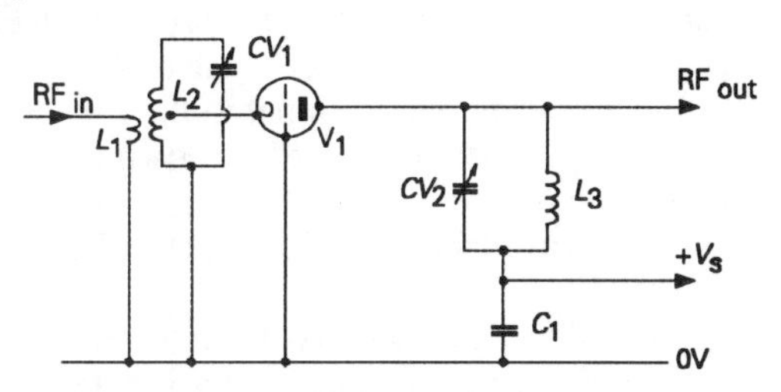

Figure 12.4 *Grounded grid RF amplifier stage*

which is in antiphase to that at the valve anode is developed at the opposite end of the *LC* circuit, and this allows the deliberate feedback, through the neutralizing capacitor (C_n), of a signal which can exactly cancel that arising from the internal feedback due to the valve stray capacitances (C_{fb}). This arrangement works quite well, but demands a tapped output coil, and also prevents the use of a ganged tuning capacitor, since the rotor of the tuning capacitor (CV_2) cannot now be taken to a point of zero RF potential. It is also, in practice, rather a tiresome task to set C_n to a value which will eliminate the tendency to oscillation over the whole required operating frequency range.

The grounded grid amplifier stage of Figure 12.4, which uses the grid electrode of a triode valve as an electrostatic screen between the input and output tuned circuits, works very well, and is often found, in a 'solid-state' version, as the RF gain stage in inexpensive FM radio circuits, where a bipolar transistor is used in a layout of the kind shown in Figure 12.5. The description 'grounded base' is normally used for this type of circuit.

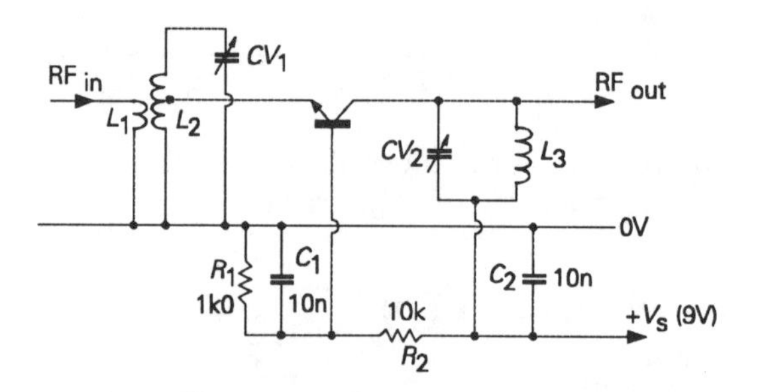

Figure 12.5 *Transistor operated grounded base RF amplifier stage*

The only problem with either of these last two circuits is that both the cathode circuit of a valve and the emitter circuit of a transistor are very low impedance points, because of the effect of internal negative feedback. This means that the input to the gain stage must be taken at a tapping point low down on L_2 to provide a low enough drive impedance, and this limits the voltage gain available from the input tuned circuit, L_2/CV_1.

In the case of valve amplifiers, an engineering solution to the problem of unwanted feedback from anode to grid, due to inter-electrode stray capacitances, was provided by the introduction of a 'screen grid', connected to a suitable low RF impedance positive potential, and mounted internally between the grid and the anode. Such screened grid valves, and their later 'RF pentode' successors, solved the problem of internal capacitative feedback very satisfactorily, and allowed a useful degree of amplification to be obtained up to frequencies at which the loss of efficiency due to electron transit times within the valve, and to the parasitic inductance of the connections to the valve from its holder, began to be a significant factor.

The contemporary solid-state equivalent to the screened grid valve is the dual gate MOSFET, shown symbolically in Figure 12.6. In this type of device a second insulated gate electrode metallization (G_2), is formed on the FET channel between the signal gate (G_1), and the drain. With contemporary dual gate MOSFETs, feedback capacitances of 0.01pF, or less, are attainable; as compared with the 3–10pF collector–base capacitance of a typical small-signal bipolar junction transistor, or the similar drain–gate capacitance of a typical junction field effect device; and this allows stable RF amplification, using a straightforward gain stage, of the kind shown in Figure 12.7, up to several hundred megahertz. Recent dual-gate MOSFETs based on gallium arsenide, rather than silicon, have extended the usable frequency range up to the GHz region.

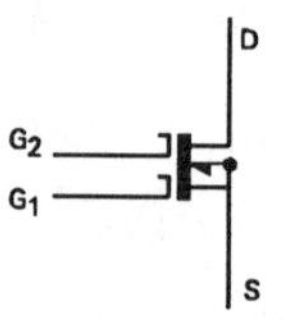

Figure 12.6 *Dual-gate MOSFET RF amplifier device*

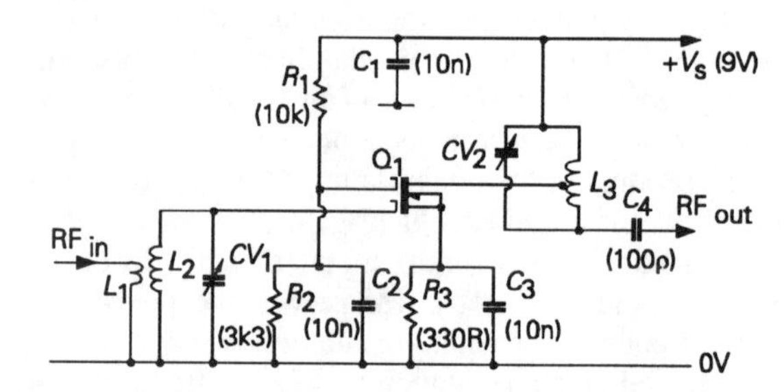

Figure 12.7 *Dual-gate MOSFET RF amplifier stage*

Unfortunately, the high circuit impedances associated with MOSFETs lead to a somewhat worse circuit noise figure in comparison with gain stages based on bipolar junction transistors or junction FETs. Circuit arrangements using standard devices which avoid the problem of feedback capacitance by virtue of the circuit layout are the long tailed pair layout shown, for bipolar junction transistors, in Figure 12.8, and the cascode layout shown in Figure 12.9, using junction FETs. In these cases the output transistor is driven by either its emitter or its source, so the the base or gate – which is taken to a low impedance point in respect of RF – is interposed between the input and output of the amplifying device. This gives a similar degree of input–output isolation to the grounded-base circuit shown in Figure 12.5, but with a higher input impedance. Both of these circuit layouts allow stable gain to be obtained up to quite high operating frequencies.

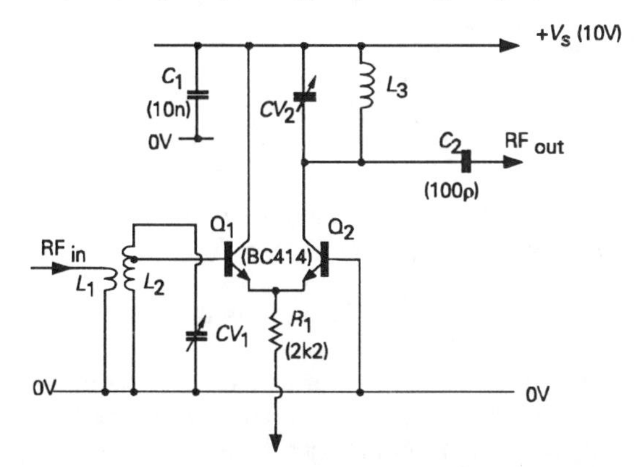

Figure 12.8 *Long-tailed pair RF amplifier stage*

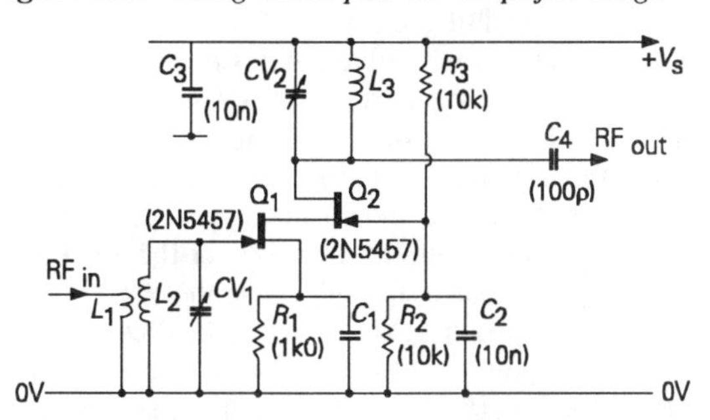

Figure 12.9 *Cascode connected FET amplifier stage*

An interesting version of the cascode circuit which combines the characteristics of both of these layouts, by the use of complementary symmetry devices, is shown in Figure 12.10. This arrangement gives the isolation of the grounded-base stage as well as the high input impedance associated with an emitter follower.

In both the cascode and the grounded-base layouts the current through the output device is almost entirely determined by the emitter/base forward potential. This gives rise to the type of relationship between collector current and collector/base voltage shown in Figure 12.11, in which the actual collector voltage has very little influence on collector current, which means that the device has an exceedingly high dynamic output impedance. This is useful since such a stage can be

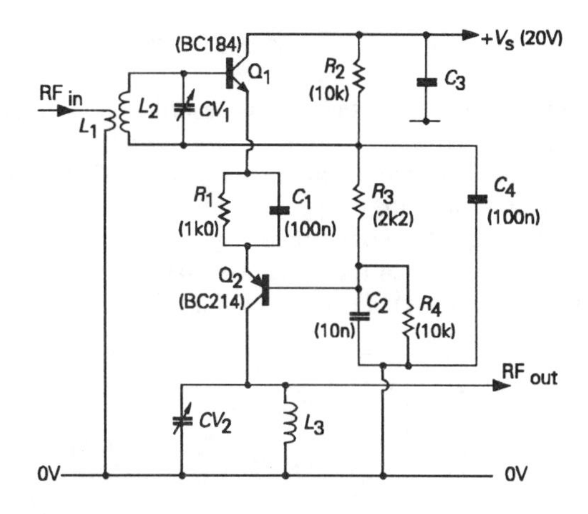

Figure 12.10 *Complementary cascode RF amplifier stage*

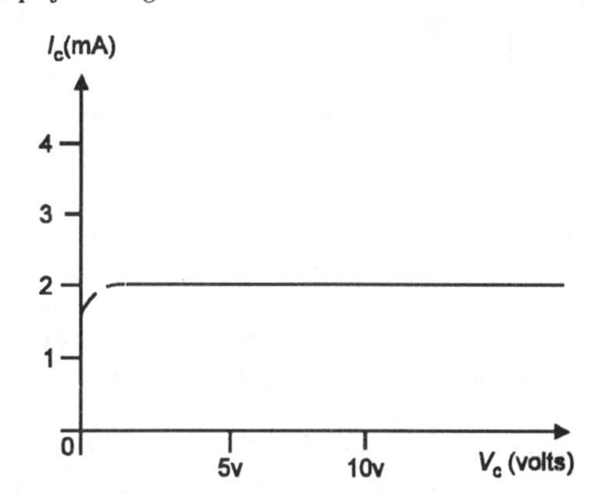

Figure 12.11 *Cascode collector current characteristics*

connected directly to a parallel resonant tuned circuit without introducing significant parallel resistance damping of its characteristics.

Both MOSFETs and junction FETs have very high input impedances (>100MΩ), and can therefore be connected directly across an *LC* tuned circuit, as shown in Figure 12.9, without lowering its Q – its voltage magnification factor – whereas a bipolar device would nearly always require that it should be driven from a tapping point rather lower down on the tuned circuit to lessen the effect of its relatively low (5–10kΩ), input resistance on the Q of the circuit.

An ingenious alternative method of neutralizing an RF gain stage – in this case a junction FET, chosen because its nearly ideal 'square law' transfer characteristics give low cross-modulation between wanted and interfering input signals – is described by Philips (*Electronic Components and Applications*, May 1980), and shown in the circuit of Figure 12.12. In this, the tuning capacitor of the input tuned circuit (CV_1), is connected effectively between gate and source, and a small inductor, in this case of 10nH inductance, is connected between the source and the 0V line.

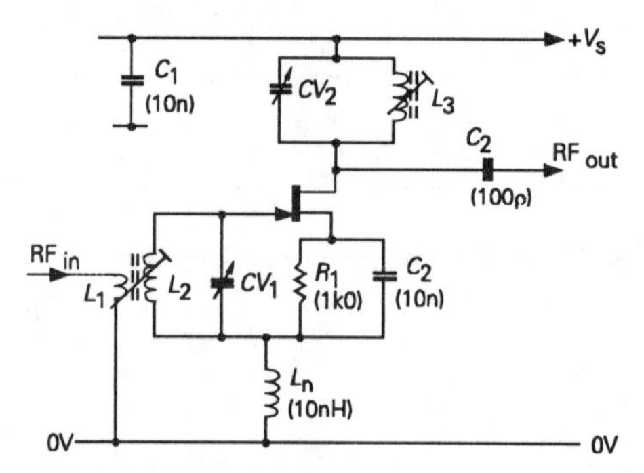

Figure 12.12 *Source inductor neutralization circuit*

Where ceramic SAW filters (see Chapter 11), are to be used as selectivity elements, they usually require an input and output load impedance of the order of a few hundreds of ohms. At this sort of impedance level a standard bipolar transistor operated feedback amplifier, such as the circuit shown in Figure 12.13, which I designed for an FM tuner IF stage, provides quite a satisfactory answer. In the case of the circuit illustrated, the amplifier provides a stable gain of approximately 50dB at 10.7MHz. Allowing for the anticipated 28dB loss in the four SAW filters this meets the design value for the residual IF gain of 22dB (12.5).

Bandwidth, noise and cross-modulation

For any given source impedance, ambient temperature and measurement bandwidth there will be an associated thermal noise voltage (E_n), defined by the equation:

$$E_n \text{ (RMS)} = \sqrt{4RkTB} \quad (1)$$

where k is Boltzmann's constant (1.38×10^{-23} J/°K), R is the circuit impedance, T is the ambient temperature in °Kelvin (the so-called 'absolute temperature', based

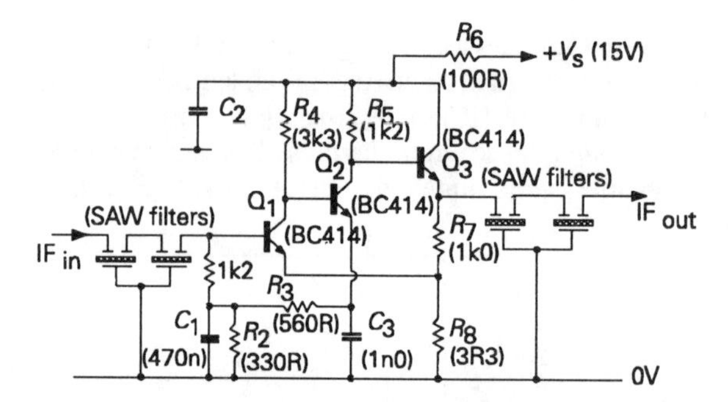

Figure 12.13 *FM receiver IF amplifier for use with SAW filters (gain = 50dB @ 10.7MHz)*

on a zero of –273.15°C) and B is the measurement bandwidth, in Hz.

With ideal components, it wouldn't matter at what stage the operational bandwidth was determined since the result, in terms of the output noise level, would be the same, and would be the RMS sum of all the noise components resulting from the input circuit impedances and cumulative voltage gains of the preceding gain stages. Unfortunately, there is always some degree of nonlinearity associated with any amplifying device, and this could lead, for example, to the type of transfer characteristic shown in Figure 12.14. If any nonlinearity is present, the curvature of the input/output transfer characteristic will cause the voltage gain of the circuit (proportional to the vertical component of the slope of the transfer curve) to be greater at some point on the input/output transfer curve than another.

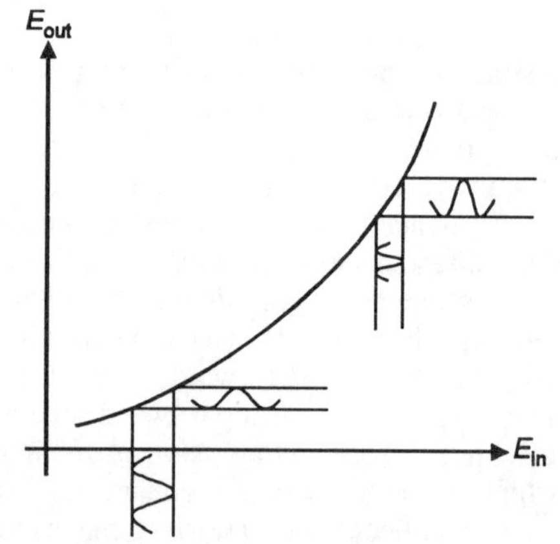

Figure 12.14 *Effect of nonlinearity in amplifier transfer characteristics*

So, if a signal is being amplified by such a nonlinear stage, and this is true whatever the nonlinear shape of the transfer characteristic, any mechanism – such as an additional signal voltage being handled at the same time by the same amplifying stage – which causes the operating point to move up and down along the amplifier E_{in}/E_{out} transfer curve, will have the effect of 'modulating' (increasing or reducing the size of) the resultant signal voltage.

When the mechanism for moving the operating point up and down is the presence of a second simultaneous signal voltage, the result will be that both output signals will be modulated by each other – a phenomenon which is called 'cross modulation'. This effect will occur in any stage which suffers from any degree of waveform distortion, which, in practice, means virtually every real – as distinct from theoretical – system, and is of particular importance in wideband amplifier stages. This is because such amplifier stages will have a higher thermal noise component – see equation (1), above – and because there is a greater probability, especially in radio frequency preamplifier systems, that one or more relatively powerful signals will be included within the amplifier pass-band. So, in the presence of such inevitable nonlinearities in the amplifier characteristics, all signals will be cross-modulated by wideband noise, a noise component which will remain as a feature of each signal and will be unaffected by subsequent bandwidth limitation (apart from considerations of sideband curtailment). Furthermore, all signals which are present at the same time will, to some extent, cross-modulate each other: an effect which cannot subsequently be removed by later improvements in selectivity.

Some measure of relief is afforded by the fact that the effects of nonlinearity in an input transfer curve are related to the size of the input signal, so that the smaller the signal, the more closely the transfer characteristic will approximate to the ideal straight line shape, and the less significant cross-modulation, of all kinds, will be. However, the fact remains that the wider the bandwidth any stage is called upon to handle, the greater the scope there will be for various operational problems.

Good amplifier design will therefore seek to combine the following requirements: that the gain of the input stage shall be sufficient to ensure that the noise contributions from all later stages will be less than that due to the input stage; that, for the same reasons, the inter-stage couplings shall be chosen to achieve the best possible transfer of signal energy from each stage to the next; that the effective bandwidth of any stage is never greater than is necessary; that all stages have the maximum practicable linearity of their transfer characteristics; and that the signal level at any point in the system is never allowed to exceed that which can be accommodated within an adequately linear part of the transfer curve. It could also be added that it is a useful design ambition to try to achieve the required performance with the minimum feasible number of stages: what isn't there won't add noise, and can't go wrong!

Wide bandwidth amplifier design

There are three main difficulties in obtaining a good high frequency performance from a simple transistor amplifier, of the kind shown in Figure 12.15, ignoring for the moment such arcane things as carrier transit time effects within the device. These are the effect of the collector–base feedback capacitance (C_{fb}), the output stray capacitance (C_s), and the input (base–emitter) capacitance (C_{in}), which is largely due to the forward biased base–emitter junction diode, an important factor which will be examined later.

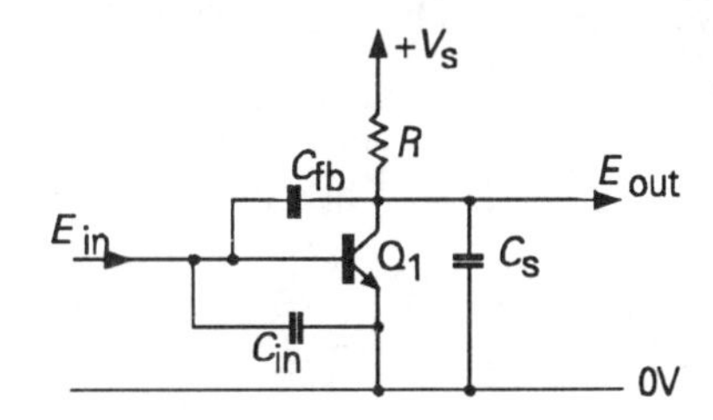

Figure 12.15

The importance of the feedback capacitance, in a simple resistance–capacitance coupled amplifier, is mainly due to the fact that since the stage is a voltage inverting one, an input voltage of y will give rise to an output voltage of $-Ay$, where A is the amplification factor of the stage. This means that the input voltage y will cause a total voltage of $-(A+1)y$ to develop across the capacitor, so that the input charging current required to allow a voltage change to arise at the input of the device will be $A+1$ times greater than if the output end of C_{fb} was just connected to the 0V rail.

This effective magnification of the size of the feedback capacitance is known as the 'Miller effect', and, for a feedback capacitance of 5pF and a stage gain of 50, this effect would make the capacitance seen at the input appear to be 255pF. If we define the Miller capacitance as C_m, then to get a useful degree of amplification, at, say, 5MHz from such a stage would demand that the source impedance (Z_S) is significantly lower than that of C_m at 5MHz, i.e. less than 127 ohms.

The effect of the stray capacitance (C_S) from the output point to earth, depends on the charge/discharge time constant of C_S and R_l. Again, for a –3dB point in the gain curve at 5MHz, for a 10pF value of stray capacitance, the load resistor cannot exceed 3k. (Both of these resistor values are derived from the equation $Zc = 1/[2\pi fC]$.)

The simplest way of minimizing the effect of Miller capacitance is to use either a grounded base, or a differential amplifier or a cascode layout, of the kinds shown in Figures 12.5, 12.8, 12.9, and 12.10, modified for use in a resistance coupled layout as shown in the circuits given in Figure 12.16.

A dual gate MOSFET could also be employed as the amplifier stage, but such a device, though excellent in

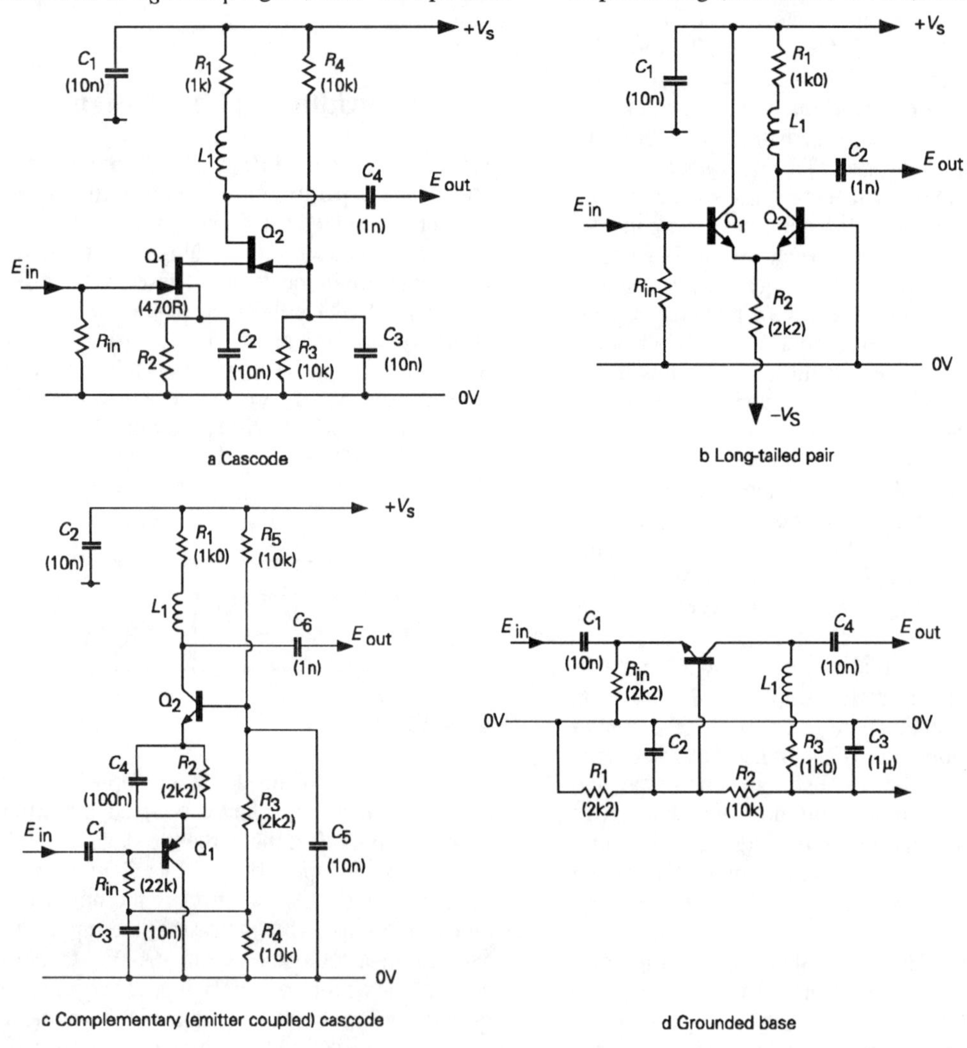

Figure 12.16 *Practical wide bandwidth amplifier circuits*

respect of its very low feedback (Miller) capacitance, would have limitations in respect of its possible drain current, which might need to be fairly high in order to achieve a useful output voltage swing with a low value collector or drain load resistor.

A further common technique, illustrated in Figures 12.16, is to incorporate a small inductor, L_1, in series with R_L. The value of this inductor will be chosen to resonate with the stray capacitance, as a parallel tuned circuit, at a frequency close to the upper −3dB gain point of the amplifier stage. This will give a boost to the gain of the stage at this frequency, and may allow higher values of load resistor to be employed – which will give a higher stage gain – for the same value of stray capacitance.

An alternative approach which is also used, where the amplifier stage is to be used to drive a known capacitative load, such as a long length of cable, is to include an inductor, L_2, as shown in Figure 12.17, in series with the output circuit, with the value of L_2 chosen to resonate with C_S as a series tuned circuit, in order to generate a higher output voltage at the HF end of the specified pass-band.

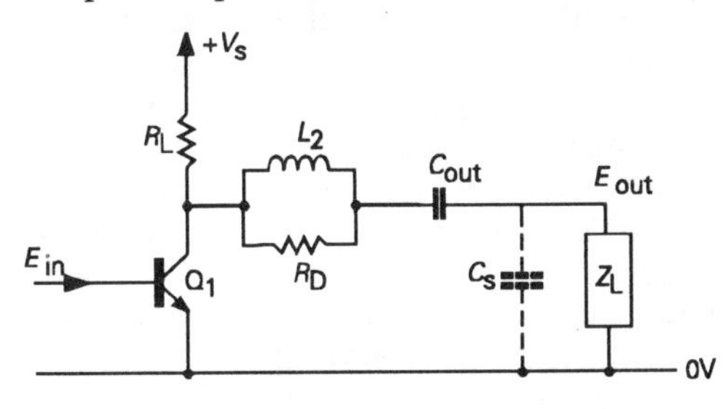

Figure 12.17 *Series inductor output RF compensation*

Another very useful way of limiting the shunting effect of the load stray capacitances on the amplifier load resistor is simply to interpose an output emitter-follower (or its FET equivalents) between the gain stage and the load circuit, as shown in Figure 12.18. This will allow an effective output impedance, as seen by the load, of less than 100 ohms, with a consequent substantial increase in the effective gain/bandwidth product of the stage. Some caution is needed, however, in the use of this kind of circuit, in that an emitter follower can break into oscillation, at VHF, with certain values of output load capacitance and inductance, due to its internal stray capacitances and carrier transit

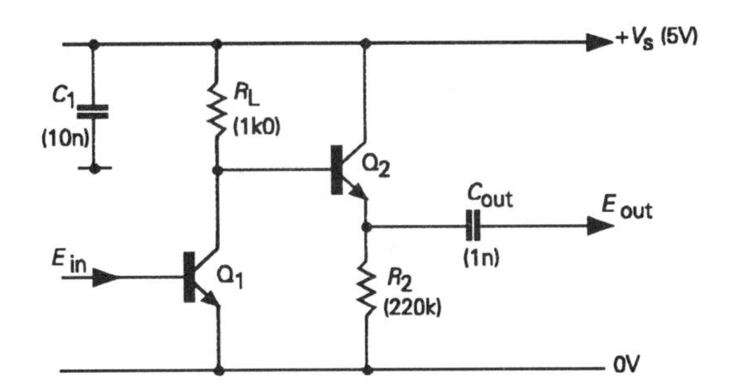

Figure 12.18 *Output emitter follower impedance converter*

times. This type of mechanism is explained more fully in Chapter 13.

For very high frequency use, the load resistor may be dispensed with entirely, as shown in Figure 12.19, so that the load inductor, L_1, acts simply as an RF choke – a load whose RF impedance increases with frequency, but whose DC resistance can be quite low – allowing relatively high values of collector current to flow, with consequently increased values of amplifier transfer conductance (g_m), expressed as milliamperes of change in output current for a unit voltage change in the applied input potential. The drawback with this arrangement, apart from the fact that the stage gain will be very low at low frequencies, is the fact that above the frequency of resonance of the choke with the load stray capacitance, the output load will appear to be purely capacitative, and the stage gain will consequently fall linearly with increasing frequency.

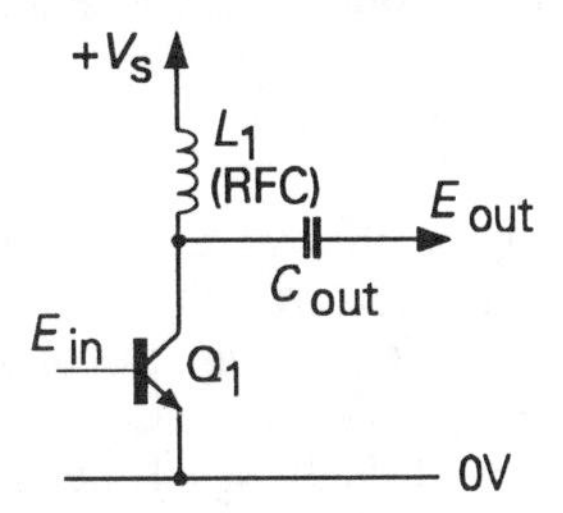

Figure 12.19 *RF choke used as load*

Although, for simplicity, I have illustrated the preceding amplifier stages as using a single transistor, designers will, in practice, more often choose to use a gain stage, such as the cascode layout shown in Figure 12.20, which will limit the effects of Miller capacitance.

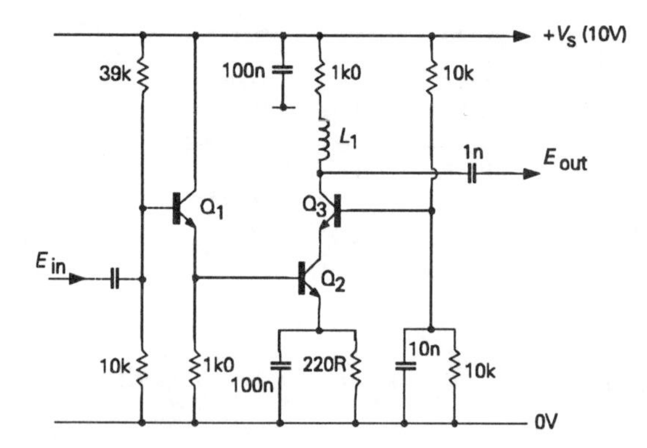

Figure 12.20 *Cascode connected wide-band amplifier stage*

Effects of junction capacitances

The capacitances within a junction transistor are composed of two separate components, the 'stray' capacitances which are simply due to the physical proximity of the connecting wires, and those which are due to the presence of the collector–base and base–emitter junction diodes.

In normal operation, the collector–base junction is reverse biased, so this diode looks like any other junction capacitor of equivalent area and doping level, and will offer a capacitance (C_j) which will vary with the reverse bias level in an identical manner to the well known 'Varicap' tuning diode, whose capacitance is given approximately by the equation

$$C_j = m/\sqrt{(V + V_d)} \qquad (2)$$

where V is the applied voltage, m is a constant depending on the junction area, thickness and doping level, and V_d is the forward diode conduction voltage – approximately 0.6V for silicon devices.

Under normal small-signal operating conditions the collector–base voltage will remain substantially constant, so that C_j can be considered simply as an internal device capacitance in much the same way as the anode – grid capacitance of a valve – the treatment I have adopted above. Unfortunately, the effect of the forward biased base–emitter junction is much more complex, because the effective capacitance for a given junction area will be much larger, and much more strongly dependent on doping level and bias voltage, especially the latter. For this reason, transistor manufacturers do not attempt to specify the effective base-emitter junction capacitance, but, instead, quote the value of the effective 'current gain transition frequency', or 'gain/bandwidth product', 'f_T', which largely depends on it. This type of specification is, in any case, rather misleading in its apparent implication, since a quoted value 'f_T = 400MHz' tends to give the impression that some useful gain might be available at this frequency, whereas, in reality, this is the frequency at which the current gain has fallen to unity in value, so, a transistor with an h_{fe} (common emitter current gain) of 400 would have a –3dB point in its gain value at 1MHz, not 400MHz!

The transition frequency value for a bipolar junction transistor is a complex factor, and depends on the geometry of the device, including particularly the junction thickness, the doping level, the carrier mobility, as well as the effective input (base–emitter) capacitance, in that this will provide an alternative path to ground for the input signal current. The internal base–emitter conducting impedance (r_e), decreases as the emitter (and base) current increases. Unfortunately, the base–emitter junction capacitance (C_{ie}) also increases, so that, for any given source impedance, the proportion of the input current which is usefully employed increases only slowly with an increase in emitter/collector current, with most small signal transistors giving their best HF performance at collector currents in the range 5–100mA, where the effective base–emitter (input) capacitance of the transistor will be in the range 10–100pF.

There is little the designer can do to minimize the effect of this input capacitance on circuit performance except that since the gain/bandwidth product, an alternative name given to f_T, remains fairly flat over a range of collector current values, whereas the base–emitter capacitance will increase over this range, as shown in Figure 12.21, it is sensible to operate bipolar junction transistor gain stages at the lowest collector current capable of giving an adequate value of f_T.

In the case of bipolar junction transistors specifically designed for use at high frequencies, efforts will have been made to keep the junction areas, and those stray capacitances of purely mechanical origin, as low as possible, and also to keep the junction regions as thin as feasible for a given maximum collector voltage, in order to keep the carrier transit times low. Because of the low circuit impedance levels usable with bipolar

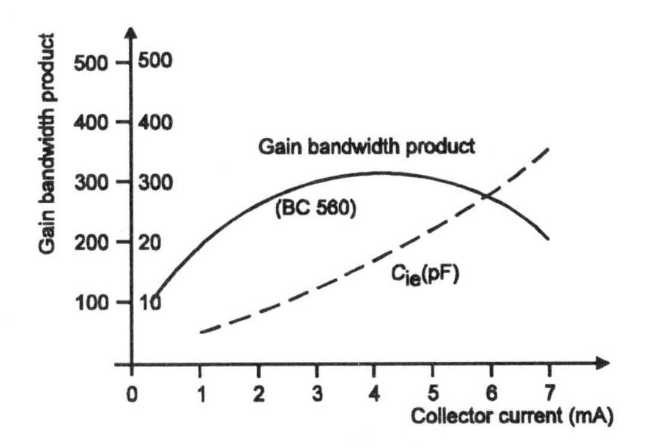

Figure 12.21 *Effect of collector current on f_T and input capacitance in small signal transistors*

junction transistors, and the very high mutual conductance values which they offer, these devices are still preferable for low noise, high gain and wide bandwidth amplifier stages, in spite of the practical difficulties in circuit design.

Input bandwidth limitation

Since the problems of input noise level and possible cross-modulation effects are bandwidth related – hence the old adage that 'the wider the window, the more the dirt flies in'– it is a sensible policy in amplifier design to limit the bandwidth to a value which does not greatly exceed that which is necessary for the purpose in hand.

Where amplification down to LF is not required, as, for example, in the case of a radio or TV receiver RF preamplifier stage, but where it is not wished to select the particular frequency to be amplified at the input to the amplifier – as might be the case in a radio receiver where the actual operating frequency is only decided at the frequency changer stage, later in the circuit – an approach which is widely used is to use an input filter circuit which combines high-pass and low-pass characteristics. In principle, this might just consist of an *LC* low-pass and a *CL* high-pass pair of filter circuits in cascade, as shown in Figure 12.22, but the performance of such a simple arrangement is unlikely to be adequate.

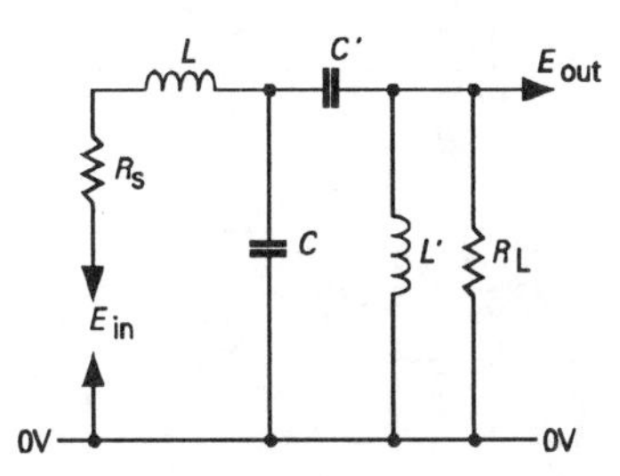

Figure 12.22 *Layout of simple band-pass LC/CL filter*

However, design tables for high performance multiple element low-pass, high-pass and bandpass filters with a flat-topped pass-band frequency response, coupled with steep cut-off bandwidth skirts, are commonly given in RF circuit design textbooks (such as the *ARRL Handbook,* 1989. pp. 2.37–2.55), and a worked example, using diode switching, and covering the frequency range 3MHz–30MHz in three bands, is shown in Figure 12.23.

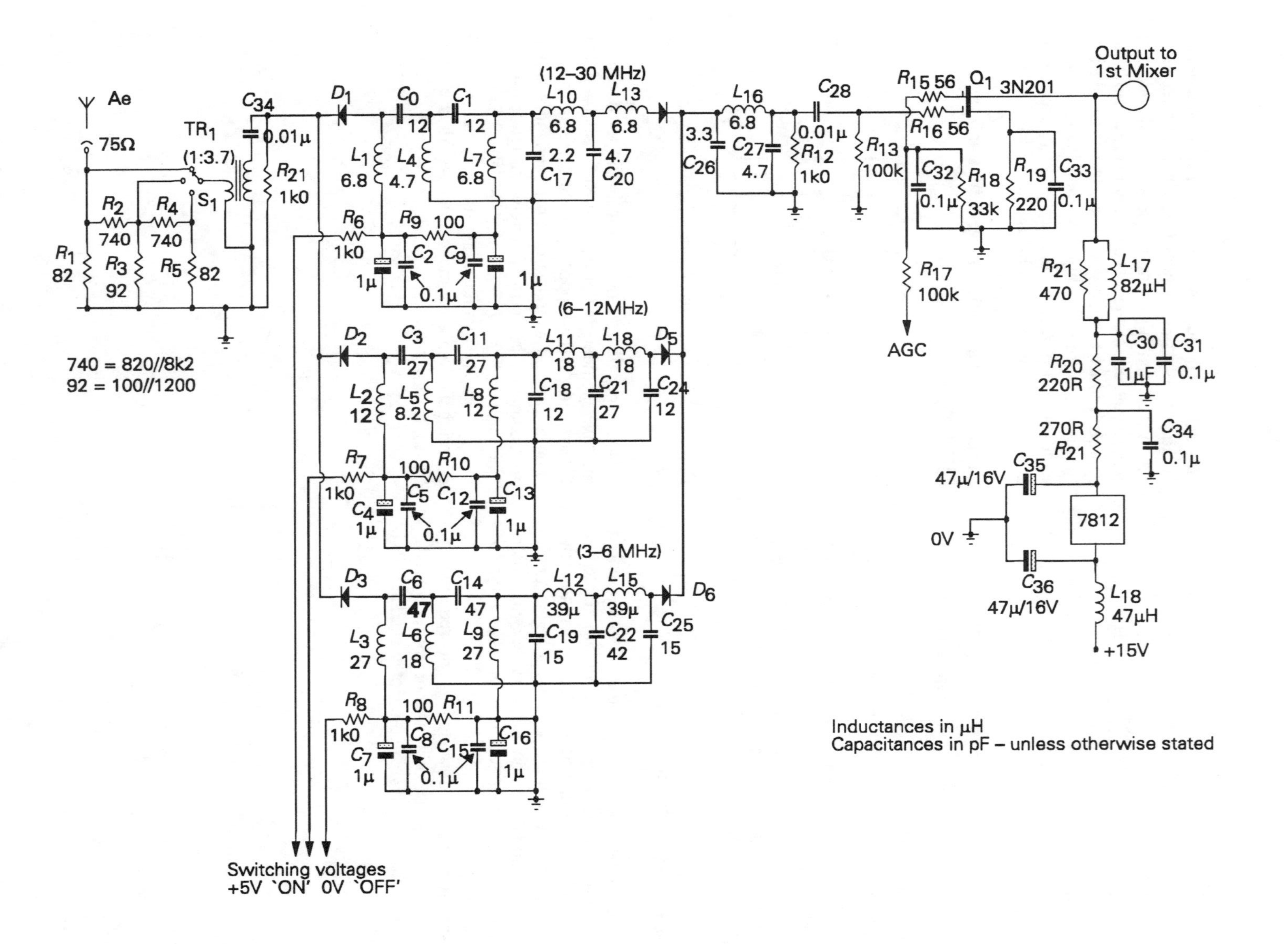

Figure 12.23 *Typical input band-pass filter for 3–30MHz communications receiver*

13

High Frequency Oscillators

Introduction

The requirements of simple industrial control mechanisms, and the exploration of non-mechanical digital computer systems would, undoubtedly, have led in due course to the utilization of the electronic emission phenomena uncovered by Edison, even if they had not been used by Fleming and Lee de Forrest, in the subsequent evolution of the thermionic valve. However, as a matter of history, the actual stimulus to the development of 'thermionics', and, following this, the exploration of electronics, as a whole, came from the commercial success of wireless telegraphy, by which means interrupted trains of high frequency alternating electric signals could be sent from a 'transmitter' to a 'receiver', without the need for a wired electrical connection between them.

In the early experiments in this field, the use of a high voltage 'spark generator', of the kind shown in Figure 13.1, to shock excite a damped train of electrical oscillations in a tuned circuit, as indicated in Figure 13.2, was quite adequate as a source of RF signals. Such a transmitter, apart from its inefficiency and noisy output signal, would inevitably generate quite a wide spectrum of RF radiation, and it would be impracticable to operate two such transmitters in the same area, because of mutual interference, or to operate a transmitter, and a receiver, simultaneously at the same location, unless their operating frequencies were very widely separated.

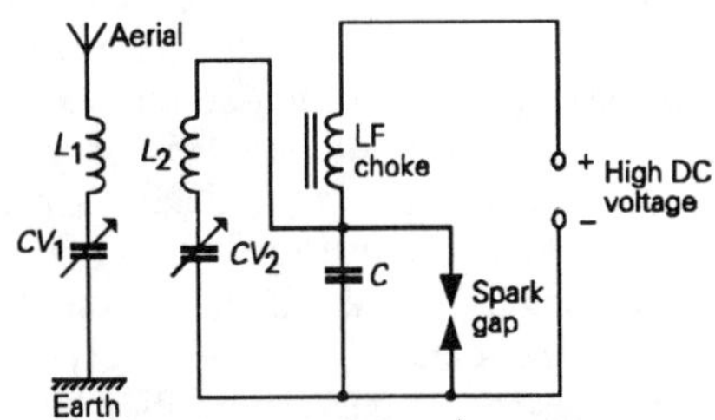

Figure 13.1 *Typical spark generator for RF*

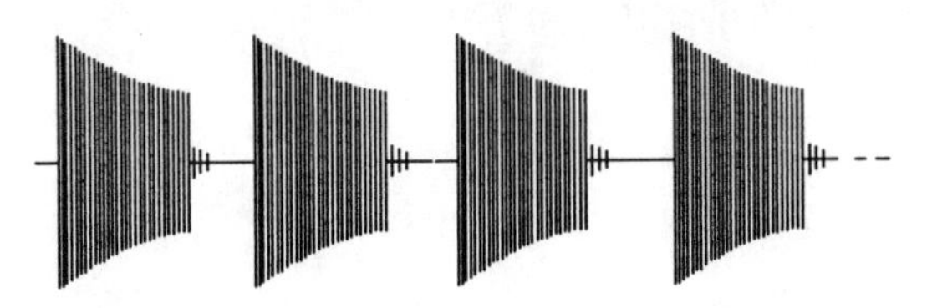

Figure 13.2 *Damped train of RF oscillations produced by circuit of Figure 13.1*

What was required was some way of generating a continuous single frequency sinewave voltage output, stable in both amplitude and frequency, and free from spurious harmonics or other unwanted signals, which could then be amplified and broadcast by the transmitter aerial. This is not an easy requirement to meet, and an enormous amount of ingenuity has been directed, over the years, to devising circuitry for this purpose. Undoubtedly, the starting point for this development was the invention of feedback, by E. H. Armstrong, in 1912: a technique in which an amplified signal was fed back into the input circuit of the amplifier, which provided an effective means of generating such a continuous oscillation. Sadly, Armstrong discovered, like many other inventors before and since, that making a useful invention and defending the title and ownership of it through subsequent, and normally unavoidable, litigation, were two very different things, and the rest of his life was plagued by the continuing need to preserve his rights in the use of the many useful ideas he had originated. However, I am happy, on behalf of my fellow electronics engineers, to acknowledge the debt we owe to him.

Although, in the beginning, oscillator circuits were mainly required as the signal source for transmitters – a need which still remains – increasingly, with the passage of time, other applications are being found, such as the use of oscillators in receiver circuitry for frequency conversion purposes, as tone generators in electronic music, as frequency standards, test signal sources and waveform generators in a whole host of applications, and as frequency control clock mechanisms for a whole range of digital circuitry.

For convenience, I have divided this chapter into three parts, covering those types of oscillator based on tuned *LC* circuits, those utilizing electromechanical effects in piezo-electric materials ('crystal' oscillators), and those based on *RC* relaxation effects – as an extension of the various multivibrator designs examined in Chapter 10.

Basic circuit designs

LC oscillators

Because of the general purity of their output waveform, these systems form the majority of the variable frequency HF oscillator designs, and are based on the use of the *LC* tuned circuit arrangement examined in Chapter 11. In all of these designs a tuned circuit is used to select the actual frequency of oscillation, and the layout of an electronic circuit containing an amplifier is arranged to provide the positive voltage or current feedback needed to sustain oscillation.

Although there are literally scores of different *LC* oscillator circuits which have been developed for various applications, there appear to be seven basic forms, from which individual circuits have been derived. Since all of these circuits were devised at a period when the only amplifying element available to the designer was the electronic valve, which acts as a high input impedance phase inverting gain stage, I have illustrated these basic oscillator designs, in schematic form, in Figures 13.3 to 13.10, using a simple high input impedance phase inverting gain block, A_1, A_2, etc., to simulate the function of the valve.

For some of the circuits, the operation of the circuit, in an elaborated form, depends on the physical characteristics of the amplifying device, as in the electron coupled oscillator, and in this case I will show either the valve circuit or one based on a contemporary solid-state equivalent device, such as a junction FET.

Hartley

In this layout, shown in Figure 13.3, the inductor (L_1) in the *LC* tuned circuit has an intermediary connection, a 'tap', usually but not necessarily somewhere near its centre point, which is taken to a zero RF potential point. This causes the instantaneous RF potential at the two ends of the coil to be in antiphase, which gives the phase inversion in the feedback path needed to meet the requirement for oscillation.

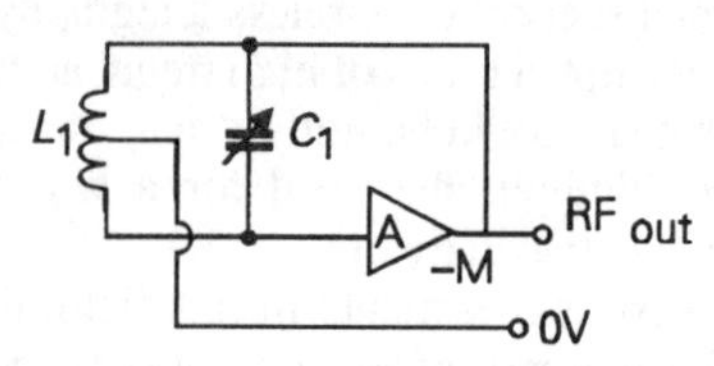

Figure 13.3 *Hartley oscillator*

Colpitts

This arrangement, shown in Figure 13.4, is very similar in its function to the Hartley circuit of Figure 13.3, except that, in this case, the tuning capacitor in the LC circuit is divided into two (C_1 and C_2), and the junction of these is taken to a low impedance point. As a matter of practical convenience, C_1 will often have the same value as C_2, but this does not need to be the case, and in some applications better performance might be obtained by the use of a different C_1/C_2 ratio.

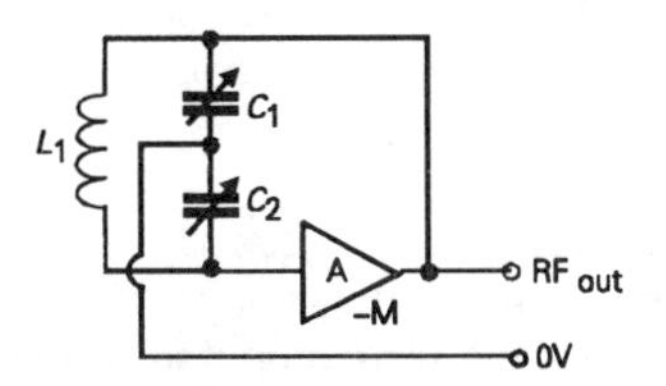

Figure 13.4 *Colpitts oscillator*

Tuned input

In this layout, shown in Figure 13.5, the tuned circuit is elaborated into a two-winding HF transformer, of which the tuned secondary is connected to the input of the gain block, and the primary feedback winding, taken from the amplifier output, is connected in such a way that the necessary positive feedback is obtained.

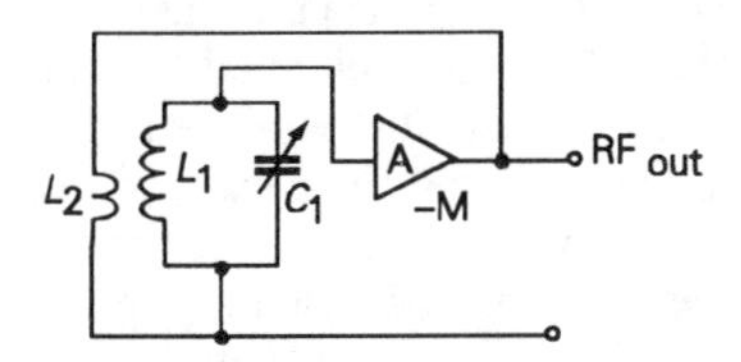

Figure 13.5 *Tuned input oscillator*

Tuned output

This circuit, shown in Figure 13.6, is basically identical in its design philosophy to that of the tuned input layout, except that, in this case, it is the primary winding of the output–input coupling transformer which is tuned. This has the advantage, where high voltage outputs are needed, that the transformation ratio can be chosen so that a large voltage swing can be developed at the output without overloading the input circuit.

Meissner

This oscillator, shown in Figure 13.7, employs a tuned

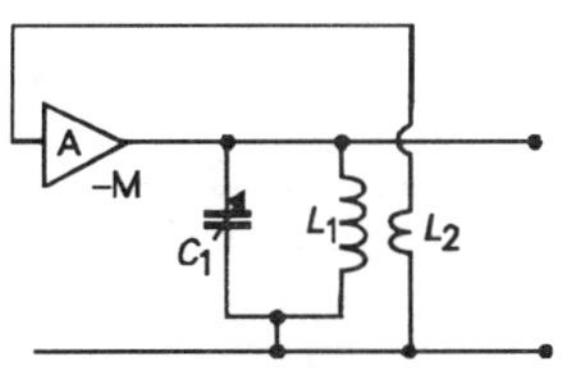

Figure 13.6 *Tuned output oscillator*

circuit in which the frequency of resonance is determined by the presence of a third resonant winding, tightly coupled, inductively, to the transformer primary and secondary circuits. Apart from the greater ease of developing a large output voltage swing across the tuned winding, this circuit offers few advantages to offset its additional complexity, and is seldom used in contemporary designs.

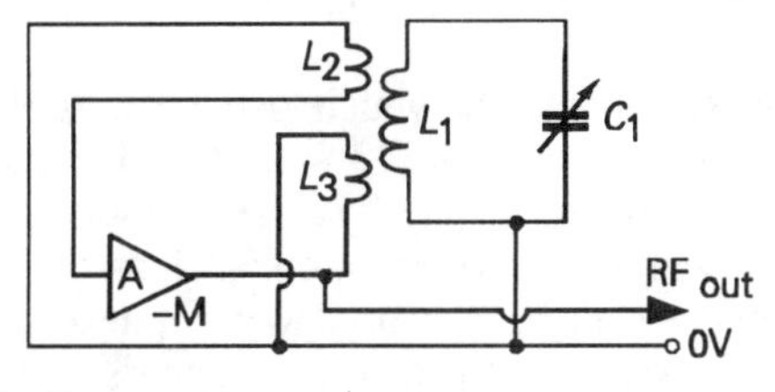

Figure 13.7 *The Meissner circuit*

Capacitive feedback

In this layout, shown in Figure 13.8a, oscillation will occur, due to the combination of capacitative voltage feedback through C_{fb} and the phase shift in the input and output circuits which occurs near resonance, when the input and output tuned circuits are tuned to very nearly the same frequency. This phenomenon was discussed in Chapter 12, in relation to the problems encountered with early tuned RF amplifier designs, which were invariably unstable in operation due to this phenomenon.

Such circuit layouts, based on triode valves, as shown in Figure 13.8b, are usually described as 'tuned anode, tuned grid' (TATG) oscillators, and are still quite commonly found in simple radio transmitter circuitry, mainly because they allow such a high RF output voltages to be developed at the valve anode. With a 300V +ve supply line, and a suitable anode coil/capacitor combination, RF voltages of 10kV RMS, or more can be generated – probably mainly limited by internal arcing within the valve or across the tuning capacitor!

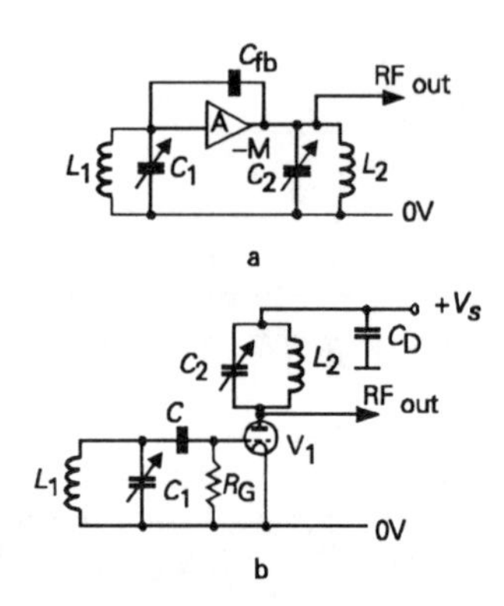

Figure 13.8 *Capacitive feedback (TATG) oscillators*

Negative impedance oscillators

In these circuits, of which one form is illustrated schematically in Figure 13.9, the tuned circuit is used to determine the frequency of oscillation, and an external gain block combination, such as the amplifier chain in the diagram, is used to generate an externally applied negative impedance characteristic. If this negative impedance is greater than, or equal to, the resistive circuit losses present in the tuned circuit, oscillation will occur at the LC resonant frequency. With a pair of valve amplifiers employed as the gain blocks, this circuit is called the 'Franklin' oscillator, and was noted for its very high stability in operating frequency, because it would operate with very 'loose' coupling (very small values of capacitors) between the tuned circuit and the external amplifier chain.

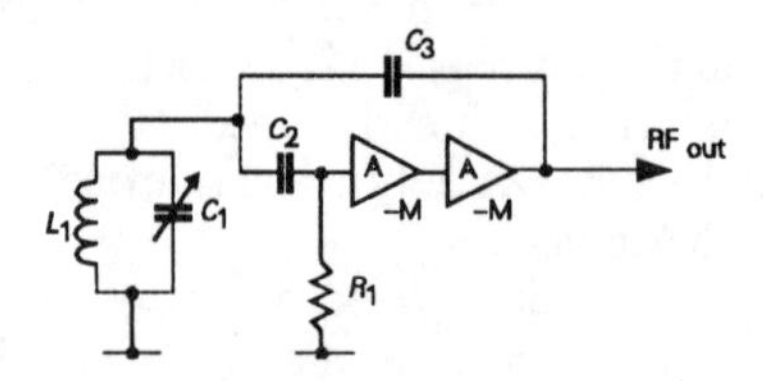

Figure 13.9 *Franklin type of negative impedance oscillator*

A modern circuit, using basically the same method of generating oscillation, is the tunnel diode circuit shown in Figure 13.10, in which the diode offers a negative resistance characteristic over part of its forward bias current range, a mechanism which was explained in Chapter 4.

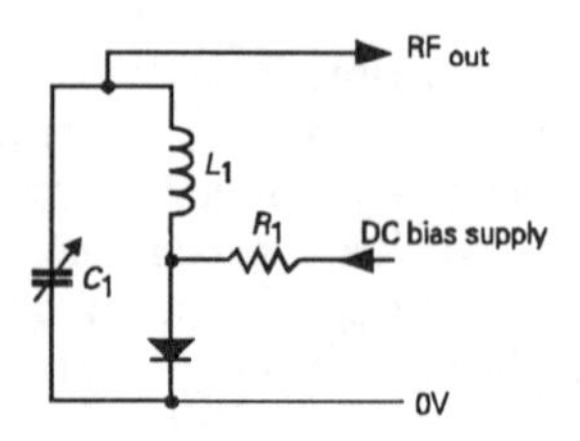

Figure 13.10 *Tunnel diode oscillator*

Practical oscillator systems

The major requirements for any oscillator are that it should be stable in frequency and output amplitude, and that it should have an output waveform which is adequately free of noise, harmonics or overtones, and spurious parasitic signals. This latter requirement is not as easy to meet as the designer would wish, and, for a high degree of purity in the output frequency spectrum, demands great care in the physical layout of the oscillator circuit, to avoid unexpected inductance/capacitance loops.

Frequency stability in a system based on an LC tuned circuit normally requires that the operating temperature of the whole system shall be held constant, and that the external amplifier or output load circuit shall not have a significant influence on the resonant frequency of the tuned circuit. It is also necessary that the inductor – usually just called a 'coil' in normal radio parlance – and the capacitor shall be stable in value, in respect of physical shocks and vibration, which usually means that they must be physically robust, and well screened from outside inductive and capacitative effects.

One way of minimizing the effect which an external load might have upon the operating frequency of the tuned circuit is to take advantage of the fact that in a screened grid or RF pentode type of valve, the screening grid can also act as a triode anode, and this allows the rearrangement of a Hartley oscillator circuit, as shown in Figure 13.11, in which this effective anode is earthed, and the output of the amplifier circuit is coupled to the tuned circuit by means of a tap on L_1, connected to the valve cathode. In this arrangement, the voltage gain needed to bring about oscillation is obtained from the step-up transformer action of the tap on L_1, driven from the low impedance cathode circuit. The electron flow through the valve to the anode is then used to couple the oscillator section of the valve to the output, which minimizes the influence which the external load may have, and thereby improves the frequency stability of the oscillator.

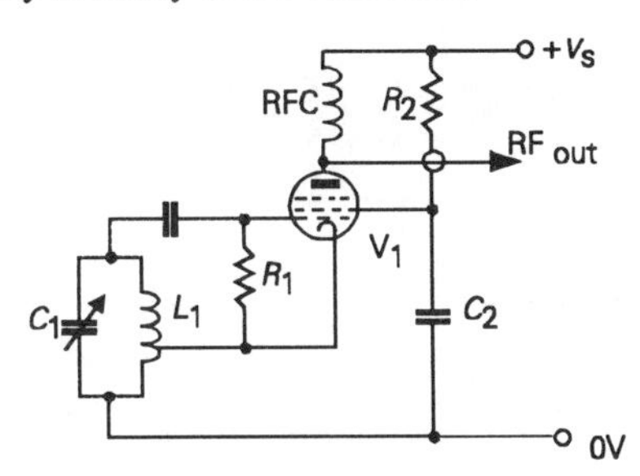

Figure 13.11 *Circuit of electron-coupled or cathode-coupled Hartley oscillator*

Because of its action, this circuit layout is called an 'electron coupled oscillator' or ECO. It is also known as a cathode-coupled Hartley oscillator.

To prevent waveform distortion due to clipping, it is necessary to provide some mechanism by which the output waveform amplitude can be limited automatically at some point below the clipping level. This can be done quite effectively in valve operated circuits, as shown in Figure 13.11, by the mechanism called 'self biasing'. This uses the input capacitor–resistor combination, C_2/R_1, to store a negative charge on C_2, each time the grid of the valve is driven positively beyond the point at which grid conduction occurs. This results in an average level of negative DC bias being developed at the grid of the valve which is proportional to the peak AC voltage present at this point, and biases the valve back to a lower gain part of its operating characteristics, which stabilizes the output voltage at some point below the clipping level.

The 'peak clipping' effect of grid conduction causes some degree of high order harmonic distortion in the output waveform – a problem which is minimized by the use of a high Q tuned circuit. A further snag with this arrangement is that the values of C_2 and R_1 must be chosen with care, in that if they are too large, in relation to the operating frequency, the problem known as 'squegging' – a continuous sawtooth shaped modulation of the output amplitude – can occur. If squegging occurs at a high enough frequency, it may not be recognised for what it is, and simply seem to be an unduly high oscillator noise level. Squegging is employed beneficially in super-regenerative receivers (another invention due to Armstrong), discussed in Chapter 14.

A version of the circuit of Figure 13.11, using a junction FET cascode layout is shown in Figure 13.12. Since the gate junction of the FET is a simple junction diode, self-biasing can also be employed, though it does not work as well as with thermionic valves, and a thermistor, of the type having a very low thermal mass, and designed for oscillator amplitude stabilization, connected in parallel across part of L_1 may be a better control mechanism.

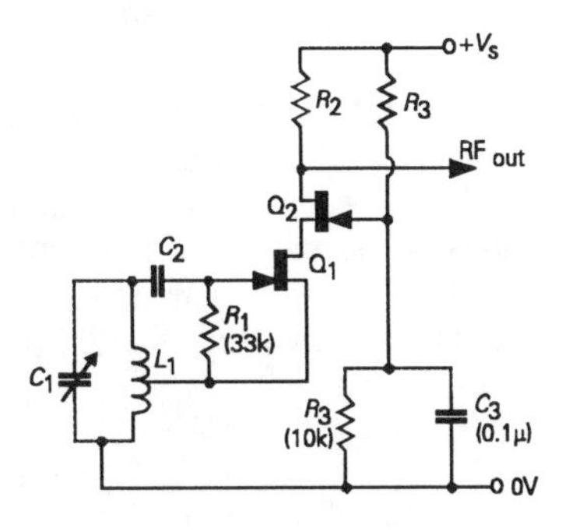

Figure 13.12 *FET cascode ECO arrangement*

The Colpitts circuit may also be used in a cathode-coupled (or source- or emitter-coupled) electron-coupled layout, as shown in Figure 13.13. In this case, because there is no DC return path through L_1 for the cathode current, an RF choke (RFC_2) is used to complete the DC circuit. A simpler circuit of the same type, though, in this case, with the RF output taken from the source electrode, is shown in a FET version in Figure 13.14.

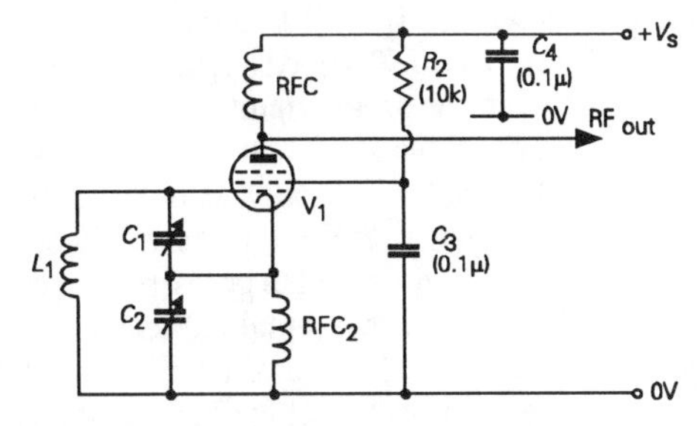

Figure 13.13 *Cathode-coupled Colpitts oscillator*

The connection of the main *LC* resonant circuit in a series, rather than a parallel form, has attracted interest as a technique offering improved frequency stability, because of the lower impedance offered to the external

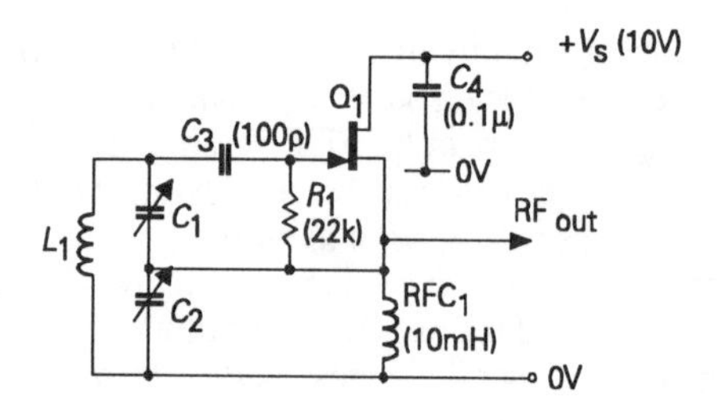

Figure 13.14 *FET source-coupled oscillator*

gain block. The best known of these layouts, using a Colpitts type feedback connection, is the 'Clapp–Gouriet' design, shown in Figure 13.15. As a matter of historical curiosity, the reason for the double-barrelled name is that the layout was invented separately and independently by two engineers, G. C. Gouriet and J. K. Clapp, in 1947–48.

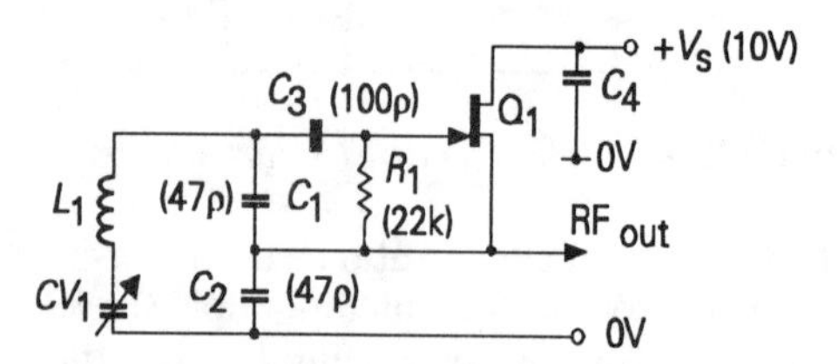

Figure 13.15 *Transistor operated Clapp-Gouriet type of oscillator*

Another layout with a similar configuration, and which also enjoys a high reputation for frequency stability, among amateur radio entusiasts, is the 'Vackar' circuit, shown in Figure 13.16, although, in this case, the source of the FET is earthed, so the RF output must be taken from the drain circuit.

I have already emphasized that nearly all of the circuits shown can be built equally well using valves, bipolar transistors, junction FETs or MOSFETs, provided that the layout is arranged to give the correct DC working conditions, and that suitable means are employed to control the amplitude of oscillation. However, in some cases, the availability of complementary polarity semiconductor devices (i.e. PNP as well as NPN) leads to particularly convenient transistor circuit layouts, such as the modified Colpitts arrangement shown in Figure 13.17. This layout offers good frequency stability, and will operate at frequencies well above 100MHz.

A variation of the Franklin circuit – a layout in which

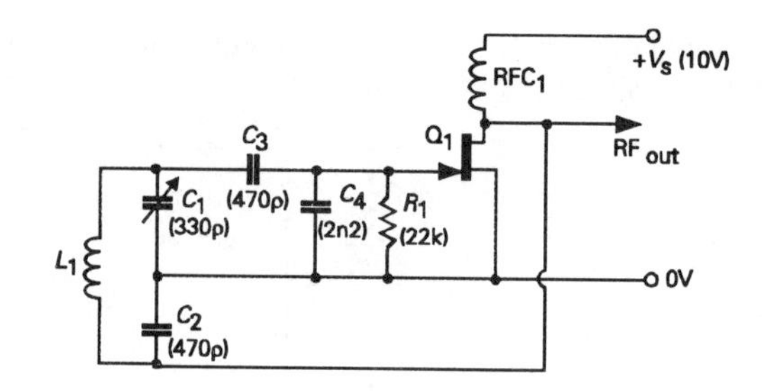

Figure 13.16 *The Vackar oscillator*

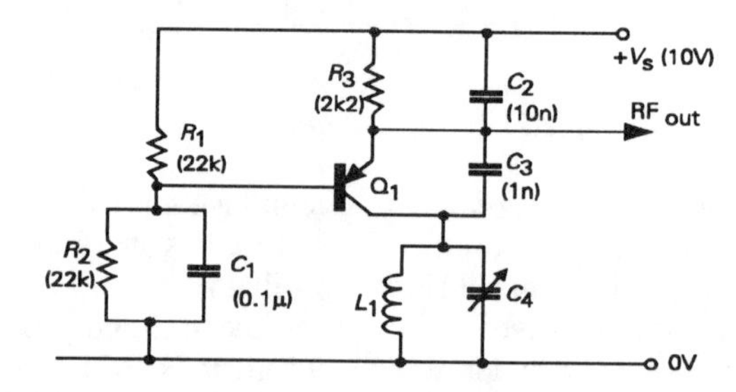

Figure 13.17 *Modified Colpitts oscillator using-PNP transistor*

the external amplifier consists of two phase inverting stages connected in cascade, so that the tuned circuit does not need to provide the in-phase feedback signal – is the source-coupled design shown in Figure 13.18. This arrangement works well with valves, FETs and MOSFETs, but is less suited to use with bipolar transistors.

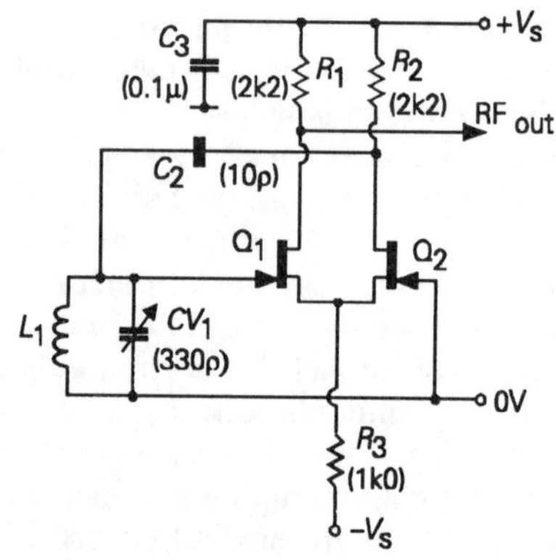

Figure 13.18 *Source-coupled Franklin oscillator*

An interesting development of this circuit is the Butler layout, where the inter-stage signal path is completed by way of a series resonant tuned circuit connected between the sources (or cathodes) of the amplifying devices, as in the FET arrangement shown in Figure 13.19. This layout shares with the Clapp-Gouriet and Vackar circuits the desirable characteristic that the RF output voltage of the oscillator remains substantially constant over the tuning range, whereas most of the systems using parallel resonant *LC* circuits as their frequency determining components tend to show an output voltage which decreases as the value of the tuning capacitor, CV_1, is increased, because this reduces both the dynamic impedance (L/Cr) and the Q ($1/r \sqrt{(L/C)}$) of the tuned circuit.

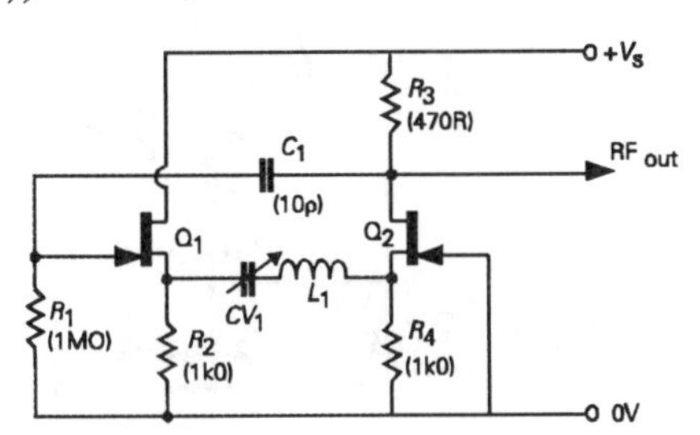

Figure 13.19 *FET operated Butler oscillator*

The use of a series resonant circuit in a modified Colpitts configuration, of the kind shown schematically in Figure 13.20 – which can be thought of as either an *LC* version of the Pierce crystal oscillator (Figure 13.23), or yet a further rearrangement of the Clapp layout – allows the tuned circuit to be isolated from the relatively low input impedance of the bipolar junction transistor, which allows improved frequency stability and increased output voltage swing. This layout was employed in a symmetrical push–pull version, in a circuit of my own, shown in Figure 13.21, which was used as a high output voltage HF bias oscillator in a tape recorder design (*Wireless World*, February 1978, p. 36).

Quartz crystal oscillators

Fundamental frequency types

The use of the electro-mechanical piezo-electric effect in quartz crystals, to provide a very high Q equivalent to a tuned circuit, was examined in Chapter 12. Where a very high degree of precision and stability in the operating frequency of an oscillator circuit is needed,

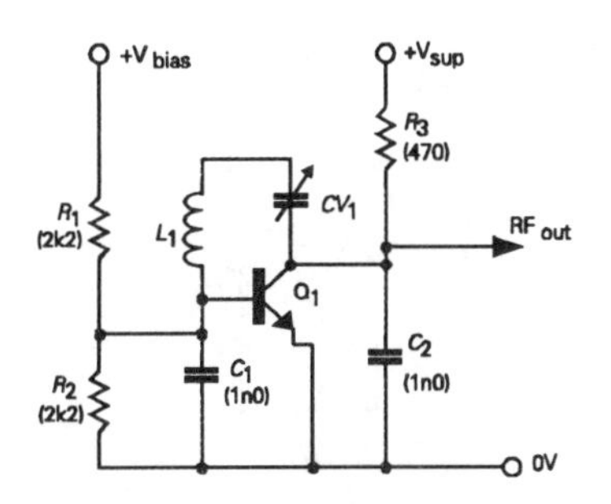

Figure 13.20 *Modified Colpitts oscillator*

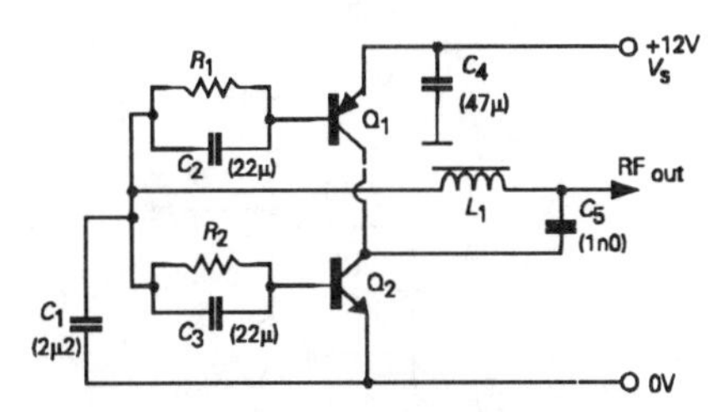

Figure 13.21 *Modified Colpitts oscillator used as tape recorder HF bias source*

the quartz crystal (or its near equivalent, the quartz surface acoustic wave resonator) is by far the best low cost option, though it imposes a need for operation at, or very close to, a fixed 'spot' frequency, determined by the physical dimensions of the crystal, and this limitation may be inconvenient.

As can be seen from its equivalent electrical circuit, shown in Figure 13.22, its principal mode of operation is that of a series resonant circuit – though it can operate as a parallel resonant circuit because of the inevitable holder, mount, and other stray capacitances (C_0). This means that quartz crystal oscillator systems will operate most happily in those kinds of circuit – such as the Clapp, or Butler or Pierce designs, shown in Figure 13.23 – which have been developed for use with series resonant *LC* layouts.

In the Pierce crystal oscillator design the capacitor, C_1, in series with the crystal is mainly used to ensure that the crystal has the recommended load capacitance for its specified frequency of operation.

A selection of typical circuit arrangements is shown in Figures 13.23–13.33, of which the first is the layout usually called the Miller or Pierce–Miller design, which is very similar in its method of operation to the tuned anode, tuned grid (TATG) design shown in Figure 13.8b. A rather more conventional Colpitts

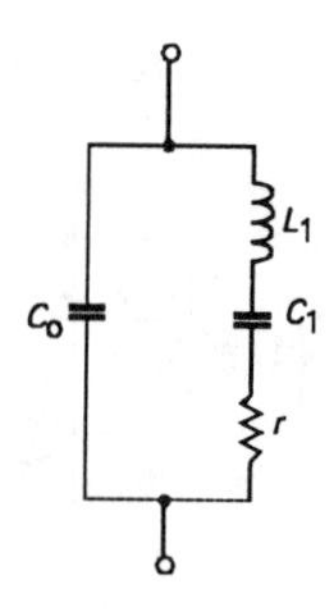

Figure 13.22 *Equivalent circuit of quartz crystal resonator*

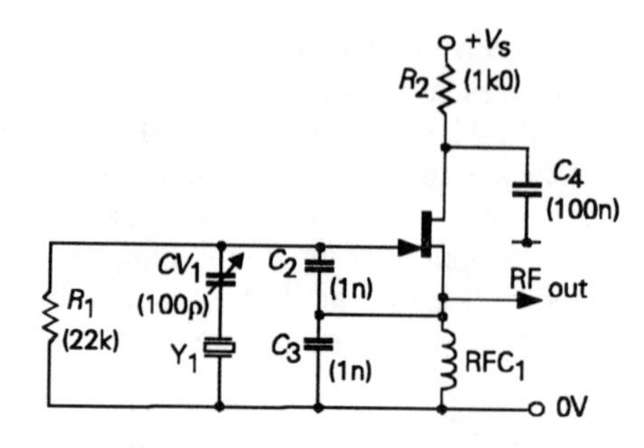

Figure 13.25 *Colpitts crystal oscillator circuit*

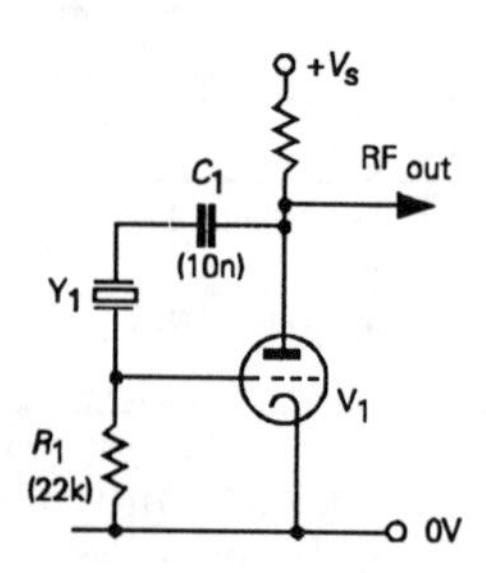

Figure 13.23 *Pierce oscillator*

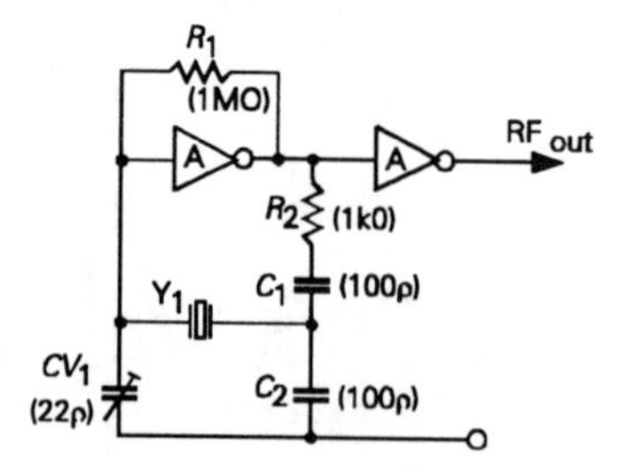

Figure 13.26 *CMOS quartz crystal oscillator*

type layout, shown in Figure 13.25, is widely recommended for use over the frequency range 100kHz–20MHz.

A type of Pierce oscillator which is well suited for use with CMOS-type inverting buffers, is the layout shown in Figure 13.26. A somewhat less satisfactory arrangement, which is, nevertheless very widely used, is the circuit shown in Figure 13.27. This has the intrinsic drawback that the circuit has overall positive feedback, and could therefore still oscillate if the crystal were to be replaced by a small capacitor. This means that there is no real guarantee that the frequency of oscillation is that due to crystal resonance, although it is probable that it will be.

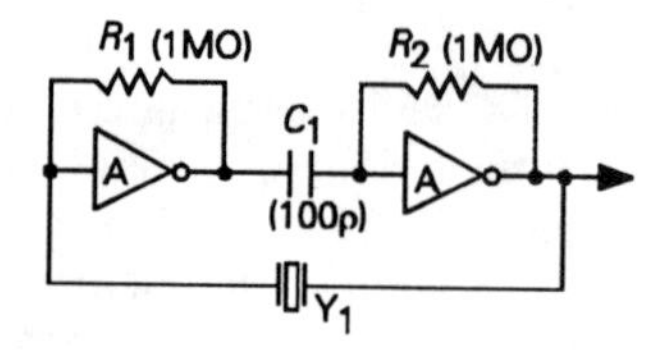

Figure 13.27 *Logic gate crystal oscillator*

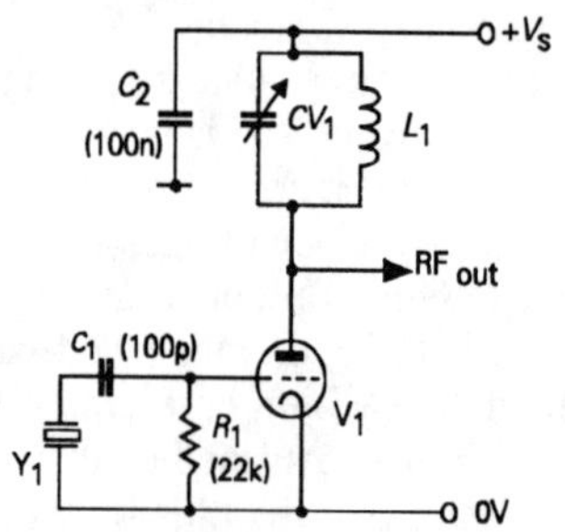

Figure 13.24 *Pierce–Miller crystal oscillator*

Overtone oscillators

Increasing the operating frequency of a quartz crystal resonator operating in its fundamental resonant mode entails increasing thinness, and fragility, in the crystal slice. It is therefore customary for crystal oscillators operating at higher frequencies to utilize overtone modes of oscilllation, based on the 3rd or 5th harmonics of the fundamental resonant frequency – a mechanism considered in Chapter 11.

The simplest version of this kind of layout is that shown in Figure 13.24, which is a TATG or capacitive feedback (Miller) type of oscillator, in which the anode *LC* circuit (L_1/CV_1) is tuned to the required crystal harmonic, though satisfactory operation may require a crystal designed for this application. The input resistor, R_1, is used to provide self biasing for amplitude stabilization.

With all crystal oscillator designs, and particularly so with TATG derivatives, care must be taken to prevent the crystal applied RF voltage and power dissipations from being exceeded, in that this may damage the crystal, as well as impairing its frequency stability.

Two similar circuits designed specifically for use with crystals made for overtone operation, up to some 105MHz, are shown in Figure 13.28 and 13.29. These are Colpitts oscillators, in which an external series resonant circuit, comprising L_1 and the stray capacitances of the crystal and holder, is used to force oscillation at the required harmonic frequency. The values of R_1 and C_2/C_3 will depend on the operating frequency range, with lower values required at higher frequencies. Like the inverting logic IC layout of Figure 13.27, these circuits suffer from the drawback that they can oscillate at a frequency not specifically determined by the crystal.

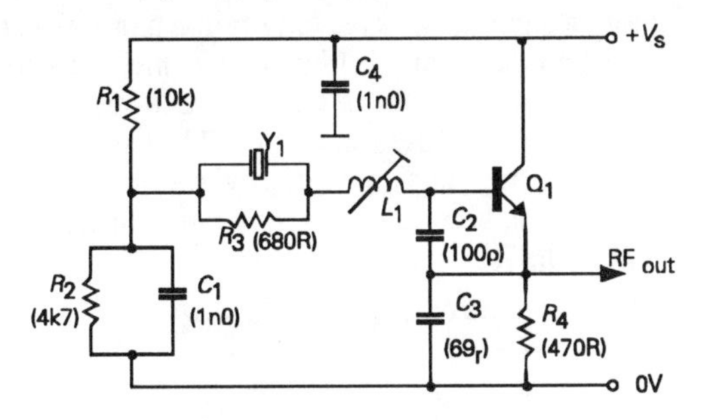

Figure 13.28 *Crystal overtone oscillator based on Colpitts system*

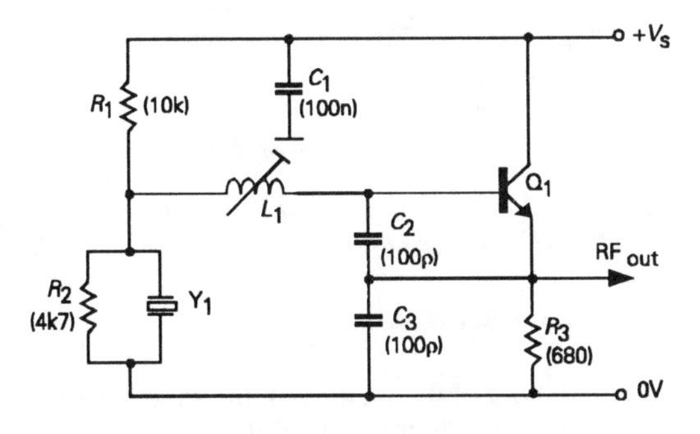

Figure 13.29 *Colpitts-based crystal overtone oscillator*

An alternative approach to the harmonic operation of a crystal, which does not rely on the use of a crystal specifically designed for this application, is just to select the required harmonic output from a crystal oscillating in its fundamental mode, as with the Colpitts layout shown in Figure 13.30, where the output circuit is tuned to the required harmonic frequency. The same approach can be used with the Butler oscillator of Figure 13.31. Two other useful overtone oscillators are the Robert Dollar design of Figure 13.32, and the grounded base circuit of Figure 13.33.

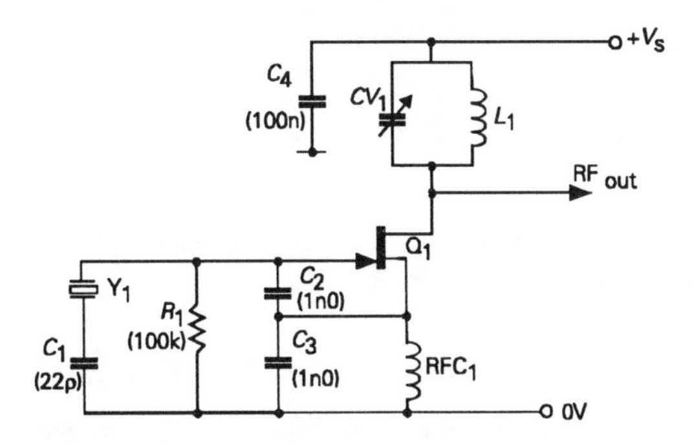

Figure 13.30 *Crystal controlled overtone oscillator*

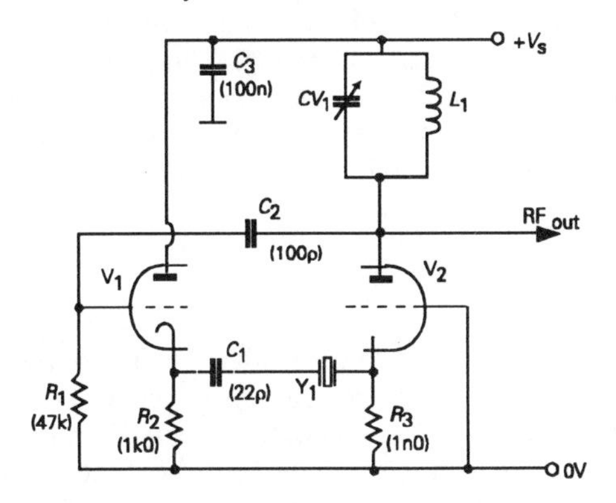

Figure 13.31 *Butler overtone oscillator*

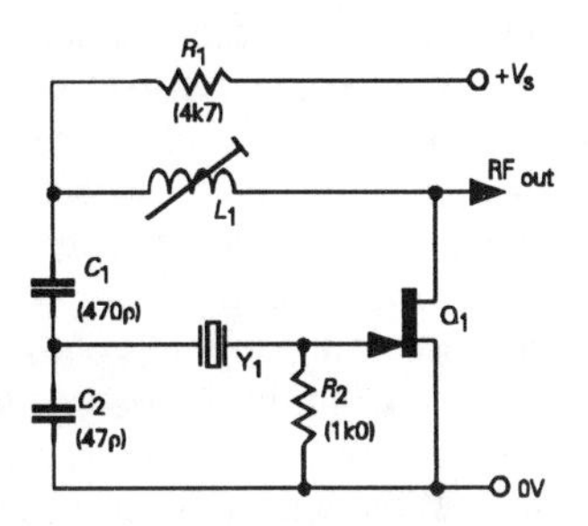

Figure 13.32 *The Robert Dollar overtone oscillator*

As mentioned in Chapter 11 there are also surface acoustic wave resonators, which offer frequency stability values which approach those of quartz crystals, but which can operate in their fundamental mode at frequencies up to 500MHz.

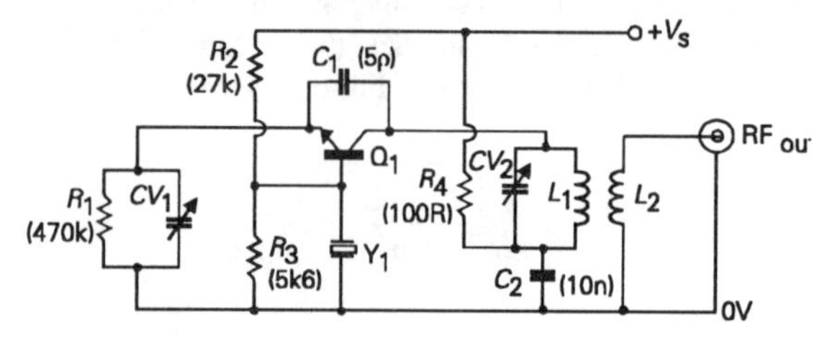

Figure 13.33 *Grounded-base 3rd/5th harmonic crystal oscillator circuit*

VXOs

Frequency trimmed quartz crystal oscillators, known as VXOs (to distinguish them from standard frequency crystal oscillators, known in radio jargon as XOs), are basically standard XO designs in which a small shift in operating frequency is brought about by a manually adjustable trimmer capacitor, usually in the range 20–100pF, connected in series or parallel with the crystal, as shown, for example, in Figure 13.25.

RC oscillators

In principle, any normal *RC* multivibrator, of the type shown in Figure 13.34, can also operate as an HF oscillator, up to a maximum frequency determined by the characteristics of the transistor and the shunting effect of stray capacitances, bearing in mind that the values of C_1 and C_2 will also need to be smaller for the shorter charge/discharge times needed for higher frequency operation. If the transistors are driven into current saturation, there will be a finite recovery time in their conduction characteristics, because of the time required for depletion layer minority carriers to be swept up by the applied field: a time which will decrease with increasing supply voltages. FETs and MOSFETs are free from this particular defect.

Although thermionic valves are also free from the carrier recombination type problems of junction semiconductors, multivibrators based on these are limited in maximum operating frequency by stray capacitances and inter-electrode electron transit times, as well as the inevitably large unwanted circuit inductances due to their connecting leads. This means, in practice, that their maximum operating frequencies are usually much lower than possible with solid-state designs.

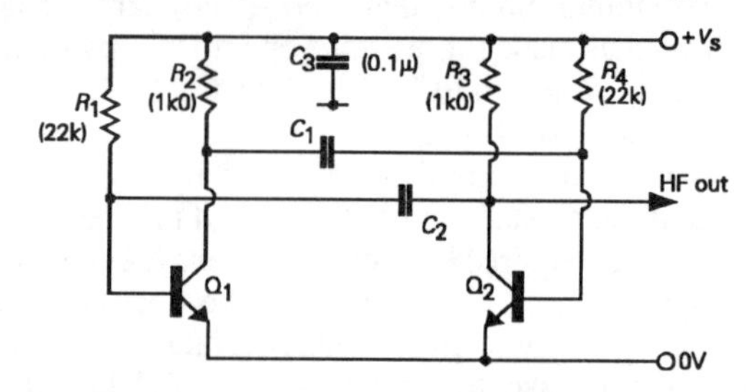

Figure 13.34 *Simple multivibrator circuit usable up to perhaps 200kHz*

Circuit designs such as the long-tailed pair layout of Figure 13.35 offer much higher possible operating frequencies, up to, say, 15MHz, because the transistors are not driven into current saturation. The basic Bowes type emitter-coupled astable multivibrator, shown in Figure 13.36, is also capable of operation at frequencies up to 10MHz.

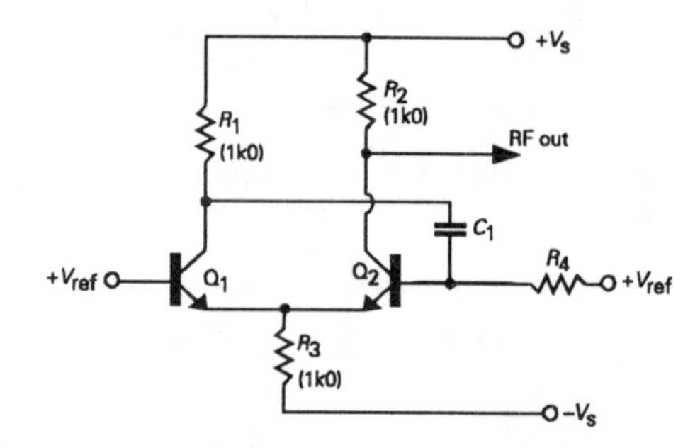

Figure 13.35 *Emitter-coupled multivibrator having better HF performance*

Problems with *RC* oscillators

There are a number of difficulties in the use of astable *RC* oscillators (multivibrators) at high frequencies, as RF signal sources. To begin with, such oscillators offer relatively poor stability in operating frequency, due to the fact that a whole range of different effects, such as transistor junction temperature, supply voltage and junction turn-on voltage, affect the duration of any on or off cycle. It also requires care to prevent the

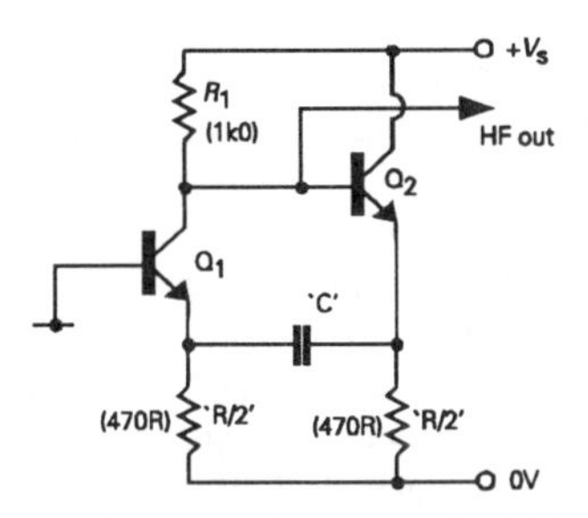

Figure 13.36 *Bowes emitter-coupled HF multivibrator*

operating frequency being affected by the output load impedance.

A more subtle problem in the operation of all *RC* relaxation-type oscillators, is that there is always a degree of statistical randomness in the actual point in time at which a transistor turns on or off, in such a circuit, due mainly to the effect of thermal voltage noise on junction potential. This leads to a relatively large frequency modulated, and a somewhat smaller amplitude modulated, noise component in the output signal. This would mean that if the output of the oscillator were combined with some other signal, as, for example, as the local oscillator in a superhet radio frequency changer stage, or added to the recorded signal, as the bias oscillator in a magnetic tape recorder, the final signal to noise ratio of the system will be worsened – especially if the amplitude of the *RC* oscillator voltage, and therefore its noise component, is large in comparison with the signal voltage to which it is added.

Voltage controlled oscillators (VCOs)

This category embraces a wide range of oscillator circuit designs, in which the operating frequency will depend on some applied voltage. With *RC* oscillator types, two general approaches are used, to alter the reference voltage to which the base of Q_2 is taken, in the circuit of Figure 13.35, which alters the voltage to which the base of Q_2 must climb or descend, through C_1, R_1 and R_3, before Q_1 and Q_2 turn on or off, which will alter the operating frequency.

The other approach is to directly control the charging or discharging current of the timing capacitor, as, for example, C_1 in Figure 13.37, through CC_1. If the resistance of CC_1 is lowered, the operating frequency of the oscillator is increased, and conversely. This is the technique which is used, in one form or another, in most of the integrated circuit VCOs, though it is often difficult to achieve a reasonably linear relationship between applied voltage and operating frequency. A design of my own (*Wireless World*, November, 1973, pp. 367–369), which gives a highly linear relationship between applied voltage and operating frequency, is shown in Figure 13.38. This circuit was used as the VCO in a phase-locked loop FM demodulator, and, with the component values shown, was designed to cover the frequency range 9–14MHz. To assist in reducing circuit stray capacitances, which helps to achieve a high working frequency range, transistors Q_1–Q_5 are part of a CA3046 monolithic transistor array. In this circuit, Q_1 and Q_2 are connected as a current mirror (see Chapter 6), with the current flow through Q_2 controlling the charge and discharge times of C_1, and, in consequence, the operating frequency of the oscillator. Increasing the value of C_1 would allow operation at lower frequencies.

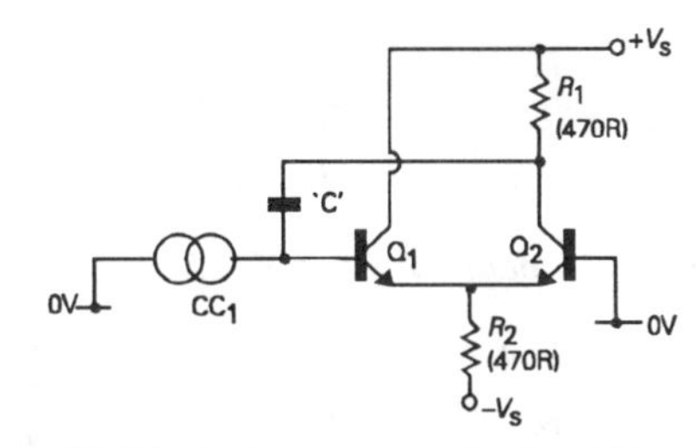

Figure 13.37 *Voltage controlled oscillator*

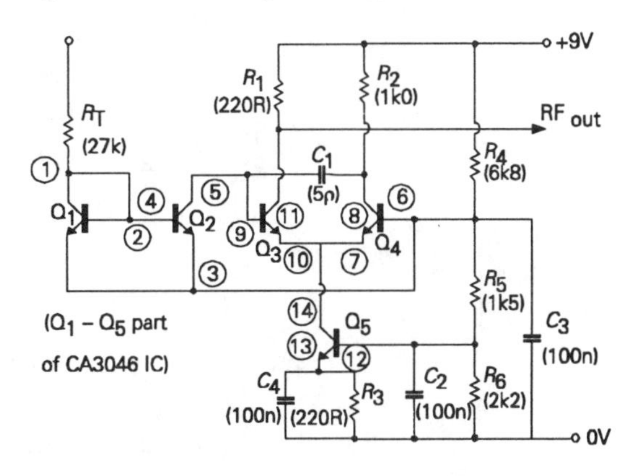

Figure 13.38 *Linear voltage controlled oscillator based on CA3046 transistor array (pin numbers shown in brackets)*

For most high frequency applications, the approach most commonly employed is just to connect a varicap diode (D_1) – a device which exploits the increase in the thickness of the depletion layer, and hence the decrease in capacitance, in a reverse biased P-N junction, as the applied voltage is increased – in parallel with a standard parallel resonant tuned circuit (L_1/C_2), as shown in Figure 13.39, and then to use a change in the bias voltage applied to the diode to alter the resonant frequency of the tuned circuit. There are, however, two problems with this type of frequency contol, of which the first is that there is a nonlinear relationship between the applied voltage and diode junction capacitance, of the general form $C = m/\sqrt{V + V_d}$, where m is a constant determined by the construction of the device, V is the applied voltage and V_d is approximately 0.6V for a silicon diode, as shown in Chapter 12 – a relationship which results in a similar nonlinearity between control voltage and tuned circuit resonant frequency.

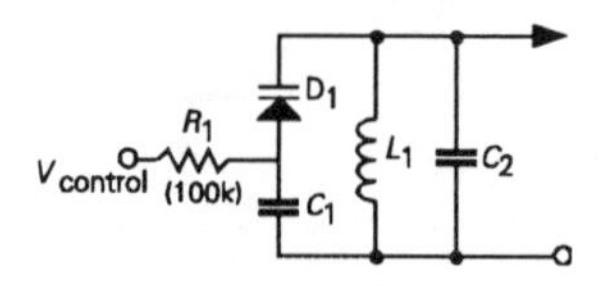

Figure 13.39 *Standard Varicap VCO arrangement*

The second problem is that the relationship between capacitance and applied voltage is temperature dependent. For some phase-locked loop frequency control systems, of the kind shown in Figure 13.40, the operation of the internal feedback loop will maintain the chosen oscillator frequency so that neither the nonlinearity nor the poor frequency stability of the varicap control system are particularly important. However, where varicap diode frequency control is used as the tuning mechanism for a radio receiver, the applied control voltage must be temperature compensated in order to minimize the extent of tuning drift. Nevertheless, the great convenience of being able to tune a number of circuits simultaneously, by means of a remotely controlled voltage, has led to the very widespread adoption of this technique in radio receiver circuitry, with suitably temperature compensated control voltages.

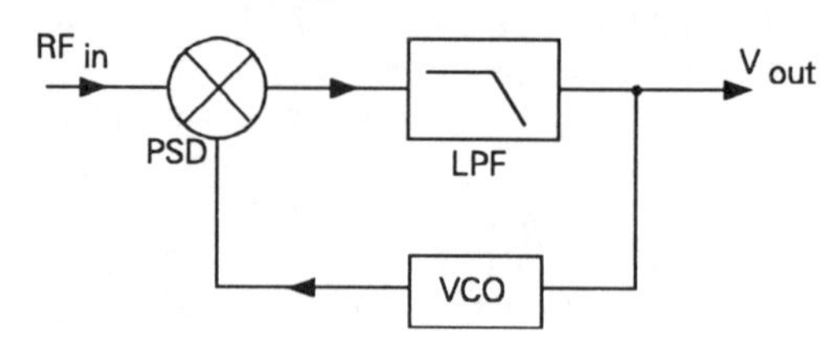

Figure 13.40 *Basic phase-locked loop system*

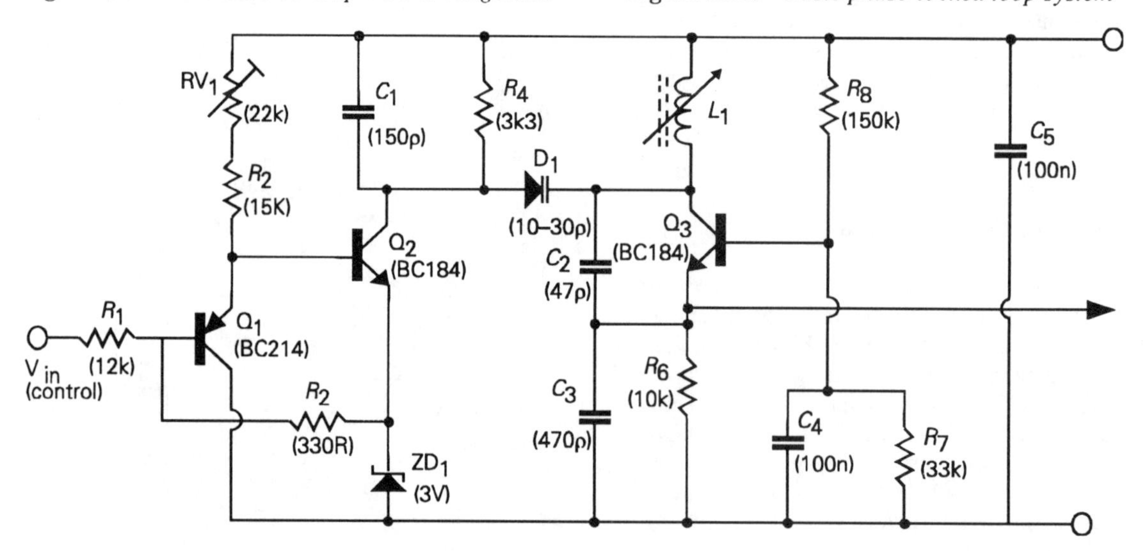

Figure 13.41 *Linear voltage controlled oscillator based on a varicap diode*

A varicap tuned oscillator which is internally corrected both for temperature effects and nonlinearity, also used in an FM tuner phase-locked loop demodulator system (JLH, *Wireless World*, September 1979, pp. 87–88), is shown in Figure 13.41. In this circuit, advantage is taken of the fact that the temperature dependence of characteristics, and the nonlinear relationship between base voltage and collector current in a bipolar transistor (in this case Q_2), is of the same form, but opposite in curvature, to that of the varicap diode D_1. This condition requires that Q_2 should be driven from a low impedance source (in this case Q_1, acting as an emitter follower), and the base-emitter DC offset of Q_2, which will also be temperature dependent, is cancelled by the opposed connection of Q_1 and Q_2. The resulting linearized voltage vs. oscillator frequency characteristics are shown in Figure 13.42.

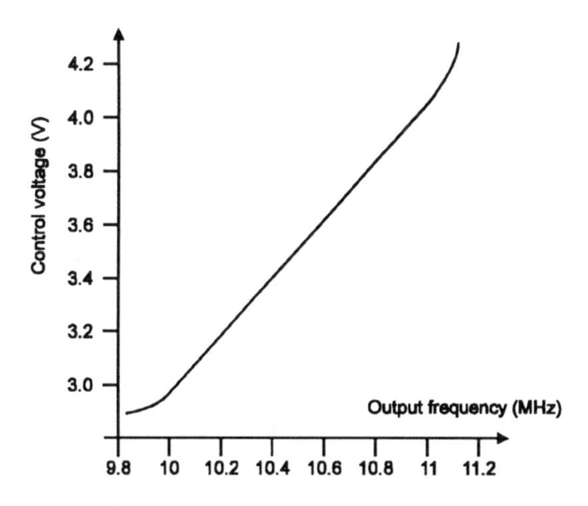

Figure 13.42 *Frequency response of circuit of Figure 13.41*

14

Radio receiver circuitry

Introduction

Although I have used the term Radio in the chapter heading, the basic techniques and systems which are used for radio reception are much the same as those employed in television, Radar, radio astronomy, and other systems used to receive high frequency electromagnetic radiations, apart from any modifications to the receiver which may be needed to meet the specific requirements of the operating frequency, or the signal bandwidth.

For me, and perhaps for many others of my generation, radio circuitry has always enjoyed a special degree of affection because it was my original introduction to the whole field of electronics, and I still think that there are two uniquely enthralling pleasures of revelation which await the practically minded – that of watching a black and white print image, of ones own, appear, for the first time, in the developing dish in a photographic darkroom, and that of hearing ones first short- wave radio signals, on headphones, on a home-built radio receiver. Of these two pleasures, that of radio has the advantage that, with care, all of the components are re-usable. One does not need to throw away ones failures, one can dismantle them, instead.

Radio receiver circuitry

Tuned radio frequency (TRF) systems

Demodulation methods

In most radio signals, the amplitude of the high frequency, (HF or RF), transmitted waveform, usually called the carrier, is modulated, (caused to vary), as a means of carrying the lower frequency information or programme content of the transmission. With a high frequency signal, the fluctuations in the amplitude of the incoming signal would be far outside the audible range, and undetectable. In order to allow the receiver to convert such an amplitude modulated (AM) RF voltage into an audible or measurable signal, the process known as demodulation or detection is used, most commonly by simply rectifying the incoming RF waveform, shown in Figure 14.1a, to give a waveform of the kind shown in Figure 14.1b. If this voltage waveform could be averaged, as shown in Figure 14.1c, the result would be a true replica of the original waveform used to modulate the carrier. Unfortunately, the easiest technique for doing this, just to use a resistor/capacitor/diode circuit, of the kind shown in Figure 14.2, leads to a substantial distortion in the recovered modulation waveform, as shown in Figure 14.1d, unless the *CR* discharge time constant is kept short in relation to the highest modulation frequency employed, and this lowers the demodulation efficiency.

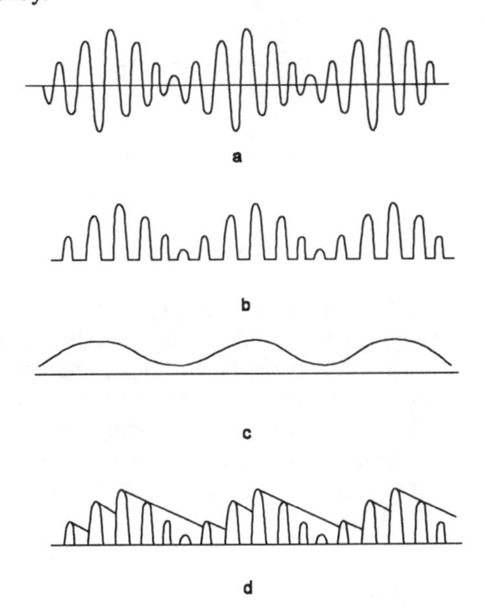

Figure 14.1 *Action of diode demodulator*

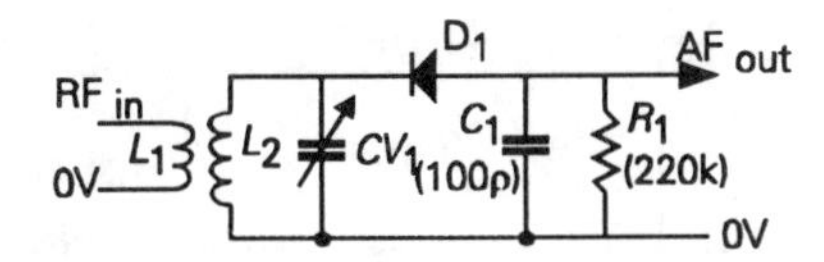

Figure 14.2 *Simple diode demodulator circuit*

A rather better way of extracting the modulation component from such a rectified carrier signal is to interpose a low-pass filter between the rectifier and the signal output point, as shown in Figure 14.3, but even this cannot entirely solve the problem of AM demodulation distortion, particularly at lower carrier, and higher modulation frequencies, so demodulation distortion remains a characteristic of all AM receivers.

The second problem with a demodulator system based on a simple diode rectifier, of the kind shown in Figure 14.2, is that low applied voltages may either not cause the diode to conduct at all, or, if a forward bias

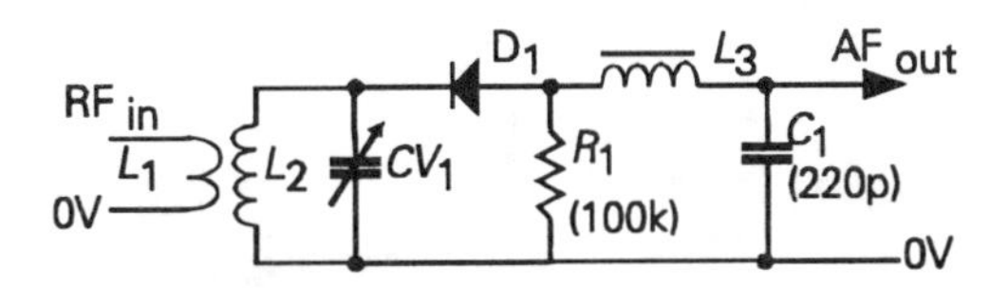

Figure 14.3 *Improved diode detector circuit*

is deliberately applied, to ensure that they do, or if a semiconductor junction is employed which does have some conduction at zero applied voltage, as shown in Figure 14.4, there will be very little difference between the diode forward and reverse direction conduction characteristics for very small input signal voltages. This leads to a very low demodulation efficiency, a characteristic which it shares with the other 'slope demodulator' circuits shown in Figures 14.5 and 14.6, which rely for their rectification ability on the curvature of the V_b/I_c, or V_g/I_d, characteristics of the semiconductor device.

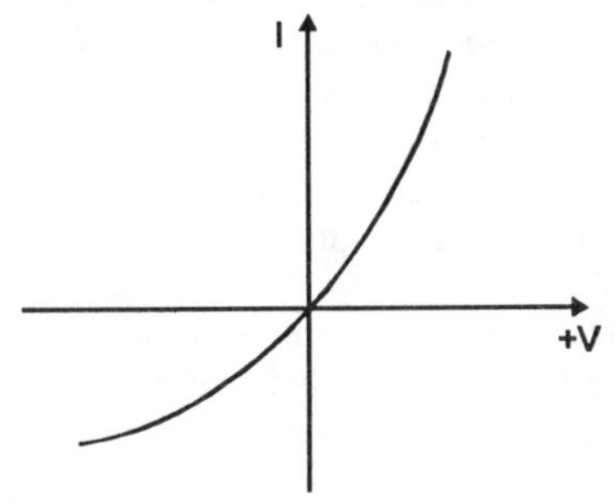

Figure 14.4 *Conduction characteristics of foreward biased diode junction*

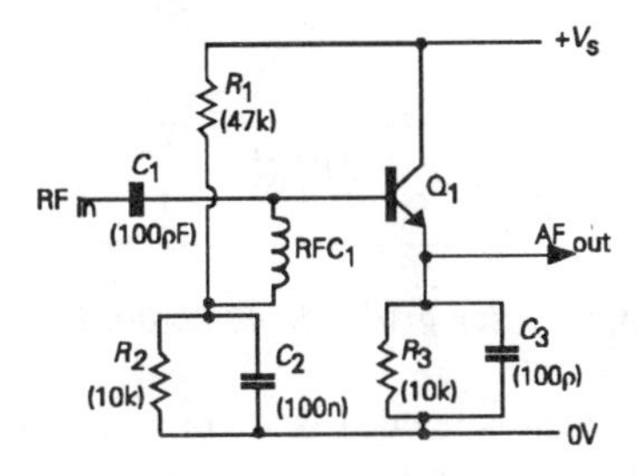

Figure 14.5

The so-called 'grid leak' detector circuit, shown for a triode valve in Figure 14.7, exploits the fact that as the positive-going peaks of the RF waveform applied to the valve grid approach zero potential, the flow of grid current into C1 will negatively bias the grid of the

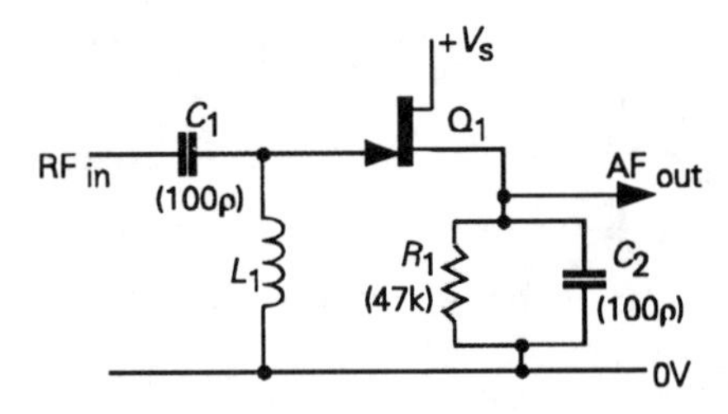

Figure 14.6

valve, which will push the mean DC level of the input signal down the I_a/V_g slope, as shown in Figure 14.8. This makes the modulation envelope lop-sided so that the average anode current of the valve fluctuates in sympathy with the carrier modulation level. This scheme could also be used with a junction FET, but, in this case, the FET gate would need to be forward biased to the point at which diode type gate conduction was about to occur. Traditionally, the grid leak resistor would be connected across C_1, to minimize the resistive damping of the input tuned circuit, (L_2/CV_1), but since R_1 would usually be a megohm or greater in value, the damping due to this cause would be negligible, and R_1 could equally well be taken to the 0V line. This type of demodulator shares with the diode detector the problem of modulation distortion due to the finite R_1/C_1 discharge time constant, as well as the problem that, for small signal levels, the demodulation efficiency is exceedingly poor.

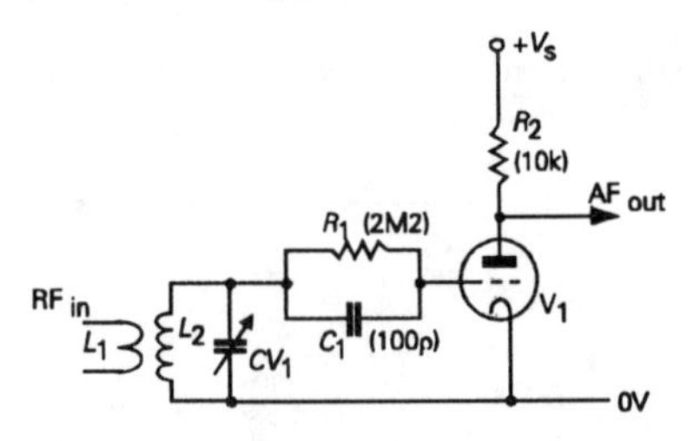

Figure 14.7 *Valve grid leak demodulator circuit*

Clearly, what is needed is some way of increasing the size of the incoming RF signal to a level at which the demodulation efficiency from any of these systems reaches a useful level.

Effect of the Q of the input tuned circuit

Some assistance in increasing the magnitude of the applied signal will be given by the action of the Q,

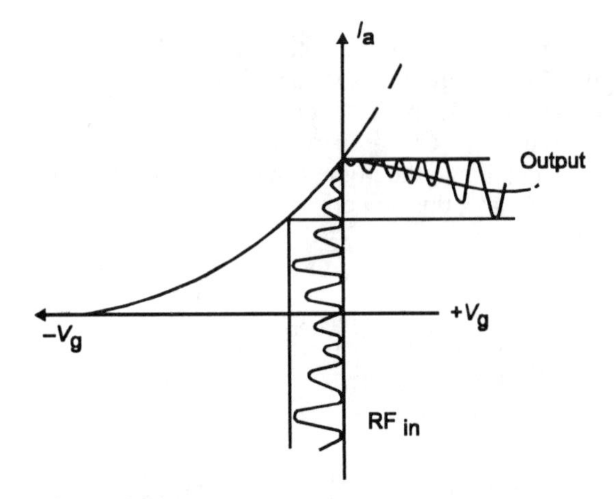

Figure 14.8

(circuit magnification factor) of the input tuned circuit, interposed between the aerial and the demodulator, in Figure 14.9, but, to prevent the low input impedance of a demodulator diode, or that of the base/emitter junction of the transistor, from lowering the Q too much, it will be necessary to take the output from the tuned circuit from a low tapping point on the inductor, (L_2), which will also lower the available output voltage.

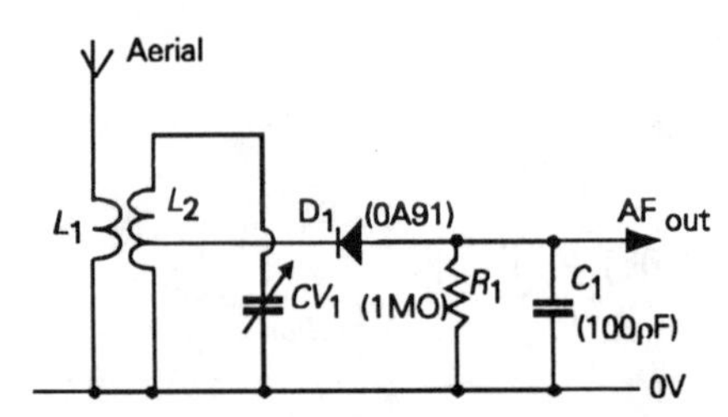

Figure 14.9 *Simple diode detector radio receiver circuit*

The FET slope demodulator of Figure 14.6 has a very high input impedance, and can therefore be connected across the whole of L_2, as shown in Figure 14.10, with a significant gain in demodulator output.

The normal engineering solution to this basic problem – that all AM demodulator circuits require an adequate RF input for satisfactory performance – is to provide some pre-demodulation RF signal amplification, using the type of circuitry examined in Chapter 12. Simple receiver layouts of this kind, using one or more tuned RF amplifier stages preceding the

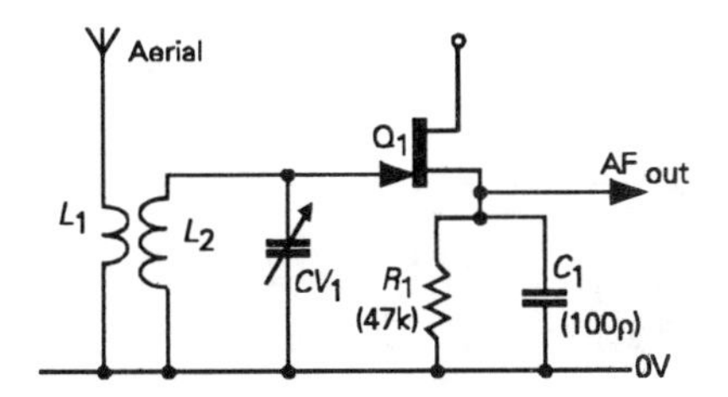

Figure 14.10 *Radio receiver circuit based on FET slope detector*

demodulator, such as that shown in a contemporary form in the circuit layout of Figure 14.11, would be described as a TRF (tuned radio frequency) receiver, and systems of this kind formed the bulk of early radio designs.

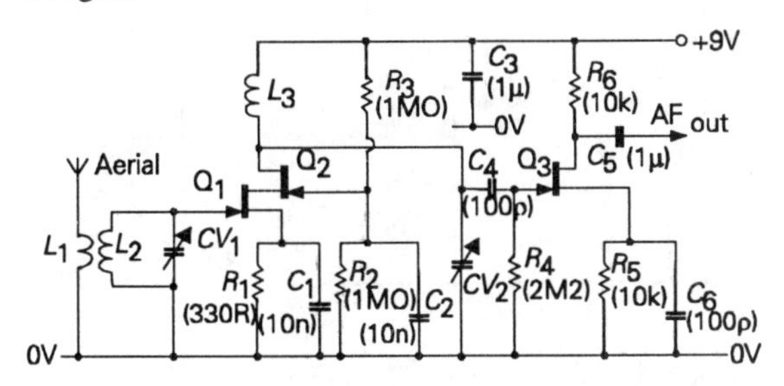

Figure 14.11 *Radio receiver circuit using cascode connected FET RF amplifier stage*

Problems with selectivity

There are a number of fundamental difficulties with this approach, of which the major one is that of providing adequate selectivity in respect of adjacent frequency signals. As seen in Chapter 11, a single tuned circuit gives a response curve of the kind shown in Figure 14.12, in which the –6dB voltage gain points occur at frequencies separated from the resonant frequency by $+/–f_o/Q$. The relative voltage output from a parallel resonant tuned circuit at frequencies on either side of resonance can be calculated, approximately, for values of Q in excess of 10, from the formula given below (Dobbie and Builder, *Radio Designers Handbook*, fourth edition, pp. 415–416):

$$\frac{E_o}{E} \approx \sqrt{1 + Q^2 \left\{ \frac{f}{F_o} - \frac{f_o}{f} \right\}^2} \qquad (1)$$

where E_o/E is the ratio of output voltage at resonance, f_o, to that at a frequency, f, on either side of resonance. This shows that the –12dB gain points occur at approximately $+/–2f_o/Q$, the –18dB points at $+/– 4f_o/Q$, and the –24dB points at $+/–8f_o/Q$. (*Note*. This type of calculation is only approximate, and shows a symmetrical gain/frequency curve, whereas, in reality, the cut-off characteristic must be somewhat steeper on the lower frequency side of resonance than the higher frequency side.)

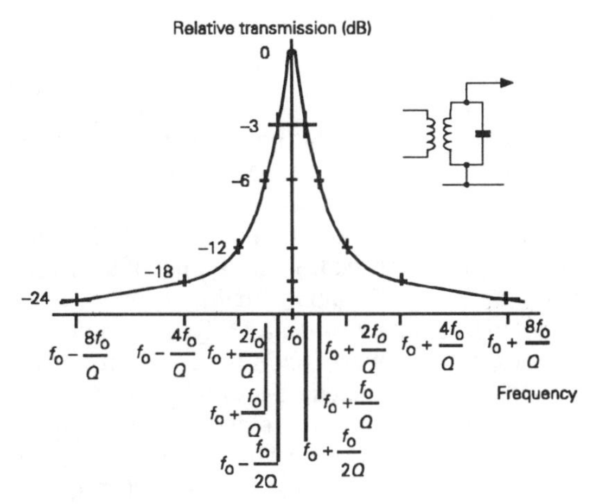

Figure 14.12 *Response curve of single loosely coupled LC resonant circuit*

Using the above data, based on an operating frequency of 200kHz, and a Q value of 100 (a good average value for a well made tuned circuit), the –6dB points would occur at 198kHz and 202kHz – a 4kHz passband: the –12dB points at 196kHz and 204kHz – an 8kHz passband: and the –18dB points at 192kHz and 208kHz – a 16kHz passband. If we accept that an 8:1 (18dB), rejection ratio for an adjacent signal is acceptable – it is clearly not very good – the ability of a receiver with a single input tuned circuit, having a Q value of 100, to provide adequate selectivity at a 200kHz operating frequency would require that adjacent stations were not closer than 16kHz in their carrier frequencies. For an operating frequency of 1MHz, these pass-bandwidth figures will be five times as great, and clearly not acceptable for the degree of crowding which exists on any broadcast band.

To improve the selectivity, one must either have more tuned circuits before the demodulator stage, or one must increase the Q of the circuits. Ganged (mechanically coupled) tuning capacitors are available to allow simultaneous tuning of more than one tuned circuit, which will help to solve the problem of selectivity, but while 2-gang tuning capacitors are inexpensive and easy to buy, 3-gang ones are costly and relatively scarce, and 4-gang types are very seldom found, and would be very dear even if they were available, so the number of variable capacitor tuned RF stages which the designer could employ in a TRF type receiver would be mainly limited by the availability of hardware.

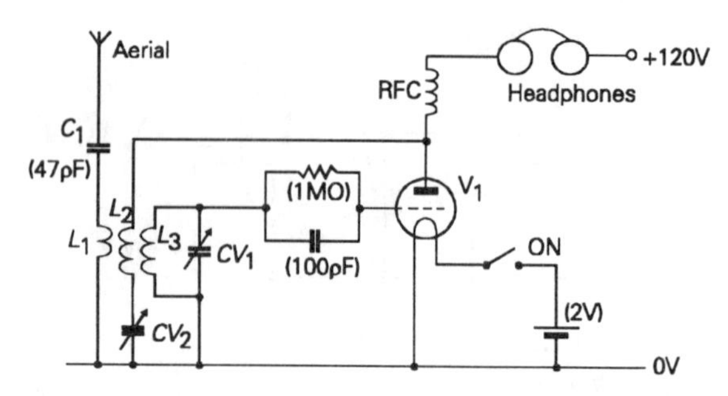

Figure 14.13 *One valve receiver using regeneration*

Regeneration or reaction

An ingenious solution to the problem of low Q values, and poor selectivity, which is particularly acute on the 'short wave' bands (approximately 2.5MHz–30MHz), is to use 'regeneration' or 'reaction'. In this technique, some energy is fed back from the output of the amplifying stage into the input circuit, in an identical manner to that employed in an *LC* type HF oscillator. However, if the amount of energy fed back is carefully adjusted, so that the circuit does not quite break into oscillation, it is possible to use the feedback signal to very nearly completely cancel the energy losses in the tuned circuit. With care, and delicate adjustment, operating Q values for the input tuned circuit as high as 50,000 can be obtained by this means. For a receiver operating at 20MHz, this would give a –18dB passband of 3.2kHz, which would be adequate to separate most wanted signals. This arrangement would also have the great advantage that the incoming aerial signal would also be magnified by the Q factor, so that a 50µV aerial signal would become a 2.5V signal at the demodulator input – a level at which efficient demodulation would occur.

Because of the simplicity and efficiency of such circuits, designs for one valve short-wave receivers using regeneration, such as that shown in Figure 14.13, were very common in the amateur magazines during the 1930s and 1940s. A comparable contemporary design, using two transistors to achieve a similar audio output power, is shown in Figure 14.14.

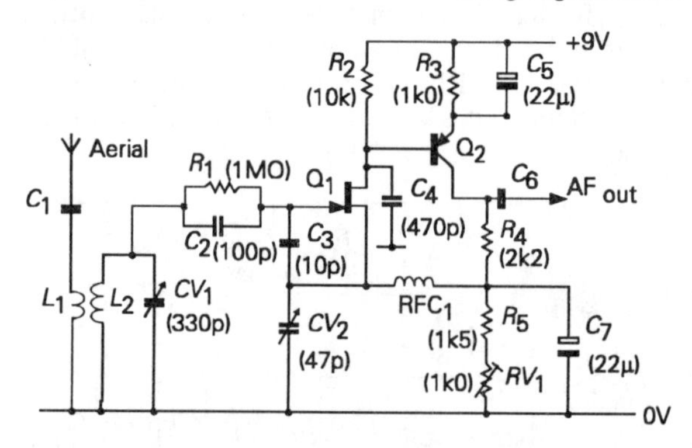

Figure 14.14 *Simple AM receiver using regeneration – usable over the frequency range 150kHz – 30MHz*

Super-regenerative receiver systems

While, with careful adjustment, regeneration provides a very simple and effective method of increasing receiver sensitivity and selectivity, it suffers from the problem that no single setting of the regeneration control will be satisfactory over the whole of the desired tuning range. This is because any alteration in the resonant frequency of the tuned circuit will require an alteration in the *L*:*C* ratio in that circuit. But the Q of the circuit is defined by the relationship

$$Q = (1/r)\sqrt{L/C} \qquad (2)$$

so altering the tuning will inevitably alter the Q value of the circuit, and, in consequence, the amount of positive feedback needed to cancel the circuit losses. This means that the user of the receiver needs to fiddle continuously with his regeneration control knob as he tunes his receiver, to keep on the correct side of the knife-edge between stable operation, and an audible 'howl', which is irritating both to the user and also to others within range of the RF signal inadvertently radiated by his oscillating receiver.

An ingenious solution to this problem, again introduced by Edwin Armstrong, in 1921, is the 'super-regenerative' system. This takes advantage of the fact that even when a sufficient amount of positive feedback is applied, around an amplifier stage containing a tuned circuit, to cause it to break into oscillation, there is a finite time lapse between the moment of applying the feedback, and the onset of oscillation. Armstrong's approach was therefore to apply an external 'quench' signal periodically to the regenerative circuit, to pull its operating point back from the point at which continuous oscillation will occur. This quench signal can be any convenient method of momentarily reducing either the stage gain or the amount of feedback – even a grid-leak/capacitor combination, whose values have chosen to make the oscillator 'squeg', will work.

However, for simplicity and reliability of operation in a super-regen. receiver, the best approach is usually just to superimpose a suitable amplitude sine-wave or square-wave on the HT supply line to the regenerative stage. In order to prevent the whistle due to the quench waveform from blotting out the wanted signal, the quench signal will usually be chosen to be at some ultrasonic frequency, which will facilitate its separation, by simple filtering, from the wanted signal.

Experience has shown that the best receiver sensitivity is obtained with quench frequencies just above the audible range. A simple modification to the regenerative receiver circuit shown in Figure 14.14, using an externally applied square-wave, of 1–2V amplitude, at 25kHz, to convert it into a super-regen. receiver, is shown in Figure 14.15. Any suitable square-wave generator, of the kinds shown in Chapter 13, will work.

Although super-regen receivers are simple and effective, especially at higher signal frequencies where other receiver systems are less efficient – and are widely used in 'Citizens Band' (CB) receivers for this reason – they suffer from several snags which have prevented their more widespread adoption. These are the annoyingly loud inter-station noise (due, it is said, to randomly occurring thermal noise impulses in the input circuit being amplified by the circuit up to full output voltage level), wide-band noise being radiated from the receiver in the neighbourhood of its tuned frequency, and the relatively poor input signal selectivity.

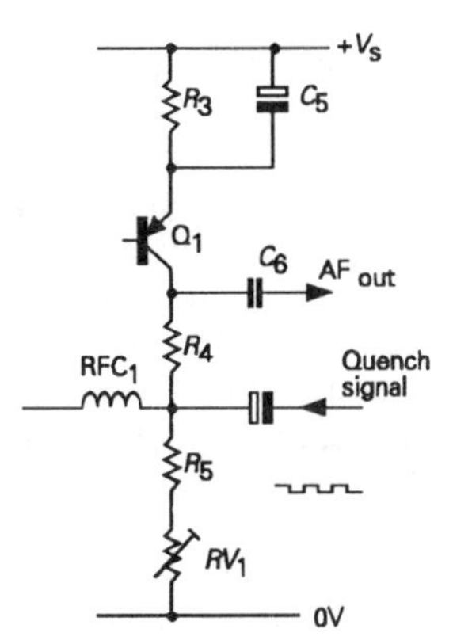

Figure 14.15 *Modification to convert circuit of Figure 14.14 to super-regenerative operation*

Effect of selectivity on the received signal

So far, all our calculations on the ability of a receiver to reject an unwanted adjacent signal have failed to consider the effect of this selectivity on the sidebands of an AM broadcast signal, which are themselves composed of the sum and difference outputs of the modulation and carrier frequencies. Too narrow a reception bandwidth in the receiver will attenuate the higher modulation frequencies, and this is an inevitable effect with any system based on either a single *LC* parallel resonant tuned circuit, or a series of these connected in cascade to improve the selectivity, simply because the output voltage from such a tuned circuit will fall immediately the signal frequency moves away from the frequency of resonance.

Tuning successive stages, in a cascaded series of tuned circuits, to slightly different frequencies will increase the receiver bandwidth, but this is a difficult practice to carry out well. A much better approach, which is very widely employed, is to use pairs of tuned circuits which are electrically coupled to one another, as shown, for example, in the inductively coupled layout of Figure 14.16.

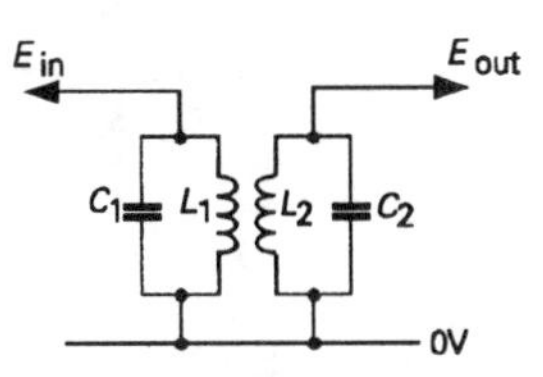

Figure 14.16 *Bandpass-coupled pair of tuned circuits*

Bandpass coupled tuned circuits

Such bandpass-coupled pairs of resonant *LC* circuits, of which a number of possible layouts were shown in Chapter 11, have the great advantage that, for appropriate values of Q and coupling factor, the pass-band has a portion, adjacent to the resonant frequency, at which the signal output will be substantially constant, as shown for a critically coupled bandpass system in Figure 14.17. Circuits of this kind can be cascaded, to improve the steepness of the skirt frequency response, without impairing the flatness of the portion of the response curve on either side of resonance, and this level response portion can be tailored, by choice of Q and coupling factor, to provide almost any desired pass-band.

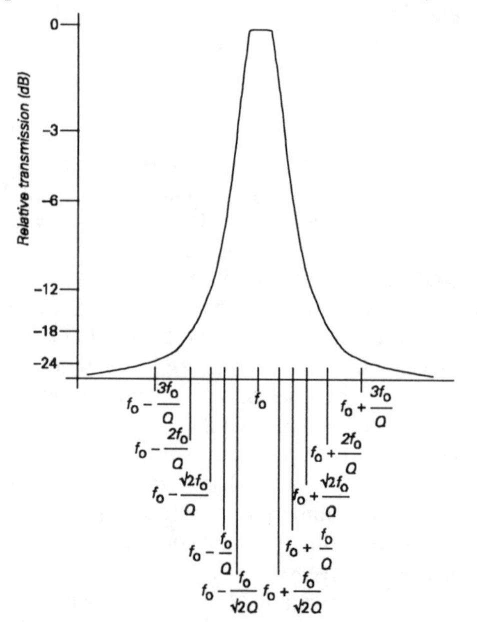

Figure 14.17 *Frequency response curve of critically coupled bandpass pair of tuned circuits*

However, these factors will depend on the operating frequency chosen, and this factor, quite apart from the impossibility of obtaining 4-, 6-, or 8-gang tuning capacitors, precludes the use of bandpass-coupled RF stages in simple TRF receivers, and has led to the almost universal adoption of the 'superhet' system.

The supersonic heterodyne, or superhet system

As with so many other useful ideas in the field of radio, this was introduced by Armstrong, in 1921, and provides an effective solution to the problem of selectivity, while avoiding the difficulty of trying simultaneously to tune a number of critically coupled bandpass-coupled circuits. The basic idea of the superhet receiver is illustrated in Figure 14.18. In this arrangement, the incoming signal is converted, in a frequency changer or mixer stage, into a signal at some other fixed, intermediate, frequency at which the bulk of the pre-demodulator, RF, amplification is performed. This frequency conversion is carried out by combining the incoming signal with the output of a local oscillator in a stage having a sufficient degree of nonlinearity as an amplifier to cause the generation of sum and difference products as a result of the interaction between the input signals. By the choice of a suitable local oscillator frequency, a composite output signal, carrying all the modulation information present on the incoming aerial signal, can be generated, and can be placed at any desired frequency. This allows the intermediate frequency to be chosen to lie at any convenient part of the RF spectrum.

For example, for an incoming signal frequency of

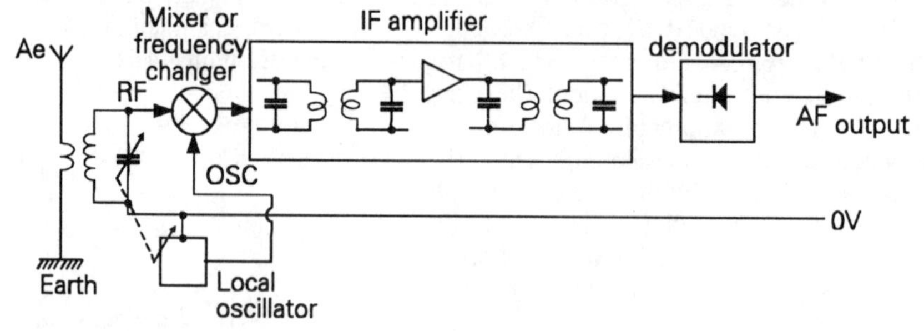

Figure 14.18 *Basic layout of superhet receiver*

1550kHz, and a local oscillator frequency of 1000kHz, sum and difference frequencies of 550kHz and 2550kHz will be generated, and will be present in the mixer stage output, along with the original 1550 and 1000kHz input frequencies.

For low to medium frequency receivers, such as are widely used for the medium- and long-wave broadcast bands, the IF frequency normally chosen is 455 or 465kHz, and this frequency band is kept free of most commercial broadcast transmissions. For earlier designs of short-wave receivers, 1.6MHz, a frequency right at the upper end of the medium-wave band, and not, at that time commercially exploited, was commonly chosen as an IF frequency, but in more modern designs, 45MHz or even higher frequencies may be used. Since it is desirable to avoid direct aerial circuit break-through at the chosen intermediate frequency, the frequencies adopted for IFs are usually those free from existing broadcast signals. It is also normal practice, because it facilitates mixer stage design, to adopt the 'oscillator high' style of operation, in which a receiver with a 455kHz IF amplifier stage, covering the frequency range 650kHz–1600kHz, would employ a local oscillator tunable over the range 1105kHz–2055kHz (i.e. above, rather than below, the received signal frequency).

The use of a fixed frequency IF amplifier conveys an enormous advantage in receiver design, in that almost any desired amount of amplification can be provided, and the adjacent channel selectivity can be chosen to give any required cut-off characteristic and passband, whether by the use of a cascaded series of bandpass-coupled, or stagger tuned, circuits, or by the use of a quartz crystal, mechanical resonator, or ladder filter, or some other surface acoustic wave type of RF passband defining system.

There are, however, also some drawbacks with the superhet system, of which the chief ones are 'second-channel' reception, mixer noise, drift of the tuned frequency, poor mixer conversion efficiency, whistles and cross-modulation effects.

Second-channel interference

Second-channel or image frequency reception can always occur in a frequency changer, for the reason which can be illustrated using the case of the receiver described above. For example, if the local oscillator is tuned to 1105kHz to produce a 455kHz IF output from a wanted 650kHz incoming signal, it will also produce a 455kHz output with an unwanted 1560kHz incoming signal, if one should, by chance, be present. The only way by which this can be prevented is by making sure that the selectivity in the RF amplification, or other tuned frequency, stages between the aerial and the mixer input is adequate to reject the unwanted image frequency signal. For signal frequencies up to a few MHz, the task of providing adequate RF selectivity to reject an unwanted signal 910kHz away from the wanted signal frequency is quite easy to accomplish. However, for a signal frequency of, say, 20MHz, an adequate degree of rejection of an unwanted second-channel signal at 20.910MHz will be much more difficult to secure. This was the reason for the choice of a 1.6MHz IF in early short-wave superhets: that the image frequency would be 3.2MHz away from the frequency of the wanted signal, and therefore somewhat easier to reject.

Double superhets and direct conversion systems

Since the efficiency of an IF gain stage will decrease as the operating frequency is increased, the 1.6MHz IF signal might then be down-converted to 455kHz for further amplification and pass-band filtering, as shown in the schematic layout of Figure 14.19. This process is termed 'double conversion', and such a receiver is called a 'double superhet' to distinguish it from the 'single conversion' method used in the simpler design of superhet receiver.

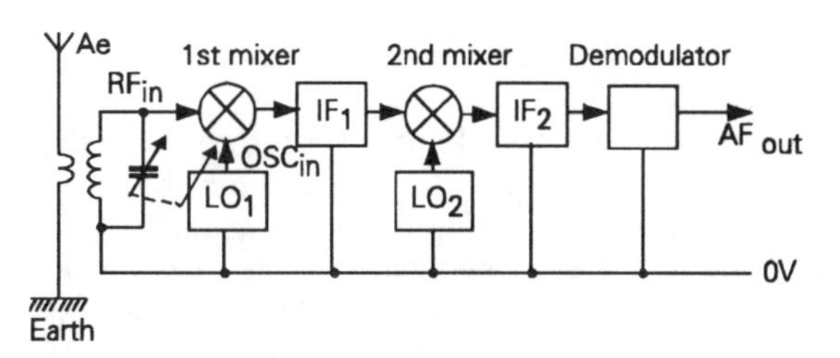

Figure 14.19 *Layout of double superhet*

The process may be extended yet further, for example to a 'triple conversion' receiver, with three IF gain blocks, and three successive frequency changer stages. In the opposite direction of development, there are what are termed 'direct conversion' receivers, shown in Figure 14.20, in which the incoming signal

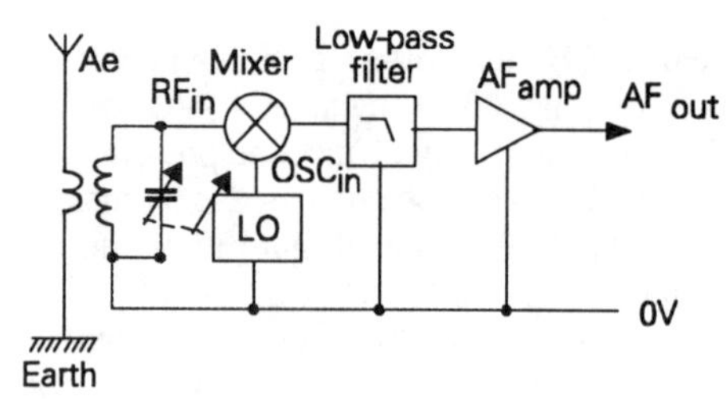

Figure 14.20 *Direct conversion receiver*

is mixed with a locally generated one, usually derived from a crystal controlled oscillator, operating at, or near, the same frequency as the wanted signal.

In the case of a transmitter whose carrier is keyed to send a Morse coded message, the output from the mixer could then be an interrupted audible difference frequency tone. In the case of a single-sideband suppressed-carrier transmission, provided that the oscillator frequency was chosen correctly, the output from a direct conversion receiver could be a normal audio band speech signal, without the need for any additional demodulator.

For a double-sideband transmission, with carrier, there would be an audibly objectionale 'howl' if the locally generated oscillator signal in a direct conversion receiver differed in frequency from the received carrier by an amount which was within the audible pass-band, and this limits the usefulness of such a system. Nevertheless, there are double-sideband receivers, in which synchronous frequency local oscillators are employed. These are called 'homodyne' or 'synchrodyne' systems, and are discussed later.

Whistles and mixer noise

A further problem with all superhets is the presence of whistles, which may occur anywhere within the tuning band. These are usually due to the presence at the input to the mixer stage of harmonics of the intermediate frequency signal – mainly generated at the demodulator stage – which then interact with incoming RF signals. Alternatively, oscillator harmonics can produce whistle generating IF frequency signals by mixing with signal harmonics.

In principle, a superhet will always have a somewhat lower signal-to-noise ratio than a simple TRF receiver, because of the additional noise introduced by the mixer stage. This noise occurs for exactly the same reasons as the thermal noise which is found in any other high gain amplifier system, and is a function of the effective conversion bandwidth, the effective input impedance of the system, and the ambient temperature. The problem is worsened by the relatively low stage gain (conversion efficiency), of the mixer stage, which means that its output noise level – bearing in mind that its output impedance may be fairly high – may be amplified nearly as much as the desired input signal.

Frequency drift

Another difficulty peculiar to the superhet is that of oscillator frequency drift, which causes a corresponding drift in the tuned frequency of the receiver. This problem is mainly caused by changes in the ambient temperature of the oscillator capacitors and inductors, and methods for minimizing this are discussed later. However, frequency drift can also be caused by changes in the input impedance of the oscillator or mixer devices, as well as by component ageing effects, or by the effects of mechanical vibration or external capacitative or inductive fields.

In VHF receivers, where high frequency IF stages, such as 45MHz, are used to reduce image frequency signal intrusion, the instability of tuning, with even a well designed variable frequency *LC* oscillator, would be quite unacceptable, and this problem is worsened because such receivers will inevitably require that the local oscillator operates at a frequency placed above the desired signal frequency. Practical receivers of this kind must therefore either use drift cancelling circuit techniques, or frequency synthesizer systems, based on a stable frequency quartz crystal reference oscillator. These techniques are explored later in this chapter.

A further associated problem is that of local oscillator frequency pulling, because of the effect of the aerial input signal on the impedance which the mixer stage presents to the local oscillator. This can largely be eliminated by good design procedures, such as the inclusion of a buffer amplifier between the oscillator output and the mixer input.

Cross-modulation

Two other associated difficulties are those of intermodulation and cross-modulation within the mixer stage. The first of these effects is due to harmonics of input signals, produced by the essential nonlinearity of the mixer, creating spurious higher frequency signal images. The second, again due to mixer nonlinearity, is the more annoying problem in which a strong incoming signal, by causing the mixer input working point to move up and down its transfer characteristic, will add its own programme modulation to that of the wanted signal.

Avoidance of these faults requires care in mixer stage design, and some limitation of the size of the signal level at the mixer input. For this reason, pre-mixer RF gain is usually chosen to be high enough to ensure that the overall signal-to-noise ratio of the receiver is mainly due to the aerial input circuit, but not so high that inter- and cross-modulation effects become noticeable.

Practical mixer circuitry

In thermionic valve operated equipment the most common frequency changer stages are the triode hexode, or heptode, valve types illustrated in Figure 14.21. These are, essentially, screened grid or RF pentode valves in which the screening grid has been separated into two sections, with an additional control grid inserted between them. This grid is internally connected to the control grid of a triode, which is contained within the same envelope and which shares the same cathode as the hexode/heptode. The internal triode section can then be employed as a separate RF oscillator whose output will modulate the electron stream flowing through the hexode/heptode section. This gives the required sum and difference signal generation, but screens the oscillator signal from the aerial input circuitry to avoid unwanted radiation of the local oscillator RF output. Unfortunately, although the triode- hexode/heptode mixer valve can give a good IM and cross- modulation performance – if the circuit operating conditions are chosen correctly – it has, by modern standards, a relatively poor noise figure.

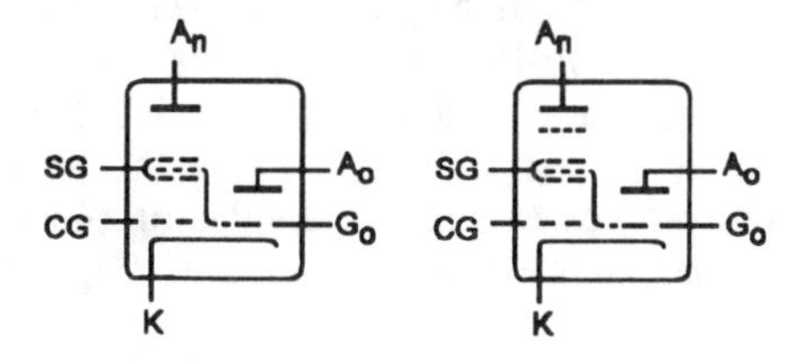

Figure 14.21 *Triode-hexode and triode-heptode frequency changer valves*

In low-cost portable transistor radio receivers, where the principle requirement of the manufacturer is to keep the total component count as low as possible, earlier circuit designs usually employed a self-oscillating mixer, of the general type shown in Figure 14.22, using a bipolar junction transistor, in spite of the fact that this type of circuit can only give a relatively poor performance in technical terms.

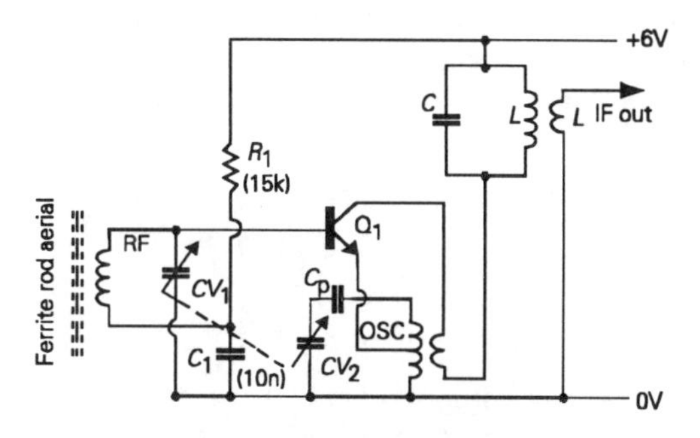

Figure 14.22 *Circuit layout of single transistor frequency changer*

In contemporary low-cost design practice the preferred approach is to use a single integrated circuit, such as the LM1868, in which all the functions of mixer, IF amplifier, demodulator, automatic gain control, and probably also those of a low power AF output stage, are fabricated on a single chip. Although the IC designers are forced to choose circuit configurations which minimize the use of hard-to-make capacitors and resistors, and this constraint can lead to some odd-looking circuit layouts, the basic electronic structures used in this type of IC will generally be similar to the circuit configurations used with discrete component designs. So such ICs, although effective in keeping the manufacturers component count down to a low level, will still usually only offer a comparably indifferent technical performance. Such single chip IC radio systems will usually alsq incorporate an FM receiver section: a method of signal transmission which is examined later in this chapter.

The best practicable performance for a 'multiplicative' mixer (one in which the output signal is the product of the aerial and local oscillator signals) is given by a system in which there is a 'square-law' relationship between the input voltage and the output (anode, collector or drain), current of the mixer device, and this condition is met most nearly by a junction FET operated at or near zero gate bias.

An efficient, low-noise, mixer circuit is provided by the circuit shown in Figure 14.23, where the aerial signal is applied to the gate and the local oscilator signal is injected into the source circuit of a junction FET, though this does not give a very good isolation of the oscillator signal from the aerial circuitry, unless an RF or other buffer stage is interposed between the aerial and the mixer. The relatively limited optimum working voltage range of such a circuit also limits the range of signal voltages which can be handled without input overload.

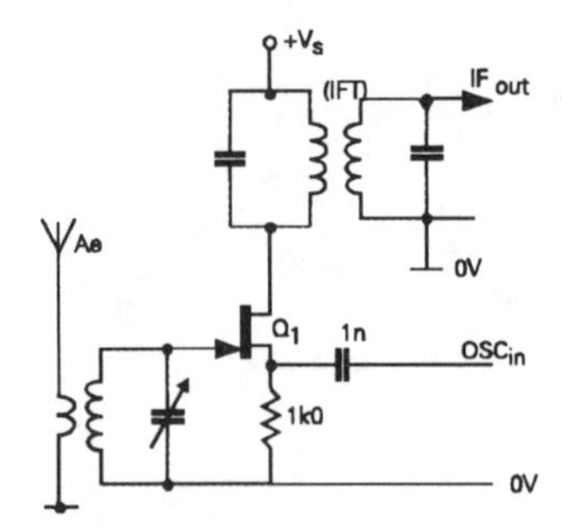

Figure 14.23 *FET mixer layout*

This circuit can be elaborated using the cascode layout of Figure 14.24, which gives a somewhat better degree of input/output/oscillator circuit isolation. These advantages are shared by the dual-gate MOSFET circuit shown in Figure 14.25, a layout which is very commonly used in medium quality discrete component superhet systems. The I_d/V_g characteristics of the dual-gate MOSFET are not quite as favourable from the point of view of avoidance of cross-modulation as those of the junction FET, though some MOSFETs are designed specifically for mixer applications, where the gate characteristics have been modified to suit this application. MOSFETs do not generally have such a low noise figure as junction FETs. They may, however, have better input overload behaviour.

Single- and double-balanced systems

All of the mixer circuits so far described are commonly known as single-ended systems, in which there are single signal and oscillator ports. This type of layout can be elaborated into push–pull systems, described as 'single balanced' and 'double balanced' layouts. These offer a better conversion efficiency –

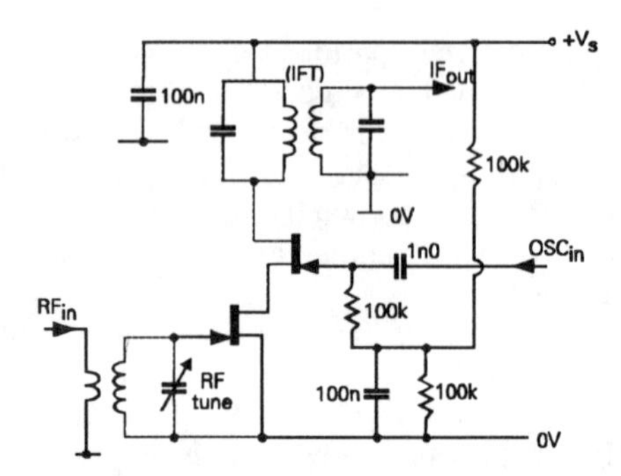

Figure 14.24 *Cascode connected FET mixer system*

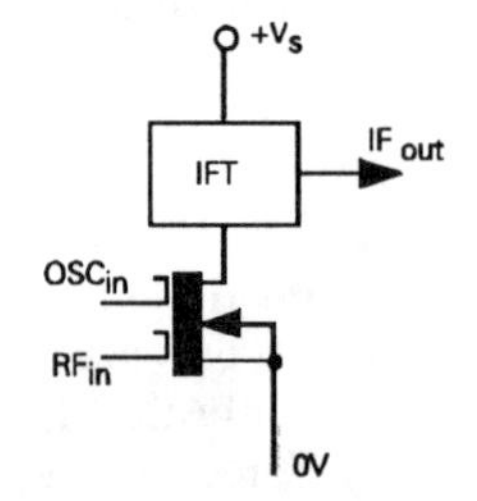

Figure 14.25 *Dual gate MOSFET mixer*

and consequently an improved s/n ratio – as well as improved port to port isolation: this factor is better with dual- than with single-balanced types.

The simplest single-balanced system is that using a diode bridge layout, shown in Figure 14.26, where the oscillator signal is injected into the centre tap of the secondary winding on a wide-band RF transformer – usually 'tri-filar' wound on a toroidal ferrite core. This layout gives good oscillator to input, but poor oscillator to output isolation. Diode mixers of this type can give excellent performance in respect to cross- modulation and input overload, but need a high oscillator output voltage at a low source impedance, and suffer from the drawbacks of both a conversion loss and a relatively poor noise figure.

A double balanced diode layout, often called a 'ring mixer', is shown in Figure 14.27. This gives good port-to-port signal isolation, and excellent IM and cross-modulation characteristics. Using 'hot carrier' or 'Schottky-type' (metal/semiconductor junction) diodes, this layout is usable up to the UHF (gigahertz) range.

Single and double-balanced mixer layouts using MOSFETs and junction FETs are shown in Figures 14.28–14.30.

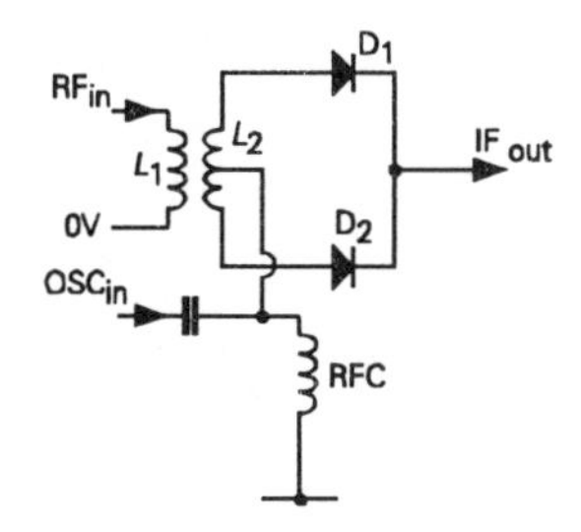

Figure 14.26 *Diode bridge single-balanced mixer circuit*

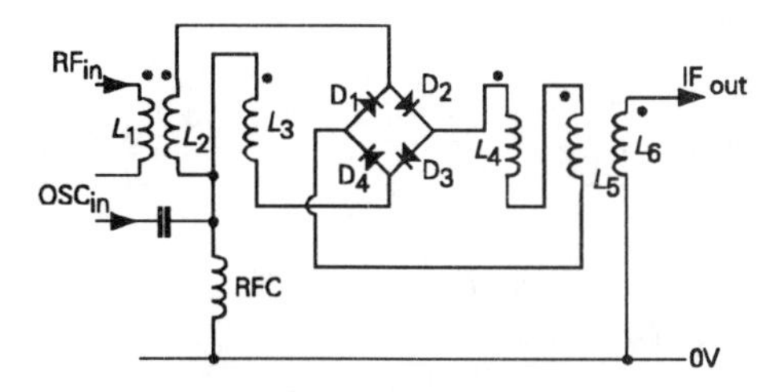

Figure 14.27 *Double-balanced ring mixer*

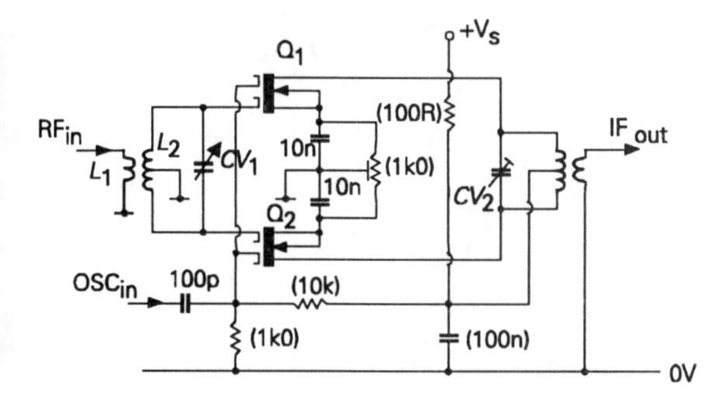

Figure 14.28 *Single-balanced mixer using MOSFETs*

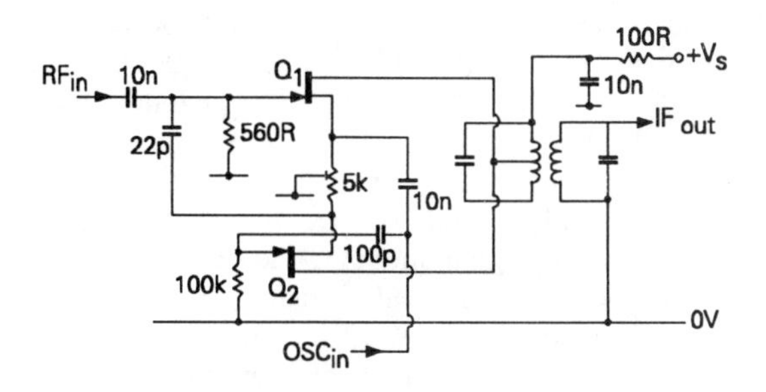

Figure 14.29 *Balanced mixer using junction FETs*

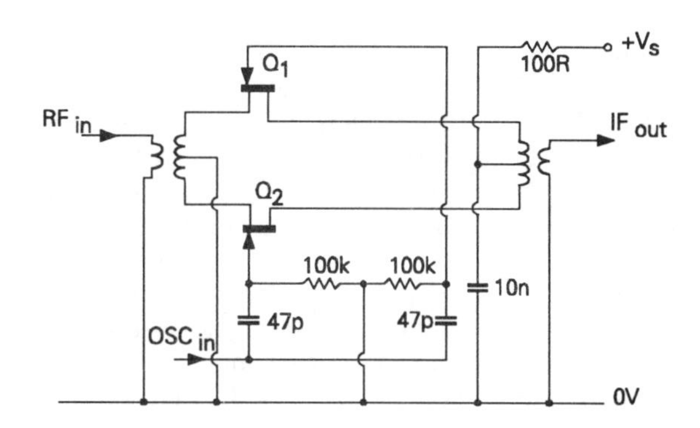

Figure 14.30 *Low noise balanced mixer based on junction FETs*

Oscillator stages

A number of circuit designs which could be used as the local oscillator in a superhet have been described in Chapter 13. The important considerations in this particular application are that they should be stable in frequency, and have a low noise and spurious signal content in their output waveform.

Temperature compensation

The importance of frequency stability is dependent upon the degree of selectivity of the receiver, and the tolerability of tuned frequency drift with time. Most *LC* tuned circuits will suffer from drift as a result of changes in the ambient temperature of the oscillator circuit, though, with low power solid-state circuitry. 'warm-up' frequency drift following switch-on is no longer a particular problem.

Some compensation for thermal drift in *LC* oscillators can be achieved by the incorporation within the tuned circuit of additional capacitors having a negative or positive temperature coefficient. These may simply be connected in parallel with the tuning capacitor, and would typically be of metallized ceramic construction. Such capacitors are usually designated N– or P–, with the coefficient specified in parts per million per degree Celsius. For example, an N–750 marking would denote a temperature coefficient of –750 parts/million/°C, and a P–100 one a capacitor with a +100 ppm/°C characteristic. Similarly a NPO marking would imply a near zero temperature coefficient for the component. Choosing the correct value and type of temperature compensation components is usually a laborious and tiresome exercise, especially if compensation over a wide temperature range is sought.

A high degree of oscillator frequency stability is particularly desirable in the case of receivers designed to receive 'CW' signals (transmissions consisting of a simple sinewave carrier, interrupted by morse or other coded keying patterns), where the signal is made audible by a beat frequency oscillator (BFO) stage. This operates by heterodyning the signal, in a further mixer stage, with an oscillator having an adjustable output frequency close to that of the IF. In this case a shift in incoming signal frequency due to oscillator drift will result in an audible change in the BFO pitch.

Oscillator frequency drift will also be embarrassing where the receiver is used to receive a 'suppressed carrier' signal, where if the reinserted carrier frequency moves away from the desired frequency, the received signal may become unintelligible.

The freedom of a local oscillator output from spurious signals is obviously necessary if whistles and the reception of signals at unwanted frequencies is to be avoided, and this demands care in the layout of the resonant *LC* circuit, to avoid inadvertent inductive loops, or inconvenient stray capacitances.

The requirement for a good signal-to-noise ratio in the local oscillator output arises because any modulation component of the local oscillator output, such as noise, will be added to the output sum and difference signals, in just the same way as the modulation present on the carrier of the wanted signal. So, if a 60dB signal-to-noise ratio is required for an input signal of 10μV amplitude then the local oscillator signal must have a signal-to-noise ratio of at least 80dB. Also, in some cases, mixer nonlinearities may exaggerate this problem by increasing the extent to which local oscillator noise is added to the composite signal.

IF amplifier stages

The basic requirements for both RF and IF amplifier stages are the same – that they should give a useful degree of gain, a high signal-to-noise ratio, an adequate input linearity (to avoid cross-modulation), and that they should be stable in operation. However, whatever the means by which adjacent signal selectivity is obtained, the fixed frequency IF stages have the advantage that they will not usually require tuning except in setting up the receiver. Any of the HF filter systems examined in Chapter 11 can be employed, and two typical IF amplifier blocks in which the selectivity is defined, in the first case by a group of four bandpass-coupled tuned circuits, and, in the second instance by a surface acoustic wave ladder filter, are shown in Figures 14.31 and 14.32.

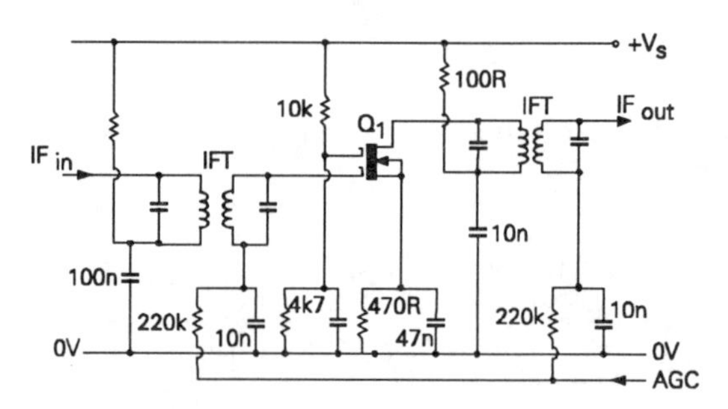

Figure 14.31 *IF amplifier stage using bandpass-coupled tuned circuits*

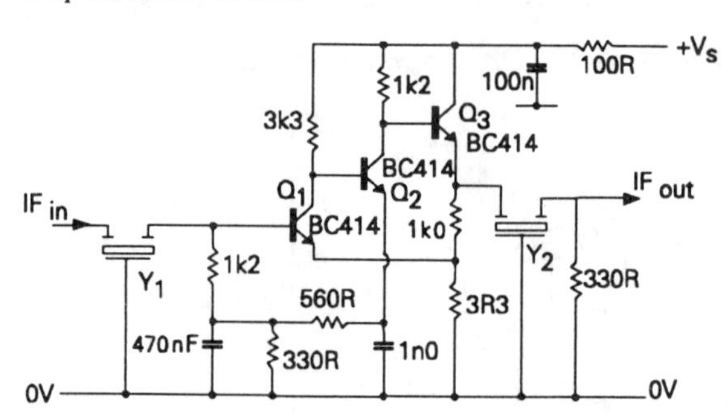

Figure 14.32 *IF amplifier using ceramic SAW ladder filters*

Automatic gain control (AGC)

An inevitable difficulty with the reception of amplitude modulated signals is that the larger the signal, the larger the demodulated audio frequency output. This means not only that strong signals will be louder than weak ones, which will necessitate continuous adjustment of the receiver output gain control as it is tuned from one signal to another across the band, but also that fluctuatinng signal strength, due to changing reception conditions, will lead to 'fading'.

Because of the very high IF gain which is usually available, it becomes possible with superhets to derive a further signal-dependent voltage, proportional to the average magnitude of the received carrier, and use this voltage as a gain control mechanism, to reduce the gain of the IF stages (and usually, also, the RF stages – if used) to try to ensure that the magnitude of the IF signal presented to the demodulator is of a substantially constant size.

A simple AGC circuit is shown in Figure 14.33, in which a negative gain control voltage, suitable for controlling the gain of a valve or MOSFET/FET operated IF amplifier, is derived from a diode demodulator. Some HF filtering is necessary in the AGC loop to avoid HF instability due to signal feedback, but the time constants for the gain control circuitry must be chosen with care, since too long a response time in the control system will reduce its ability to correct rapid 'flutter type' fading, while too short a time constant will attenuate the lower modulation frequencies, which will be interpreted by the AGC system as unwanted fluctuations in signal strength.

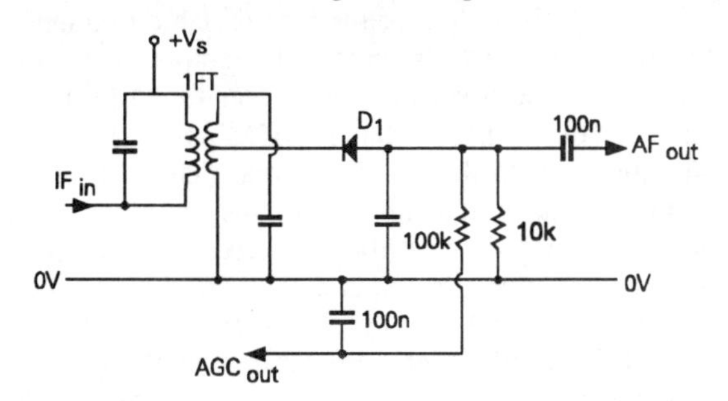

Figure 14.33 *Automatic gain control voltage generation circuit*

It must be remembered also that the gain control system is a closed loop servomechanism, and if the accumulated phase shifts in the control loop approach 180°, the whole system may become unstable, leading to the problem known, from its sound, as 'motor-boating'.

To avoid needlessly reducing the gain of very weak signals, it is common to arrange the design so that there is a lower threshold level in the AGC control system, below which the AGC voltage will not be applied. This is usually termed 'delayed AGC'. Similarly, to increase the effectiveness of the AGC system, designers may sometimes employ a DC amplifier in the AGC loop. A low distortion demodulator system, using both amplified and delayed AGC, is shown in Figure 14.34. (JLH, *Wireless World*, October 1986, p. 17).

Signal demodulation

Most of the demodulator systems described above in relation to TRF type receivers will operate quite satisfactorily as the demodulator, or 'second detector' in a superhet. However, since there is almost always an adequately large IF output signal available, and the designers of AM receivers are seldom concerned about minimizing demodulator distortion, a simple diode demodulator, of the kind shown in Figure 14.33, is nearly always used.

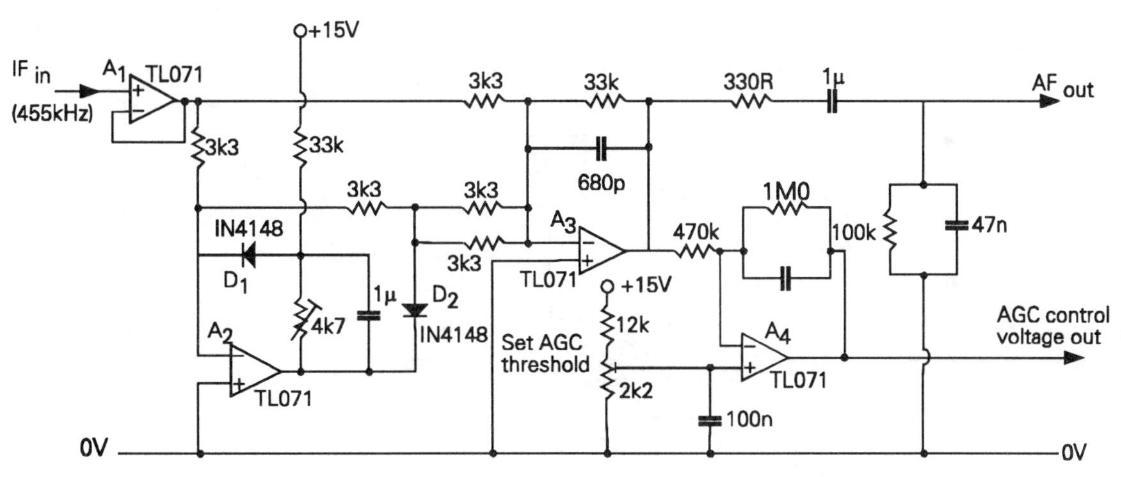

Figure 14.34 *Low distortion demodulator incorporating amplified AGC*

Stable frequency oscillator systems

The use of high signal frequencies in receiver or oscillator systems exaggerates the problems of frequency drift, simply because the same proportional drift becomes a higher frequency error at higher frequencies. For example, a 0.01% drift at 100kHz will be 10Hz – a negligible amount in most applications. However, the same proportional drift at 100MHz will be 10kHz – an error which would be unacceptable in almost every case. A relatively simple solution to this problem, provided that the coverage of the receiver is relatively limited, is to use a fixed frequency, quartz crystal controlled, local oscillator, coupled with a tunable IF. This would, however, suffer from the drawback that if the chosen, tunable, IF was low enough not to suffer from frequency drift, there would be a problem due to second-channel breakthrough, due to inadequate RF selectivity, while if the chosen IF was high enough to avoid this problem, the IF stage – which would also be a superhet receiver – would also suffer from frequency drift.

A number of systems have been evolved to reduce the extent of this problem, of which the two most interesting, in that they represent entirely different, and original, design philosophies, are the Barlow–Wadley drift cancelling loop, and the phase-locked loop frequency synthesizer system.

The Barlow-Wadley drift cancelling system

The way this system works is shown, in schematic form, in Figure 14.35. In this the aerial signal, amplified by a conventional tuned RF stage, is fed to a first mixer, along with the output signal from a good quality *LC* tuned local oscillator, operating at a frequency which is above that of the incoming signal. The resulting first IF signal is then amplified at an IF frequency of, say, 45MHz and fed to a second mixer stage. Because of the inevitable frequency drift in the first local oscillator stage, the 45MHz IF output will also suffer from drift. However, attached to the oscillator system is a stable frequency quartz crystal oscillator, operating at, say, 1MHz, and the output from this is fed to a harmonic generator circuit, to produce an array of output frequencies, at 1MHz intervals, extending up to, say, 100MHz or more. This harmonic series is then mixed with the output of the first variable frequency LC oscillator, and the composite output from this mixer is fed to a selective amplifier tuned to a frequency which is higher than the 45MHz first IF, by an amount equal to that of the second (say, 3MHz) IF amplifier. The output signal from this will suffer from an identical frequency drift, and in the same direction, as that of the first IF output signal, and the resultant errors will cancel, giving a drift-free input to the second IF amplifier/demodulator combination.

As an elaboration of this basic principle, the first IF

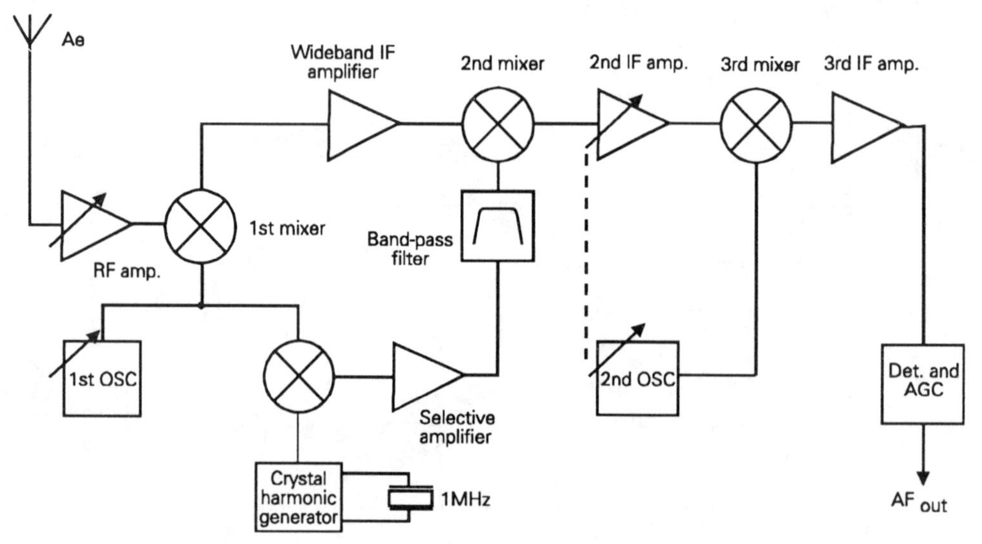

Figure 14.35 *The Barlow–Wadley loop drift cancelling oscillator system*

can be chosen to have a fairly wide, flat-topped frequency pass-band, and the second IF stage can then be made tunable, over say, the 2–3MHz range, to provide a band-spreading facility for the receiver – since frequency drift in a 2–3MHz receiver will seldom be a major problem. The first variable frequency oscillator can then be used to select the incoming signal frequency in 1MHz blocks.

Frequency synthesizer techniques

These can be divided into 'partial' synthesis and 'full' – i.e. 'digital' – synthesis designs. The partial synthesis method combines a quartz crystal oscillator with a standard *LC* variable frequency oscillator (VFO), as, for example, in the simple circuit arrangement shown in Figure 14.36. In this a receiver designed to cover the frequency range 28–30MHz, with a first IF frequency of 10.7MHz, employs a VFO tunable over the frequency range 1.7–3.7MHz. The output from this is heterodyned with that from a 37MHz quartz crystal, and passed through a 38.7–40.7MHz selective amplifier, before being mixed with the aerial signal in the receiver. This technique is adequate for coverage of a limited and specified waveband, but would require that a range of crystals were available to allow tuning over a wider input signal range. Full frequency synthesis techniques rely on the use of a phase-locked loop (PLL), for the control of the oscillator, so I propose to explain this system first.

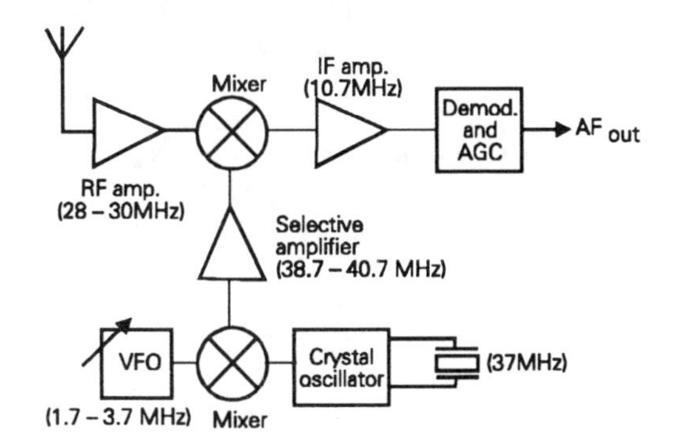

Figure 14.36 *28–30MHz receiver system*

The phase-locked loop

This circuit arrangement, shown in schematic form in Figure 14.37, provides a powerful technique for forcing a variable frequency voltage controlled oscillator (VCO) to operate at a specified frequency, as well as a means for deriving a DC voltage output related to this frequency, if this should be needed.

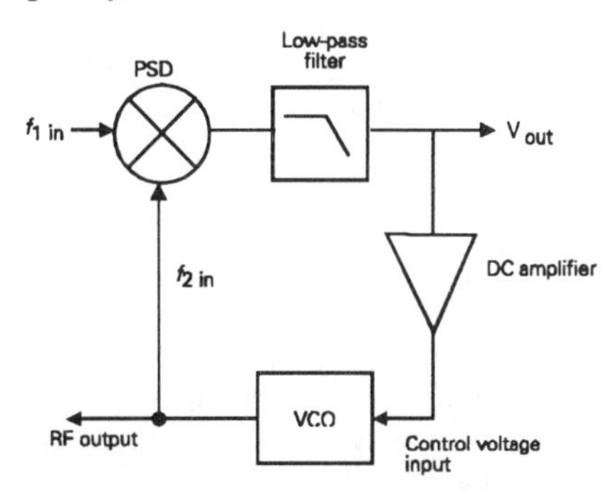

Figure 14.37 *The phase-locked loop*

In the simple circuit arrangement of Figure 14.37, the PLL consists of just four elements, an input mixer or 'phase sensitive detector' (PSD), a low-pass filter, a DC amplifier, and a voltage controlled oscillator, whose output frequency is determined by the control voltage applied to it. If the free-running frequency of the VCO is close to that of the input signal, and if it is postulated that, momentarily, the input frequency (f_1), is the same as the VCO output frequency (f_2), the PSD output will be a DC voltage which is related to the phase difference between these two signals. If this is amplified, and applied to the input of the VCO, it will cause this oscillator to speed up, or slow down, over part of a cycle, in such a manner that f_1 and f_2 are brought into phase and frequency synchronism with one another. If the free-running frequency of the VCO is truly that which requires a zero DC control voltage, then f_1 and f_2 will also be at phase quadrature, since that is the condition at which the PSD will have a zero voltage output, and if the loop amplifier has a high enough gain, a near-quadrature phase relationship between the signal and the VCO will also be the condition for non-zero VCO control voltages. Moreover, it is also found that if f_1 and f_2 are not initially identical, then, provided that the difference frequency is within the pass-band of the low-pass loop filter, known as the capture range, then the VCO will be drawn into, and will remain, in synchronism with the input signal. This is described as the PLL being 'in lock'.

If a frequency divider is incorporated in the VCO/PSD loop, then the VCO can be forced to oscillate at a multiple of the input frequency. Similarly, if a frequency divider is introduced into the control frequency path (f_1), then the VCO can be caused to oscillate at a sub-multiple of f_1.

The choice of filter bandwidth is dictated by the required capture range, and the required purity of the VCO output, since increasing the bandwidth of the filter increases the amount of wideband noise which will be added to the control voltage.

Digital frequency synthesis techniques

The complete superhet receiver shown in outline in Figure 14.38 uses an elaboration of the frequency multiplication, frequency division possibilities of the phase-locked loop circuit to generate a fully variable, but crystal-controlled, local oscillator frequency.

In this, a crystal controlled oscillator, operating at a frequency f_o, is used as a reference frequency generator, whose output is frequency divided by a factor of n, before introduction as the control signal, f_1, to the PSD. The output from a high frequency voltage controlled oscillator is also divided before being applied to the PSD, and the PSD output is filtered and amplified before application as the control voltage to the VCO. Then, when the PLL is in lock, the output from the VCO, f_3, will be determined by the crystal frequency according to the relationship

$$f_3 = f_o(m/n)$$

and will be very stable in frequency. By the use of a microprocessor to control the two division ratios, the required input signal frequency of the receiver shown in Figure 14.38 can be controlled by receiver front panel push-button selection, and the selected signal can be shown, for convenience, on a digital frequency display.

A number of manufacturers now offer single-chip frequency synthesizer ICs, which greatly facilitate the use of this technique.

Synchrodyne and homodyne receiver systems

The difficulty of achieving adequate selectivity with TRF type receivers, and the many technical problems associated with superhet systems, has prompted exploration of the possibilities of direct conversion receivers, of the kind discussed above. The advantage of a direct conversion receiver, if the local oscillator frequency can be controlled so that it is truly identical

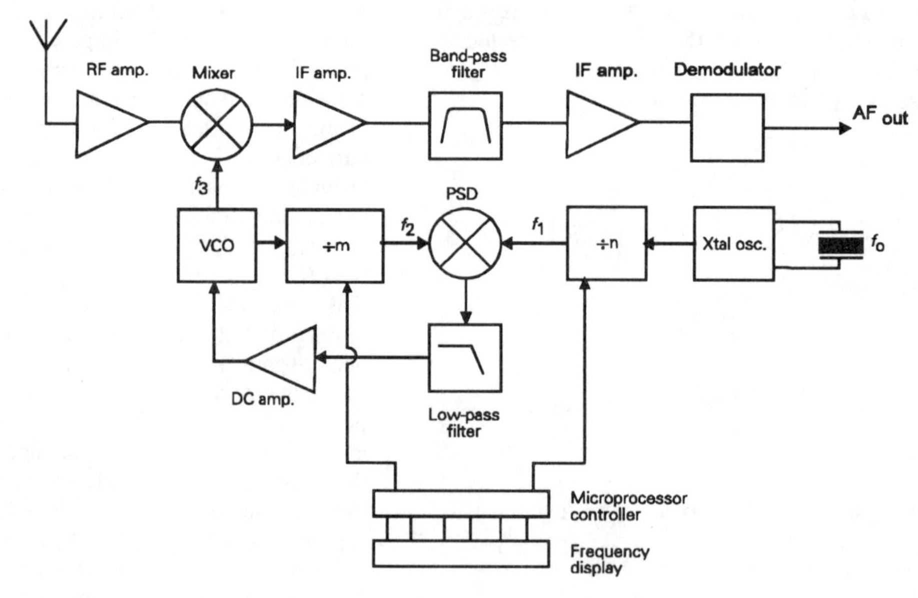

Figure 14.38 *Complete superhet receiver using frequency synthesized local oscillator system*

to that of the carrier of the received signal, is that demodulation of the incoming signal will occur without the need for a specific demodulator circuit: the input sum and difference frequencies produced by the amplitude modulation of the carrier will be directly transformed into an audible signal. This avoids the problems of poor demodulator sensitivity and linearity. Moreover, since the sidebands of adjacent interfering signals will also be transformed into audio signals, but of a much higher pitch, the receiver selectivity can be provided by post-mixer AF filtering, for which there are a number of useful low-pass filter circuits, as shown in Chapter 8. The only problem with this system is that if the pre-mixer selectivity is inadequate, cross-modulation can occur at the input to the mixer, to introduce an interfering signal which is proof against subsequent attempts at removal.

The difficulty, obviously, is to generate a local oscillator signal which is truly in synchronism with the required aerial signal. In the homodyne circuit, best implemented as the demodulator system for a broad selectivity superhet, a solution to this problem is attempted by using the carrier of the desired signal for this purpose, as shown in Figure 14.39. In this, a very narrow bandwidth IF stage, in parallel with the normal IF amplifier, is used to generate a signal, derived from the input signal carrier, but amplitude limited to remove its modulation component, which can be employed to synchronously demodulate the incoming signal. This system does work, but small changes in the signal tuning will alter the relative phases of the two signals applied to the mixer, and this will substantially alter the magnitude of the audio output from the mixer.

In the synchrodyne receiver, it is also required that the local oscillator signal fed to the mixer is synchronous with the wanted signal. An attempt is usually made to satisfy this requirement, in simple systems, by introducing a small amount of the aerial signal into the local oscillator circuit. If the local oscillator frequency should drift away from the signal frequency, the loss of synchronism becomes audible as a loud whistle, whose pitch is determined by the difference frequency, and this fact, especially noticeable when tuning from one signal to another, coupled with the difficulty of maintaining synchronism of the local oscillator, has prevented the use of this technique on anything other than a laboratory scale. However, a synchrodyne receiver design in which phase-locked loop techniques were used to force the local oscillator into phase and frequency synchronism, and in which inter-station muting was incorporated, is shown in outline in Figure 14.40. A practical receiver based on this structure was described in *Wireless World*. (JLH. Jan. 1986, pp. 51–54, Feb. 1986, pp. 53–56 and March 1986, pp. 58–61).

Frequency modulation (FM), systems

In this type of transmission system, the frequency, rather than the amplitude, of the transmitted signal is modulated to carry the programme content. This idea, also, was due to the same remarkable Major Edwin Armstrong who invented the concepts of positive feedback, regeneration, the super-regenerative and the superhet types of radio receiver.

As a broadcasting technique, FM has many advant-

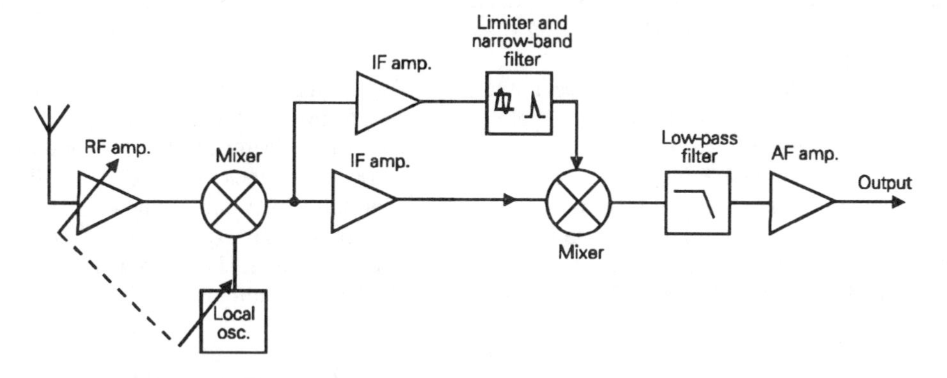

Figure 14.39 *Circuit arrangement of homodyne receiver*

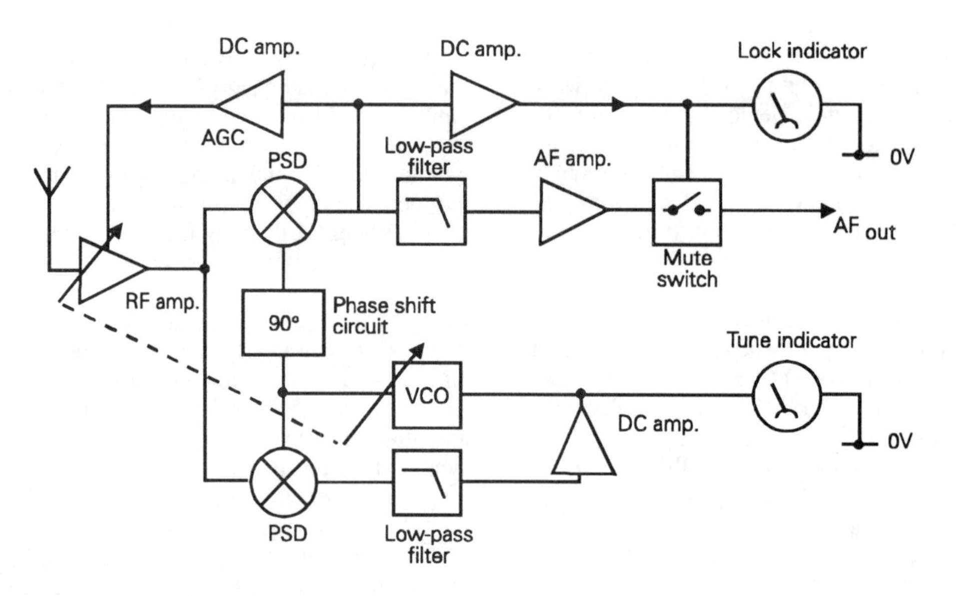

Figure 14.40 *Synchrodyne receiver*

ages, of which the major ones are that, since the amplitude of the received signal is now no longer important, all of the incoming signals can be amplified sufficiently to allow their amplitude to be limited by a clipping stage, such as the simple back-to-back diode limiter shown in Figure 14.41. This has the immediate practical benefit that signal fading is eliminated, and that all incoming signals are received at the same volume level. In addition, 'frequency discriminator' demodulation systems are, at least potentially, very much better in terms of linearity, and freedom from signal distortion, than comparable AM demodulators. They will also, in good designs, give a high degree of rejection of 'impulse type' noise, such as that caused by electrical switch contact interference and motor vehicle ignition noise. Rejection of interfering adjacent channel signals is also assisted by the 'capture' effect, a feature of the demodulator system employed, in which the presence of a stronger signal, at the demodulator, will completely suppress a somewhat weaker one: the extent to which this occurs is known as the 'capture ratio'. Good FM demodulator circuitry can offer capture ratios better than 1dB.

The major drawback of FM broadcast systems is that, for optimum modulation bandwidth and linearity they require a very wide transmission bandwidth. Peak modulation bandwidths of +/–75kHz are typical of contemporary FM broadcast transmitters, for which

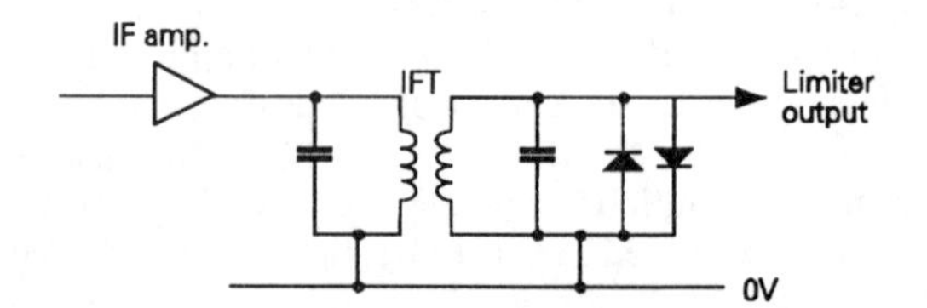

Figure 14.41 *Simple diode amplitude limiter circuit*

good quality reception requires a receiver bandwidth of at least +/–120kHz. Clearly, this amount of broadcast bandwidth is not available in the existing and crowded long- or medium-wave commercial transmission bands, so a part of the frequency spectrum between 88MHz and 108MHz has now been set aside, by international agreement, for domestic FM broadcast transmissions. The use of this part of the RF spectrum offers both advantages and disadvantages, in that the general absence of significant ionospheric reflection means that reception will only be possible, normally, over line-of-sight distances. So, for general reception, a network of local stations is necessary, and the choice available to the listener is usually restricted to the programmes broadcast on his or her own national network. On the other hand, the limited range of transmissions in this frequency band eliminates overcrowding, and the likelihood of unwanted adjacent channel interference.

FM receiver design

The design of receiver used for FM reception usually follows conventional radio practice, only modified as necessary to suit the 88–108MHz signal spectrum, and the required IF bandwidth. The frequency normally used for the IF is 10.7MHz, and a wide range of piezo-electric ceramic surface acoustic wave (SAW) filters is available, allowing substantially flat-topped response curves, with a range of bandwidths, for different applications.

Contemporary 'tuner head' design practice favours either dual-gate MOSFET, or neutralized junction FET stages, with similar devices for the mixer stage. Since varicap diode tuning allows additional tuned circuits to be added without practical difficulties, it is common, in better class tuners, to find three or four tuned circuits preceding the mixer, to reduce cross modulation effects, and improve the effective s/n ratio.

A typical layout for a varicap tuned FM RF/mixer circuit is shown in Figure 14.42, and a 10.7MHz IF stage, using ceramic SAW filters, is illustrated in Figure 14.32.

FM demodulator systems

The slope detector

Various methods have been adopted, over the years, as a means for converting a constant amplitude RF signal which varies in frequency into a varying DC output voltage level. Of these, the simplest, and most straightforward, is to feed the signal to a diode detector, coupled to a tuned circuit, whose natural resonant frequency is displaced somewhat to one side of the signal frequency, as shown in Figure 14.43. This will somewhat distort the demodulated signal, as shown, because of the curvature of the tuned circuit resonance curve. It will also offer no immunity from AM signal breakthrough or impulse-type interference.

The Round–Travis circuit

An improvement in this arrangement is offered by the 'Round–Travis' layout, shown in Figure 14.44, in which two such demodulator circuits are connected with their outputs in series, with one tuned somewhat

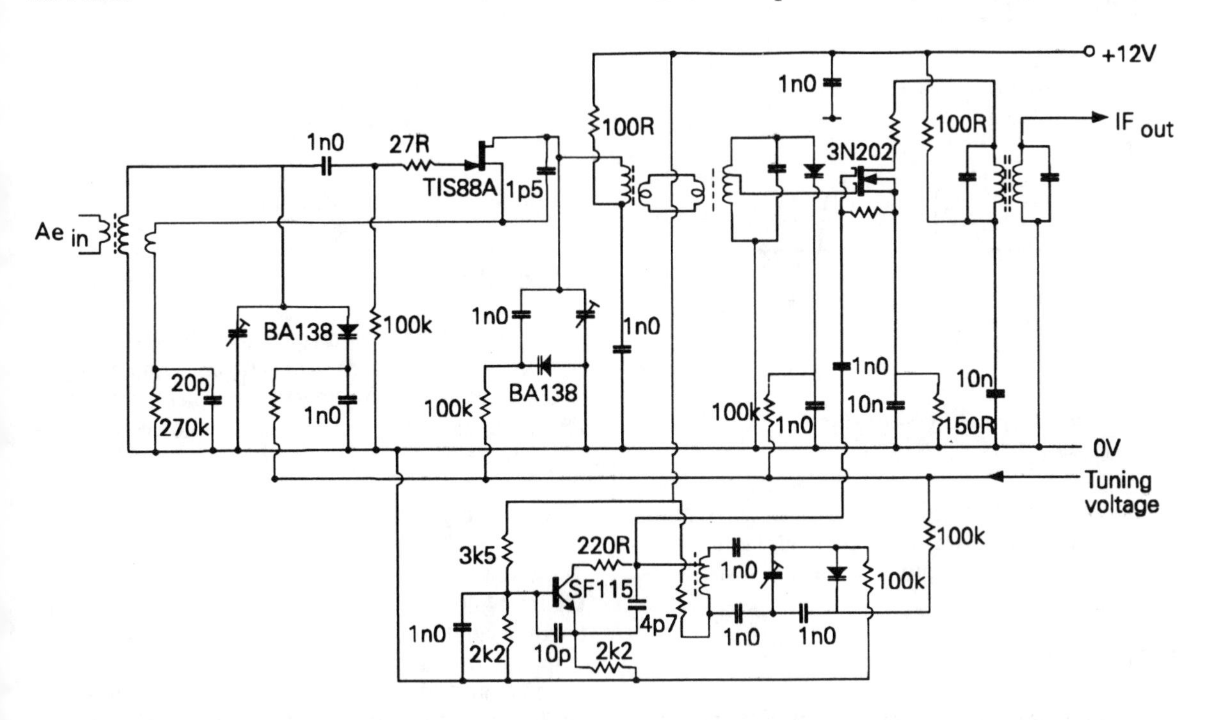

Figure 14.42 *Commercial FM tuner head design*

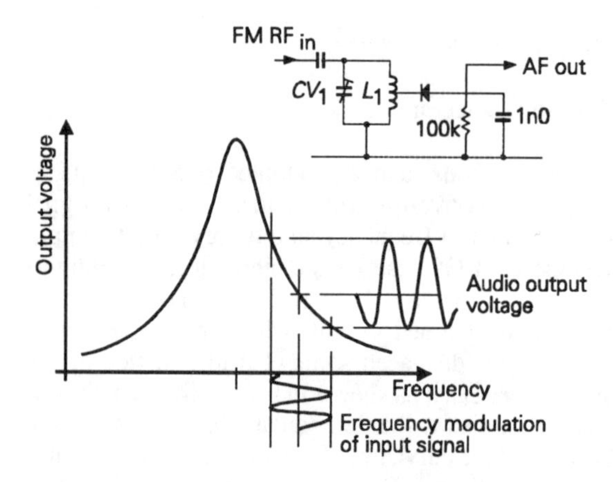

Figure 14.43 *Simple FM 'slope' demodulator*

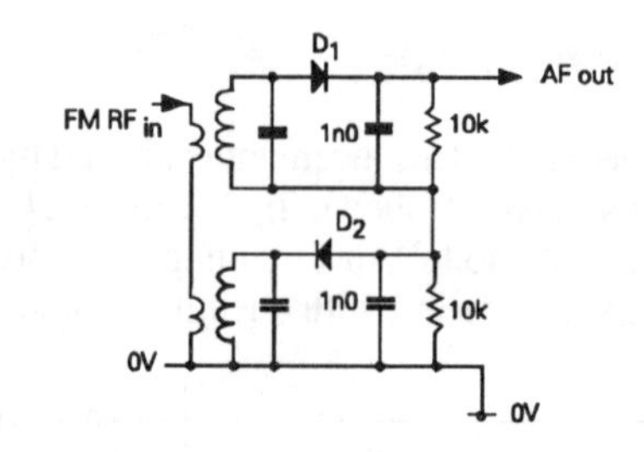

a The Round–Travis FM demodulator circuit

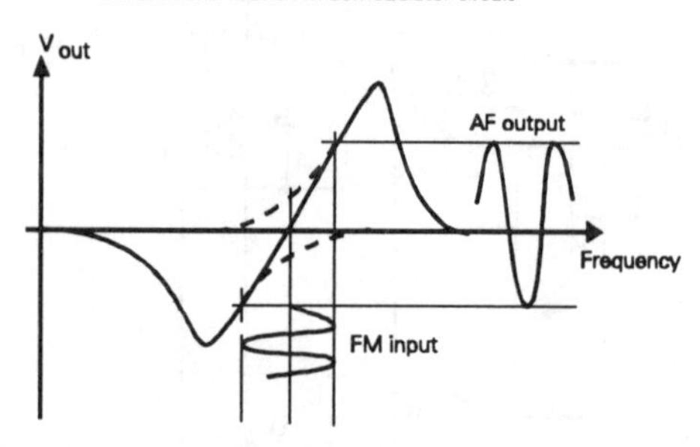

b Output voltage from Round–Travis circuit

Figure 14.44

above, and the other tuned somewhat below the mid-point signal frequency. The effect of adding the two resonance curves is to improve the overall demodulator linearity.

The Foster–Seeley discriminator

The most important improvement in the effectiveness and linearity of the FM demodulator was the introduction of the 'Foster–Seeley' circuit shown in Figure 14.45. The operation of this is based on the fact that the phase of the voltage developed across a tuned circuit, which is loosely coupled to an input winding, will be at quadrature (90° or 270°), to that of the input signal when the tuned circuit is at resonance. This leads to the possibility that, if an input signal is injected into the centre tap of the resonant circuit, it will either add to the output voltage or oppose it, depending on the input frequency, and this circuit will give a very linear output voltage/input frequency relationship for relatively small input signal voltages.

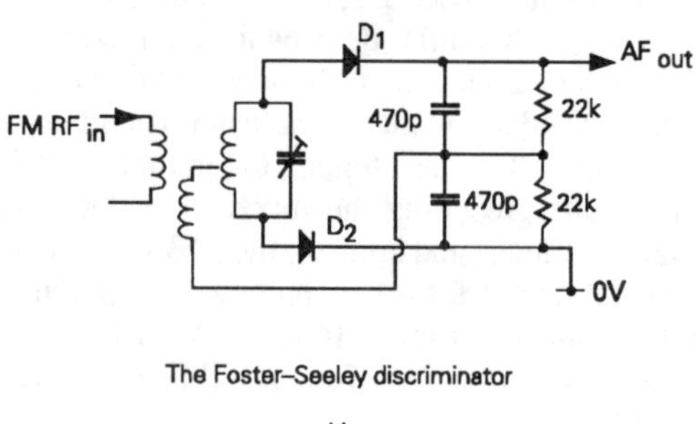

The Foster–Seeley discriminator

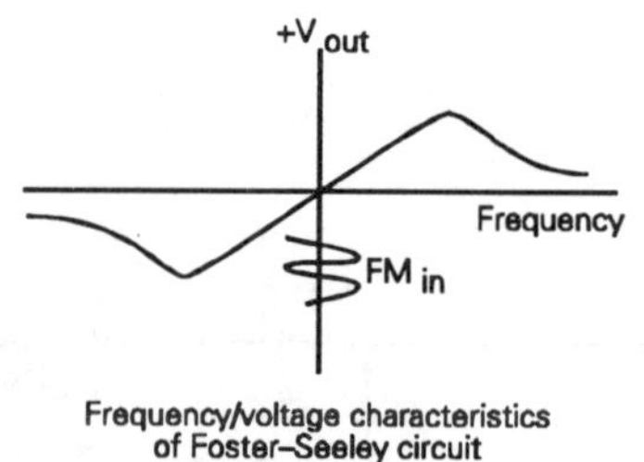

Frequency/voltage characteristics of Foster–Seeley circuit

Figure 14.45

The major practical drawback with this layout is that it offers very little discrimination against impulse-type AM breakthrough, a persistent problem with VHF reception due to the prevalence of motor vehicle 'ignition noise'.

The ratio detector

The circuit known as the 'ratio detector', shown in Figure 14.46, minimizes this problem by connecting the two demodulator diodes in opposition, so that the output is the ratio between the two signal voltages rather than their sum. It is not, however, either as efficient or as linear in its demodulation characteristics as the Foster–Seeley discriminator.

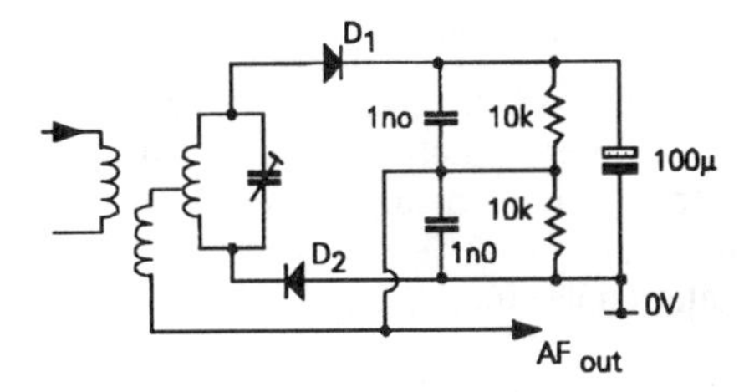

The ratio demodulator

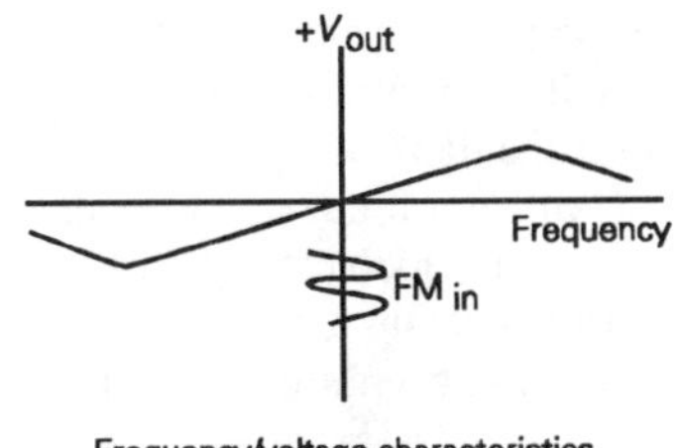

Frequency/voltage characteristics of ratio demodulator

Figure 14.46

The gate coincidence demodulator

Contemporary design practice favours, almost exclusively, the use of a quadrature coil 'gate coincidence' demodulator (GCD) combination, shown in Figure 14.47, because this can easily be fabricated as part of a monolithic IC structure. Since these ICs, such as the RCA CA3189, also offer a host of other useful functions, such as gain control, automatic frequency control, signal strength indication and off-station noise 'muting', it is understandable that they are widely used.

The GCD is a true phase-detector, in that there will be no output voltage change if an input signal voltage is applied to either of its signal ports, on it own, or if two input voltages are applied in phase quadrature – which would be the case if the quadrature coil, L_1/C_1, is resonant at the incoming frequency. If the input frequency is altered, there will be a change in the output voltage level which, over a limited voltage range, will be as linear as the output from a Foster–Seeley discriminator, but with improved AM rejection. Typical THD values for a single quadrature coil GCD design lie in the range 0.8–1.5% THD for a +/–75kHz modulation level. By comparison, an AM demodulator would give 2–5% THD for a 50% modulation level.

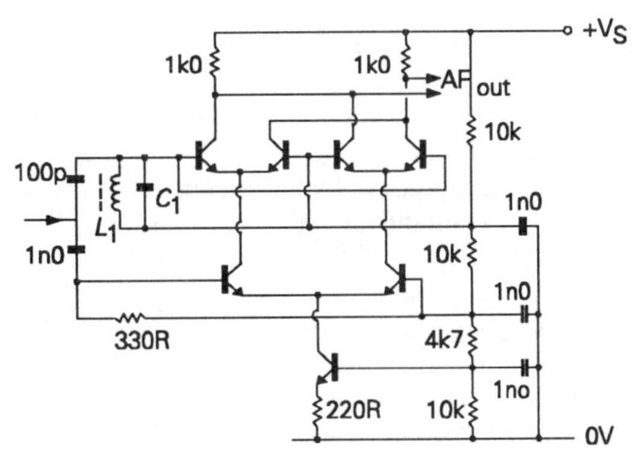

Circuit layout of gate-coincidence demodulator system

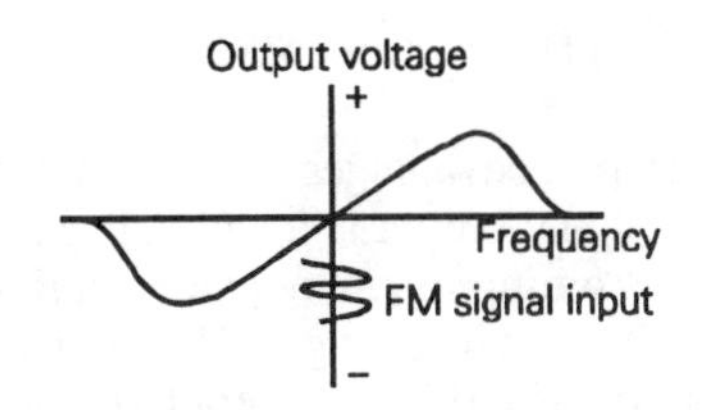

Performance of gate-coincidence demodulator

Figure 14.47

It is claimed by the manufacturers of gate coincidence demodulator ICs that the linearity of the demodulator system can be improved to allow better that 0.1% THD distortion levels, by coupling a further resonant tuned circuit to the quadrature coil. This configuration remains more of a theoretical possibility than a commercially valuable system because the output distortion given by this arrangement depends critically upon the coupling factor between these two tuned circuits. If they are 'under coupled' little benefit is given, while if they are 'over coupled', the demodulator output will show a pronounced crossover distortion type kink at the mid-point frequency (JLH, *Wireless World*, March 1991, p. 220).

Phase-locked loop demodulator systems

The circuit layout shown in Figure 14.37 can also be used to provide a very efficient, low distortion FM demodulator system, in that if the loop is locked, the frequency of the voltage controlled oscillator will be identical to that of the input signal. If this varies in frequency, the DC output from the loop (point 'B' in the figure) will be that which is required by the voltage controlled oscillator to cause it to oscillate at that frequency. If there is an accurately linear relationship between the VCO control voltage and its output frequency, then an input FM signal, applied to point 'A', will also be demodulated, by this means, at point 'B', with very little modulation distortion. Such a PLL demodulator will also give very good AM rejection and an excellent capture ratio. As a matter of personal interest, I have designed a number of experimental FM receivers using this system, of which the most recent was offered as an amateur constructional project in 1987 (JLH, *Electronics Today*, March 1987, pp. 34–38).

Automatic frequency control (AFC)

One of the inherent advantages of any FM demodulator is that the signal output is a DC voltage which is related to the incoming signal frequency. If this varies, the output voltage will change. With tuner circuits using varicap diodes, this DC output voltage can be applied to the oscillator tuning system so that unwanted drifts in the tuned frequency can be corrected. This is a particularly valuable feature in receivers operating at such relatively high signal frequencies, so that most commercial tuner head designs provide an AFC input control point, for direct connection to the AFC output from the demodulator IC.

Inter-station noise muting

With FM tuner designs in which a high signal amplification level is coupled to an amplitude limiting clipping stage, all signals, including the thermal noise input from the aerial, or that due to the input stages to the receiver, will be amplified to the clipping level. Moreover, since the thermal noise will have a random distribution in both amplitude and frequency it will be demodulated into a wideband (white), noise signal, and this is a disconcerting feature in inexpensive FM receivers without automatic inter-station muting facilities.

Various techniques are used to mute this noise, of which the simplest is to monitor the IF signal strength, at some point prior to the limiter stage, and arrange the circuit to switch off the audio output if the measured signal level falls below some predetermined value. A circuit for this purpose is shown in Figure 14.48.

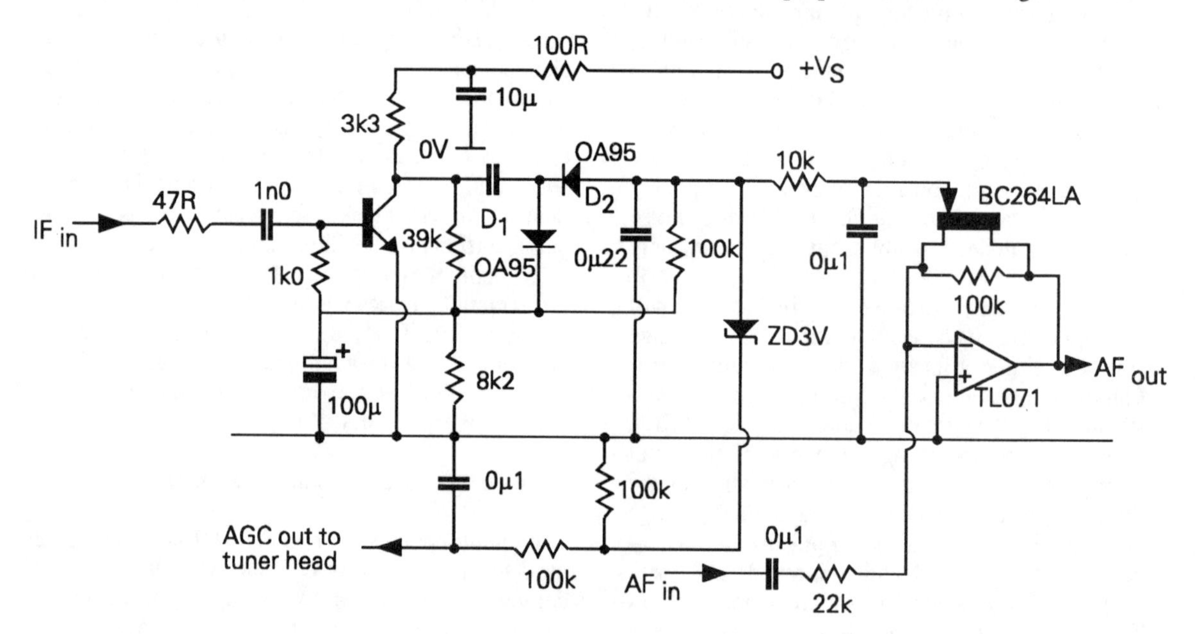

Figure 14.48 *Inter-station muting circuit for FM receiver*

A rather more sophisticated approach, employed by RCA in their CA3189 IC, is to monitor the extent of signal frequency deviation, and mute the output if this exceeds a certain value, as it will do with a wideband noise input.

FM Stereo broadcast transmissions

The extra modulation bandwidth available on the Band II region of the VHF allocated for FM broadcasts has been exploited to allow stereo broadcasting, using the GE-Zenith 'pilot tone' system, for which the bandwidth distribution is as shown in Figure 14.49. In this a normal 'L+R' mono signal is transmitted, using the 30Hz–15kHz part of the modulation spectrum, while an additional 'L–R' signal is broadcast as a double sideband supersonic signal based on a carrier frequency of 38kHz, giving a maximum modulation for the composite (L+R/L–R) signal of 90% of the permitted +/–75kHz deviation. This 38kHz carrier is suppressed before transmission to avoid unwanted breakthrough into the normal audio passband, and is then reconstructed, in the 'stereo decoder' of the receiver, from a phase synchronous 19kHz pilot tone broadcast at 10% effective modulation depth. This technique is 'mono compatible' in that, on FM receivers without stereo decoding facilities, the signal is received as a normal mono broadcast, whereas, on suitably equipped receivers, a pair of separate, low distortion, 'L' and 'R' channel outputs, having up to 40dB channel separation, are provided. Decoding this composite signal can be effected by treating the L–R signal as a normal LF radio signal and, after separating the carrier from the L+R signal, demodulating it in a conventional way to give a L-R output, from which the separate L and R signals are obtained by subtraction or addition. Alternatively, the L and R outputs can be obtained by synchronously switching the composite signal between the two channels, at a 38kHz frequency – a process which effectively does the same job.

These alternative decoding techniques are illustrated schematically in Figures 14.50 and 14.51.

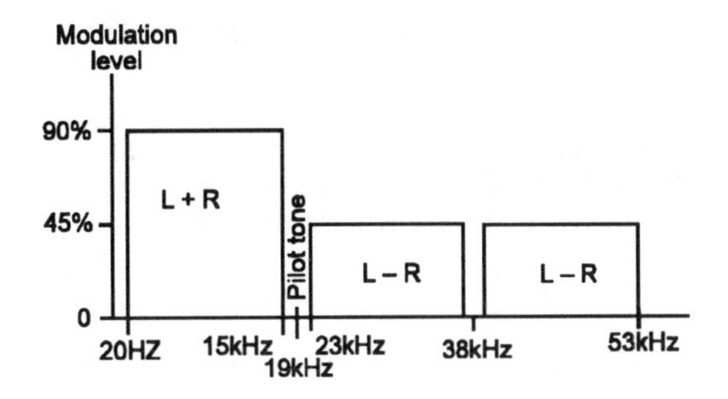

Figure 14.49 *The carrier modulation characteristics used in the GE-Zenith FM stereo transmission system*

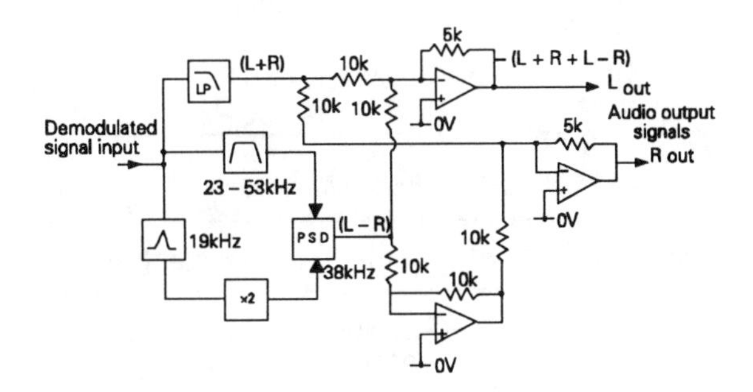

Figure 14.50 *Matrix stereo decoder system*

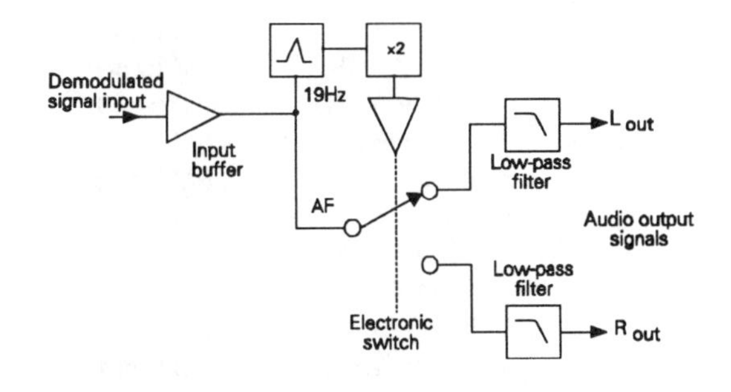

Figure 14.51 *Switching type stereo decoder*

Appendix. Transmission and reception data

There are a number of terms which are conventionally used in the classification of radio broadcast signals. Those relating to the frequency spectrum are listed in Table 14.1.

Table 14.1 *Classification of radio frequency bands*

VLF	3–30kHz
LF	30–300kHz
MF	300kHz–3MHz
HF	3–30MHz
VHF	30–300MHz
UHF	300MHz–3GHz
SHF	3–30GHz

Table 14.2 *Broadcast band allocations*

Waveband	Allocation	Wavelength
LW	150–285kHz	
MW	525–1605kHz	
SW	1.8–2MHz	160M band *
	3.5–-3.8MHz	80M band *
	5.95–6.2MHz	49M band
	7–7.1MHz	40M band *
	7.1–7.3MHz	40M band
	9.5–9.775MHz	30M band
	11.7–11.975MHz	25M band
	14–14.35MHz	20M band *
	15.1–15.45MHz	19M band
	17.7–17.9MHz	16M band
	21.0–21.45MHz	13M band *
	21.45–21.75MHz	13M band
	25.5–26.1MHz	11M band
	28.0–29.7MHz	10M band *
	50.0–54.0MHz*	
	70.025–70.7MHz*	
	144.0–148.0MHz*	
	430.0–440.0MHz*	
	1215.0–1325.0MHz	23cm band *
	2300.0–2450.0MHz	13cm band *

(* denotes amateur transmitter frequency allocations).

Within these frequency bands segments have been allocated, by international agreement, for domestic and amateur usage, as shown in Table 14.2.

(*Note*. It is observed that not all national/commercial broadcasting authorities adhere strictly to these frequency allocations.)

In addition to these allocations there are other VHF/UHF band designations, listed in Table 14.3.

Table 14.3 *VHF/UHF Frequency band designations*

Waveband	Frequency allocation.
I	41–68MHz
II	87.5–108MHz
III	174–223MHz
IV	470–585MHz
V	610–970MHz
VI	11.7–12.5GHz

Recent changes in the internationally agreed transmitter frequency separation in the LF and MF bands have imposed constraints on the broadcast bandwidth, in order to limit adjacent channel sideband interference. VHF broadcasts are not subject to these constraints, and can therefore provide a wider programme frequency range.

Details of the British broadcast standards, published by courtesy of the BBC, are listed below.

Audio bandwidths
LF/MF(AM)

40–5800Hz	+/–3db	(cut-off rate >24dB/octave beyond 6kHz)
50–5000Hz	+/–1db	

VHF(FM)

30Hz–15kHz	+/–0.5dB	(Cut-off rate >24dB/octave beyond 15kHz)

Distortion
LF/MF(AM)

<3%THD @75% modulation
<4%THD @100% modulation

VHF(FM)

<0.5%THD @ 80% modulation

Modulation depth
LF/MF(AM)

Up to 100% over range 100–5000Hz

VHF(FM)
Peak deviation level (100% modulation), corresponds to +/–60.75kHz deviaton

15

Power supply systems

Introduction

Active components operate by transforming energy supplied by a power source into some other desired form. Because of this, the way they operate must always be dependent, to some extent, on the quality and purity of the power supply by which they are energized, although one of the aspects of circuit design can be to minimize, where possible, the influence of the power supply on the behaviour of the circuit.

Most electronic circuitry will be designed to be powered by some form of DC supply arrangement, although it is possible to design some kinds of equipment which will work satifactorily with a raw AC voltage supply rail, and this may be done where low cost is the major design objective. Of the DC sources, the commonest forms are 'primary' (expendable), or 'secondary' (rechargeable), batteries, AC mains powered DC supplies, and solar cells.

In general, the choice of supply type will depend on the total energy requirement, and whether the equipment needs to be fully portable. Solar cell power supplies are restricted to use in medium to high light level environments, and with systems, such as electronic calculators, which require very little power, though output efficiency improvements in such cells are changing this situation.

Batteries (groups of electro-chemical cells, connected in parallel or series) allow a higher output power, but are either expensive to replace or require recharging at intervals. Their use also carries with it the possibility of damage to the equipment with which they are used through leakage of corrosive electrolytes, though improved battery design has led to 'leak free' systems.

AC (mains) supply line operated power supplies will allow an almost unlimited range of available output powers, but they may be heavy, and they will usually restrict the portability of the equipment. They also lead to the possibility of mains hum intrusion into the electrical output of the circuit, either because of the presence of residues of mains frequency voltage ripple on the DC output rails, or by interaction between the alternating magnetic field associated with the mains transformer in the power supply and the internal electrical circuit connections in the equipment with which it is used.

Battery power supplies

Depending on the relative costs and service requirements the choice will lie between primary cells – those types which are used until their output voltage falls below the minimum satisfactory level, but must then then be removed and discarded – and secondary cells, which can be repeatedly recharged.

Primary cells

Of the primary cells, the choice is between cost and performance – in terms of their output power to weight ratio, their constancy of voltage output as a function of time, electrical load, and ambient temperature, their output voltage, and their resistance to leakage of electrolyte. The major contemporary cell types are listed below.

Leclanché

A wide range of 'voltaic' cells was devised, during the early part of the 19th century, based on the combination of various pairs of metals with some acid or alkaline electrolyte, following Alessandro Volta's discovery in 1799 that zinc and silver metal discs, separated by a layer of moist cloth, would generate an externally measurable electrical potential. Most of these cells, such as those due to Daniell, Grove and Bunsen, remained little more than laboratory curiosities, but the system devised, in 1866, by Georges Leclanché which used an electrode pair of carbon (+ve), and zinc (–ve), in an ammonium chloride electrolyte, proved exceedingly successful, and has, with some improvements, been adopted as the basis for the vast majority of inexpensive 'dry' batteries.

In its simplest form, the construction adopted is as shown in Figure 15.1, in which a U-shaped thin-walled tube of zinc, is lined on its inside with blotting paper, and is used to hold either an ammonium chloride or a zinc chloride electrolyte, or a mixture of these, with some liquid absorbent or gelling agent, to reduce the likelihood of leakage. Also contained inside the tube is a cloth bag holding the positive electrode, which consists of a carbon rod, surrounded by a mixture of graphite, to help conductivity, and manganese dioxide, which operates both as a 'depolarizing' agent – to prevent the evolution of gaseous hydrogen at this electrode – and as a source of oxygen to activate the cell. The cell is completed by crimping a tin plated brass cap on the top of the central carbon rod electrode, sealing the gap between the rod and the outside tube at the top of the cell with pitch, and wrapping a paper label around the outside of the zinc tube. Such cells are cheap to make, have a storage life of up to a year, require little exotic manufacturing technology, and have an adequate output power to weight ratio.

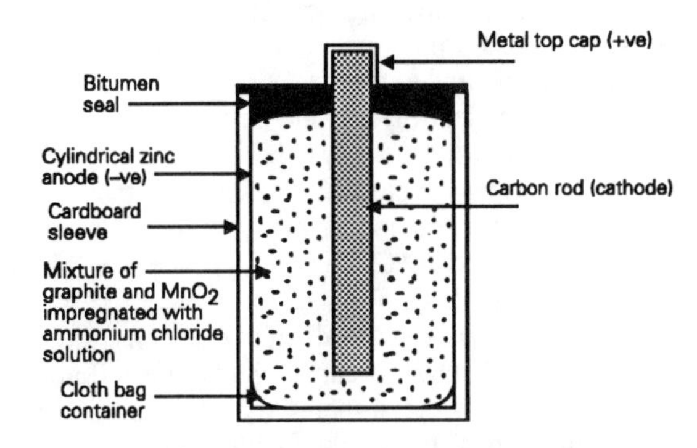

Figure 15.1 *Simple dry battery (Leclanché cell)*

The initial output voltage for a new cell will be in the range 1.55–1.6V, which will gradually fall, during use, to 1.1–1.2V, by which time its service life must be considered to be at an end, since any further discharge beyond this point will cause the output voltage to collapse rapidly to a near-zero level. In more recent cell technology, zinc chloride has largely come to replace the ammonium chloride electrolyte, because it gives a somewhat greater cell capacity: a simple modification which allows the makers to charge a higher price. During the discharge process, the manganese dioxide is reduced to manganese hydroxide, and the metallic zinc electrode is either oxidized to a basic zinc manganate, or combines with the ammonium chloride to form a diammine chloride. If a zinc chloride electrolyte is used, the zinc is mainly converted into zinc oxychloride. In either case the cell becomes discharged when most of the accessible metallic zinc is consumed, or when the bulk of the manganese dioxide is reduced to an unreactive state. Similar processes occur in all of the other cells using a zinc anode, except that those using a potassium hydroxide electrolyte will ultimately convert the metallic zinc into zinc hydroxide or oxide.

The major problem with the simple Leclanché type of cell is that since zinc is consumed during the discharge process, the outer zinc tube will, in due course, become eroded, and will perforate. This is hazardous if the cell is allowed to remain *in situ* in a discharged condition, since a perforated outer shell will allow the corrosive electrolyte to creep into other parts of the equipment in which it is used, and this can be exceedingly destructive if not caught in time.

A more modern type of construction, shown in Figure 15.2, uses an extruded plastic tube, crimped over tin plated iron top and bottom caps, both to contain the cell, and to form a short-term barrier to the leakage of the electrolyte if the zinc tube should perforate. With improved purity in both the metal used for the zinc tube and the electrolyte chemicals these cells have a much greater storage life – up to 3–4 years at temperatures up to 20°C. The useful range of working temperatures for this kind of cell is effectively between 5° and 30°C, since with all of these cells the life expectancy and retained capacity are lowered above 30°C, while both the ampere/hour capacity and the maximum output current obtainable from the cell are greatly reduced below 0°C.

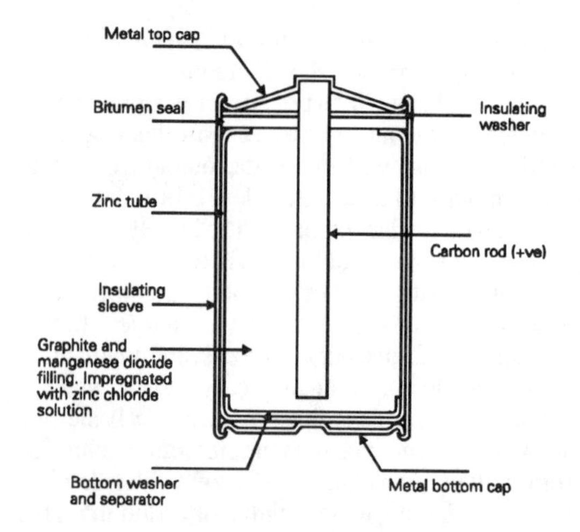

Figure 15.2 *Improved form of dry cell. The zinc chloride cell*

Another common style of this type of battery is the layer cell type, shown in Figure 15.3, in which a stack of cells – usually of four or six – made in pancake form, and individually contained in a plastics (i.e. an insulating synthetic polymer such as polypropylene or rigid PVC) tube, are mounted within a plastic lined sheet metal jacket, and internally connected to give an output voltage of 6V or 9V. This type of battery was very popular as a source of power for portable radios, but has been rather overtaken by circuit design improvements which allow operation from one or two of the less expensive cylindrical form cells.

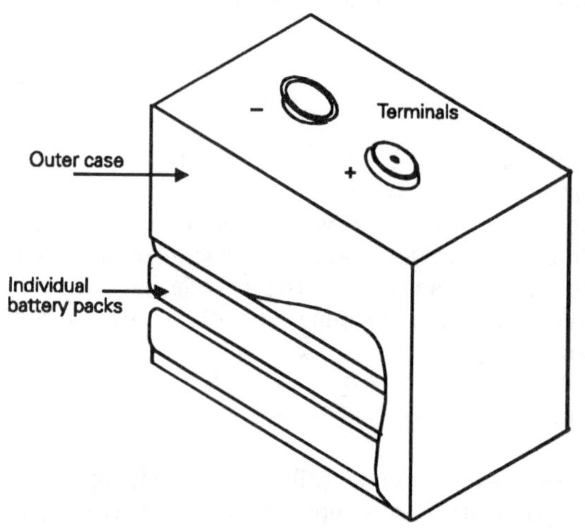

Figure 15.3 *Layer cell battery*

Alkaline manganese cells

While still a zinc/carbon/manganese dioxide system, with a nominal output voltage of 1.55V, this style of cell, shown in cross-section in Figure 15.4, gives a substantial step forward in performance in that it gives almost double the storage capacity and shelf life expectancy of the normal zinc/carbon Leclanché cell.

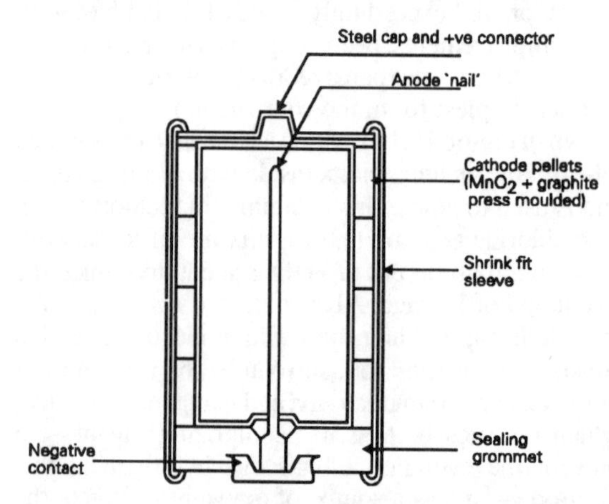

Figure 15.4 *Alkaline – manganese cell*

To the user, the most conspicuous visual difference is that the cell construction is reversed, with the positive terminal being a protruding dimple formed on the outer metal case, while the negative contact is a cup-shaped disk, insulated from the steel outer case by a plastic sealing washer.

In its construction, the extruded zinc container of the simple Leclanché cell is replaced by a thin sheet steel tube inside which, and in intimate contact with it, is a stack of cylindrical cathode 'pellets', formed under pressure from a mixture of graphite and manganese dioxide. An inner sleeve of porous insulating material separates this from the anode electrode, which is a filling of zinc powder, formed around the zinc current collecting 'nail', and made into a paste with the potassium hydroxide solution electrolyte.

From the point of view of the user, a major advantage, which offsets its greater purchase price when used with expensive electronic equipment, is that the outer steel case is resistant to the electrolyte, and the cell is therefore leak proof under all normal conditions of use. Also, because of the greater efficiency and conductivity of the potasssium hydroxide electrolyte, the internal resistance of the cell is lower, which allows greater peak output currents to be drawn for brief periods. However, apart from its greater ampere/hour capacity for a given cell size, the characteristics of the alkaline manganese cell are very similar to that of the Leclanché.

Button cells

Mercuric oxide systems

Both the zinc–carbon systems and the nickel–cadmium rechargeable cells, examined later in this chapter, tend to be relatively bulky. This is not a particular disadvantage in the case of hand torches, or most of the cassette recorders, cordless telephones or portable radios with which they will normally be used. However, the advent of miniaturized hearing aid systems, electronic wrist watches, credit card sized pocket calculators, and electronic exposure and shutter control systems in small cameras, created a demand for cells of very much smaller dimensions, and with a much more constant voltage output. Of these, one of the earliest to be exploited commercially was the mercuric oxide–zinc cell, of which the basic system had been invented in 1886 by Aron, and further developed in the 1930s by Samuel Ruben, although Ruben's interest was in the relative constancy of the output voltage of this kind of cell, rather than its capacity for miniaturization.

The method of construction of a contemporary cell of this type is shown in Figure 15.5, and consists of a cathode of mercuric oxide, with some graphite to increase its conductivity, compressed into a small flat pellet, pressed into the base of a tin or cadmium plated steel can, and separated from a high purity zinc–mercury amalgam anode by an absorbent pad containing the potassium hydroxide electrolyte. This kind of cell has an output voltage of 1.3–1.33V, depending on output current, which will remain constant up to the point at which it is almost completely discharged. This gives the sometimes disconcerting characteristic, for example in electronic wrist watches, that when they do stop, nothing short of replacing the cell will cause them to re-start.

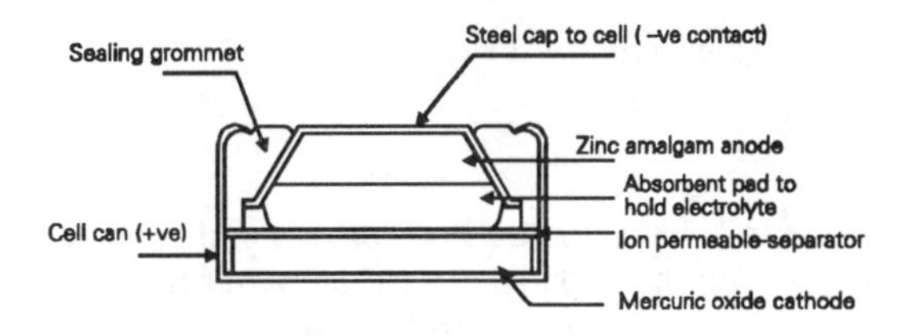

Figure 15.5 *Mercuric oxide cell*

Silver oxide-zinc cells

This is very similar in construction and application to the mercury cell, and was also derived from a design originally patented in the mid-1800s (in this case by Clarke) developed in the early years of the 1939–45 war by André, as a miniature voltage source for military applications. In this cell, the cathode is a compacted pellet of silver oxide and graphite – added to increase conductivity. Its output voltage is 1.55V, and, like the mercury button cell it has a very flat output voltage vs. discharge characteristic.

Zinc–air cells

By using atmospheric oxygen, adsorbed onto a catalytic carbon composite layer, as the cathode, more of the internal cell space can be used for the consumable zinc amalgam anode, leading to a greater power capacity for a given volume. The basic construction, shown in Figure 15.6, is similar to that of the mercury cell, apart from the fact that the bottom of the cell case is perforated to provide access for air. During storage these perforations are covered with a layer of impermeable tape, which must be removed to activate the cell. However, once the cell has been activated it will continue to discharge slowly, so long as these access holes are uncovered – a characteristic which makes such cells more suited for applications requiring continuous rather than intermittent use. The output voltage of this type of cell is 1.2–1.4V depending on load current, and it has a discharge voltage characteristic which is similar to that of the mercury cell.

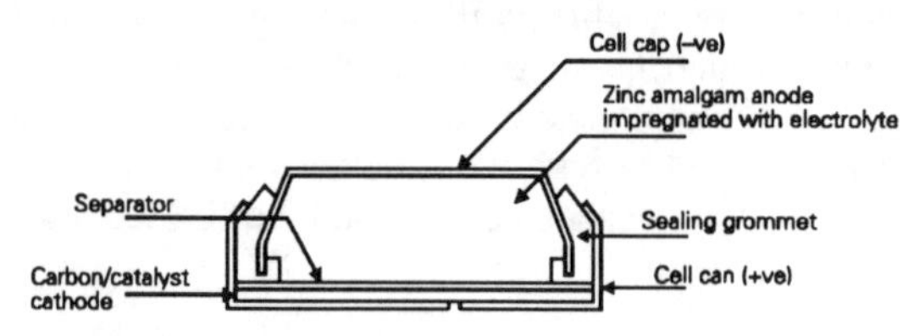

Figure 15.6 *Zinc-air cell*

Lithium cells

The search for very high power-to-weight ratios in primary cells has led to the exploration of a number of exotic electro-chemical combinations, of which the most successful have been those based on metallic lithium in combination with a wide range of oxide, sulphide or fluoride cathodes. Because lithium will react vigorously with water, it is necessary to find some non-aqueous solvent for the electrolyte, such as dimethoxyethane.

A commercially available cell (the Ever Ready type 2016 button cell) employs a manganese dioxide cathode and a lithium perchlorate electrolyte to give a 3.2V output voltage, and good storage and discharge voltage characteristics. Similar cells, such as the Crompton 'Eternacell', are available which employ a thionyl chloride electrolyte.

All of these Lithium cells are characterized by a high energy to weight ratio – up to three times better than the alkaline manganese cell – a wide operating temperature range (typically –50°C to +60°C), and an exceedingly long storage life, which can be in excess of ten years. This feature is exceedingly useful where a permanent voltage source is required in applications where only a very low output current demand is likely, such as, for example, as a back-up voltage source for 'volatile' computer memory banks, to protect against data loss during a power failure.

It is essential to avoid reverse (charging) current flow through these cells, an action which can lead to the cell exploding, so the circuit shown in Figure 15.7 is normally used in back-up applications to prevent this happening.

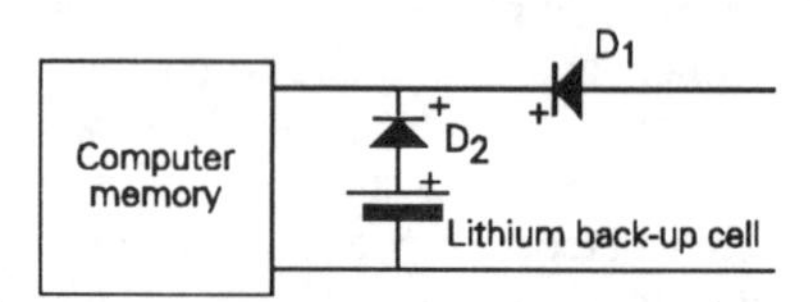

Figure 15.7 *Lithium cell computer memory back-up circuit*

Secondary cells

Those types of cell which can be recharged to restore their initial output voltage and stored energy capacity are known as secondary cells, to distinguish them from the use and throw away types, known as primary cells. Although it is possible to recharge standard Leclanché type dry cells, provided that they are not too fully discharged (see *Wireless World*, August 1981, p. 70, and Feb. 1982 p. 46), the results are not as satisfactory as in the case of cell systems specifically designed for this purpose.

Of the secondary cells which have achieved more general popularity, the major types are the lead–acid, the nickel–iron, or NiFe, and the nickel–cadmium, or NiCad cells. Of these, the lead–acid types are by far the most common, in that they are the standard source of power for starting and lighting motor vehicles.

The specification of secondary cells generally includes both their nominal output voltage and their discharge capacity, in milliampere or ampere hours. In all cells, this capacity is dependent on the effective surface area of the reactive materials on the plates, so means are taken to make this as large as practicable, so far as this is compatible with mechanical robustness and impact resistance. The discharge capacity rating is not usually quoted for primary cells, mainly because it depends on so many factors, such as operating temperature and discharge rate, but, as a general rule, the ampere hour capacity of a D size sealed lead-acid cell would be of the order of 2–2.5Ah, as compared with 4Ah for a NiCad cell, 7Ah for a zinc chloride electrolyte Leclanché type, and 15Ah for an alkaline manganese cell, when discharged intermittently at a 30mA output current.

Lead–acid systems

In the lead–acid type of cell, the electro-voltaic couple is between lead dioxide (PbO_2) and metallic lead, using a dilute sulphuric acid electrolyte, of which the specific gravity, SG, when fully charged = 1.26. This gives an output voltage – when fully charged – of about 2.25V/cell, decreasing to about 1.9V/cell when approaching the discharged state.

In a typical open cell system the positive plate electrode consists typically of a hollow lead matrix, of rectangular or honeycomb type cavities filled with lead peroxide, with a negative plate held out of direct electrical contact with the positive one by means of a porous 'separator', having a pocketed or spongy surface. The construction of such a cell is shown, schematically, in Figure 15.8. The chemical reaction which occurs during discharge is, theoretically, that the lead dioxide is reduced to metallic lead releasing oxygen, while the lead (negative) electrode is oxidized to lead sulphate, releasing hydrogen. This action consumes part of the sulphuric acid, and releases water by electrolytic recombination, so the specific gravity (SG) of the electrolyte falls during discharge, to a nominal SG when fully discharged, of 1.1. If the cell is allowed to stand for long in a discharged condition, both of the plates may become covered in lead sulphate, and the electrolyte thus further depleted of acid has an increasingly high resistance, which restricts the possible current flow, and makes recharging difficult. This is described as the battery being 'sulphated'.

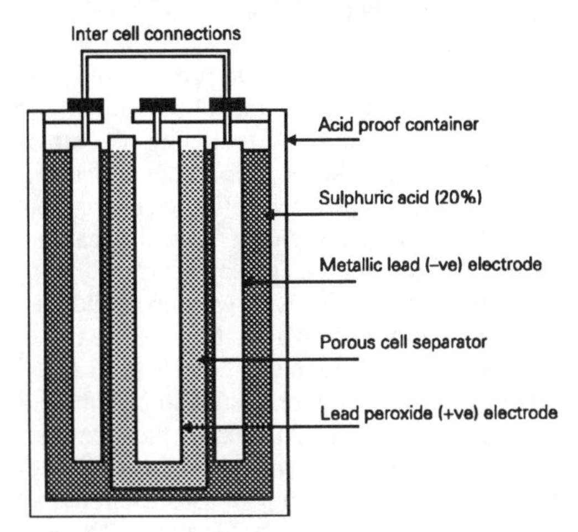

Figure 15.8 *Basic lead–acid cell*

In normal use, the ampere hour capacity of lead–acid cells is gradually reduced by the loss of the lead dioxide coating on the positive plate. However, the normal failure mechanism for lead–acid cells is that this lead dioxide, when shed by the positive plate, and then partially reduced to metallic lead, forms a conductive bridge between the plates which permits the battery to discharge even under no-load conditions. In the case of automobile batteries, one might suspect that the design of the battery is chosen to permit, or even encourage, this type of failure mechanism, to prevent the cells from lasting too long!

Ideally the level of current used to recharge a lead acid battery should be between C/5 and C/20, where C is the ampere hour rating of the cell. These limits would be referred to, respectively, as 'five hour' or 'twenty hour' charging rates. In emergencies, higher currents can be used, up to 2C, but both high charging and high discharging rates tend to accelerate the loss of lead dioxide from the positive plates. Since both the charge and discharge mechanisms are less than completely efficient, 'outgassing' (the electrolytic evolution of gaseous hydrogen and oxygen) occurs during both the charge and discharge cycles, especially at high current levels. This is unimportant in 'open-cell' car battery systems, where provision is made to allow such gases to vent to the atmosphere, with the lost water being replaced as necessary. At high charge/discharge rates the electrolyte may be lost as spray, and simple topping up with distilled water will gradually cause the electrolyte to become increasingly dilute.

In 'sealed' lead–acid cells, the electrolyte will either be held in some absorbent filling material, or as a gel, in order to lessen the possibility of leakage. The type of construction adopted will also be chosen to promote electrolytic recombination. It is probable also that the manufacturers will recommend a maximum charge/discharge current to help prevent too rapid a rate of evolution of gases.

Lead–acid cells have the advantage of a relatively high output voltage and, when new, a very low internal source resistance, but worries about possible electrolyte leakage have tended to make designers choose NiCad types, whose immunity form this type of problem has been more fully established, even though improved sealed lead–acid systems have now been shown to be exceedingly reliable and trouble-free.

NiFe and NiCad systems

The exploration of systems based on an alkaline hydroxide electrolyte, by Waldemar Jugner in Sweden, and Thomas Edison in the USA, led to the use of both nickel hydroxide/iron and nickel hydroxide/cadmium couples in open-cell batteries, with Edison's NiFe cells being employed first in the early 1920s. These were relatively inexpensive and exceedingly robust – electrically – indeed they had the reputation of being virtually indestructible, but had poor charge efficiency, which led to rapid loss of electrolyte through outgassing, and a relatively low energy density – only about 25% of the theoretical utilization of active material being practicable at that time. In recent open-cell NiFe types better energy densities of up to 40–50Wh/Kg have been obtained. The output voltage discharge curve of such cells falls from 1.3 to 1.15V at a C/6 discharge rate. A very popular application for such cells was in the 120V 'Milnes' radio HT unit, popular in the 1930–40s as a source of anode voltage in battery operated radio sets. The arrangement of the cells in this unit – individually contained in small glass jars in a wooden case – was such that they could be switched into parallel groups for recharging from a standard 6V car battery, or back again into series to provide the 120V output.

The use of this type of system in sealed cells was not easily possible, and this factor, together with the improved NiCad energy capacity, led to the greater popularity of the nickel/cadmium types, which can be designed so that there is very little evolution of gas in normal use. The construction of these cells is shown, schematically, in Figure 15.9. In this, both the negative (cadmium) and the positive (nickel) plates are formed from a microporous mass of powdered metal, with the nickel electrode being largely oxidized during manufacture to basic nickel oxide (NiOOH). The chemical reaction which occurs during discharge is that this oxide is reduced to the hydroxide ($Ni(OH)_2$), while the released oxygen diffusing through the cell oxidizes the cadmium negative plate to its hydroxide, ($Cd(OH)_2$) – a process which is fully reversible. Although water, from the potassium and lithium hydroxide solution which forms the electrolyte, is involved in this process, it is not consumed in normal use, so, in sealed cell manufacture only the amount judged to be theoretically necessary is provided.

If charging is continued beyond the fully charged condition surplus cadmium on the negative plate is able to convert the oxygen evolved on the nickel plate back to water, only heat being evolved. The warm-up of the cells can be used as an indication that they are fully charged.

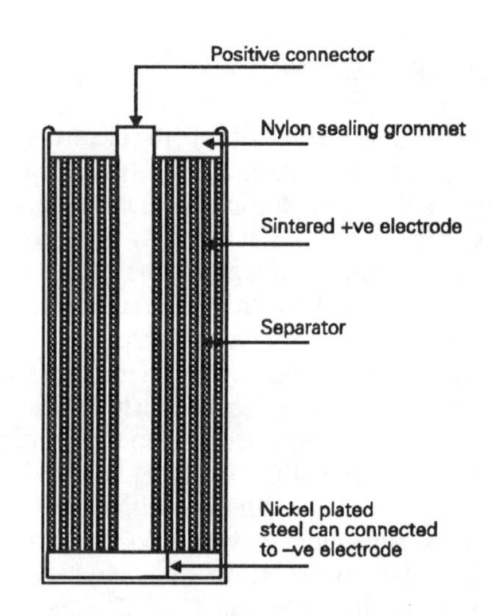

Figure 15.9 *Cylindrical form Nickel–Cadmium cell*

The major failure mechanism in NiCad cells is the loss, by electrolysis, of the available water in the system as a result of inadvertent reverse charging – a process which can happen, all too easily, if current is drawn from a partially discharged battery which contains one cell which has become more completely discharged than the others, but can also occur if the cell is excessively over-charged.

A further mechanism to which more gently used NiCad cells seem particularly prone is the growth of fine threads of conducting metallic cadmium – known as 'dendrites' because of their tree-like structure – through the porous separator between the plates, which not only causes the cell to discharge, but may also prevent its being charged again because of the internal shortcircuit which is present.

These failure mechanisms were examined in detail by Cooper (*Wireless World*, May 1985, pp. 61–63, June 1985 pp. 60–63, and July 1985, pp. 32–36.), and the means of restoring internally short-circuited cells to working order, by subjecting them to bursts of charging current large enough to fuse the cadmium dendrites is examined by Johnson (*Wireless World*, February 1977, pp. 47–48). On the credit side, such cells are relatively light and mechanically robust, largely trouble free in normal use, and are capable of very high output currents over brief periods, if needed. Indeed, a very successful range of portable soldering irons has been marketed, powered by NiCad cells housed in the handle, which supply a current to the low resistance heater element which is high enough to bring the bit up to operating temperature within a few seconds.

Various charging systems have been suggested for NiCad cells, but the most straightforward is to recharge them separately at a fixed voltage of 1.42V/cell, when damage due to overcharging is not possible. Where fixed voltage charging is not feasible, current charging rates in the range C/5 to C/10 are usually recommended – preferably combined with a cell temperature rise cut-out mechanism.

The output voltage discharge curve is very flat, falling from 1.22V to 1.19V at the threshold of discharge, at a discharge current of C/5. Immunity to self discharge is reasonably good in modern cell designs, though less satisfactory in sintered plate systems.

Other rechargeable cells have been devised, based on, among other possibilities silver–zinc (1.5V), silver–cadmium (1.1V), and Nickel–Zinc (1.8V) types, but these remain relative rarities.

I have illustrated in Figure 15.10 the relative output voltage discharge curves for the more common primary and secondary cell types, scaled to represent the performance which would be given by cells having an equivalent physical size, and discharged at the same C/*x* current flow.

I would like to express my gratitude to the Ever Ready and Duracell battery companies for their generous provision of technical information on this subject.

Mains operated power supplies

Simple transformer–rectifier systems

The need to provide a stable, reasonably noise and ripple free DC output from a 50 or 60Hz AC power supply line is capable of being met by a wide range of circuit layouts, but, in general, the designer must first decide how good a quality of output DC is needed by the system to be powered, and to what extent a more expensive or complex power supply layout would be justified by circumstances.

In its simplest form, a DC output can be provided by any of the transformer/rectifier/capacitor layouts shown in Figure 15.11. A point which must always be borne in mind, in any rectifier/reservoir capacitor layout, is that, on light load, the output voltage will be 1.414 times the RMS AC output voltage provided by the transformer secondary winding. This means that the energy output for which the transformer is rated, in volt–amperes, usually abbreviated to VA, must be down rated by a factor of, at least, 1.414 to allow for the fact that the output current from the power supply is being drawn, effectively, at a higher voltage than the secondary RMS output value. This means that, in the case of the simple circuit layout of Figure 15.11a, if the secondary winding is rated at, say, 1A, the maximum output current which can be drawn from the supply will be 0.707A, even though, under load, the resistive and core losses in the transformer, and the inevitable forward (conducting), voltage drop across the rectifier diode, will mean that the voltage actually developed across the reservoir capacitor, C_1 will actually be a good bit less than 1.414 V_{out}. Also, in

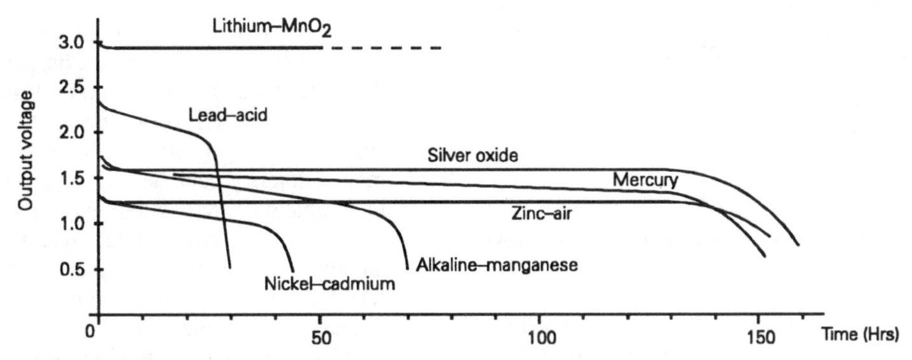

Figure 15.10 *Voltage output and energy capacity for common cell types*

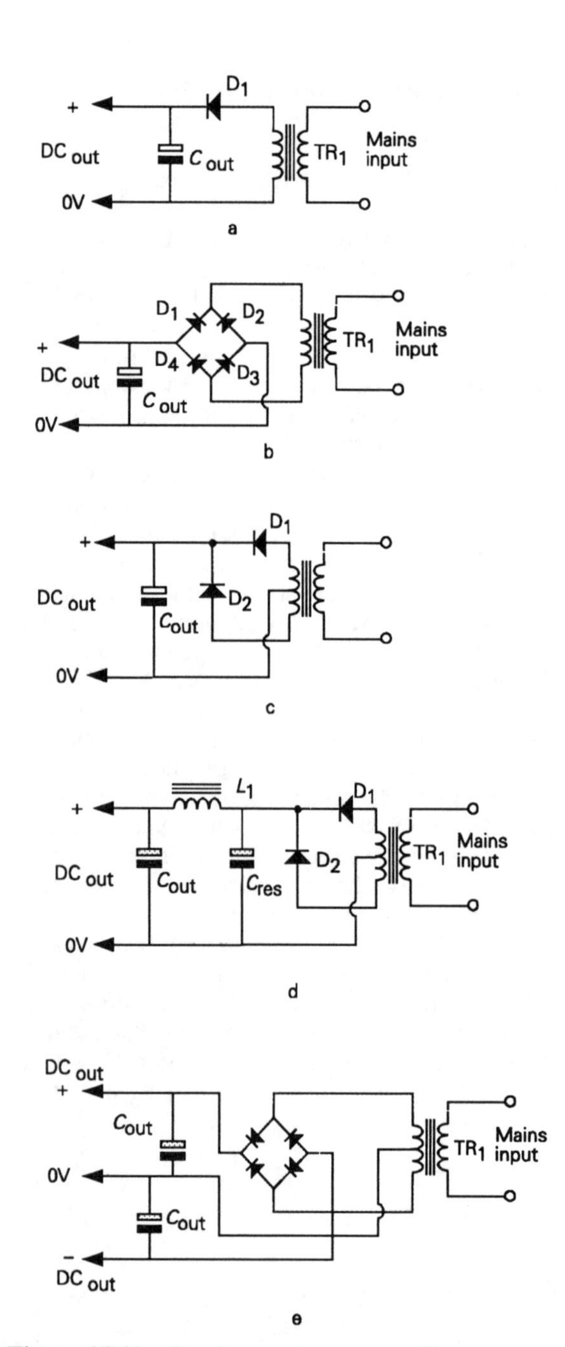

Figure 15.11 *Simple transformer–rectifier power supplies*

practice, the maximum output current will be reduced even further because the current drawn from the secondary winding of the transformer will take the form of short duration high-current pulses, as shown in Figure 15.12. It must be remembered that the core and winding resistance losses are proportional to I^2 rather than I, so the high output current pulses drawn from the secondary winding will cause a disproportionately high transformer dissipation.

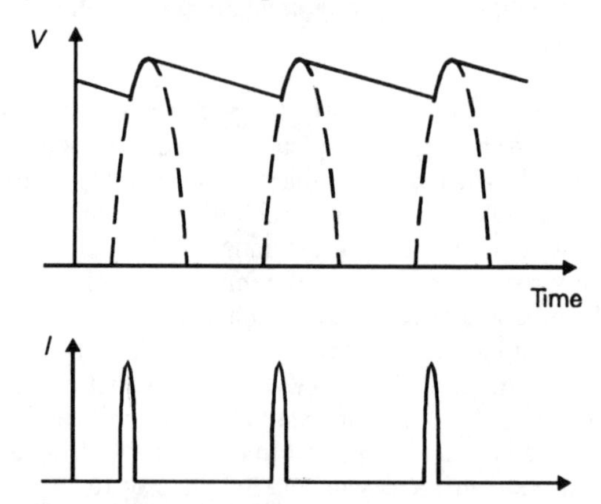

Figure 15.12 *Voltage and current waveforms – half-wave rectification*

The allowance which must be made for this effect cannot be precisely specified because it depends on both the transformer core characteristics and the size of the reservoir capacitor employed – the larger the capacitance of this the shorter in duration and the larger in magnitude the repetitive charging current pulses will be. A working rule of thumb is that the DC output current (I_{dc}) should not exceed 0.65 I_{ac}. This leads to the need for a design compromise, in that, if the DC output is to be drawn directly from the reservoir capacitor, as is increasingly the case with modern transistor and IC power supplies, the desire to keep the output ripple level low would urge the use of a large value reservoir capacitor, but, because this will cause the current drawn from the mains transformer secondary to be reduced to progressively shorter reservoir capacitor charging pulses, this worsens the transformer efficiency and heat dissipation. However, the inevitable secondary winding resistance, coupled with core saturation effects, means that, in practice, there is an upper limit of reservoir capacitor size, beyond which no further benefit will be obtained.

The 'choke-input' filter system shown in Figure 15.13 makes much better use of the potential power output from the transformer, but the DC output voltage developed across the reservoir capacitor, C_1, will always be somewhat less than the transformer RMS secondary voltage. Also, because of the need for a gapped core 'swinging' choke – a somewhat rare type of component, which will add to the bulk and component cost of the circuit – this type of power supply system is rarely found in low to medium output power designs.

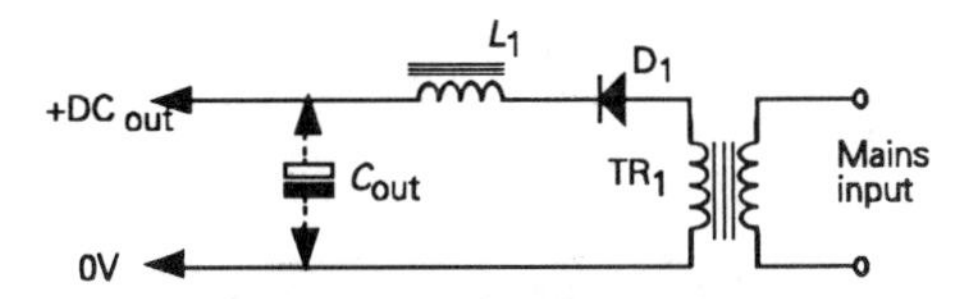

Figure 15.13 *Choke-input filter system*

Where just a single line or a pair of +/– supply lines are needed, the most common types of circuit are those shown in Figures 15.11b, c, d and e, where 'full-wave' rectification is employed to double the frequency of the charging pulses fed to the reservoir capacitor, and nearly halve the magnitude of the residual sawtooth supply line ripple. A comparison which is illustrated in Figure 15.14.

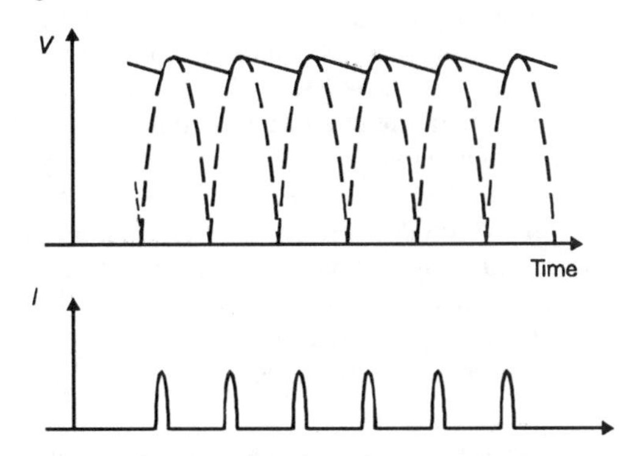

Figure 15.14 *Voltage and current waveforms. Full-wave rectification*

Even with full-wave rectification, and a large value reservoir capacitor, there will always be a significant amount of (100/120Hz) ripple on such a DC supply output, and it is inevitable that the mean output voltage will fall and the ripple content will worsen as the output current demand is increased. However, such simple systems are very widely used and, with care in the design of the circuitry to be used with them, are often quite adequate for their purpose.

In valve operated electronic systems, where relatively high DC supply line voltages were used, and the DC current demand was therefore proportionately lower, it was conventional practice to insert a 'smoothing choke' (L_1), between the reservoir capacitor (C_1), and the output 'smoothing capacitor' (C_2), as shown in Figure 15.11d. This converts L_1/C_2 into a –12dB/octave *LC* type low-pass filter and substantially reduces the amount of residual sawtooth HT ripple present in the output circuit: a ripple voltage which is caused by the way in which the rectifier/reservoir capacitor charging mechanism acts.

However, with high current, low voltage, power supplies the additional DC resistance introduced into the output path by a series choke of adequate inductance to be useful would be undesirable, and the use of such series connected chokes has become a rarity; particularly since output voltage smoothing can be provided in a much more compact and economical way, if needed, by the use of a voltage regulator circuit.

Voltage regulator systems

It is often necessary to supply the operating circuitry with a substantially constant, and ripple free, DC voltage – a requirement which cannot be met by any simple mains transformer/rectifier layout. Such improved supply stability can be provided by the use of some kind of stabilizer or regulator circuit, and these can be divided into 'shunt' and 'series' systems.

Shunt regulator layouts

The simplest example of this type of system is the zener or avalanche diode circuit shown in Figure 15.15. (*Note.* Although true zener diodes are only those with a turn-over voltage of about 5V or below, the term zener has come to be used as a generic term for all reverse breakdown regulator devices.)

Such a layout is called a shunt regulator because the voltage sensitive component/s are connected so that they will draw current in parallel with the load, and will increse their current flow if the load current is reduced or the output voltage tends to increase.

In the circuit of Figure 15.15, the input supply voltage, V_{in}, the value of the series resistance, R_1, and the power handling capacity of the zener diode, D_1, are chosen so that the required output voltage will be

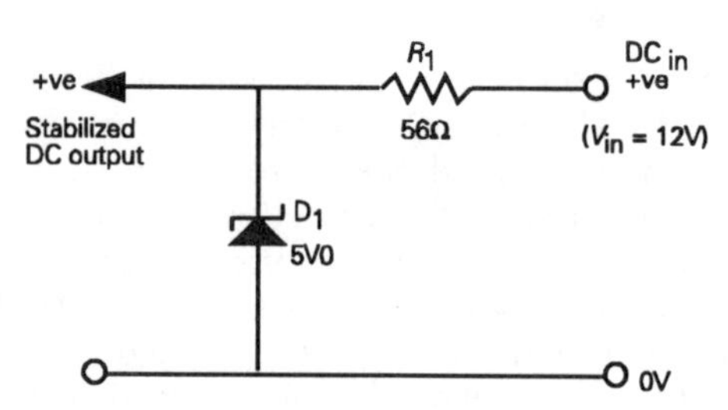

Figure 15.15 *Simple zener diode shunt regulator system*

provided, without exceeding the permitted dissipation of the diode, at zero output current demand, but so there will always be an adequate current flow through the diode at the maximum likely load current. The output source resistance is determined by the slope of the diode voltage, V_d, when plotted against diode current, I_d, as shown in Figure 15.16. For example, if the diode characteristics are such that an increase in current of 100mA causes an increase in the voltage developed across the diode of 0.1V, the effective source resistance will be one ohm, the attenuation of the input voltage ripple, for the circuit values quoted in Figure 15.15, will be 50:1, and the required dissipation for D_1 will be 0.6W. Unfortunately, for reasons examined in Chapter 4, both zener and avalanche diodes generate a small amount of wideband noise, so it is probably worthwhile to connect a capacitor, C_1, in parallel with the diode output to reduce somewhat this noise contribution.

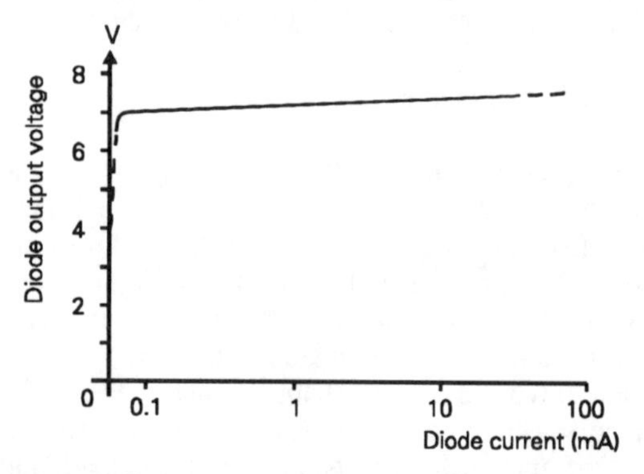

Figuer 15.16 *Typical zener diode regulation characteristics*

For low voltages a simple forward biased silicon junction diode is usable as a reference source, and will give an output voltage between 0.55V and 0.65V, depending on diode current, with a very low electrical noise contribution. Several such diodes could be connected in series, as shown in Figure 15.17, to give, for example, a 2.4V DC source.

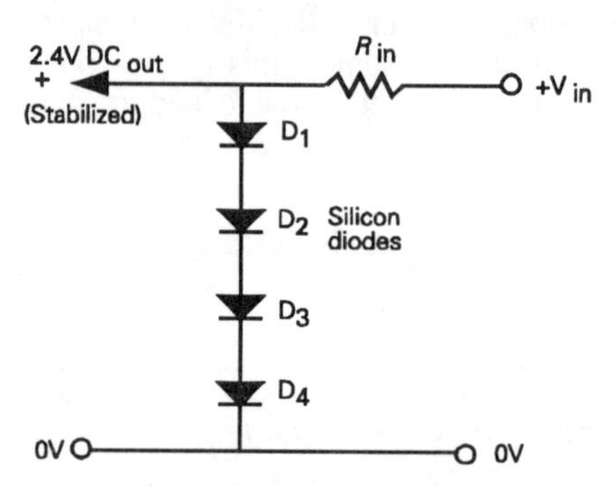

Figure 15.17 *Forward biased diode chain*

The output characteristics of the simple zener diode circuit of Figure 15.15 can be substantially improved by connecting the diode in the base circuit of an amplifying transistor, as shown in Figure 15.18. This also allows a low–power diode to be used, which is useful since medium to high power transistors are usually cheaper than zener diodes having the same dissipation ratings. Also, since voltage regulator diodes, with turn-over voltages above some 5V, have a positive temperature coefficient of output voltage – depending somewhat on diode current for a given junction area, as shown in Fig 15.19, – while normal silicon junction diodes, such as the base–emitter junction of a transistor, have a negative temperature coefficient, connecting these in series, as in Figure 15.18, can partially compensate for the positive temperature coefficient of the diode.

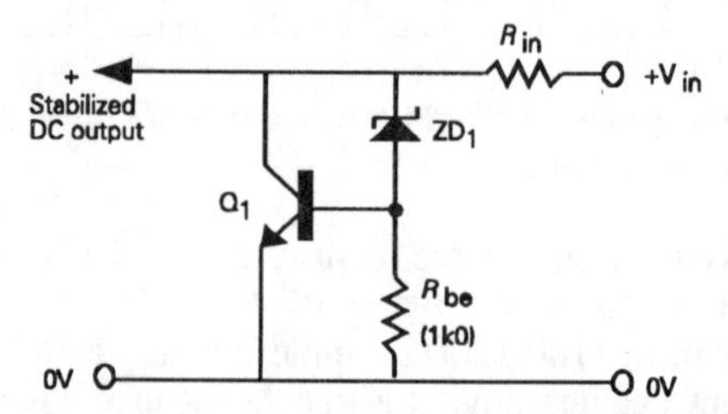

Figure 15.18 *Improved zener diode shunt regulator*

Another simple shunt regulator layout is the so-called 'amplified diode' circuit shown in Figure 15.20. In this the transistor conducts if the voltage developed

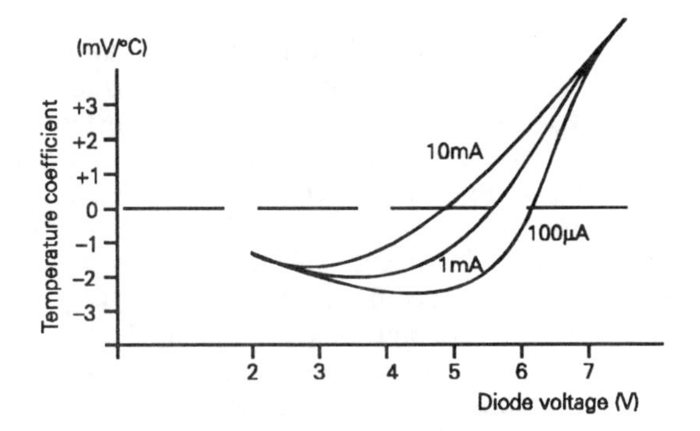

Figure 15.19 *Temperature coefficient of zener diode regulator*

across R_3 exceeds some 0.55–0.6V, giving an output voltage of $V_{be}(R_2+R_3)/R_3$, which is adjustable, within limits, by the choice of values of R_2 or R_3. This type of layout is commonly used as a method of stabilizing the forward bias on the output transistor pair in class AB audio amplifiers, where Q_1 will often be mounted in thermal contact with the output transistor heat sink, in order to provide a measure of operating temperature compensation.

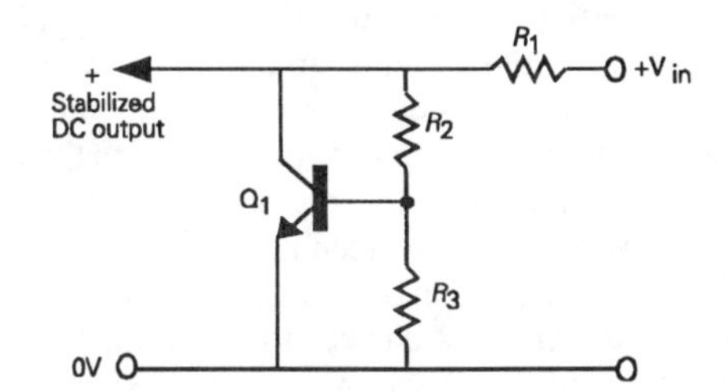

Figure 15.20 *The amplified diode circuit*

More elaborate shunt regulator systems, of the type shown schematically in Figure 15.21 – in which the output voltage is compared with a precison voltage reference source, and the difference voltage is amplified and used to control a shunt transistor load – can be used to provide improved regulation characteristics, but this layout is only likely to be used where the difference between the available input supply voltage and the required output voltage is so small that a series regulator circuit would be impracticable.

Series voltage regulator systems

This type of circuit layout is one in which a series

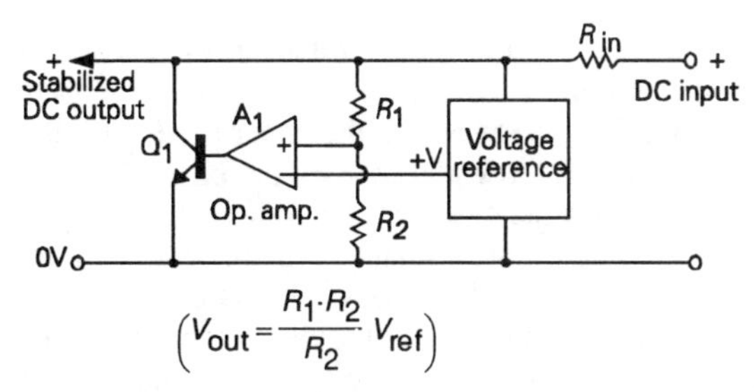

Figure 15.21 *High stability shunt voltage regulator circuit*

transistor, or equivalent device, is interposed between the input voltage source and the output point, so that, if the output voltage tends to rise above, or fall below the target figure, then the current flow through the series (pass) transistor can be adjusted by some external circuit so that the required output voltage level is maintained.

The simplest feasible circuit for this purpose is one in which the base of the pass transistor is fed from a fixed voltage source, such as from a simple zener diode voltage regulator, as shown in Figure 15.22. If the output voltage falls, the base–emitter voltage applied to Q_1 will increase, and so will the current through Q_1. If the output voltage tends to increase, then both the forward bias on Q_1, and the current flow through Q_1 will decrease.

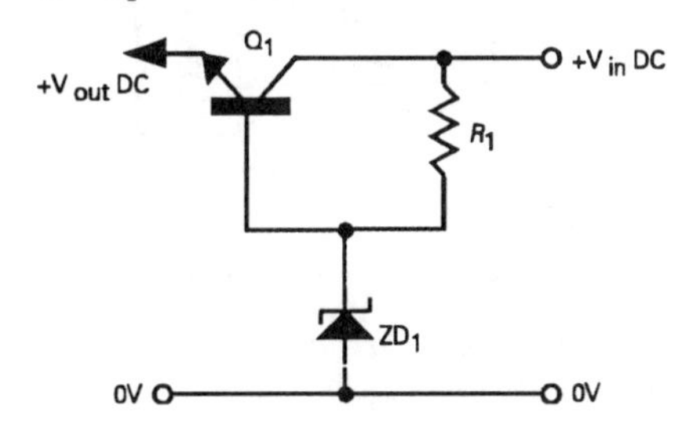

Figure 15.22 *Simple zener regulated series stabilizer*

There are three main snags with this circuit, of which the first is that R_1 must be able to pass a sufficient current to meet the base current demand of Q_1 at the required maximum output current demand, plus whatever minimum current is required to cause the zener diode to operate. Secondly, the value of R_1 cannot be made too low, or otherwise the desired attenuation of the input noise and ripple will be greatly lessened, and this means that the input voltage must exceed the required output voltage by an adequate amount. Thirdly, the voltage across the zener diode will drop as more current is withdrawn by the base of the pass transistor, and the base–emitter voltage drop of Q_1, as the output current demand is increased, will also become larger. Both of these effects operate in the same sense to reduce the output voltage provided by the circuit, and this gives the circuit a relatively high output (source) impedance.

A more conventional series regulator circuit will normally use the circuit layout shown in Figure 15.23, in which the base current of the pass transistor is derived from the output of a voltage amplifier, whose input circuit is connected so that it is able to sense the difference between its input voltage, – usually related to the regulator output voltage – and some stable and noise-free voltage reference source. If the output voltage from the regulator should vary, the amplifier will operate to oppose this variation. By the use of an error amplifier having sufficient gain a high degree of output voltage stabilization, and a very low output impedance, typically less than 0.02 ohms, at frequencies up to a few kHz, can be obtained.

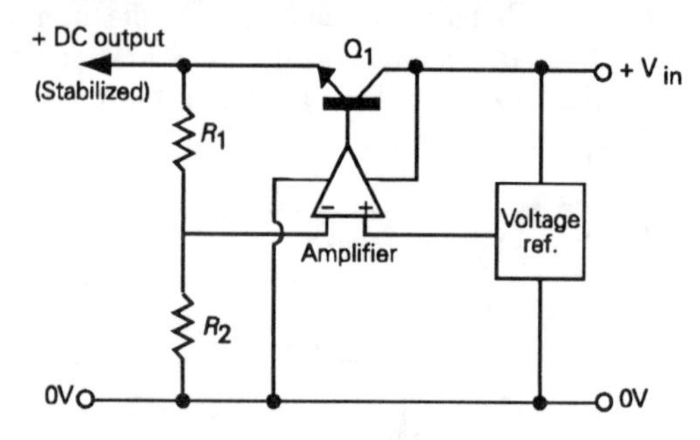

Figure 15.23 *Schematic layout of conventional series stabilization circuit*

The major drawback with this type of layout, in which the pass transistor is connected as an emitter follower, is that enough voltage difference must exist between the available input potential and the required output voltage for the amplifier to be able to drive the pass transistor adequately. The minimum input–output voltage requirement is commonly termed the 'drop out' voltage, and, with a well designed circuit can be as low as 2–3V.

The problem of the need to allow enough voltage drop in the circuit to provide forward bias for the transistor can be lessened by using the transistor in an inverted mode, in which the output current is drawn from its collector, so that the forward operating bias is derived from the potential difference between the input supply lines, as shown in Figure 15.24. By comparison with the circuit design of Figure 15.23, the intrinsic output impedance of this layout is higher because the load is fed from the collector of the pass transistor, which is a high impedance point. By the use of a high system gain, however, output impedances below 0.05 ohms can still be obtained.

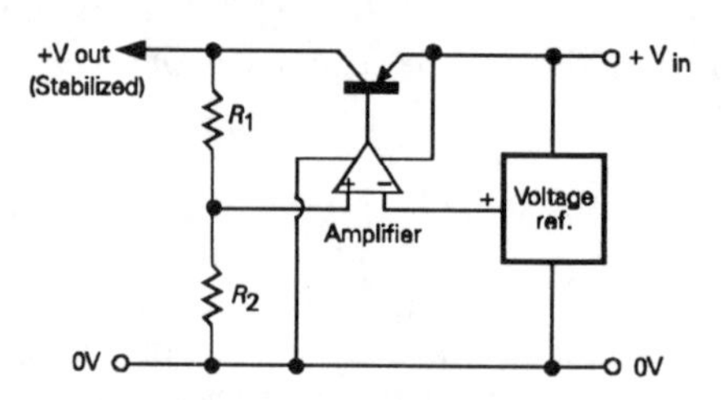

Figure 15.24

Over-current protection systems

A problem which is inherent in any series regulator system is that, in the event of an output short circuit, a very heavy instantaneous current could flow from the power supply reservoir capacitor through the pass transistor into the output. This will nearly always destroy the series device, since no simple fuse could operate rapidly enough in practice to rule out this type of failure. The simplest answer to this problem is to incorporate some rapid acting current-sensing circuit in the feed path to the pass device, which can be arranged either to 'steal' the input current to this transistor, as shown in Figure 15.25a or to disconnect its drive, as shown in Figure 15.25b. Frequently, such a current limiting circuit will be arranged to be 're-entrant' in its operating characteristics, as shown graphically in Figure 15.26, so that the output current which the pass device is allowed to deliver is related to the voltage existing across this device, which will limit the thermal dissipation of this device in the event of a sustained short circuit. This will also prevent the possibility of 'secondary breakdown' – a failure mechanism which operates if the boundaries of the permitted area of current flow and collector voltage, shown for a bipolar power transistor in Figure 15.27, are exceeded.

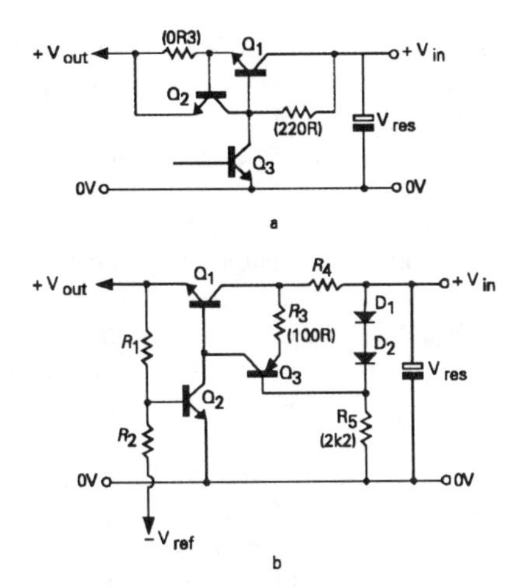

Figure 15.25 *Over-current limiting circuits (I_{max} = 2A)*

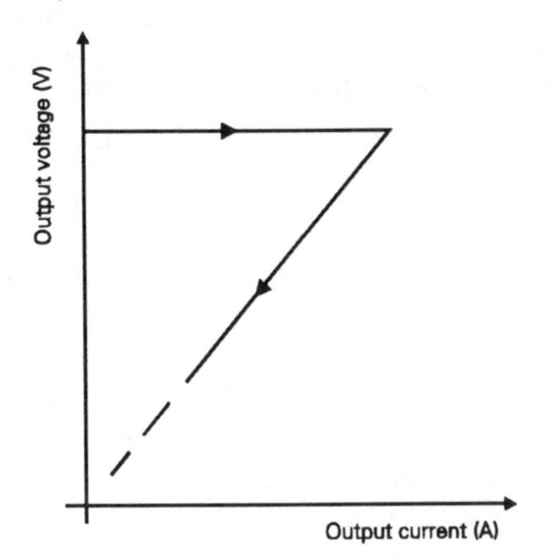

Figure 15.26 *Re-entrant over-current protection*

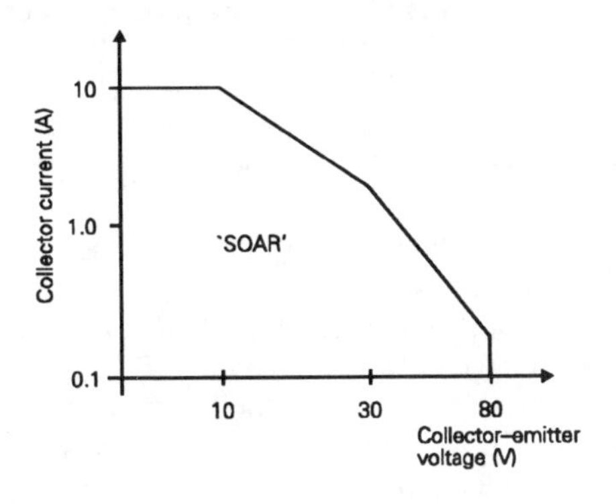

Figure 15.27 *Typical power transistor safe operating area rating (SOAR)*

Practical series regulator circuitry

Some simple series voltage regulator circuits, which could, if necessary, incorporate current limiting systems, are shown in Figures 15.28–15.31. All of these require some sort of voltage reference arrangement with which the output voltage from the regulator circuit can be compared. The simplest traditional type of layout used for this purpose is that shown in Figure 15.28a, where an amplifier transistor, Q_1, is connected with its base fed from a potential divider (RV_1 and R_1), connected across the output of the regulator, while the emitter of Q_1 is held at a fixed potential by the zener diode, D_1. If the output voltage tends to increase, the current flow through R_2 will increase, and the base voltage of Q_2 will be lowered: an action which will reduce the regulator circuit output voltage again.

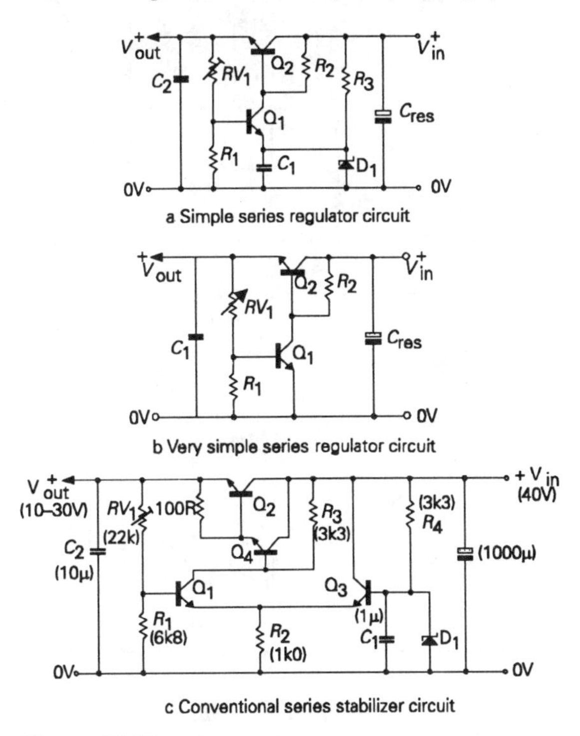

Figure 15.28

The most immediate practical problems with this layout are that, while the output voltage can be adjusted by altering RV_1, the lowest possible output voltage is $V(D_1) + V_{be}, (Q_1)$, and it is not possible, with any value of RV_1, to adjust the output voltage down to

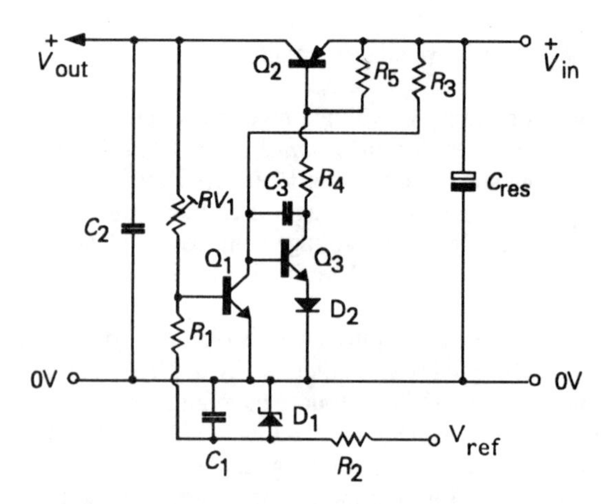

Figure 15.29

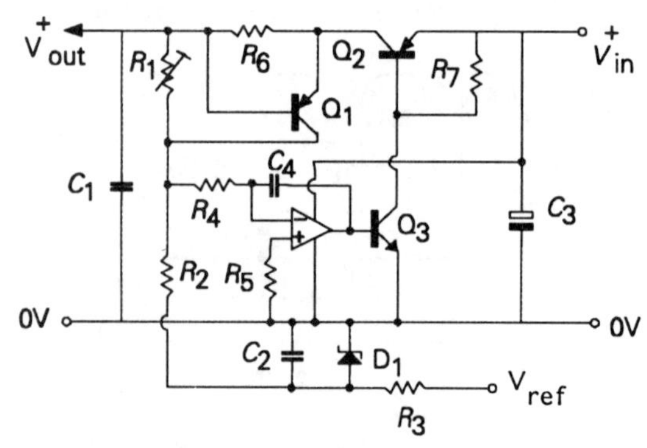

Figure 15.30

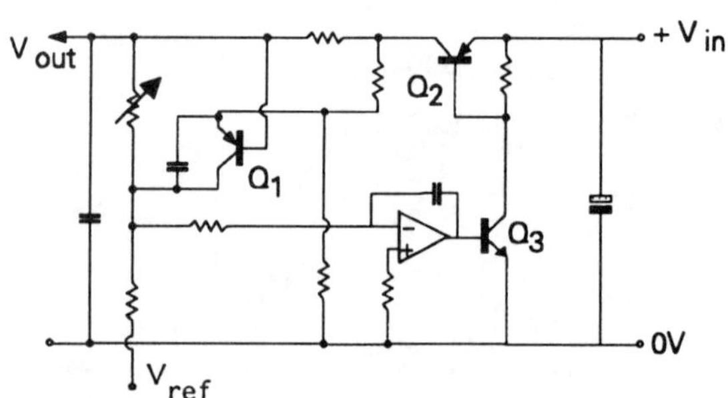

Figure 15.31 *Elaboration of circuit Figure 15.30 to provide re-entrant current limiting*

zero. This circuit can be further simplified by dispensing entirely with the zener reference source, as shown in Figure 15.28b, where the forward base–emitter potential of Q_1 is used as the reference. This arrangement works quite well, and allows the output voltage to be adjusted down to about 0.6V, but its regulation characteristics are poor because output voltage errors are also attenuated by the output potential divider (RV_1 and R_1), which limits the ability of the circuit to correct these.

The output voltage control performance of the circuit can be improved by elaborating the input error comparator into a long-tailed pair configuration, as shown in Figure 15.28c, where the zener diode has only to supply the base current of Q_1, and the emitter impedance of Q_1 is lowered by the emitter follower action of Q_3.

All of the circuits shown in Figure 15.28, suffer from the snag that the total base current required by Q_2, which is $I_{max.out}/H_{fe}$, and can be quite large, must be supplied by R_3. This leads to a voltage drop across R_3, which, together with the necessary base–emitter potential for Q_2, determines the minimum drop-out voltage for the circuit. If R_3 has a large resistive value, this drop-out voltage will be large. If, on the other hand, R_3 is made small to lessen this problem, the gain of the amplifying transistor, Q_1, will be lowered, and the output voltage regulation factor, and the output impedance of the circuit, will be worsened. Moreover, at low or zero output currents from the circuit, all of the current flowing through R_3 must be drawn off by Q_1, so the current handling capacity of Q_1, and its thermal dissipation characteristics, must be adequate. In the case of the simpler circuit of Figure 15.28a, under zero output current drain, the bias current not required by Q_2 must also pass through the zener diode, so it too must have adequate current handling capability.

The use of a Darlington connected pair, as shown in Figure 15.28c, will reduce the current handling capacity necessary for the control voltage handling stages, which will improve their gain, and the performance of the circuit. The snag, here, is that this will also tend to increase, somewhat, the drop-out voltage of the pass transistor.

An alternative layout, using the pass transistor in its inverted mode, is shown in Figure 15.29. In this circuit, the input potential divider, RV_1/R_1, is taken to a negative voltage reference source (R_2,D_1,C_1), so that, if the output voltage from the regulator circuit is inadequate, Q_1 is cut off, and the current flow through R_3 causes both Q_2 and Q_3 to be turned hard on. Since the output voltage under these conditions will exceed that which is required, the potential now present at Q_1

base will cause Q_1 to conduct, and draw off current which would otherwise flow through R_3 into the base of Q_3, causing the output voltage to be held at the required level.

Although the pass transistor is now operated in its common emitter, rather than its common collector (emitter follower) mode, and therefore offers a relatively high output impedance, all three of the transistors are operated as amplifiers, which gives a high regulator circuit open-loop gain, which reduces the circuit output impedance.

Advantages offered by this type of circuit are that the drop-out voltage can be very low, since the forward bias for the pass device is drawn from the 0V line; that the zero output current flow through Q_3 (and through the base–emitter path of Q_2) will be zero, so that the dissipation of Q_3 needs only to be that required by the base current of Q_2 to provide the regulator circuit output current; and that the output voltage can be adjusted down to some 0.6V, as in Figure 15.28b, but without the penalty of the relatively poor regulation factor offered by this layout.

An elaboration of this layout, due to the author (JLH. *Wireless World*, Jan. 1975, pp. 43–45), uses an operational amplifier in place of Q_1, which allows a further improvement in the precision of the output voltage control, and is shown in Figure 15.30. In this the output voltage is adjustable down to within a few millivolts of zero volts, and the drop-out voltage – the minimum practicable regulator voltage drop between supply and output potentials – is less than 1V. This circuit also uses Q_1 in a current limiting arrangement, in which, on excessive output current demand, the power supply output voltage is progressively reduced, making it operate as a constant current source, but with the same degree of output noise and ripple rejection as it would possess in its normal constant voltage mode. The circuit can be made to operate in a re-entrant mode, if required, by connecting Q_1 emitter to a tap on a further potential divider, as shown in Figure 15.31, or, in the case of a more conventional circuit, as is shown in Figure 15.32.

Voltage reference systems

The use of a simple zener diode, or merely the forward base–emitter turn-on voltage for a junction transistor, have been employed, in the circuits so far described, as the reference voltage against which the regulator

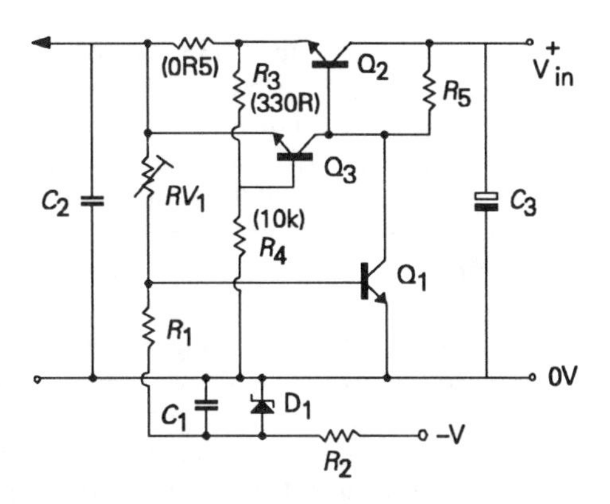

Figure 15.32 *Elaboration of circuit of Figure 15.28b to provide re-entrant current overload limiting*

output voltage is compared to establish the required output level. All of these, however, are somewhat temperature sensitive, though it is possible to temperature compensate an avalanche-type zener diode (i.e. one having a breakdown voltage in excess of some 6V), which will have a positive temperature coefficient, by connecting it in series with one or more forward biased silicon junction diodes, which will have a negative temperature coefficient. Indeed, temperature compensated zener diodes, such as the 1N821/1N827, are available, with a typical reference voltage of 6.2V, and a temperature coefficient as low as 0.001%/°C.

An alternative approach, and one which is used to provide a temperature stable voltage reference in almost all of the integrated circuit voltage stabilizer blocks, is to use a 'band-gap' arrangement.

The band-gap voltage reference circuit

The use of a temperature compensated zener diode, with an output voltage of, say, 6.2V, would lead to some design difficulties in voltage stabilizers which were required to have an output voltage of, say, 5V. Also, all zener diodes have a significant noise component in their output voltage, and this would also appear in the output voltage of the regulator. For these reasons, the voltage against which the regulated output voltage is compared is normally derived from a band-gap reference system. This type of circuit derives its name from the energy-band gap of the semiconductor material, at 0°Kelvin, which, for silicon, is 1.205V.

If, as in the circuit of Figure 15/33, two identical transistors, Q_1 and Q_2, are operated at substantially different currents – perhaps so that that through Q_1 is ten times greater than that through Q_2 – there will be a voltage developed across R_2, in the emitter circuit of Q_2, which is equal to the difference between these two base–emitter voltages (defined as ΔV_{be}). This potential difference has a positive coefficient of voltage against temperature, and provides a means for generating a reference voltage which has a near zero temperature coefficient. The way by which this can be done can be seen from Figure 15/33. If the current gain of Q_2 is sufficiently high that its base current flow can be neglected, the voltage developed across R_3 will be $(R_3/R_2)\Delta V_{be}$. The total voltage across the circuit, from Q_3 collector to the 0V. line, will be, if Q_3 is identical to Q_1 and Q_2, $V_{out} = V_{be} + (R_3/R_2)\Delta V_{be}$ and if these two component voltages – one with a positive and one with a negative temperature coefficient – add up to the band-gap potential of 1.205V, then the output voltage will have a zero temperature coefficient.

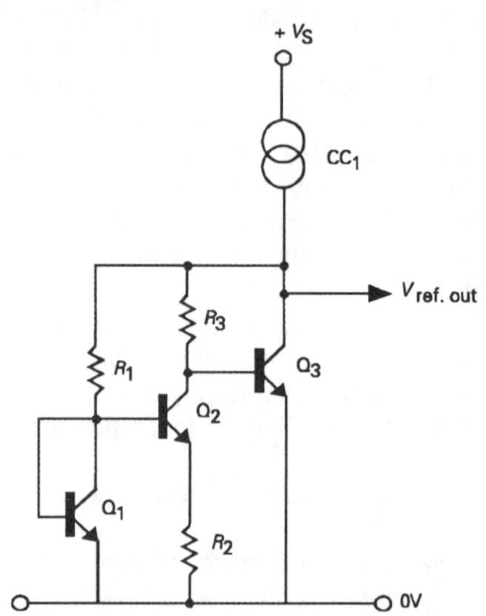

Figure 15.33 *Band-gap voltage reference circuit*

IC voltage regulators

The requirement for a zero temperature coefficient reference source is particularly critical in IC voltage regulators, since the voltage reference circuit will be contained within the IC package, and this package may become hot in use. A further point is that since all the silicon junctions are forward biased they will generate very little electrical noise. Also since the breakdown voltage of a zener diode is critically dependent on the doping level, and a selectively doped region of an IC chip would be inconvenient to arrange, and difficult to control to the required degree of precision, from the point of view of IC manufacture the more predictable behaviour of circuitry using only forward biased junctions is much to be preferred.

The actual circuit layout used in most positive output IC voltage stabilizers is of the general form shown in Figure 15.34 – which can be compared with the discrete component layouts illustrated in Figure 15.28. However, various circuit artifices are used to obtain perfomance refinements, such as the use of a low voltage drop constant-current source to feed the Darlington-connected output pass transistor pair to reduce the drop-out voltage, and to reduce the regulator IC output current if the chip temperature approaches its maximum permitted working level.

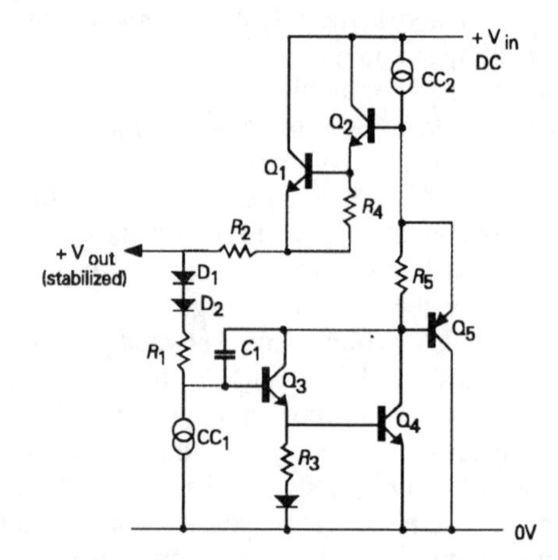

Figure 15.34 *Typical +ve input IC voltage regulator schematic circuit*

For those who are interested, a detailed explanation of the way in which these internal circuit arrangements operate in a typical IC voltage regulator is given in *Wireless World*, March 1982, pp. 41–44.

The complete circuit of the LM 109 positive output three-terminal voltage regulator IC is shown in Figure 15.35. Although this IC is designed to provide a fixed +5V output, its output voltage can be varied, above this figure, up to some +22V, by the use of the circuit shown in Figure 15.36 – an arrangement which is

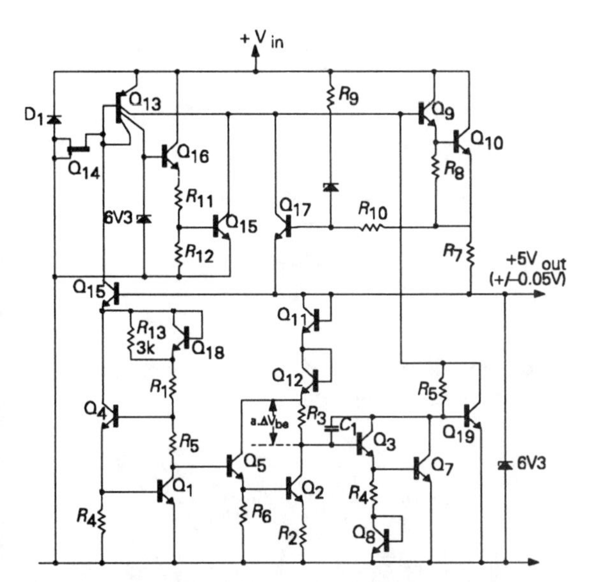

Figure 15.35 *Circuit layout of LM109 positive voltage regulator*

commonly used in adjustable voltage regulators, such as the LM317.

The use of adjustable output voltage regulator ICs in normal power supply applications, as distinct from their use as zener diode substitutes, has been made largely unnecessary by the availablity of both +ve and –ve output fixed voltage regulators covering the range 5–24V, of which the most common types are those

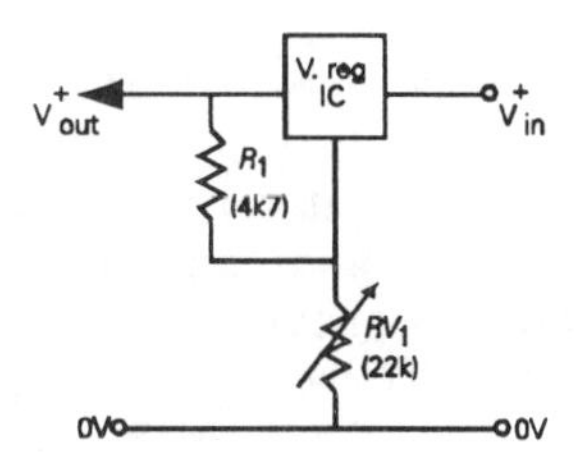

Figure 15.36 *Method of adjusting output voltage of fixed voltage regulator IC*

derived from the Fairchild 78xx and 79xx series. These have now become so widely adopted as power supply components that discrete transistor voltage regulators are now only justifiable where output voltage and current requirements are outside the range available from IC systems.

Because of the difficulty of fabricating good quality PNP transistors in ICs, the negative output voltage regulators, such as the μA79xx devices, employ a pass transistor layout which is similar to that shown in Figure 15.29, with the output taken from the pass transistor collector. The band-gap voltage reference circuit still uses NPN transistors, referred to the –ve line, with a DC amplifier (Q_6, Q_{11}, Q_{14} and Q_{17}) employed to invert the DC control voltage for application to the pass devices. The circuit of the μA79xx voltage regulator is shown in Figure 15.37. The use of the modified pass transistor configuration gives the 79xx type regulator an improved drop-out voltage (1.1V) as compared with that ot the 78xx positive rail

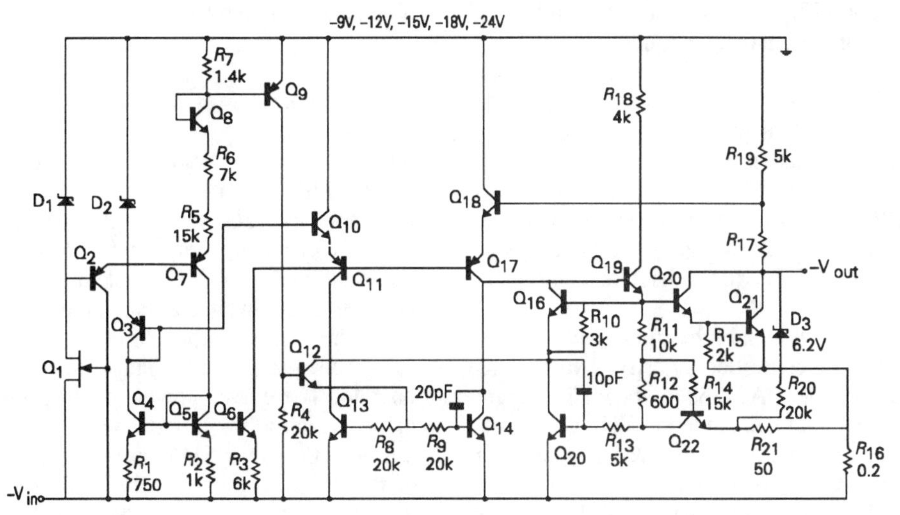

Figure 15.37 *Fairchild 79xx negative voltage regulator circuit*

regulator IC. The output noise figure (125–375μV) is, however, worse than that of the 78xx device (40–90μV), with the lower output voltage ICs having better noise figures than the higher output voltage equivalents. If a very low output noise figure is desirable, it is worth remembering that the low output current ICs, such as the 78Lxx and 79Lxx devices have a noise output of only about one third that of their higher power equivalents.

Increased voltage and current ratings for IC regulators

The normal IC voltage regulator has an input voltage limit of 35V – though the 24V output μA7824 and μA7924 ICs allow a 45V maximum input voltage and there are some higher voltage devices available – and this may be inadequate for some applications. In this case, the easiest way to improve the voltage and current ratings is by hybridizing the IC with a few external discrete components.

For example, the maximum current rating can be extended by causing the IC to control a parallel connected power transistor, as shown in Figure 15.38. This layout takes advantage of the fact that an IC voltage regulator draws very little current when its output voltage slightly exceeds the predetermined output voltage level. So by connecting the compound emitter follower circuit, Q_1 ,Q_2, in parallel with a resistor (R_1) in series with the IC, the parallel high current transistor will only be forced into conduction when the IC input current reaches a high enough level to cause a 0.6V drop across R_1. The output voltage can be adjusted, if necessary, by inserting a network (R_4, RV_1), in the reference limb of the IC. This gives almost exactly the same precision of output voltage control, and input noise and ripple rejection as would be provided by the IC on its own. Over-current limiting can be provided for this circuit in exactly the same way as shown, in Figure 15.25a, for a discrete component system.

A circuit allowing both a higher input and a higher output voltage to be controlled by a fixed voltage IC regulator, such as the μA7815, is shown in Figure 15.39. In this, the input voltage to the 7815 is controlled by the potential divider circuit (R_5, R_6), and the emitter follower, Q_2, while output current (and input current) is only drawn from the IC if the emitter voltage of Q_1 falls below 15V, which allows the output

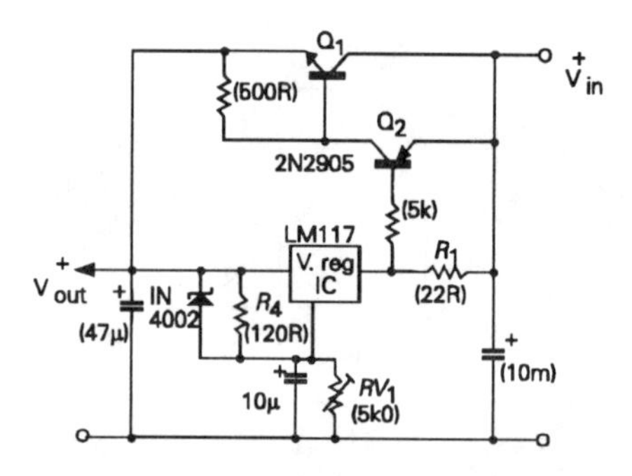

Figure 15.38 *Method of extending output current capacity of IC voltage regulator*

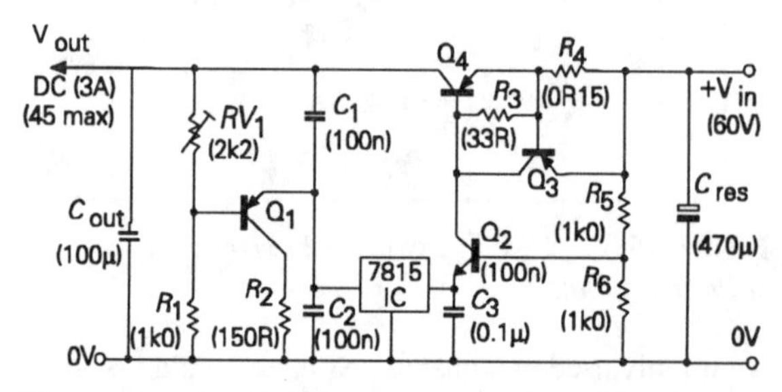

Figure 15.39 *Circuit providing increased current and voltage capability from IC regulator*

voltage to be adjusted down to about 14.5V by the setting chosen for RV_1. Over-current limiting is provided by the transistor Q_3, which steals the drive current from the pass transistor if the voltage across R_4 exceeds some 0.6V.

Switching and Switch-mode voltage regulators

All of the voltage regulators examined so far have come within the category of DC to DC converters, in that they have taken a DC input; usually neither particularly clean, in terms of freedom from electrical noise and mains induced ripple, nor stable in voltage; and converted this into a stable, much more noise free, but rather lower output voltage, since the regulator required some internal voltage drop in order for it to operate. There are, however, other useful arrangements which allow, for example, an input source of one polarity, say +15V, to be converted into an output of the opposite polarity, say –15V, or, alternatively, which will allow a lower voltage, say +5V, to be converted into a higher voltage, say +15V. Since all of these systems rely on the transfer of the energy stored in a capacitor or an inductor from the input to the output circuit, by some repetitive switching action, these circuits are generally called 'switching regulators'.

There is also a specific category of switching power supply system in which the input DC source is just an input rectifier/reservoir capacitor circuit. This provides a high power, but rather rough DC supply straight from the raw AC line, and this is then converted into a stable DC output, isolated from the mains input, by a high frequency oscillatory switching circuit, coupled to a small high frequency transformer. This type of circuit is employed because great savings in size, weight, and heat dissipation can be made by the use of high frequency, rather than 50–60Hz, voltage transformation techniques. The term 'switch mode' is usually used when referring to systems of this kind.

DC to DC reverse polarity converters

One of the simplest DC–DC converter systems, for which circuits frequently appear in the 'Design Ideas' columns of electronics magazines, uses a free-running square-wave generator, such as, say, a 555 type IC, connected as shown in Figure 15.40, to produce an output in which the charge stored in C_1 on each +ve going half cycle, can be rectified by the diodes D_1 and D_2, and 'pumped' into the output circuit. By the choice of the polarity of D_1 and D_2, the DC output from the circuit can be either +ve or –ve.

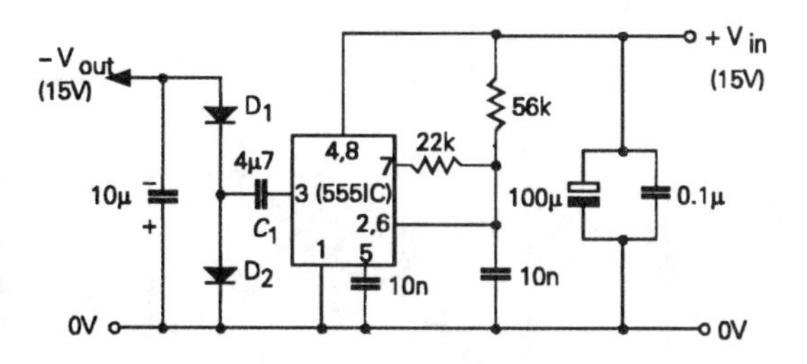

Figure 15.40 *Negative voltage generator circuit*

A more elegant switched-capacitor converter circuit, using the same philosophy, is that shown in Figure 15.41, based on a MC7660 IC. This gives a polarity-reversed output voltage which is very close to that of the input, for supply voltages up to 10.5V, and for output currents up to 10mA. The overall conversion efficiency of this circuit is typically about 95%. Similar types ofdevice are available from Texas Instruments, National Semiconductor, and many other manufacturers.

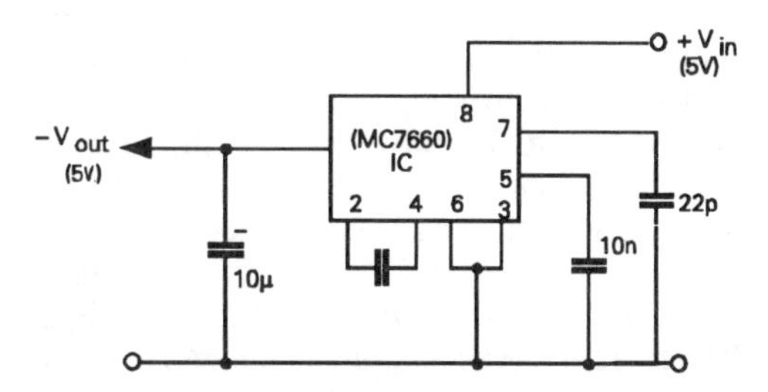

Figure 15.41 *Switching polarity inverter circuit*

A point which must be remembered in any switching regulator is that it is capable of injecting a significant amount of noise back into its own supply line, so the voltage feed to the circuit must be adequately decoupled.

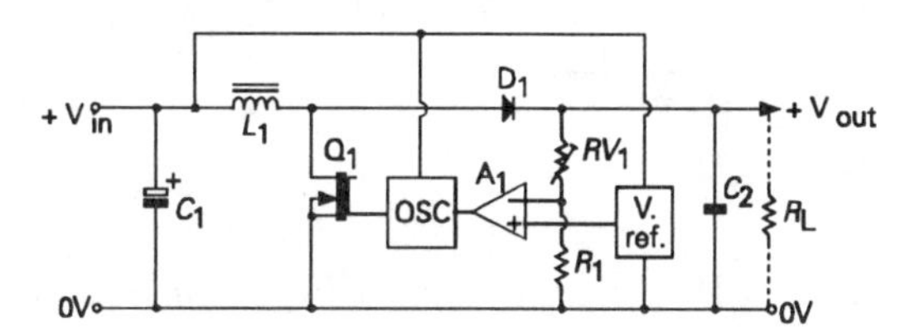

Figure 15.42 *Simple switch-mode regulator circuit (DC–DC)*

A much better type of circuit is the kind in which the energy store is an inductor, in that, for a given physical size, the current transferred to the output can be much greater. A typical example of this kind of arrangement is shown, schematically, in Figure 15.42, in which a switching transistor Q_1 is used, alternately, either to permit or to interrupt the current flow from the input DC supply through the inductor, L_1, to the 0V line. When the current flow through L_1 is interrupted, the voltage at the collector of Q_1 will rise, very rapidly – a phenomenon known as the 'back EMF', or, more popularly, as the 'flyback effect'. This causes the diode, D_1, to conduct, and causes current to flow into the output load, R_L. The output voltage provided by the circuit is then compared, through the voltage divider network, RV_1, R_1, with a reference voltage, and the amplified difference used to control the mark-to-space ratio of a high frequency oscillator. This, in turn, controls the switching characteristics of Q_1, to regulate the output voltage.

Although bipolar junction transistors can be used, as Q_1, in this application, power MOSFETs are more commonly used because they have a more rapid switching action and their high input (gate) impedance makes them easier to drive. By reversing the positions of L_1 and Q_1, as shown in Figure 15.43, so that when the current flow through L_1 is interrupted the voltage at its live end suddenly swings negative, the output polarity can be reversed, to allow a –ve output voltage to be derived from a +ve input supply line. As in Figure 15.42, the output voltage can be controlled, quite precisely, by comparing the output level with a reference source, and using the amplified difference to control the on-to-off time duration (the mark-to- space ratio), of the switching transistor, Q_1.

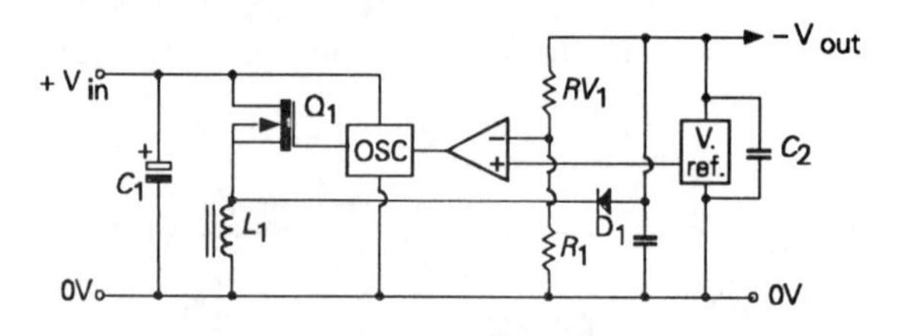

Figure 15.43 *Negative output voltage switch-mode regulator circuit*

Since the switching frequency can be made quite high – say 100kHz – the value of the output smoothing capacitor, C_2, does not need to be very high to reduce the residual switching ripple to the required low level. Both of these switched inductor circuits can provide an output voltage which is greater than the input supply line level – say, –15V out from a +5V in – and typical efficiencies can be as high as 90–95%, so that the system runs cool.

The control of the mark-to-space ratio of the waveform used to operate the switching transistor, in order to adjust the output potential provided by the circuit, can be done by the kind of circuit shown in Figure 15.44. In this a free-running 'triangular' or sawtooth waveform generator is 'sliced' by a high gain voltage comparator IC, which gives an output waveform in which the relative durations of the +ve and –ve parts of the cycle can be altered with reference to one another. If necessary, this output waveform can then be sharpened up still further, to improve the efficiency of switching, by reducing the length of time occupied in the on to off transistions, by interposing a CMOS logic gate IC between the output of the comparator IC and the switching device.

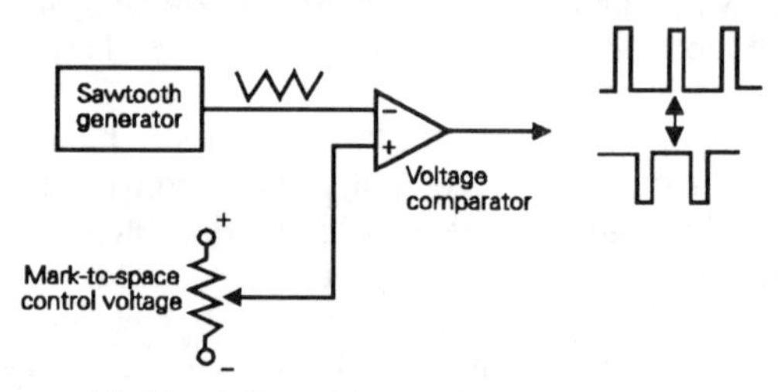

Figure 15.44 *Adjustable mark-to-space ratio oscillator circuit*

Switch-mode power supply (SMP) systems

A typical layout for a switch-mode power supply is shown in Figure 15.45. In this, the mains line AC voltage supply is rectified by the diode bridge (D_1), and charges a capacitor (C_1), to provide a DC source of either 163V or 340V, depending on whether the input supply line is 115 or 240V. Either this, or some alternative DC supply is used to power a pulse-width modulated oscillator, operating normally in the range 25kHz–200kHz, depending on the designer's choice, which controls the conduction of the switching transistor Q_1. In the past, this would normally have been a bipolar junction device, but the growing availability of higher switching efficiency MOSFETs has made these the more attractive choice.

It is nearly always essential to provide full electrical

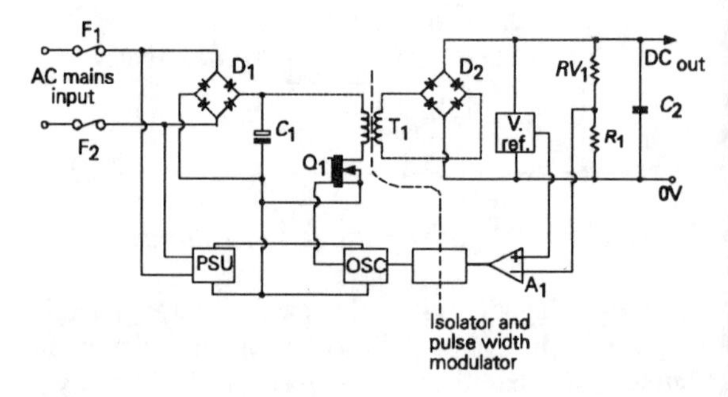

Figure 15.45 *Complete AC–DC switch-mode power supply*

isolation between the mains input and the DC output lines in any power supply, for safety in operation, and to allow freedom in the choice of interconnections in the load circuit. In the case of an SMP supply, this isolation is given by the transformer T_1. Because of the high operating frequency of this system, this transformer need only have a small core cross-sectional area – in practice this core will often be a 'ferrite' toroid – and the windings need only few turns, so that they can be wound with relatively thick wire which gives a low winding resistance and a high overall transformer efficiency.

The rectified output voltage is used to power a DC voltage reference source, against which the output DC level can be compared. As in the previous systems, shown in Figures 15.42 and 15.43, the error voltage detected by the amplifier (A_1) is used to operate a pulse-width modulator circuit which controls the conduction cycle of the switching transistor, Q_1. However, in the case of an SMP, the control circuit must be chosen so that a further isolation element, usually either an 'opto-coupler' or a small pulse transformer, can be interposed in the control path between the error amplifier and the switching transistor.

The DC operating supply for the switching oscillator, and other low voltage electronic circuitry shown on the left-hand side of the isolating transformer, is most easily provided by a simple, very low power, transformer–rectifier circuit of conventional form, although various ingenious schemes have often been used in commercial designs to provide a short burst of DC to this circuit, on switch-on, to allow the oscillator power supply, once the oscillator has started working, to be derived from a tertiary winding on T_1.

Once again, since the 'ripple' frequency of the rectified supply is so high, only a small value reservoir capacitor (C_2) is needed to reduce this to an acceptable level. However, since the impedance of an inductor increases with frequency ($Z_L = 2\pi fL$), even a low value inductor, with a very low DC resistance, interposed between the secondary rectifier bridge (D_2) and the output capacitor (C_3) and load, will give a substantial reduction in the output noise and ripple.

Because the SMP is so compact in size, and has such a low heat dissipation, this kind of power supply is now very widely used in 'desk-top' personal computers. Also, because of the popularity of this type of circuit, both pulse-width modulator systems, and switched capacitor and inductor converters, as well as complete switch-mode regulator circuits are available in IC form, from a wide range of manufacturers, such as Harris, Hitachi, National Semiconductors, SGS, Siliconix, Silicon General and Texas Instruments.

16

Noise and Hum

Introduction

Most linear circuit design is based on the convenient, and simplifying, assumption that the system will be noise free – that the end result of any amplification, filtering, or other signal handling processes will be the signal on its own, attenuated, enlarged, or modified as a result of the stages through which it has passed, but free from any alien signals not presumed to be present in the input, and not wanted in the output.

Sadly, in reality, this desirable situation can never exist, and wanted signals are always contaminated, to some extent, by unwanted electrical intrusions of one kind or another. For example, this pollution of the desired signal could be due to hum pick-up from the 50 or 60Hz mains power supply network, almost always found within buildings – even when a mains voltage supply is not used to power the equipment concerned – or to unwanted feedback of signal components; usually via the supply lines; from the output into the earlier stages of the system. In addition, all electrical input signals, and the output of all electrical signal handling circuitry, will be accompanied by noise. This may be externally generated, from a variety of sources, such as atmospheric electrical discharges, or from man-made, and usually impulse type, interference arising from such things as fluorescent lighting installations, thermostat switch-gear, or petrol engine ignition systems. Noise will also exist within the system itself as a fundamental characteristic of the basic physical nature of the signal source and all the associated electronic circuitry.

All of these unwanted noise signals can, to some extent, be lessened, or made less obtrusive, by care in the circuit design and in the layout and construction of the equipment, and it is therefore desirable for the designer to keep in mind the nature of noise sources so that he can come closer to the ideal result of a noise free output signal.

The physical basis of electrical noise

Thermal or Johnson noise (resistor noise)

At all temperatures above absolute zero, (0°K or –273.15°C) the electrons present in any conductor are in a state of restless random movement, which leads to the statistical probability that, at any given moment, there will be more electrons present at one end of a conductor than the other, a situation which causes a fluctuating electrical voltage to be developed across the conductor. This noise voltage will increase as the temperature is raised, since this increases the degree of agitation of the electrons, and as the resistance of the conductor is increased – because the electrostatic attractions which tend to oppose the movement of charges away from their rest condition of uniform distribution are lessened by the isolating effect of electrical resistance – and as the bandwidth over which the noise signal is measured is increased.

The mathematical relationship between these effects, originally defined by Johnson – hence the term 'Johnson' noise – is usually expressed in the form

$$e_n \text{ (rms)} = \sqrt{4kT\Delta fR} \qquad (1)$$

where e_n is the mean noise voltage, normally defined as volts per $\sqrt{\text{Hz}}$, k is Boltzmann's constant (1.38×10^{-23}), T is the absolute temperature, in °K, Δf is the measurement bandwidth, and R is the resistance of the conductor, in ohms.

The probability that there will be an imbalance in the charge present between the two ends of the conductor is an essentially random one, which leads to the wideband noise characteristic of this voltage. Similarly, the instantaneous magnitude of this imbalance is also entirely random in nature, which means that the noise voltage distribution has a 'Gaussian' characteristic curve, as shown in Figure 16.1.

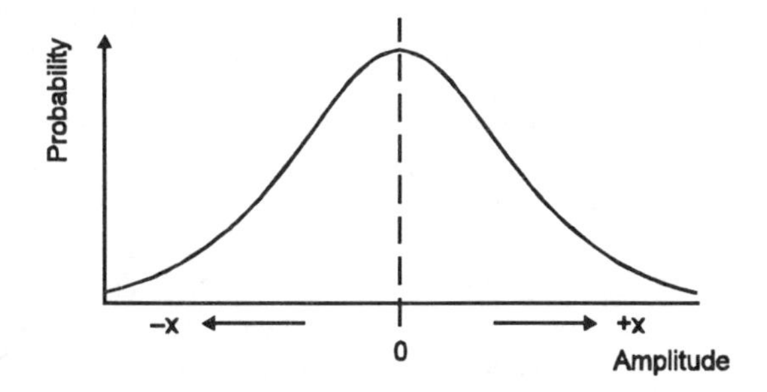

Figure 16.1 *Gaussian noise peak amplitude distribution curve*

It can also be seen from equation (1) that the energy distribution of thermal noise, as a function of frequency, is constant, per unit bandwidth – a characteristic which is given the descriptive term 'white' noise. By the use of this formula, it is a simple matter to plot a graph of the thermal noise level associated with any value of bandwidth, ambient temperature and source resistance. A typical set of curves relating noise voltage to source resistance for various bandwidths (20kHz is included as the nominal bandwidth of the audio spectrum) and an ambient temperature of 25°C (298.15°K) is shown in Figure 16.2.

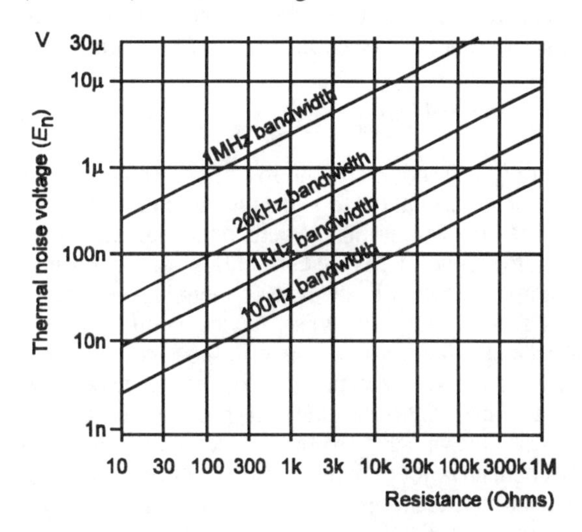

Figure 16.2 *Relationship between thermal noise voltage and circuit resistance (at 25°C)*

There are other sources of electrical noise, but Johnson noise is an invariable feature of any system which contains resistive elements, whether these are associated with the signal source or with the circuitry and active devices which are connected to it.

Shot noise (Current flow noise)

Another important source of noise which is an inherent feature of all electrical circuitry is 'shot' noise, which is characteristic of all current flow. Whether current flow is due to the motion of ions or electrons, it is, by its nature, discontinuous and particulate, with units of charge arriving individually at their destination in a random pattern. In very large current flows, this random time interval separating the arrival of any one unit of charge from the next will be less than that in the case of smaller current flows, but it will still be particulate in its nature, like lead shot falling into a receptacle. This means that the total noise current will increase with current flow, but the scatter of the individual noise pulse amplitudes will increase as the current decreases. Like thermal noise, this is white noise, with a constant energy per unit bandwidth, but it is, in this case, related solely to current flow, and is independent of both ambient temperature and the resistance of the conductor.

The relationship between current flow and shot noise (current) can be expressed by the equation

$$\bar{i}_n = \sqrt{2q \cdot I_{dc} \cdot \Delta f} \qquad (2)$$

where $\bar{i}_n$ is the noise current, in amperes (independent of source impedance), q is the charge on the electron (1.6×10^{-19} coulombs, where one coulomb is an ampere second), I_{dc} is the current flow, also measured in amperes, and Δf is the measurement bandwidth.

A further source of noise, having similar characteristics to shot noise, is found in those components, such as tetrode or a pentode valves, in which the internal current flow is offered a choice of destinations. This is known as "partition noise'. This particular mechanism is not of particular importance, in the case of solid state components, except that some integrated circuit constructions employ, as a matter of circuit convenience, multiple collector transistors, which will, because of the existence of partition noise, be significantly noisier in operation than single collector devices. However, in addition, majority and minority carrier recombination effects in bipolar transistors do give rise to a similar type of noise mechanism.

Flicker, excess or 1/f noise (defect noise)

These are terms given to a further type of noise which will arise within a resistor or a semiconductor, or any other conducting material, when a potential applied across it causes a current to flow. This effect is due to the inevitable existence of irregularities and imperfections within the material from which any conducting component is made, and was first observed as a 'flicker' type variation in the electronic emission from the surface of the cathode of a thermionic valve. Like partition noise, these defects offer a choice of paths for current flow, but with the difference that both the total resistance and the transit time for these paths will fluctuate, from one instant to another, giving rise to an additional noise voltage superimposed on the current (shot) noise, and the existing thermal noise of the resistive path.

Although a mean value of 'excess' noise can be determined, for any given component type, by laboratory measurements made on a sufficiently large number of nominally identical devices, it is impracticable to define it by any simple mathematical relationship which would be applicable to all circumstances, although in general form the power spectral density of this noise component follows the relationship

$$P_f(\omega)(\text{rms}) \approx K \cdot I^{\alpha}/\omega^{\beta} \qquad (3)$$

where K is some constant appropriate to the manufacturing process and materials used, ω is the frequency in radians/s, $\alpha \approx 2$, and $\beta \approx 1$.

This type of noise is a low frequency phenomenon, which is principally noticeable at frequencies below 1kHz, and which increases as the frequency is lowered – a characteristic which has given rise to the common description '1/*f*' noise. (This type of noise frequency distribution is known as 'pink' noise, to distinguish it from the wideband, uniform energy distribution 'white' noise)

This kind of 1/*f* noise effect is found in resistors, as a result of mechanical imperfections in their manufacture, such as variations in the uniformity of the conducting path, or in the integrity of the attachment of the connecting leads to the resistor body. This is the principal reason for the lower noise figure of metal film resistors in comparison with carbon composition types.

Contact noise has similar 1/*f* characteristics to that of excess noise, and is due to the breakdown of insulating films of surface contaminants present on contacting conductive faces, or to the interruption of minute current paths within the material. This is the reason for the lower noise of welded or soldered contacts, in comparison with 'crimped' end-caps, on, for example, wire wound resistors – welded or soldered connections are not practicable with film-type components.

Typical mean (rms) values of excess noise in different resistor types are shown in Table 16.1, in $\mu V/V$ (i.e., measured at a current flow of 1mA through a resistor value of 1kΩ), with a measurement bandwidth of 30–300Hz. (The measured values have been corrected to allow for thermal noise.)

A similar type of noise occurs in junction transistors, due to surface recombination effects within the base

Table 16.1 *Excess noise in different styles of resistor*

Resistor type	Noise voltage
Carbon composition, rod type, crimped ends	0.2–5μV
Carbon composition, rod type, soldered ends	0.1–2μV
Carbon film	0.05μV
Metal film	0.04μV
Metal glaze	0.03μV
Metal oxide film	0.03μV
Thick film	0.02μV
Wirewound, crimped end caps	0.03–0.1μV
Wirewound, welded end connections	0.01μV

region, and is one of the reasons why a PNP type transistor (which has an N-type base region), will usually have a lower noise figure than an NPN one. $1/f$ noise also occurs in junction FETs, and MOSFETs, due to reverse gate–source leakage currents. Most data sheets for low-noise transistors and ICs will show graphs of the $1/f$ noise characteristics of the devices in question.

A typical noise characteristic for a low-noise op. amp. IC, illustrating this typical $1/f$ behaviour, is shown in Figure 16.3.

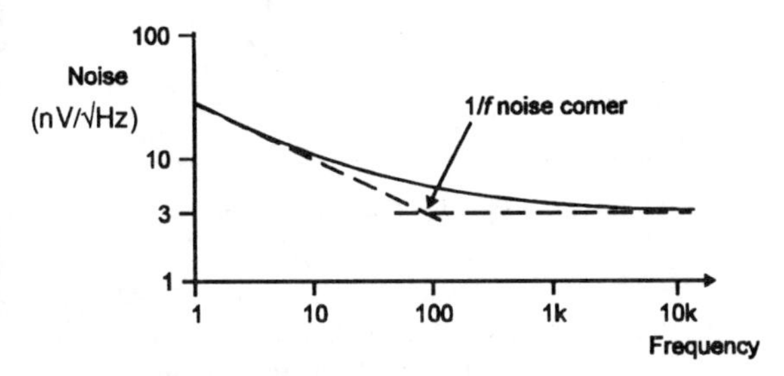

Figure 16.3 *Flicker noise characteristics of typical low-noise op. amp.*

Noise figure or noise factor

There are several ways by which this can be defined, but probably the simplest is to regard it as the extent to which the imperfections of the component or circuit in question degrade the signal/noise ratio of the system – measured in dB. This avoids the intellectual pitfall of thinking – as, for example, in the case of the data quoted for the BC322, a low-noise PNP transistor, whose 'noise figure', as a function of generator resistance, is illustrated in Figure 16.4 – that a lower system noise could be obtained by choosing a low collector current and a high source impedance. In reality, the only bargain that is offered is that if the generator resistance is made high enough, the thermal noise of the source will swamp the thermal, shot and flicker noises due to the transistor – which flatters the performance of the transistor but actually frustrates the designer's original intentions.

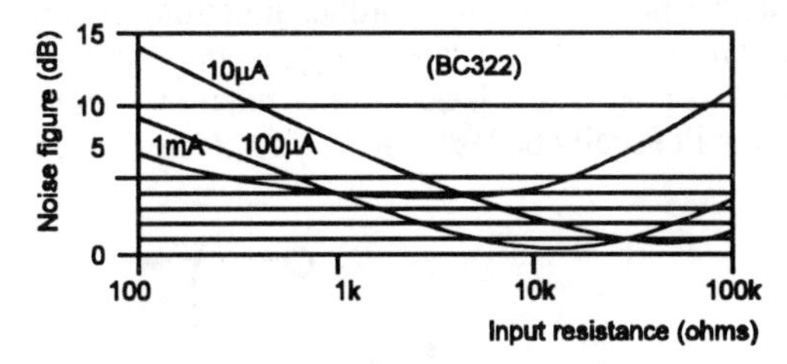

Figure 16.4 *Noise figure for low-noise PNP transistor as a function of input circuit resistance and collector current*

If the generator impedance is, or can be made, low enough that the thermal noise from this source is already very low, the best design approach is to choose a transistor whose noise figure is acceptably low at a level of collector current which is appropriate for the chosen input impedance.

Low noise circuit design

Noise models for junction transistors and J-FETs

The easiest way to visualize the practical effect of the various noise components which can arise in an amplifier circuit is to assume that one has an ideal noiseless device, in which the noise signals, due to the various physical mechanisms operating within the device, are represented by voltage or current noise generators connected to its input. The magnitude of these noise signals, referred to the input, is then measured by dividing the noise output voltage by the stage gain.

For example, in the case of the bipolar transistor circuit, illustrated in Figure 16.5, the voltage noise, e_n, referred to the input, will be the RMS sum of the thermal noise generated by the resistive components present in the transistor base circuit, as well as the shot noise resulting from current flow through the base–emitter region, and the flicker noise due to lateral current flow within the base region. These input resistive components consist of the base spreading resistance, R_{bb} (that resistive component, effectively in series with the base connection, due to the resistance of the lateral current path from the base connection to the active part of the base–emitter junction), and the base–emitter 'bulk resistance', R_{ee}, (representing the bulk resistivity of the base–emitter junction region – which will usually be less than R_{bb} by at least a factor of 10).

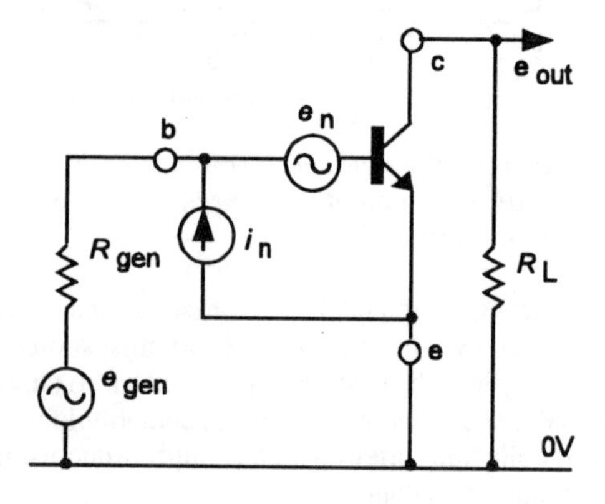

Figure 16.5 *Junction transistor noise model*

In Figure 16.5, the sum of these voltage components is represented by an idealized zero impedance voltage generator, e_n, in series with the input circuit. In practice, shot noise in the transistor will be reduced, as seen in equation (2), if the transistor is operated at the lowest practicable collector current, and flicker noise will be minimized if the transistor is operated at the lowest sensible collector voltage.

There are also input noise current components, represented in Figure 16.5 by the current generator, i_n, assumed to offer an infinitely high source impedance, connected in parallel with the input base–emitter circuit. These noise current sources are the shot noise components of the base current, and the modulation of this base current by flicker type effects. These noise sources can again be minimized by operating the transistor at its lowest practicable collector current, and by choosing a transistor with a high current gain, since both of these actions will reduce the base current of the device.

From the point of view of amplifier performance, very little can be done to minimize the effect of any voltage noise, e_n, which is due to the device, since this will appear in series with any input signal voltage. The main options open to a designer seeking the best possible s/n ratio are to choose the best device type which is economically available, or to arrange that the input generator voltage is large in relation to the input voltage noise of the amplifier.

Current noise, i_n, is important because the flow of this current through the input (generator) impedance will give rise to a noise voltage. If the generator impedance is high, and cannot be reduced, a better noise figure can, perhaps, be obtained by using a junction FET as the amplifying device, as illustrated in Figure 16.6, since this type of transistor has a very low gate leakage current, and therefore a low value of input noise current, i_n.

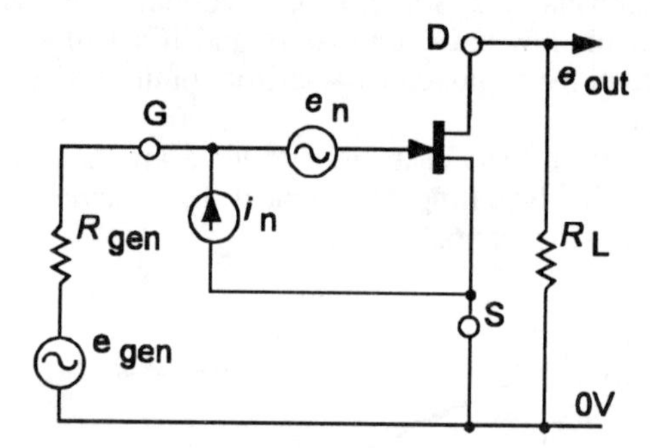

Figure 16.6 *Junction FET noise model*

(*Note.* Thermal noise is a feature of the real (i.e. resistive) part of a circuit impedance. However, in the case of a noise voltage which arises because of the flow of a noise current through an input impedance, the reactive (i.e. capacitive or inductive) parts of the input impedance must also be considered.)

On the other hand, junction FETs usually have a relatively large voltage noise figure, e_n, because the conducting channel length of a typical small-signal J-FETs, and therefore its effective thermal noise resistance, is substantially greater than the emitter–collector path length of a bipolar junction transistor. This noise voltage due to the FET channel can be defined by the equation

$$e_n(\text{rms}) = \sqrt{4kTR_n\Delta f} \quad (4)$$

where Δf is the measurement bandwidth, k is Boltzmann's constant, T is the absolute temperature, °K, and R_n is the effective channel resistance, and in typical devices is approximately defined by

$$R_n = 0.67/g_{fs} \quad (5)$$

where g_{fs} is the FET common-source forward transconductance.

In the flicker noise region, the mean value of the noise component e_n is approximately defined by the equation

$$e_n(rms) = \sqrt{4kR_n \cdot \Delta f\,(2 + f_1/f^{\beta})} \quad (6)$$

where f_1 is the upper corner frequency of the flicker noise, and β has a value between 1 and 2, depending on the process by which the device was made, and the purity and homogeneity of the semiconductor slice. The combined effects of these noise sources, at low frequencies, is illustrated in the graph, due to Siliconix (Application Note 74–4), shown in Figure 16.7.

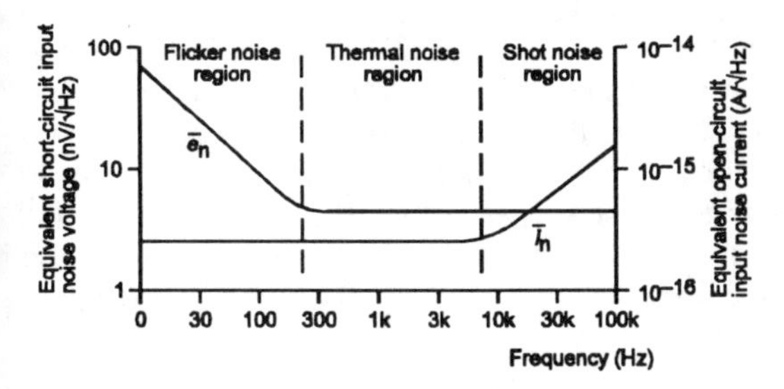

Figure 16.7 *Junction FET noise characteristics*

Since the channel resistance is related to the transconductance of the device, the noise voltage, e_n, can be reduced by increasing the forward transconductance of the device, g_{fs}, which can be done by parallel connecting two or more identical FETs – in numerical terms, this will reduce e_n by $1/\sqrt{N}$ if N devices are connected in parallel.

A similar result is given if the FET gate area is increased, which increases the channel area, and also reduces channel resistance. However, both of these measures increase the input (drain–source), and the feedback (drain/gate), capacitance of the FET.

FETs also suffer from a type of input noise current, called 'burst' or 'popcorn' noise (because of the supposed similarity of the noise produced in an audio system to that of corn popping during roasting). This occurs particularly at low frequencies (10Hz or below) and is presumed to be a type of contact noise affecting the aluminium gate metallization (Observation has shown that it is made worse if the surface of the semiconductor die is known to have been contaminated prior to the metallization process.) Although still present to some extent in all junction FET amplifiers, modern devices are much less affected by this defect.

Increasing the reverse bias of the FET gate/channel junction will reduce the input gate capacitance, but will – up to the point of cut-off – somewhat worsen both flicker and popcorn noise, as well as the thermal noise voltage associated with the channel, since the resistance of this will also be increased by increasing the reverse bias.

The operational amplifier model

The op. amp. circuit, shown schematically in Figure 16.8, presents a very similar situation to that of a junction bipolar transistor or FET circuit, since the op. amp. input stage devices will be either transistors or FETs. However, because the input stage will almost invariably consist of an input long-tailed pair configuration, of the kind shown in Figure 16.9, if the two input transistors are physically identical, as is usual,

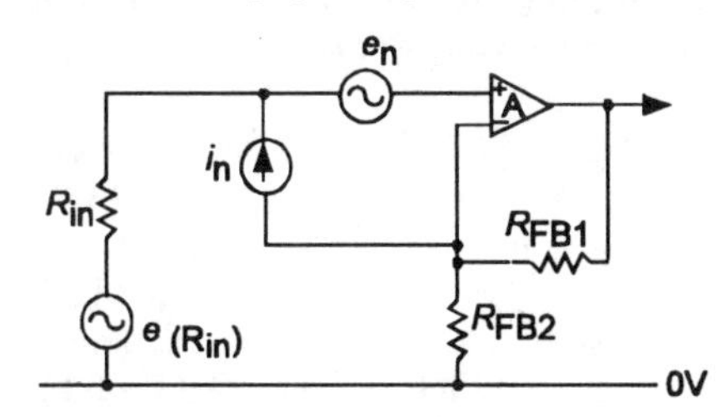

Figure 16.8 *Operational amplifier noise model*

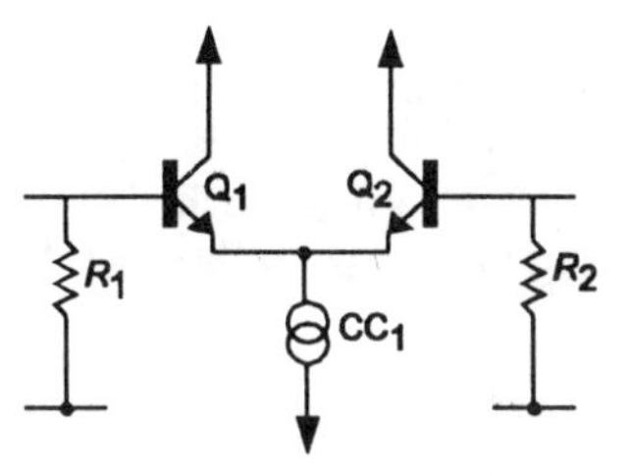

Figure 16.9 *Input long-tailed pair layout*

all the input device noise contributions will be worsened by a factor of $\sqrt{2}$ (3dB), by comparison with the single-ended layouts shown in Figures 16.5 and 16.6.

The equivalent input noise resistance due to the source and feedback resistors, shown in Figure 16.8, and associated with any practical op. amp. circuit layout, can be determined from the equation

$$R_{eq} = \sqrt{R_{in}^2 + [R_{fb1} \parallel R_{fb2}]^2} \qquad (7)$$

in which the symbol $\parallel$ denotes parallel connection (i.e. $R_{fb1}R_{fb2}/(R_{fb1} + R_{fb2})$). The equivalent input noise resistance defined by equation (7) is also applicable to single-ended stages where negative feedback is applied, for example, from the output to the emitter or source of a junction transistor or FET, but with the complicating factor that R_{fb2} will, in any case, modify the noise figure and stage gain of the device.

MOSFETs

These have very much worse noise figures at low frequencies than either junction FETs or bipolar transistors, mainly due to flicker noise and popcorn noise phenomena. It seems probable that the major cause of MOSFET flicker type noise is the continuous creation and recombination of charges from the neutral charge-pairs constituting the conductive channel, but some of this noise may be due to the on-chip protective zener diodes.

Because of their poor LF noise figure, MOSFET devices are seldom used for DC amplifier or low-frequency low-noise applications, except where the extremely high input impedance typical of a MOSFET is an essential requirement – as for example in ionization chamber amplifiers. (In those MOSFETs in which the gate is not protected by an integral zener diode, input resistances of the order of 10^{15} ohms are obtainable.)

(*Note.* It should be remembered in this context that small signal junction FETs can offer an input resistance which is typically greater than 10^{12} ohms, and are also free from proneness to damage due to electrostatic breakdown.)

Amplifiers for very low impedance systems

Bipolar transistor systems

The growth of interest in low-noise amplifiers for use with low impedance, low output voltage sources, such as strain gauges, thermocouples, and, in the audio field, moving coil gramophone pick-up cartridges (whose output voltages, for a 1cm/s. groove excursion velocity, may be as low as 50–500μV), has encouraged the semiconductor manufacturers to develop very low noise devices, frequently packaged as matched dual transistor pairs, specifically intended for use with input configurations of the kind shown in Figure 16.9.

In order to achieve low amplifier noise, all of the transistor internal noise components must be of a low level, and, in particular, its thermal noise resistance, referred to the input, should be no greater than that of the signal source. The effective internal resistance of the transistor can be lowered either by choosing a device with a large base/emitter junction area, such as a medium power device, or by parallel connecting a number of low-noise small-signal transistors. However, in practice, such transistors must either be closely matched in their base–emitter current threshold (diode) voltage or an emitter resistor (R_e) must be inserted as shown in Figure 16.10, to ensure that the total collector current is equally shared – and these added resistances will worsen the thermal noise figure of the circuit.

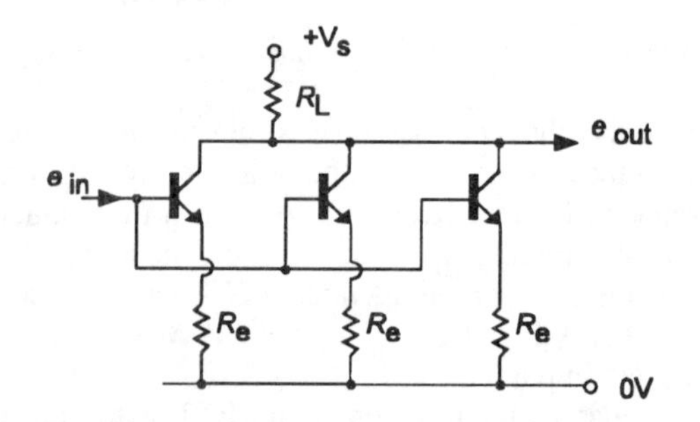

Figure 16.10 *Use of bipolar transistors connected in parallel to reduce input noise resistanceof circuit*

One of the earliest semiconductor designs aimed at the solution of this problem was the National Semiconductors LM194/394 'super-match pair'. This is essentially a multiple device integrated circuit in which two groups of parallel connected, and notionally identical, transistors have been fabricated, in a manner which gives a random distribution across the semiconductor chip to even out any differences between one group and the other.

Because of the multiplicity of parallel connected transistors, the base spreading, $R_{bb'}$, and the bulk, $R_{ee'}$, resistances are very low, at 40 and 0.4 ohms respectively, which means that the thermal noise voltage, due to the internal resistance of each half of the transistor pair, is kept to a very low level.

The optimum noise figure for any transistor amplifier requires that the collector current, I_c, should be chosen to give the best performance for the known input (generator) resistance. In general, this value of I_c would be determined by reference to the manufacturer's data sheet, or by the choice of a transistor for which this value was known.

Typically, the relationship between the noise figure, collector current and source resistance will be defined by the use of a diagram of the kind illustrated in Figure 16.4. However, for a transistor, such as the LM194/394, whose behaviour approximates very closely to the ideal model, the base current shot noise, i_n, can be fairly accurately predicted from the equation

$$i_n\,(\text{rms}) = \sqrt{\frac{2q \cdot I_c}{h_{fe}}} \qquad (\text{A}/\sqrt{\text{Hz}}) \qquad (8)$$

and the optimum collector current, for each half of the long- tailed pair, can be derived from the relationship

$$I_c\,(\text{optimum}) = \frac{kT}{q} \cdot \frac{\sqrt{h_{fe}}}{r_s} \cdot (\text{A}) \qquad (9)$$

where q is the charge on the electron, and h_{fe} is the small-signal, common emitter current gain of the transistor, and r_s is the generator resistance.

With any real-life device, there will be also some flicker noise at frequencies below about 1kHz, and this will be a function of the manufacturing process, and therefore unique to the device in question. That for the LM194 is shown in Figure 16.11.

In recent designs, improved manufacturing techniques have allowed even better noise figures to be obtained than those given by the original LM194/394

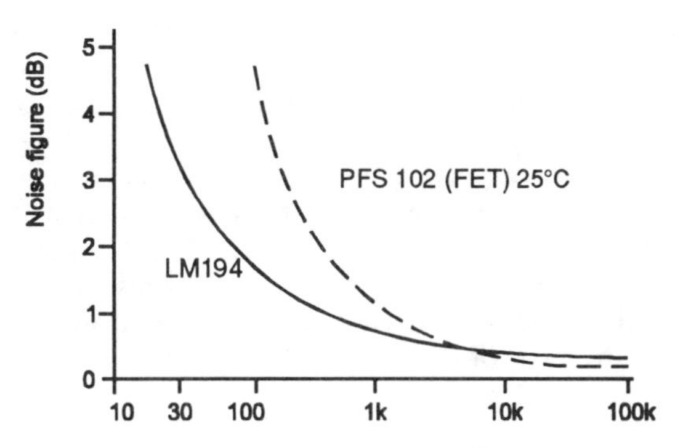

Figure 16.11 *Noise figure of LM194, compared with that of a low-noise FET as a function of input resistance*

super-match pair, with components such as the PMI, MAT02, and more recently still, the SSM-2220 and SSM-2221, PNP and NPN matched pairs, from the same manufacturer, all of which have noise voltages (with zero external input resistance) of less than 1nV/√Hz over the frequency range 10Hz to 100kHz.

A typical ultra-low-noise circuit design, intended for use as a strain-gauge DC amplifier, is shown in Figure 16.12 (National Semiconductors Application Note AN-222-7).

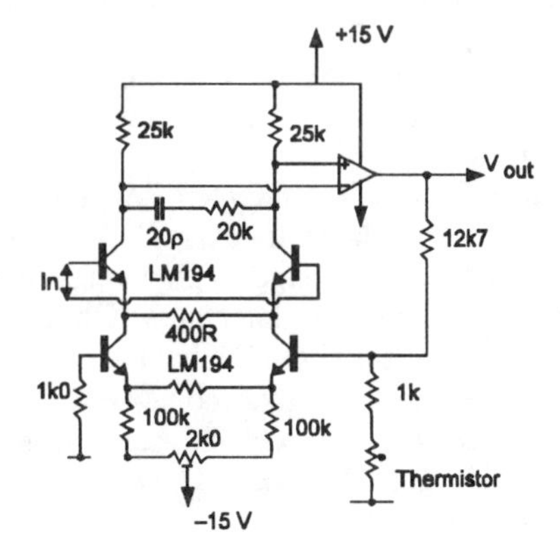

Figure 16.12 *Strain gauge DC amplifier design using LM194 super-match pair transistors*

Junction FET designs

As noted above, junction FETs will, by their nature, generally have a higher input thermal noise resistance than bipolar junction transistors and will therefore be of greatest value in systems designed for use with higher source (generator) impedances where the noise current associated with the amplifying transistor is more important than the device thermal noise voltage.

With the best available devices, the 'crossover' input impedance, above which the junction FET will offer a lower noise figure than the best available bipolar transistor, is of the order of 1k ohms. Above this figure, the better linearity, and greater tolerance of input overload of the J-FET may offer useful performance advantages.

The effect of the distribution of stage gains

In multiple stage circuitry, where every stage will contribute some noise to the final total, it is desirable to arrange that as much signal gain as possible is obtained in the early stages of the system, so that the signal level at later stages of the circuit will be high enough to swamp the noise voltages due to these stages. The effect of stage gain on overall circuit noise can be expressed by the equation

$$E_n = \sqrt{[N_1\cdot(a\cdot b\cdot c\cdots x)]^2 + [N_2\cdot(b\cdot c\cdots x)]^2 + [N_3\cdot(c\cdots x)]^2 + \cdots[N_X\cdot x]^2} \qquad (10)$$

where E_n is the total noise voltage at the output of a multi-stage system, N_1, N_2, N_3 ... N_x are the noise voltages, referred to the inputs, of stages 1, 2, 3 ... to X, of the system, and a, b, c ... x are the stage gains of these stages.

If the magnitudes of the early terms in this series can be made sufficiently large, for example, by making the stage gains of the early stages high enough, the noise voltages contributed by the later parts of the circuit can be ignored. Since it is assumed that the design of the circuit will be chosen so that the noise figure of the early stages is as low as possible, care must also be taken to avoid signal loss, such as by input attenuation prior to amplification.

Summary

In order to attain the lowest thermal noise figure in a given stage, the circuit should be chosen to employ the lowest practicable values of circuit resistance (and input impedance, where there is a significant input noise current) compatible with satisfactory circuit performance in other respects. However, one must bear in mind that it is not profitable to design for a low amplifier input resistance, in the pursuit of the lowest possible level of circuit noise, if this results in a mismatch to the generator output impedance, and a consequent loss of input signal voltage. So make sure, prior to deciding on the circuit design, that the optimum load resistance is known.

Where a very low noise level is essential, constructional techniques and components of the best available type and quality should be used, and if active devices are to be employed, then the maker's data sheets should be consulted to discover the best values of operating current and voltage for these devices, for the intended input circuit resistance.

With junction transistors, if the data sheets suggest a range of operating currents which apparently give similar device performance for a given source impedance, choose an operating value at the lower end of this range, since it will be associated with a lower level of base current shot noise.

Also choose transistor types having a high value of current gain, since they will have the lowest level of base bias current for a given collector current, and consequently the lowest level of input current noise, and they should be operated at the lowest sensible level of collector voltage commensurate with the likely level of output voltage swing that the stage will be required to handle, because this will minimize the leakage currents, and the flicker noise level associated with these.

With J-FETs, in which, at low frequencies, the input noise currents are usually exceedingly low, the best noise performance is likely to be obtained with circuits in which the input resistances are relatively large (of the order of 1kohms or greater). Unlike bipolar transistors, J-FETs show lower device noise figures at the upper end of the normal drain current range – because this is associated with lower channel resistance values, and consequently lower levels of the input thermal noise voltage associated with this. Obviously, it is desirable to keep the operating temperature low, since thermal noise is associated with circuit temperature. However, there is not usually a lot that the designer can do about this, and, in any case, the performance, of most semiconductor devices worsens, either in respect of gain figures or in terms of leakage currents, if they are used at temperatures outside the normal ambient temperature range.

The final point to remember is that all noise problems are worsened if the system bandwidth is increased – remember the old adage that 'the wider the window, the more the dirt flies in' – so make sure that the working bandwidth is not wider than necessary for proper circuit operation. Judicious filtering or bandwidth limitation can be greatly helpful in preserving a good signal to noise ratio.

Noise from external sources

With any high gain amplifier, especially if it has a high input impedance, and a wide bandwidth, there is a tendency for externally generated wide bandwidth impulse-type noise to break through into the signal channel. There are many mechanisms by which this can happen, of which the most common is by direct pick-up by the amplifier unit itself, or by its input leads, if these are inadequately screened, or if the screening (usually provided by a woven metallic braid on the outside of the cable, or by the use of a conductive, usually metallic, enclosure housing the input components) is not connected to the common 0V line by a sufficiently low impedance path.

Alternatively, electrical impulse-type noise signals may be carried into the enclosure housing the amplifier – even when this is properly screened – by the mains power cable, and by the power supply transformer primary winding, particularly if the transformer is unscreened. Direct pick-up of impulse noise signals by the internal input connections can occur from both of these sources, and it may be necessary either to employ some form of mains input low-pass RF filtering, such as that shown in Figure 16.13, or to house the power supply unit within a separate enclosure, to reduce interference pick-up to an accepable level.

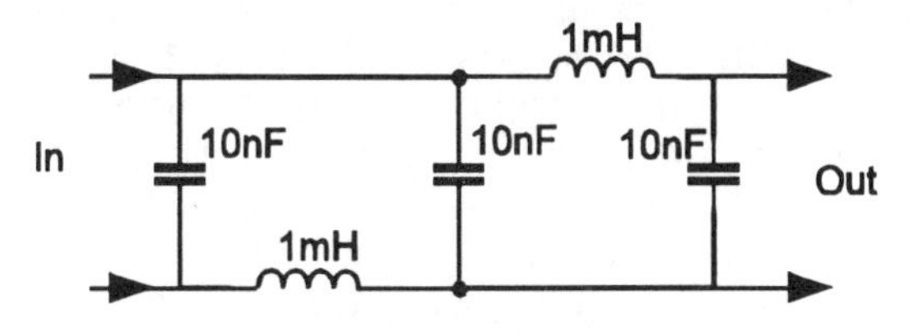

Figure 16.13 *Mains input RF filter*

Breakthrough from the mains input on to the internal DC power supply lines can occur, but is unusual, since these DC rails are nearly always well decoupled to the common 0V line by low impedance by-pass capacitors. The sensitivity of all electronic systems to interference pick-up is related to the pass-band of the system. Once again, the wider this is, the worse the problem will be.

A particular problem with amplifiers employing negative feedback is that these can show an exaggerated sensitivity to impulse-noise pick-up, if, by malfunction, or by an error in design, the feedback loop has an substantially reduced stability margin, and a consequent tendency to sporadic oscillation – especially if this occurs at some relatively high frequency.

Radio pick-up

This is an annoying phenomenon which can often occur in prototype systems, and is most noticeable in audio amplifiers, where even a low-level radio signal break-through can be very obtrusive when present under 'zero signal' conditions. This problem can have many causes, but is usually due to pick-up on input connecting leads – particularly if imperfectly screened or if the screening is inadequtely earthed – and to needlessly wide bandwidth input stages.

It is often suggested that poor quality soldered joint connections can provide the necessary (Schottky diode type) rectification of an input RF signal to allow the LF modulation to be separated from the RF carrier of a radio signal, which allows these LF components to be amplified as a normal audio signal, but since most contemporary audio amplifiers use input devices which are based on P–N rectifying junctions, there seems little need to seek additional rectifying mechanisms.

The problems which need to be solved are, firstly, how an RF signal, of sufficient amplitude to cause trouble, has found its way into the input circuit, and, secondly, since it is inconceivable that a spurious RF signal, at this stage, will have an adequate magnitude to drive the available P–N junctions into rectification, why the input stage has sufficient bandwidth, as an amplifier, for it to be possible for the intruding signal to pass through this stage and overload later ones. Unfortunately, the experience of many constructors suggests that there is no universally applicable solution to this problem, and that amplifiers (even those of respected commercial origin) which are satisfactory in one location may be prone to RF pick-up troubles in another.

The remedies are to ensure the integrity of the screening on all cables: to ensure that there is only a single connection between the circuit 0V line and the chassis: that this 'earthing point' shall be as close as possible to those input connections which have the highest subsequent gain: and that the bandwidth of the input stages shall be no greater than is necessary for the proper functioning of the circuit.

Mains hum

The intrusion into the signal path of 50/60Hz or 100/120Hz ripple voltages, directly derived from the mains AC power supply lines, is seldom a problem with contemporary semiconductor circuit designs – except where very high levels of voltage amplification are employed. This is partly because of improved DC power supply components and techniques, and the growing use of electronic voltage stabilization circuits to provide substantially noise and ripple free DC supply rails, and partly because of a better understanding of the ways by which supply line ripple rejection can be improved.

For example, in the amplifier circuit layout shown in Figure 16.14a, while there might be 30–40dB of negative supply line ripple rejection, that from the positive supply rail would be negligible, because of the relatively low resistance current path through R_2. However, by the use of a very high impedance constant current source (Q_3/Q_4), as the load for Q_2, as shown in Figure 16.14b, the positive supply line rejection can be increased to some 40dB or more. (The application of overall loop negative feedback around the amplifier will, of course, improve things still further, but it is good design practice to reduce performance defects as far as possible before the application of NFB.)

A major source of hum intrusion, particularly when a mains operated power supply is incorporated within the same housing as an amplifier system, arises because of the very heavy currents which flow during the conduction cycle of the rectifier diodes (D_1/D_2), in capacitor-input DC supplies of the kind shown in Figure 16.15a. Since there will always be some path resistance between points A and B in this circuit, the pulsating voltages which can arise, along what appears to be a direct conducting path, because of the high

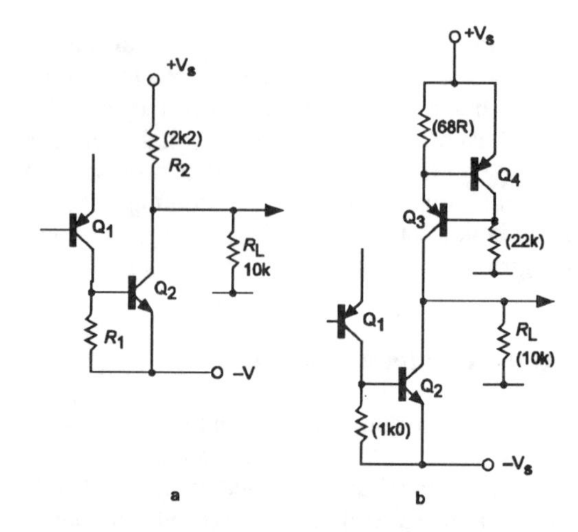

Figure 16.14 *Comparative amplifier circuits having different degrees of immunity to hum or noise break-through from the DC supply lines*

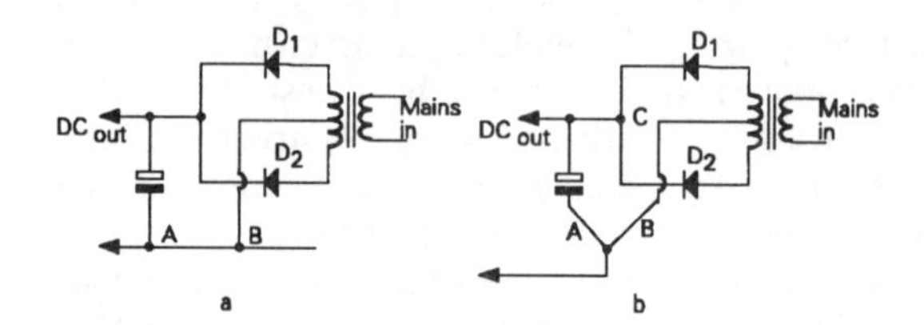

Figure 16.15 *Capacitor input power supply layouts*

reservoir capacitor charging currents at the 100/120Hz diode conduction frequency, may well be significant in relation to the operation of the circuit. A more insidious problem is that the high level pulsating current flow through part of the 0V line may mean that, for small signal return paths, the 0V line may be anything but this!

A type of wiring layout which was commonly used in high quality amplifier systems, to avoid this and similar difficulties (before the widespread adoption of copper-clad printed circuit boards made this type of layout rather more difficult to implement), was that called 'single-point earthing', in which a common connection point was chosen as the 0V anchor point, and all of the 0V connections were taken directly to this. In the particular case of the circuit layout shown in Figure 16.15a, trouble can be minimized if points A and B are the same connection, which is then joined to the 0V line, as shown in Figure 16.15b. However, there will still be substantial current flows through the loop enclosing points A, B and C, and this will act as the single turn primary winding of an air-cored transformer, and will induce hum voltages in any wiring in the vicinity, as will any stray magnetic field from the mains transformer itself.

Modern practice, in high quality amplifier designs, favours the use of power supply transformers built around 'toroidal' cores, because the tight magnetic circuit of this type of core minimizes the transformer stray hum field. A practical point to keep in mind is that the magnitude of all such induced hum voltages will increase as the physical sizes of the primary (hum generating) and secondary (hum pick-up) loops are increased (because the larger loop area increases the degree of relative coupling of the primary and secondary of this unintended 'transformer') so that layout designs which ensure that all such conducting loops are as short as possible will help minimize the induced hum voltages. This is also important in respect of the AC mains hum field present in all buildings containing power service wiring, and in the case of very sensitive amplifier stages, the more compact the input/output signal loop can be made, in any signal stage, the lower the likelihood of hum pick-up.

Screening

A simple conducting screen will usually serve to provide electrostatic isolation from one circuit to another and minimize the unwanted effects of direct capacitative coupling. However, a plain metallic screen is relatively ineffective in reducing the extent of magnetic coupling, because such a conducting housing will act as a short-circuited single-turn secondary winding. This means that currents will be induced in this loop, in the presence of an external magnetic field, and these currents will themselves generate an external magnetic field – and so on.

Much better results are obtainable if such a screen is made from a material of high magnetic permeability, such as 'Mumetal' or 'Permalloy', or others of the kinds listed in Chapter 2, Table 2.4. This is standard practice for the screening cans of microphone transformers, which are likely to employ a large number of turns of wire on their secondary windings (leading to a high sensitivity to stray magnetic fields), and are also usually situated at the inputs of high-gain amplifier units.

Balanced cables

The avoidance of hum pick-up, due to the interaction of stray external magnetic fields with low-level long path length signal cables, has led to the use of twin 'balanced' feeder systems to interconnect, for example, sound studio microphone lines.

The basic system is illustrated in Figure 16.16, in which a signal, symmetrically balanced in respect to the screening of the cable, is taken to a differential amplifier, which amplifies only the difference voltage between the two incoming lines. By this means, problems due to hum pick-up on the cables – which will run in parallel, closely spaced paths, and will therefore have identical hum voltages induced in them – will be cancelled.

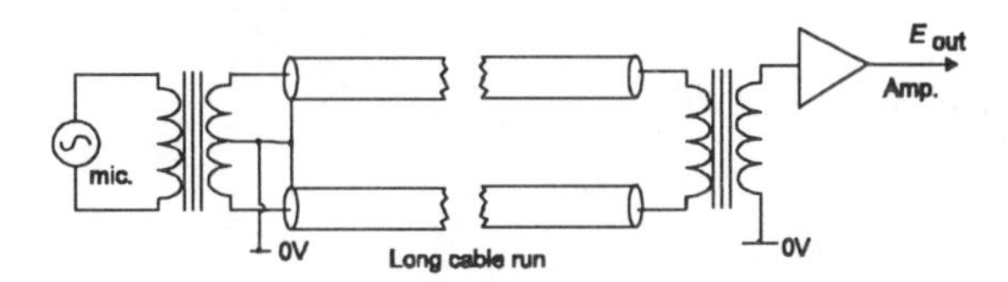

Figure 16.16 *Typical layout for balanced cable connections to minimize hum field interference*

Microphony

This is the name given to the irritating type of spurious circuit noise which occurs when the mechanical vibration of a sensitive component generates an unwanted electrical signal. This was a common problem with thermionic valves (tubes), where oscillatory motion of the grid windings, in relation to the cathode or anode of the valve, would lead to 'booming' or 'ringing' noises, but is seldom a source of trouble with semiconductor components, which are physically more robust and shock and vibration resistant. Care should be taken, however, in high gain amplifier systems, in the mounting of both components and PC boards, to limit the extent to which they can move in relation to one another.

A particular form of microphony is that which can arise in the screened cables used to interconnect very low signal level systems, particularly where there is a DC potential difference between the inner core and the outer screening braid. In this case, when the physical separation between these conductors is altered by flexure of the cable, or, perhaps, because it has been walked upon, the change in local cable capacitance will generate a noise voltage. Cables intended for use in this application, usually called 'microphone cables', employ an intermediate conductive screening layer, such as a coating of graphite or conductive plastic, bonded in intimate contact with the outer surface of the insulating core of the cable. Because this is at the 0V line potential, relative movement of the outer screening braid, in relation to this, will not generate unwanted noise voltages.

17

Test and measurement equipment

The importance of measurements, and the need for testing

I suppose it is true of any field of human endeavour that nothing goes right all the time, and that much care is needed even to make sure that things go right most of the time. Certainly, my own experience in the practice of electronics, and my observation, over the years, of the various problems which have been encountered by my colleagues in this field, has taught me to recognize the potential fallibility both of human beings and of the materials they choose to employ. In particular, however skilful the designer may think he has been in his initial concept, and however thorough he feels he has been in his calculations, there is always a possibility that he may have overlooked some relevant aspect of the system which he has designed. It is also possible that he may have been aware of, but dismissed as unimportant, some factor which was more significant than he had realized.

Similarly, when a prototype is built to a design which is, in itself, entirely satisfactory, there is always some chance that a constructional fault or a component defect will cause a circuit malfunction. In most cases the existence of a problem will be immediately obvious, and investigations will be made to find its cause and put it right. However, if the fault, due either to design or to component error, is not a particularly conspicuous one, it may remain unseen, resulting in an unexpectedly poor performance of the system. The only really satisfactory way of avoiding the problem of unseen malfunctions is to take very little for granted, and always to make at least a few appropriate measurements of the static and dynamic characteristics of the circuitry. In this task the better the test equipment available, and the more skilled the engineer in its use, the easier it becomes to check whether things are working as they should do – and, if they aren't, why not.

By static tests, I mean those which relate, in the main, to measurements made on the components in use, such as their resistance, capacitance or inductance (and if there are any doubts about component characteristics, it is usually much easier to check these before they are installed in the circuit), or to the quiescent conditions which exist in the equipment under test, such as measurements of its DC voltage and current levels, and, perhaps, its operating temperatures.

Dynamic tests are those measurements which are made on the system when it is in operation – usually with a known input test signal – and include output voltage or current swing, frequency and transient response, waveform distortion, voltage or current gain, input or output impedance as a function of frequency, signal to noise level, and, in the case of signal sources, output frequency and frequency stability.

Although I am sure that most 'proper' design engineers would frown on such a suggestion, the use of appropriate test equipment can save a lot of time in calculation, in showing which system works the best, if there is a choice between several, and also, at the prototype stage, in checking that the optimum component values have been chosen for the system. One must always remember, however, that component values may not always be what their markings indicate, and also that components will vary anyway, so the skilled engineer should always seek to design for a median value, to make sure that the system will still work well with component characteristics which depart somewhat from the ideal value. So, don't 'tweak' a system for perfection, if that perfection then rests on the edge of a performance precipice, or you may find that while the prototype works excellently, all its successors are duds.

Static testing (voltage, current and resistance measurement)

Voltmeters

Like some of the other test instruments, these come in two basic kinds, 'analogue' and 'digital', and there is also a division between 'direct reading' and 'electronic' instruments – digital meters will, obviously, be electronic in their nature. Each of these types have their own advantages and snags.

An ideal voltmeter should present an infinitely high shunt resistance when connected across the circuit under test. Similarly, an ideal current meter should cause a zero voltage drop when it is inserted in a circuit to measure direct current flow.

Electronic measurement instruments can come much closer to this ideal than any direct reading meter, but require an external power source – usually a battery, which leads to the disadvantage that it will run down if the meter is inadvertently left switched on – and may suffer from zero or calibration drift with time or changes in temperature.

Digital instruments can give a much greater degree of precision in reading (usually limited to +/–1 in the least significant digit), particularly in measurements of circuit resistance, and are particularly useful in determining the differential potential existing between non-zero points, such as across the load resistor (points 'x' and 'y') of the simple amplifier circuit shown in Figure 17.1. However, they do not usually approach the ideal zero voltage drop condition in current measurement applications as closely as, for example, an electronic analogue meter, and they are virtually useless for measurements of changing voltage or current levels, which makes them very awkward to use as a peak or a null detector during circuit adjustments. There are, indeed, some digital voltmeters (DVMs) which have a ten or fifteen segment 'bar-graph' display, in addition to the digital read-out, but these offer nowhere near the delicacy of adjustment possible with even a cheap analogue meter.

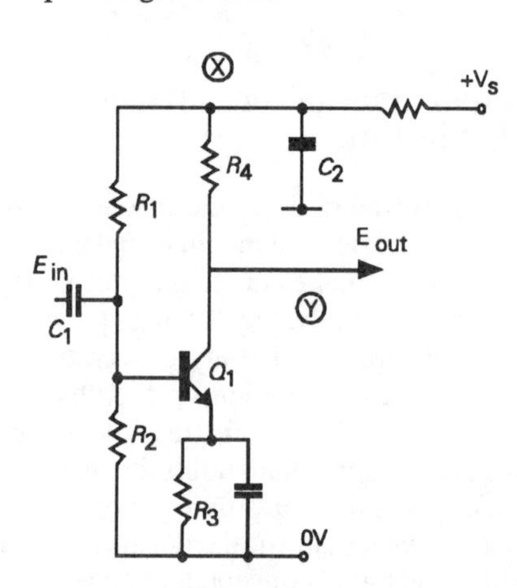

Figure 17.1

Most direct reading analogue displays are based on the use of a 'moving coil' indicating instrument, sometimes called a d'Arsonville movement, after its inventor. The construction of these is shown, schematically, in Figure 17.2. In this, a coil of wire, wound on a thin rectangular former, is suspended in the gap between a pair of magnetic pole-pieces, and pivoted so that it is free to swing about its axis. A pointer is attached to this movable coil, and the whole movement is caused to return to some predetermined zero position by means of a pair of spiral wound springs. These are mounted in opposition, so that very little torque is required to cause the movement to swing about its axis. The displacement of the meter movement away from its zero position is due to the interaction of the magnetic field due to the permanent magnet and that due to passage of current through the coil to which the external meter circuit is connected. The resulting movement of the coil is related to the strength of the permanent magnet and the smallness of the air gap in the magnetic circuit. This is reduced by inserting a soft iron collar into the gap between the permanent magnet poles, and the profile of this can be adjusted, if required, to improve the linearity of the display. Ideally the instrument should be both sensitive and robust. Unfortunately, these requirements are in conflict, in that high sensitivity of indication implies a light coil wound with fine wire, suspended by delicate springs on very low friction bearings in a narrow magnetic gap, while robustness demands the opposite. The skill of the instrument manufacturer lies in resolving these conflicting requirements, especially if, as is common, the movement is to be used in a multi-range test instrument which may have a rough life. A fairly common safety precaution with high sensitivity moving coil (m/c) meters is to incorporate a 'cut-out' mechanism which will interrupt the electical circuit to the meter movement if the pointer travels beyond its full-scale reading, or if the acceleration of the pointer exceeds some predetermined value.

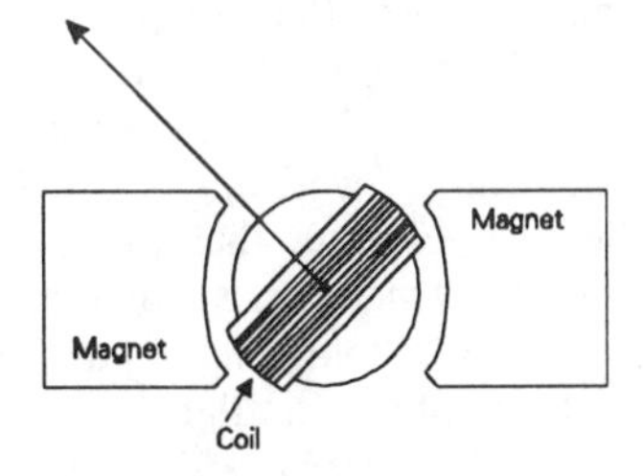

Figure 17.2 *The d'Arsonville m/c meter movement*

If a moving coil instrument is to be used solely as a DC voltage indicator, a high resistance, multiple turn coil can be used, the upper limit to the number of turns, and the consequent coil resistance, being set by the lowest full-scale deflection (FSD) voltage likely to be required from the movement. If, on the other hand, it is to be used solely as a direct current indicator, a low resistance coil would be preferred, with the lower limit to the number of turns being set by the minimum current required for a FSD display.

Typical commercial instruments are offered with FSD current ratings of 25, 50, 100, 250, 500μA, and 1mA, with display scale lengths of 3.8–15cm. The accuracy of indication is generally greater in those movements with longer scale lengths, and ranges, typically, from +/–2.5% to +/–0.25%

To use a current indicating meter as a voltmeter, an additional resistance, called a 'multiplier', is usually connected in series with the coil, as shown in Figure 17.3a. The value of this resistance is determined by the relationship

$$R_M = R_t - R_c$$

where Rc is the total meter coil resistance, RM is the value of resistance required for the multiplier, and Rt is the total circuit resistance needed to satisfy the condition that

$$V/I = R_t$$

where I is the meter movement FSD current and V is the required FSD voltage display, in amps and volts respectively.

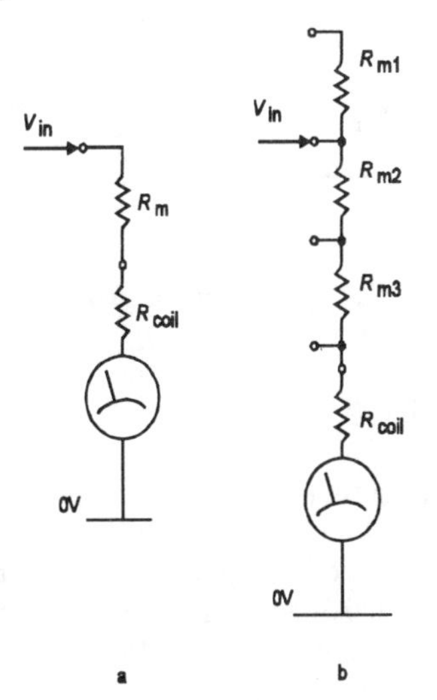

Figure 17.3 *Connection of multipliers in a moving coil voltmeter circuit*

It is apparent that the meter can be made to display as high a FSD voltage value as required by this means, provided that the multiplier resistors are able to support the voltage present between their ends. In normal practice, high voltage multipliers are arranged by connecting a number of lower value resistors in series, and if the value of the multiplier resistors can be altered by switching, as shown in Figure 17.3b, a multiple range indicating instrument can be made.

Current meters

In principle, the FSD current reading given by a meter can also be changed by connecting other resistors, known as shunts, in parallel with the coil, as shown in Figure 17.4a. However, this method of operation tends to demand very low value resistors to give correct operation at high FSD current ratings. As a solution to this difficulty, the 'current multiplier' circuit layout shown in Figure 17.4b is normally used – particularly in multi-range test meters – because this effectively adds a voltage multiplier to the meter circuit, at higher FSD current taps, and allows the shunt resistors to have somewhat higher values, which are easier to make.

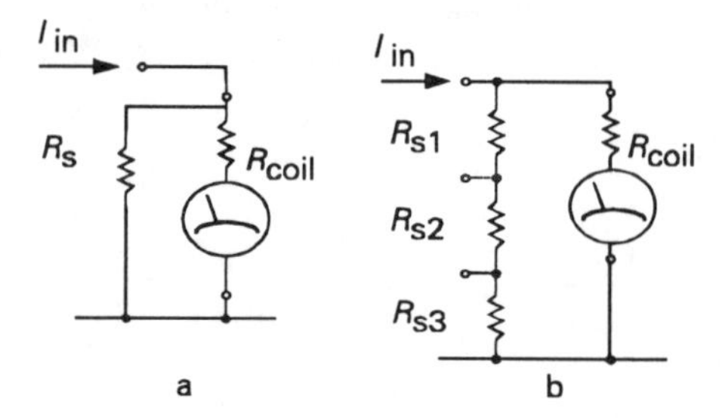

Figure 17.4 *Use of moving coil movement as current meter circuit*

Resistance meters

In essence, a resistance meter (ohm meter), is simply a current meter to which a voltage source, B1, can be connected by way of the 'unknown' resistor, as shown in Figure 17.5a. An adjustable resistor, RV1, is also connected in the circuit to allow the full-scale reading to be correctly set for a zero resistance value. The higher the voltage provided by B1 and the lower the FSD current of the meter, the higher the value of the

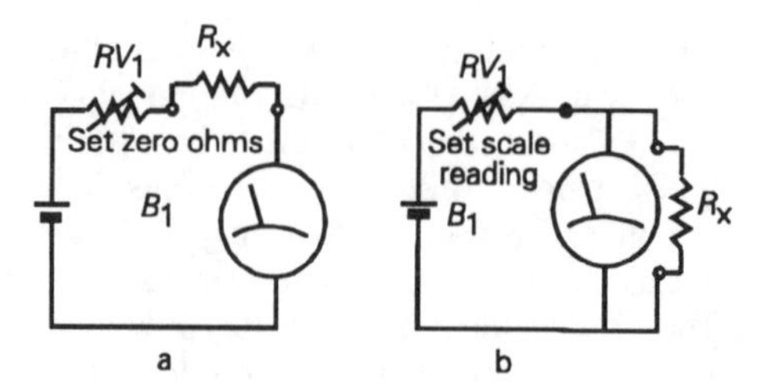

Figure 17.5 *Use of moving coil meter as a resistance measurement circuit*

unknown resistor for which a measurement can be made.

For measurements on lower values of resistor, a simple design approach is to shunt the m/c movement so that it has a greater FSD current rating. Another method is to use the circuit shown in Figure 17.5b, in which the unknown resistor is connected in the position of a current shunt. This layout has the advantage that the resistance reading is in the correct direction – i.e. with the zero resistance value at the LH end of the indicator scale.

Meter sensitivity

As mentioned above, there is a conflict between the requirements of ruggedness and sensitivity in any simple direct-reading analogue instrument, simply because the amount of current which the instrument needs to draw, in order to generate the magnetic field needed to displace the pointer from its rest position, increases as the solidity of the coil and the strength of the springs is made greater. Recognizing this fact, the sensitivity of a moving coil voltmeter is normally expressed as 'ohms per volt'. A low-sensitivity, and robust, instrument might well be specified as 1000 ohms/volt, which would imply that it was based on a basic 1mA FSD current meter. A 10 volt range on such a meter would imply a total meter plus multiplier resistance of 10,000 ohms, similarly a 100V range would imply a total meter resistance of 100,000 ohms. If a voltage measurement was made using such an instrument it would be inaccurate to the extent that the current drawn by the instrument, which would be in the range 0–1mA, would reduce the potential at the point of measurement. If the circuit being measured was a low resistance one, the error could be tolerable, but in a high resistance circuit the error could be quite large, and, perhaps, in a valve or FET circuit, with high circuit resistances, the change in DC potential which occurred, when the voltmeter was connected, might be so great that the circuit would no longer work correctly.

A more sensitive, and probably more delicate, instrument could well be rated as 20,000 ohms/volt, which would indicate that the fundamental meter movement was of a 50μA FSD type. With such a meter, a 10 volt range would present a meter circuit resistance of 200,000 ohms.

There are, of course, always snags and a basic 50μA movement would probably require 300mV to be applied to the coil to produce a full-scale deflection. This would determine the minimum voltage drop which would be produced by the meter in use as a current meter, which should, ideally be zero.

Moving iron meters

These are not very often found in small power electrical and electronic circuitry, but are still moderately common in higher power systems. These consist basically of a fixed coil through which the passage of current generates a magnetic field which attracts a pivoted magnetic 'tongue' away from some spring-loaded rest position. These movements do not offer very high sensitivity, but are robust and can be made quite linear. They can offer comparable accuracy to m/c designs.

Electrostatic movements

These instruments make use of the electrostatic attraction which exists between two oppositely charged plates – or between a charged plate and an earthed plate – to produce a deflection in a spring-loaded pivoted movement similar to that of a moving coil meter. However, in this case, instead of a coil twisting in a magnetic field, the pivot carries one or two suitably shaped plates which are arranged so that they may be drawn, by electrostatic attraction, into the gap between an oppositely charged pair. The layout used is shown, schematically, in Figure 17.6.

Meters of this type have the advantage that they draw a negligible amount of current from the circuit under test, and since they are insensitive to polarity

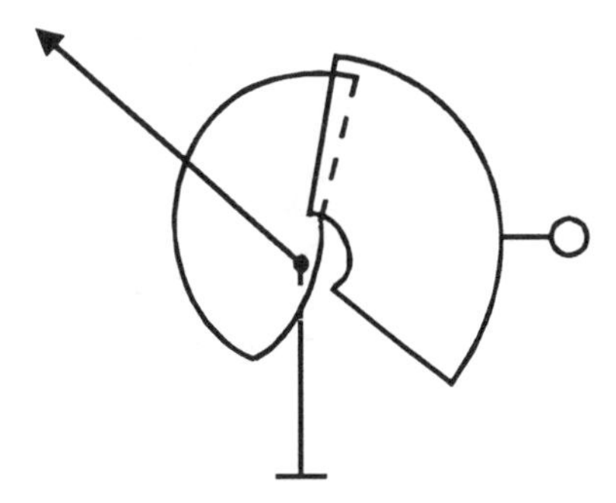

Figure 17.6 *Schematic arrangement of an electrostatic meter movement*

they can be used to measure AC as well as DC. However, they are not very robust, and they are seldom available for FSD voltages of less than 500V (2.5kV–10kV are more common ratings), and their meter scale is very non-linear, being very cramped at the low voltage end.

Electronic analogue meters

A whole new range of test instruments can be made by combining an electronic amplifier with a moving coil milliammeter movement, and these can combine high sensitivity with robustness and freedom from damage through input overload. In the earliest of these instruments, a typical circuit design made use of a balanced long-tailed pair, of the kind shown in Figure 17.7 for an arrangement based on discrete transistors. In this type of layout, the display meter is connected between the drains of the two J-FETs, Q1 and Q2, and the voltmeter input is connected, via a suitable voltage multiplier chain, to the base of Q1. A preset trimmer potentiometer connected between Q1 and Q2 sources allows a 'set zero' adjustment.

With contemporary components, a similar result could be achieved, at a much lower level of static battery drain, by the use of an op. amp. connected as a unity gain voltage-follower, as shown in Figure 17.8a. In this circuit, the minimum FSD voltage will be that which would be required to drive the m/c movement to full scale, on its own, but with an input resistance which could be almost as high as one wished, particularly if one used an FET-input op. amp. type, such as a TL071 or an LF351.

Greater sensitivity can be obtained by connecting

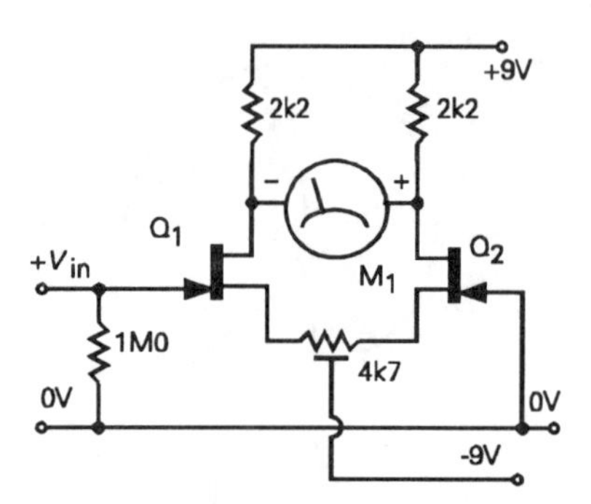

Figure 17.7 *Simple electronic voltmeter*

the op. amp. as a DC amplifier, as shown in Figure 17.8b. In both cases, a meter zero adjustment can be arranged, if needed, by the use of the op. amp. 'offset adjust connections' (usually pins 1 and 5 in the 8 lead pin-out configuration) provided for this purpose.

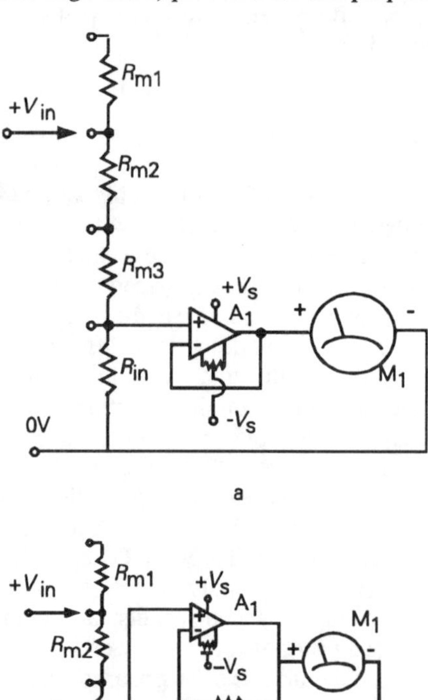

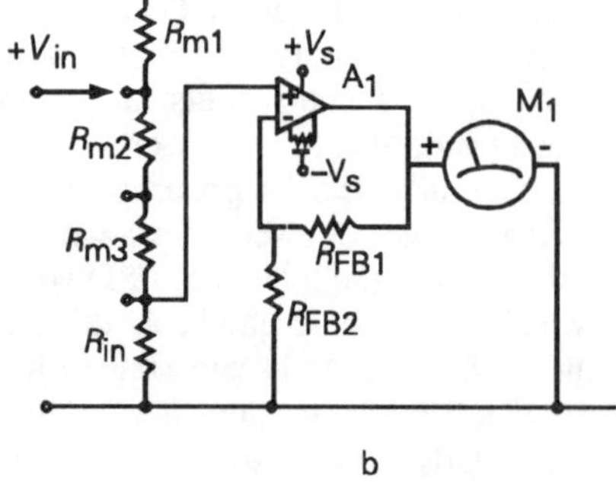

Figure 17.8 *Improved electronic voltmeter circuits*

A somewhat more elegant, and certainly more versatile, circuit layout is shown in Figure 17.9, in which the m/c meter movement is connected in the op. amp. negative feedback path. In this case, the internal resistance of the m/c movement is almost irrelevant, provided that its FSD current demand is less than the maximum permitted output current of the op. amp., and the meter FSD is determined solely by the chosen setting of the preset potentiometer RV1. This situation arises in this type of circuit layout because the feedback loop will operate so that the potential difference between the + and – (non-inverting, and inverting) inputs will be the op. amp. output voltage (the voltage required to drive the meter to its appropriate scale deflection) divided by the op. amp. open-loop gain. For example, if the op. amp. has a DC gain of 100,000, and the meter requires 300mV for FSD, the differential input voltage required between the +in and –in connections will be 300/100,000 mV, or 3μV, which is negligibly small.

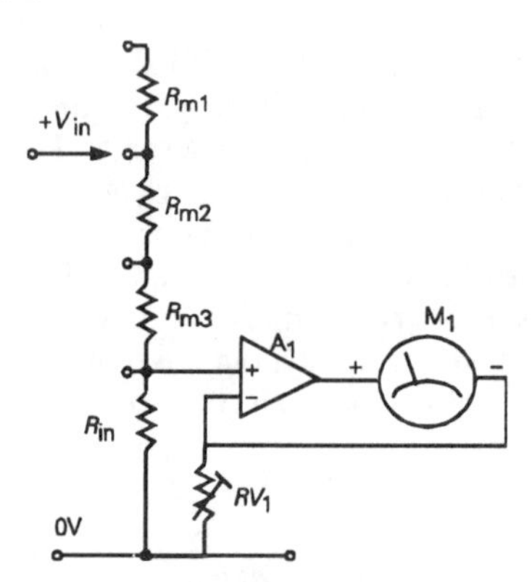

Figure 17.9 *Unity voltage gain*

In addition to the ability of such a circuit to operate as an exceedingly high input impedance voltmeter, this kind of layout would also allow the construction of a test instrument which had an exceedingly low voltage drop when used as a current meter. For example, the circuit would require an input voltage differential of only 1mV for a meter FSD, if one employed a 500mA m/c display movement and RV_1 was set to 20 ohms.

This kind of circuit arrangement would also allow the construction of a very high sensitivity ohmmeter, when used in either of the circuit layouts shown in Figure 17.5.

Digital meters

In general digital instruments supplement rather than supplant their analogue predecessors, since both systems have advantages and snags, as noted above. A variety of techniques are used to derive a numerical read-out proportional to the input DC voltage, of which the two most common are the 'double ramp' system, and the 'successive approximation' method.

A typical digital voltmeter circuit layout using the double ramp system is shown, in schematic form, in Figure 17.10. In this, the input (analogue form) signal is amplified, as required, by an input buffer stage, and used to generate an output current which is directly proportional to the input voltage. This current is then caused to charge a high quality low leakage fixed capacitor for a precisely controlled period of time. At the end of this time, an electronic switch is actuated which causes the ramp output voltage to fall linearly towards zero, through a constant current load circuit. This switch also resets the pulse counter so that it will display the number of clock pulses which occur during this downward portion of the voltage ramp, in the time between the ramp voltage starting its downward slope and reaching the zero level once more. A single cycle of operation is shown in Figure 17.11.

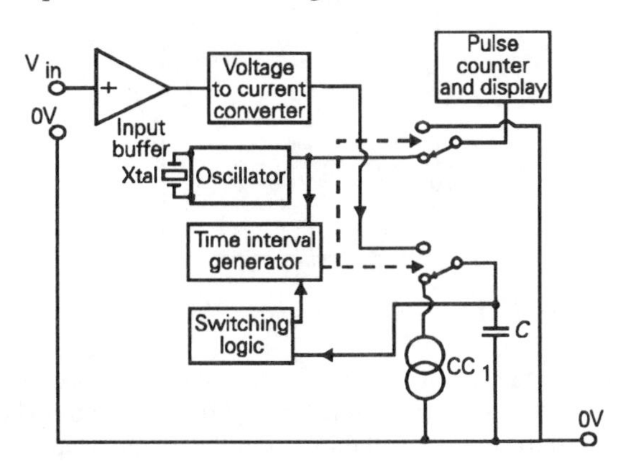

Figure 17.10 *Schematic layout of digital voltmeter*

Because such digital signal conversion systems usually have a fixed full-scale input voltage range, such as 0.1999V or 1.999V, and the use of input voltage gain stages will increase the risk of zero drift, digital current meters usually have a greater insertion voltage drop, sometimes called the voltage burden,

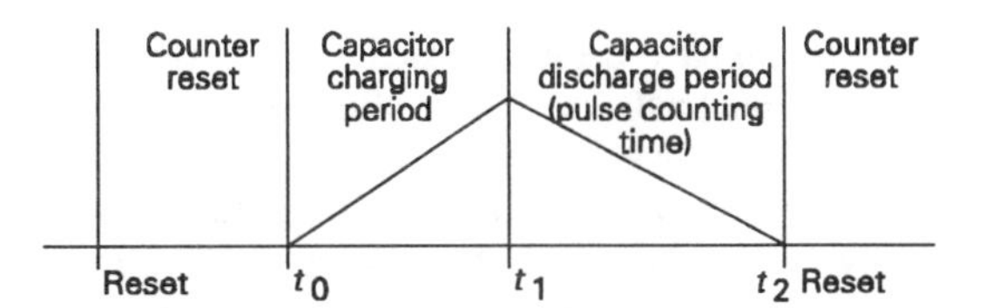

Figure 17.11 *Operating cycle of double ramp digital voltmeter*

than electronic analogue instruments. However, with this proviso, it is just as simple to construct a multi-range digital volt/amp/ohm meter using a digital display meter as it is to do so with an analogue one, using the types of circuit arrangement shown above for analogue meters. The AC bandwidth of inexpensive digital multimeters (DMMs), is usually limited to a few kHz by the characteristics of the coupling transformer/rectifier circuit used to derive a DC output voltage from the AC input signal.

The successive approximation technique uses the internal electronic 'clock' to generate a binary coded sequence of decreasing magnitude voltage steps, as shown in Figure 17.12. At the end of each step, the output from a voltage comparator is used to sense whether the sum of the 'step' voltages is greater or less than the input voltage, and then either to reject (logic 0), or to accept (logic 1), the last binary encoded voltage step from the binary staircase sequence. This provides, directly, a digitally encoded transform of the input voltage level at the instant of sampling.

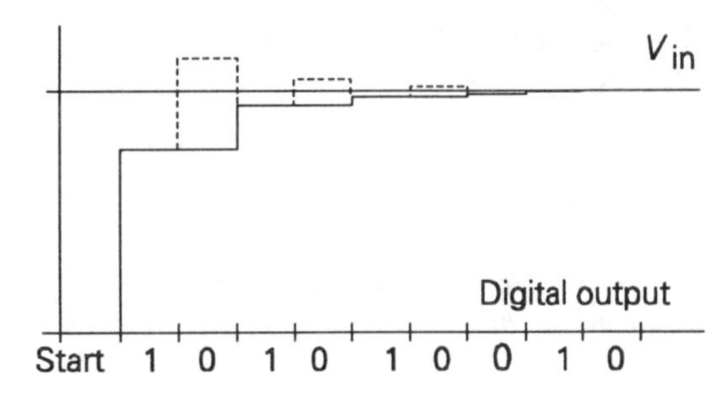

Figure 17.12 *Operation of successive approximation type of digital voltmeter*

Inductance and capacitance measurement

Various techniques are used to make these types of measurement, depending largely upon the relative importance of convenience or precision. Of the various systems which have been proposed or used, the most common, in the past, was one or other of the 'bridge' arrangements shown in Figure 17.13, in which the resistance or impedance of one or more of the arms of the bridge would be adjusted to give a 'null' reading on some adequately sensitive indicator mechanism. If the adjustments of the reference arms are precisely calibrated, bridge measurement systems can give high accuracy, but achieving a null reading may require the sequential adjustment of several controls, which is usually time consuming.

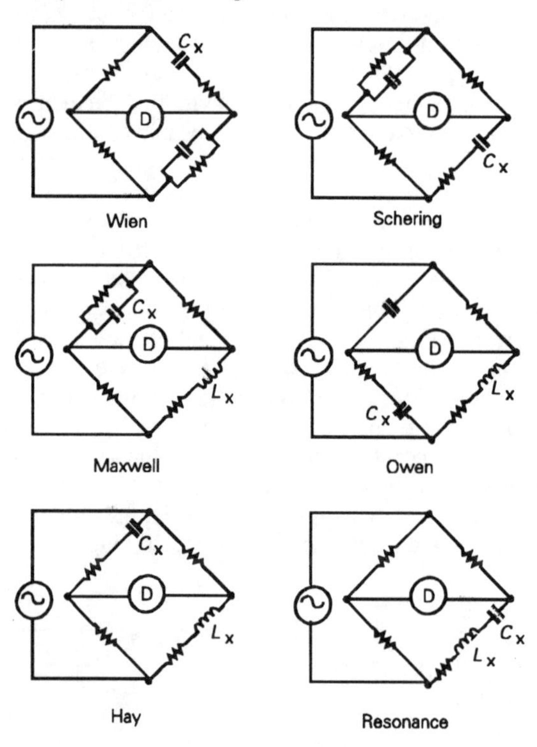

Figure 17.13 *Some of the common bridge circuits used for inductance and capacitance measurement*

Alternatively, the value of the component under test can be measured by connecting it in a circuit whose performance is determined by the inductive or capacitative impedance of the unknown component, perhaps as part of the frequency determining network in an LC or RC oscillator, from which the value of the L or C can be discovered by noting the resultant resonant frequency of the system.

However, for a simple indicating meter, the most straightforward approach is just to use the kind of layout shown in Figure 17.5a, but using an AC current measurement circuit, with a source of AC voltage in place of B1. One can then make use of the fact that, for any input frequency, there will be a precise relationship between alternating current flow and the value of the capacitor or inductor, in Farads or Henries. Appropriate AC current meters are shown in the section covering dynamic measurements.

Measurement of Q

A particular type of measurement which is of interest where inductors are intended for use in resonant circuit applications, more particularly at high frequencies, is that of the circuit magnification factor, or Q. A variety of techniques have been proposed for this purpose, but many of these are variations of the simple layout shown in Figure 17.14, in which the magnitude of the AC voltage developed across a tuned circuit, with a standard value, high quality capacitor, in series with the unknown inductor, is determined when the system is forced into resonance by an input signal from a constant output voltage AC signal generator.

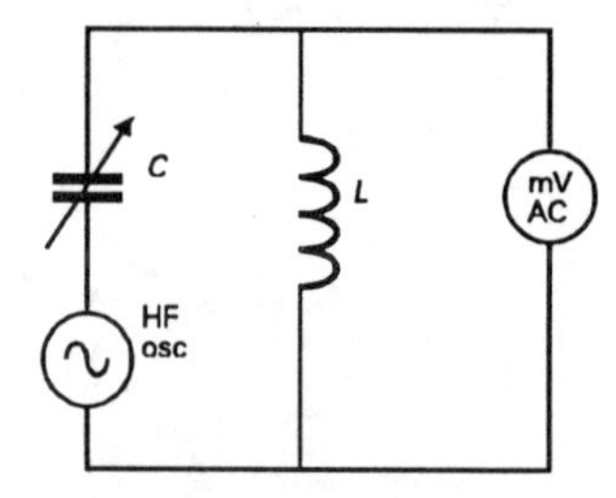

Figure 17.14 *Simple Q meter*

The measurement of LC resonant frequency will also allow the values of the capacitor or inductor to be determined, by calculation, from the relationships

$$L = (1/4\pi^2 f_r{}^2 C) \quad \text{and} \quad C = 1/(4\pi^2 f_r{}^2 L)$$

where fr is the frequency of resonance, and L and C are the values of capacitance and inductance of the components under test, provided that the signal generator output frequency, and either the value L or C is known.

Chopper type DC amplifier systems

Because of the difficulties associated with voltage drift in electronic amplifiers, due to thermal effects and component ageing, the measurement of very low voltage levels presents particular difficulties. A commonly used method for avoiding or minimizing this problem is the use of a chopper system of the kind shown, schematically, in Figure 17.15. In this circuit, an input switch, S1, is used to connect the input of a gain stabilized AC amplifier alternately to the input DC signal voltage and to the signal 0V line. If the input switch is rapidly reciprocated between these two input points, the incoming DC potential will be converted, at the input to the amplifier, into an equivalent alternating rectangular waveform, and this can then be amplified. If the amplified AC output signal is then rectified, it will provide a DC output which is directly proportional to the input DC voltage, but unaffected by any drift in the amplifier DC levels. In early systems, the chopper mechanism was commonly used to operate switches working in synchronism at both the input and the output of the amplifier, in order to provide the necessary output AC to DC conversion, but, with contemporary equipment, electronic input switching systems and AC/DC conversion techniques – see below – are more commonly employed.

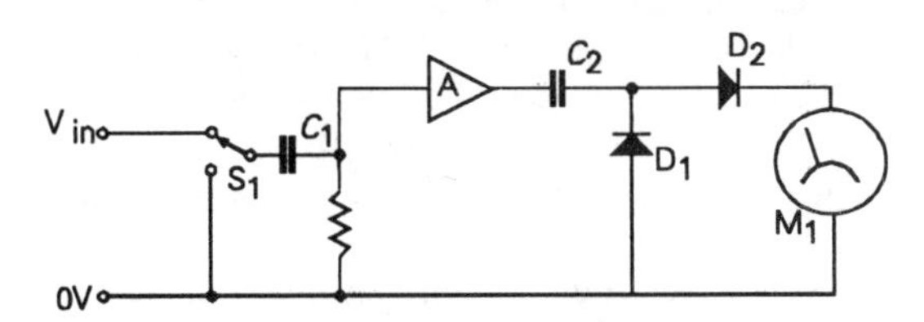

Figure 17.15 *Simple chopper-type DC amplifier*

The simple system shown in Figure 17.15 would only be able to operate with steady state, or only slowly varying, input signals – depending on chopper speed. This has encouraged the development of chopper stabilized systems, in which a conventional op. amp. is used as the amplifier, but with a 'nulling amplifier' used to monitor and cancel the main amplifier DC drift. Because of the great value of such chopper stabilized amplifiers for low level DC signals, complete circuits of this kind are now available as IC packages, such as the LCT1052, the ICL7650 and the ICL7652 devices. These offer effective input DC drift levels as low as 100nV/√month.

Dynamic testing

AC voltage and current meters

Direct reading instruments

The simple d'Arsonville type moving coil current meter can be used to indicate the magnitude of an applied AC voltage or current by rectifying the input voltage waveform, using one or other of the rectifier circuit arrangements shown in Figure 17.16, but all of these arrangements will suffer, to some extent, from the fact that all solid-state rectifiers have the less than ideal kind of conduction characteristics shown in Figure 17.17. The principal problem which results from this type of forward/reverse conduction characteristic are that a finite forward voltage must be applied before the rectifier will conduct at all, and that there will also be some reverse conduction.

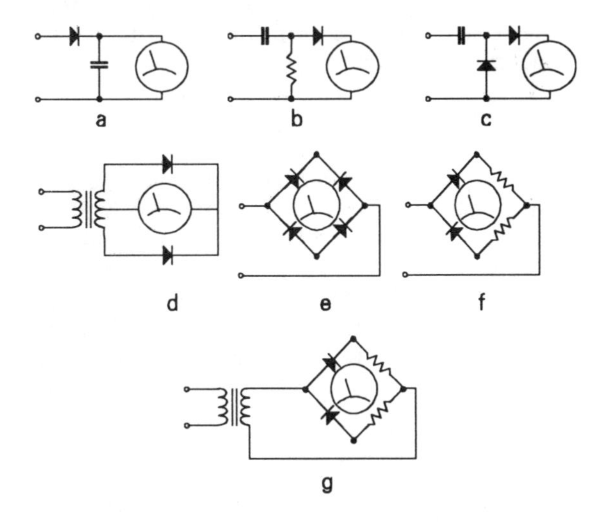

Figure 17.16 *Meter rectifier circuits*

Those types of rectifier, such as the silicon junction diode, which have low reverse leakage currents, at voltages below their reverse breakdown potential, require a forward potential of 0.55–0.6V before they will conduct at all, while those rectifying diode systems, such as germanium P–N junction diodes, or copper/copper oxide rectifiers, which have relatively low forward threshold potentials, suffer more from reverse

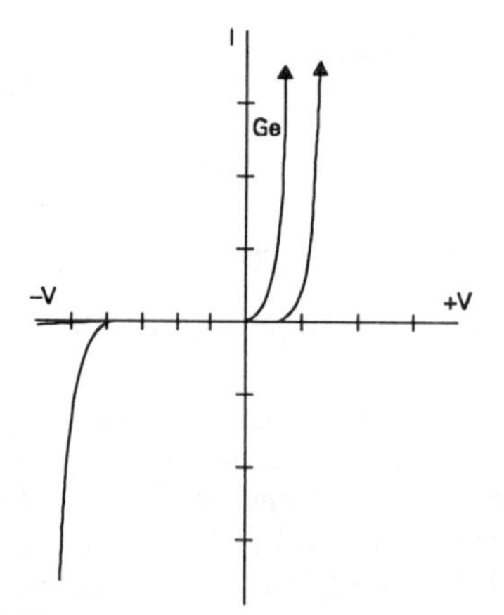

Figure 17.17

leakage. This means that any simple meter/rectifier circuit will tend to be rather nonlinear in its response to low voltage inputs: a problem which can be minimized by including a resistor in series with the meter movement, of a value which will swamp the variation in the rectifier resistance with voltage in its forward (conducting) direction.

Waveform sensitivity

A further problem with any alternating voltage indicating instrument is that its indication will be influenced by the waveform of the input signal – as shown in Figure 17.18 – a circumstance which will be of little importance with a symmetrical sinusoidal input waveform, but will influence the reading quite a lot if the input signal is non-sinusoidal: a factor which must be borne in mind when making measurements.

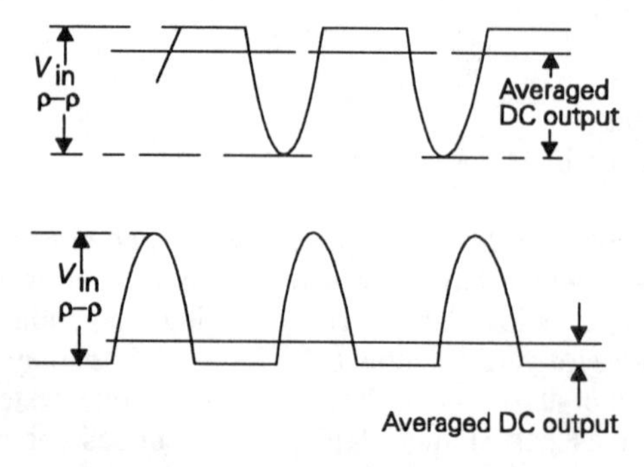

Figure 17.18 *Influence of input waveform on average DC output voltage*

A simple half-wave rectifier circuit such as that shown in Figure 17.16a, will, with a sinusoidal input voltage waveform, give an output which is approximately half the RMS value. If the meter is calibrated to take this factor into account, it would give an erroneous reading – either too high or too low – if the input voltage waveform was non-symmetrical. The particular circuit illustrated would also suffer from the problem that any DC component, associated with the AC voltage to be measured, would also affect the meter reading, as would also be the case in the full-wave rectifier circuits of Figure 17.16e, and 17.16f.

This problem can be avoided by using an input DC-blocking capacitor, as shown in Figures 17.16b and 17.16c, but with the penalty that the meter would lose sensitivity at lower measurement frequencies, unless C1 was made adequately large. Since an alternating input voltage would be applied to the rectifier circuit, a polar (i.e. electrolytic) capacitor type would be unsuitable, and large value non-polar (i.e. polyester or polycarbonate film type) capacitors are bulky.

The current transformer input arrangement shown in Figure 17.16d provides a better solution to this problem, in that, in addition to isolating the meter/rectifier circuit from any incoming DC potentials, if it is connected as a voltage step-up component, it will allow the voltage present across the meter/rectifier circuit to be increased, thereby reducing the importance of rectifier nonlinearity. Such an input transformer will, of course, also impose some limits on the bandwidth over which the meter will give an accurate reading.

Full-wave rectifier systems such as the voltage-doubler layout of Figure 17.16c, and the bridge circuits of Figures 17.16e, and 17.16f, tend to be less affected by input waveform asymmetry, because they tend to average the readings from the two halves of the waveform. However, while a sinusoidal input will give a meter reading which approximates to a true RMS value (0.707Vin) in the case of a sinusoidal signal; a rectangular waveform will give a peak value reading, and both triangular and sawtooth waveforms will give readings which approximate to half of the peak value.

All of the circuits shown in Figure 17.16 can be made to give an approximate peak voltage indication by putting a capacitor, C2, across the meter, as shown in Figure 17.16a.

Because of the need to insert a swamp resistor in series with the meter movement, to lessen the nonlinearity of the scale reading due to the rectifier characteristics, it is common practice in commercial instruments to use the layout shown in Figure 17.16f, where a pair of swamp resistors are used in place of the other half of the rectifier bridge, either as shown or, more commonly, in combination with a DC-blocking input transformer, as shown in Figure 17.16g.

Electronic meter systems

The use of electronic amplification techniques provides an elegant means of increasing meter sensitivity, and minimizing any nonlinearities due to meter rectifier characteristics, as well as avoiding errors due to the presence of DC components in the signal being measured.

A typical AC voltmeter circuit based on an op. amp. and a moving coil meter/bridge rectifier combination is shown in Figure 17.19. In this circuit arrangement, the fact that the meter/rectifier combination is placed in the feedback path of the amplifier means that the closed-loop negative feedback arrangement acts to generate a feedback voltage across R_2 and C_2 which is closely similar to the input voltage present across R_1. During those parts of the voltage swing across the meter/rectifier combination in which the silicon junction diode rectifiers are open circuit the negative feedback loop around the amplifier is disconnected, so that the amplifier output voltage (as seen at point 'A') is driven, with the full op. amp. open-loop gain, to 'jump' across the rectifier open circuit voltage band, as shown in Figure 17.20a. The voltage developed across R_2 is as shown in Figure 17.20b and that across the meter as shown in Figure 17.20c.

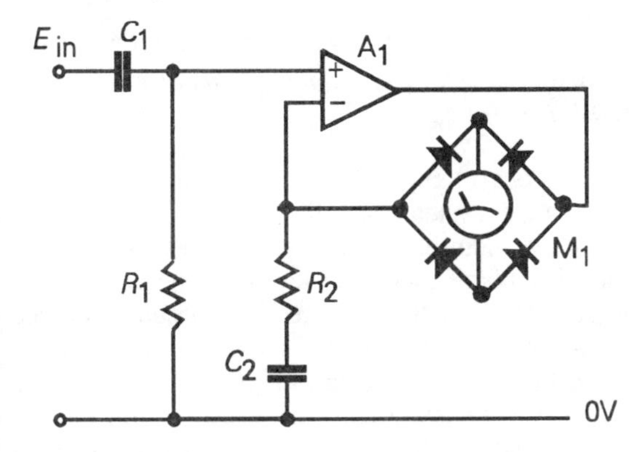

Figure 17.19 *Electronic AC voltmeter circuit*

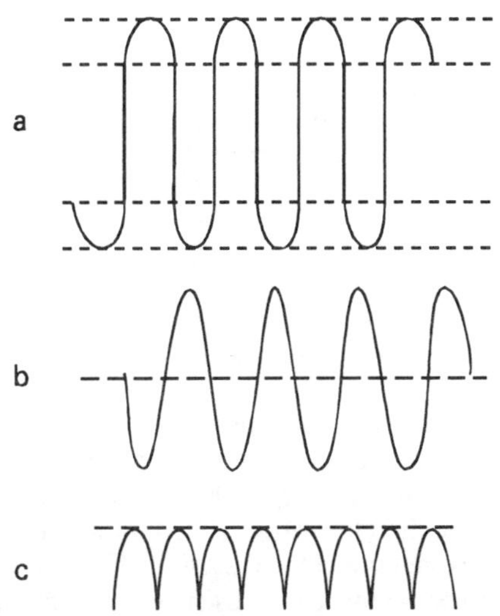

Figure 17.20 *Voltage waveforms in circuit of Figure 17.19*

Because there is no DC path through R_2 and C2, the circuit will be insensitive to DC voltages present across R_1, even were these not, in any case, blocked by C_1. In this circuit, the meter FSD will be ImR2, where Im is the meter FSD current, so that, for example, if R_2 = 100 ohms, and I_m = 100µA, a full scale deflection of the meter would be given by a voltage of 10mV. The sensitivity of the circuit can be increased, as required, by increasing the overall AC stage gain of the circuit. The basic layout of such a multi-range AC millivoltmeter, is shown in Figure 17.21. If the design specification for such an instrument was that it should have a bandwidth of 3Hz–1MHz, at its –3dB points, and a scale range of 1mV–100V FSD, this order of bandwidth and sensitivity could be achieved with the basic meter/rectifier circuit without difficulty. However, there are problems which must be resolved in the design of the input attenuator, in respect of the desired input impedance (which should be high, to limit the error caused by connecting the meter to the circuit under test), and the (input open circuit) noise level at zero input signal, which is due to the thermal noise of the input attenuator resistors, and should obviously be as low as possible.

The designer has two options in this circuit, of which the first is to use a low resistance input circuit, such as the constant impedance attenuator network of the kind shown in Figure 17.22, which offers an identical input

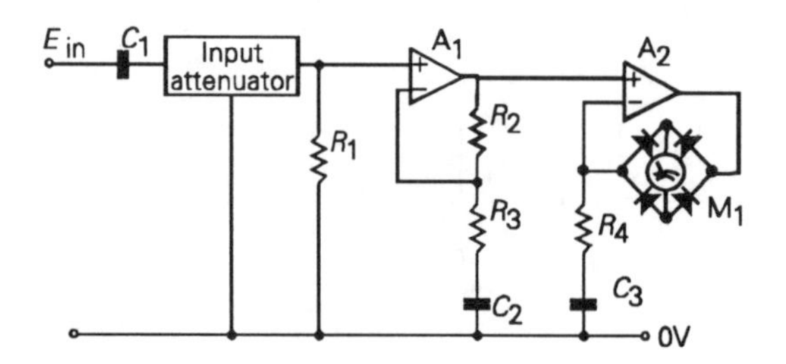

Figure 17.21 *Multi-range AC millivoltmeter*

and output resistance at all range settings, and however long the attenuator chain. In this kind of circuit layout, if K is the attenuation, R_t is the necessary value of resistance required to terminate the line, at both ends, and R_c is the required characteristic resistance of the attenuator (i.e. the value of resistance which will be seen by an ohmmeter between the 0V rail and any tapping point on the network), then the value of a will be given by

$$a = R_c \cdot (K^2 - 1)/K$$

b will be given by

$$b = R_c \cdot (K + 1)/(K - 1)$$

and Rt will be given by

$$R_t = R_c \cdot (K + 1)/K$$

so that if the desired attenuation from one step to the next was 10, and Rc – the input impedance of the meter circuit – is to be 10kohms, then a will be 99k, b will be 12.22k, and Rt will be 11.0k.

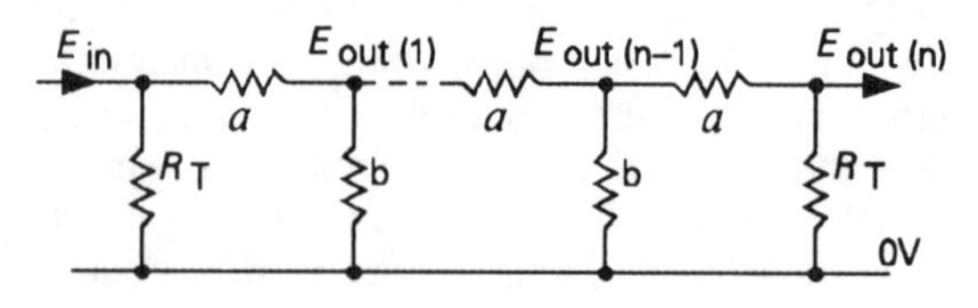

Figure 17.22 *Constant impedance attenuator network*

Assuming that this level of input resistance was acceptable in a general purpose multi-range instrument, and it is somewhat on the low side, nevertheless it would still be associated with a background thermal noise voltage, over the 1MHz instrument bandwidth, of some 12μV. – equivalent to just over 1% of the full-scale reading, which would cause a constant zero offset on all voltage ranges. This would probably be just about tolerable since the offset could be trimmed out by the meter zero adjustment, and the consequent scale error would not be great. On the other hand, increasing the characteristic resistance of the attenuator network to 100k, which would be more suitable for a general purpose instrument, would increase the background meter reading, due to thermal noise, to some 40.6μV, or 4% of FSD, and this would certainly not be acceptable.

Using the simple resistor chain attenuator arrangement shown in Figure 17.23, in which the input resistance at the maximum sensitivity position was 1k, would reduce the zero-input-signal noise threshold to 4μV, or 0.4% of FSD, but this circuit layout would need some very high resistance values – difficult to obtain with any precision – for the 10V and 100V input ranges. Also, if it is required that the basic HF –3dB response of the meter remains the same, regardless of the range setting, then care must be taken to keep the stray capacitances of the attenuator network as low as possible, so that their reactive impedances (Z_c = $1/(2\pi fC)$), are high in relation to that of the attenuator multiplier resistors. This would be feasible with the low impedance attenuator arrangement of Figure 17.22, but not with that of Figure 17.23.

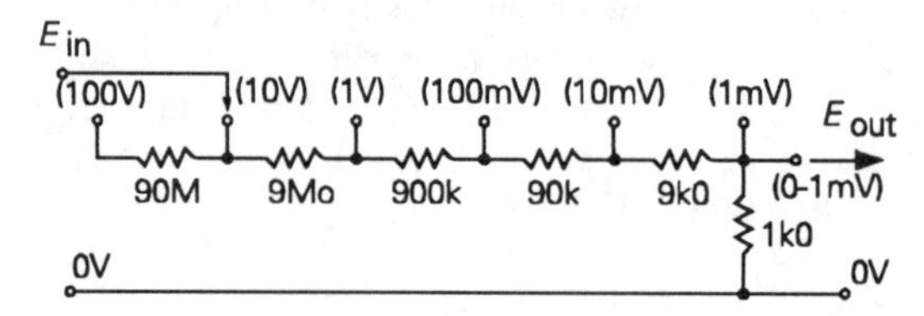

Figure 17.23 *Resistor chain attenuator circuit for 100V–1mV range*

A practical compromise, easily achieved with electronic meter systems, is to use an input attenuator to select the required FSD sensitivity over the ranges 1mV to 100mV, and then to adjust the gain of the amplifier feeding the meter/rectifier circuit to provide the 1–100V range settings, as shown in the final instrument design illustrated in Figure 17.24. The only penalties with this arrangement are that on the the 1mV FSD range, the meter will have a low (1k) input resistance and that the low frequency –3dB point is increased to 30Hz on this range. On all range settings above 100mV FSD the meter would accept an input voltage overload of up to 300V without damage. The variable resistor, RV1, is used, when the instrument is initially assembled, to set its FSD sensitivity to the required full-scale values, using an AC voltage reference standard.

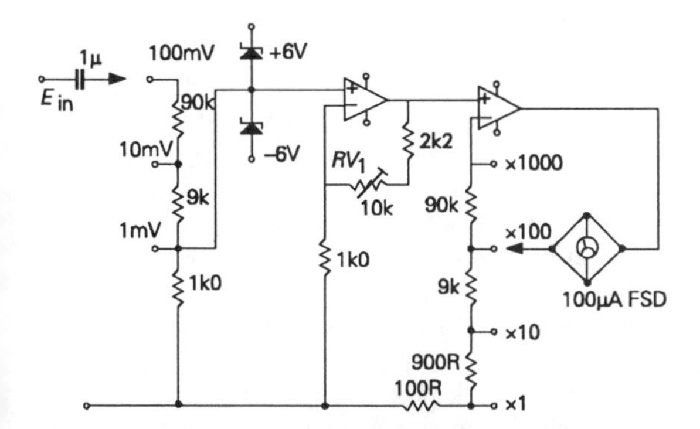

Figure 17.24 *Wide range millivoltmeter with two stage attenuation*

Waveform distortion measurement

It is often desirable that the waveform at the output of an AC signal processing or amplification system is the same as the waveform at the input. If the signal is non-sinusoidal, this kind of waveform distortion, when it is present, can usually be seen on a cathode-ray oscilloscope display, provided that the change in waveform shape is big enough. However, even when the distortion can be seen, it may still be very difficult to define the effect in numerical terms.

With the rather more restricted category of sinusoidal (single tone) signals, a variety of techniques are available which can define, and quantify, the waveform distortions introduced by the system. These types of test may not be so informative about system defects or malfunctions, but are widely used simply because they do give numerical values by which the linearity of the system can be specified. These methods fall, generally, in two categories, 'notch filtering' and 'spectrum analysis'.

THD measurement by notch filtering

This technique for measuring total harmonic distortion (THD) relies on the fact that a 'pure' sinusoidal signal will contain only the fundamental frequency, f_0, free from any harmonics at $2f_0$, $3f_0$, and so on. If the signal under test is fed to a measuring system which has a notch in its frequency response, as shown in Figure 17.25, and if this notch is sufficiently sharp that it will remove the signal at f_0 completely, but give negligible attenuation at $2f_0$ and higher frequencies, then a measure of total harmonic distortion can be obtained by comparing the magnitude of the incoming signal, measured on a wide-bandwidth AC millivoltmeter, in the absence of the notch, with that which remains after the fundamental frequency has been 'notched out'.

There are a variety of circuit arrangements which can be used to provide a notch in the frequency response of an AC measuring instrument, such as the parallel tuned *LC* tuned circuit shown in Figure 17.26a, the Wien nulling network shown in Figure 17.26b, or the parallel-T notch network shown in Figure 17.26c. Of these layouts, the tuned circuit would only be useful, in practice, at frequencies above some 50kHz. Below this frequency the size of the inductor would be inconveniently large. There could also be a problem of hum pick-up due to the inductor itself. Both the Wien and the parallel-T arrangements are practicable layouts for use in the <3Hz–>100kHz range, though they will normally need to be somewhat rearranged to allow the use of negative feedback around the loop to sharpen up the frequency response of the system.

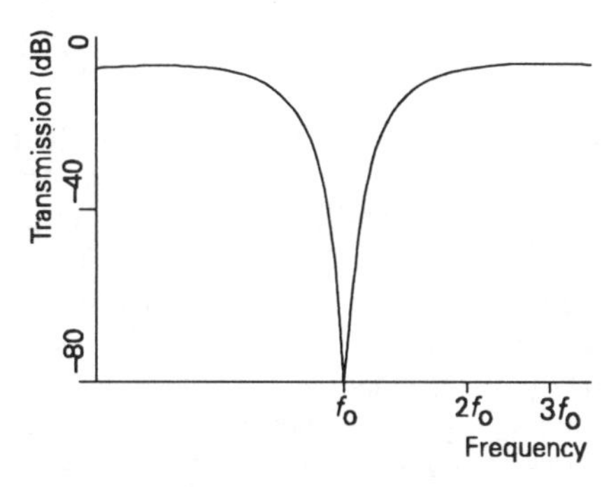

Figure 17.25 *Frequency response notch*

The principal advantage of the parallel-T notch network is that it can be connected directly between the incoming signal and the measurement amplifier (e.g. an AC millivoltmeter) so that no additional noise or waveform distortion is introduced by any input buffering or amplifying stages. Two practical notch type

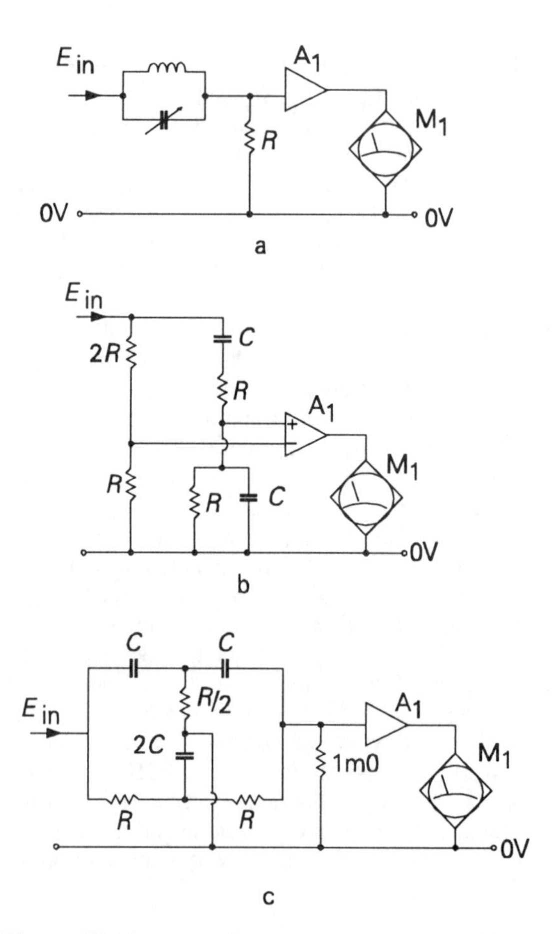

Figure 17.26 *Possible notch circuit arrangements*

distortion measuring systems of this kind, in which the sharpness of the notch is improved by the use of negative feedback around the notch-filter/amplifier loop, are shown in Figures 17.27 and 17.28. These are capable, respectively, of measuring harmonic distortion residues as low as 0.008% and 0.0001% of the magnitude of the fundamental frequency, but this measurement of signal residues will also include any hum and noise voltages present on the input signal – since these will still be left after the test signal, fo, has been removed by the notch circuit – and this snag often limits the sensitivity of the system.

The Wien network notch type of distortion meter is easier to use, on the bench, once the display sensitivity has been set, because it only requires two-knob adustment (frequency tuning and bridge balance adjustment), whereas the parallel-T network generally requires three independent adjustments to achieve a precise null setting. If the signal being measured is at, say, 400Hz and above, reductions in the meter reading error, due to the presence of 50/60Hz and 100/120Hz mains hum, can be made by introducing a steep-cut high-pass filter, with a turn-over frequency at, say, 400Hz, giving, perhaps, a –50dB (316x) attenuation at frequencies of 120Hz or lower.

Similarly, the error in the true THD reading due to the presence of the wide bandwidth white noise voltage present in the output can be reduced by restricting the measurement bandwidth by including a sharp cut-off low-pass filter in the measurement circuit – preferably with a switched choice of turn-over frequencies, at, say, 10kHz, 20kHz, 50kHz and 100kHz.

Intermodulation (IM) distortion measurements

One of the principle problems caused by nonlinearities in an amplifier transfer characteristic is that the magnitude of the output signal from such a circuit will vary according to whereabouts it sits on the transfer curve, as shown in an exaggerated form in Figure 17.29. This effect is always present to some extent in any amplifier which has not got a ruler-straight input/output characteristic. In practice, this means that when two signals are present simultaneously – let us assume that one is large and one is small – the output voltage swing produced by the larger one will sweep the smaller up and down the transfer curve, modulating the amplitude of the smaller signal by so doing. This will lead to the generation of sum and difference output frequencies, so that if the two signals are f_1 and f_2 repsctively, there will be present at the output, in addition to f_1 and f_2, signals at f_1+f_2 and f_1-f_2. Putting some numbers to this effect, if two signals, at 60Hz and 5kHz, were introduced to a distorting amplifier, the resulting 60Hz modulated 5kHz output voltage will contain spurious frequency components at 4940Hz and 5060Hz. An IM distortion test based on signals at these frequencies, at a 4:1 amplitude ratio, has been recommended by the SMPTE (the USA Society of Motion Picture and Television Engineers).

A practical measurement instrument for the SPMTE type test is shown, in outline, in Figure 17.30. In this, the composite signals, after transmission through the system under test, are passed through a high-pass filter which removes the 60Hz component. The modulated 5kHz signal is then demodulated, which will result in

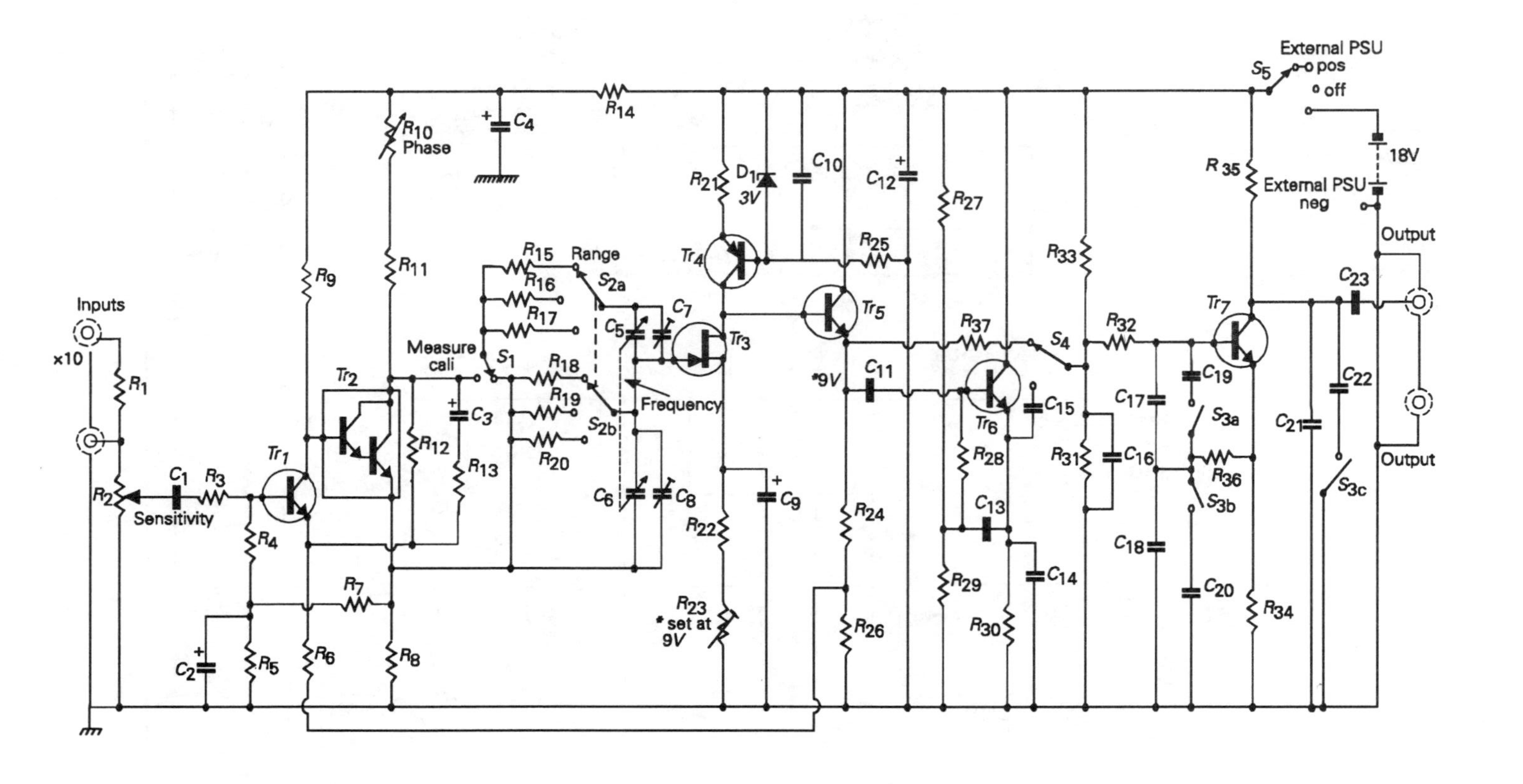

Figure 17.27 *Practical distortion meter based on Wien bridge notch arrangement*

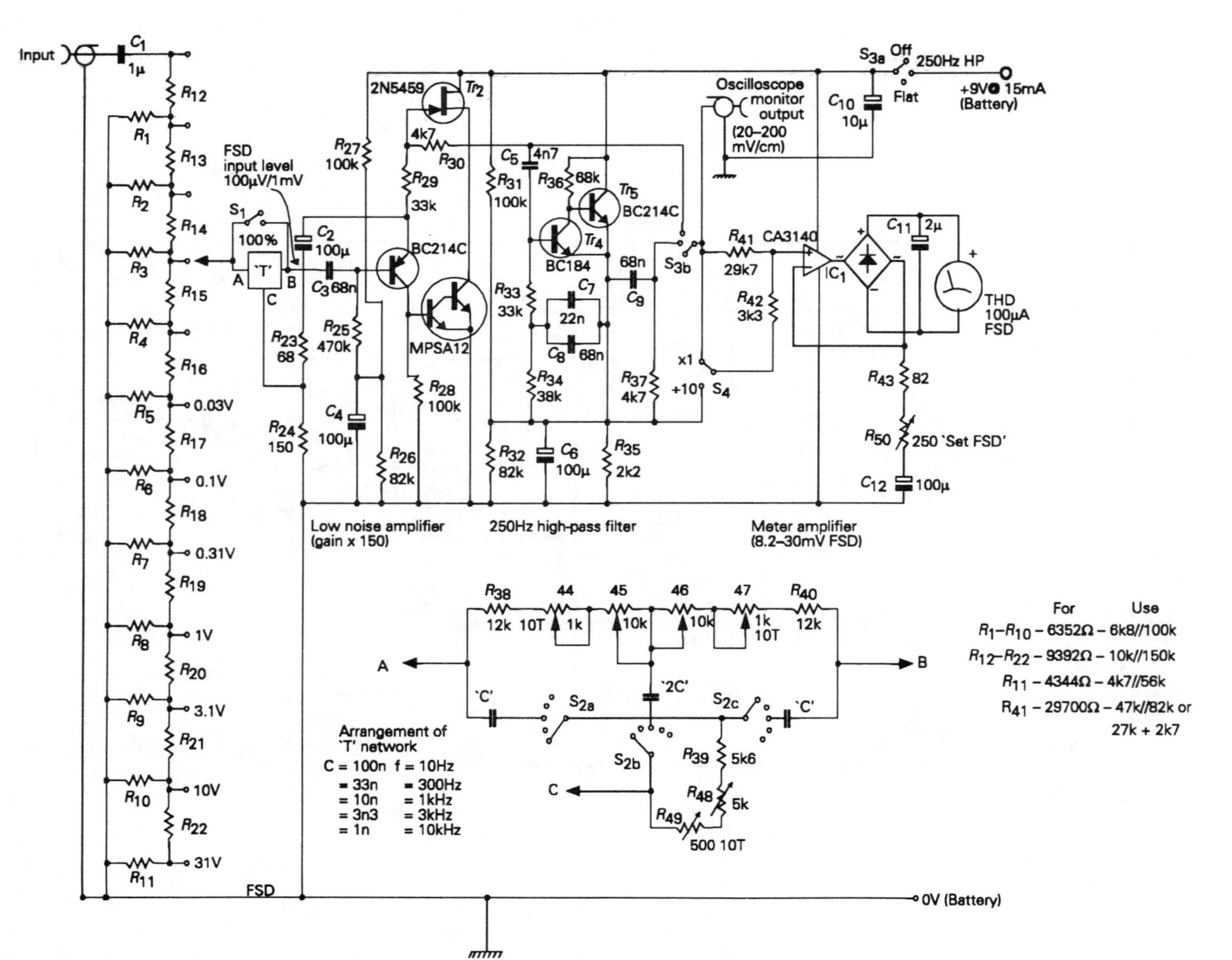

Figure 17.28 *Practical distortion meter circuit based on parallel T notch layout*

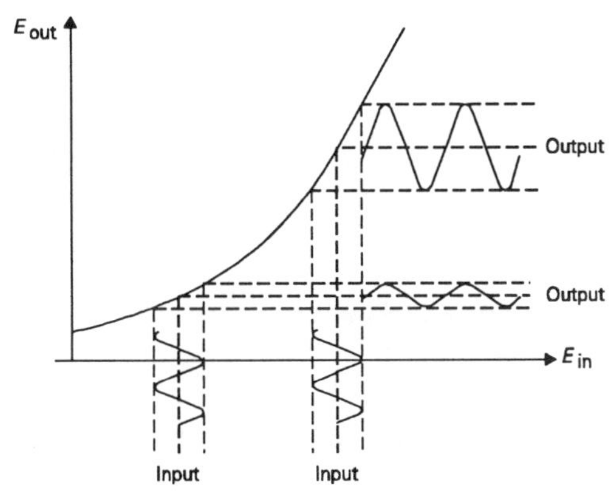

Figure 17.29 *Intermodulation distortion due to nonlinear amplification characteristics*

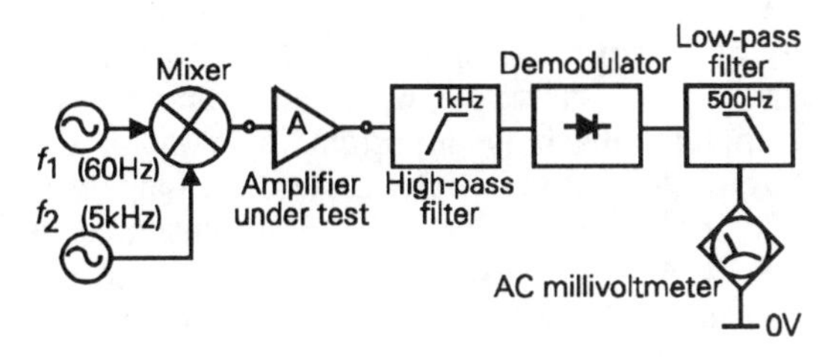

Figure 17.30 *The SMPTE intermodulation test arrangement*

the re-emergence of a 60Hz signal, and passed through a steep-cut low-pass filter to remove the 5kHz signal. The magnitude of the residual 60Hz signal, as a proportion of the 5kHz one, is then defined as the IM distortion figure.

An alternative IM distortion measuring technique has been proposed by the CCIF (the International Telephone Consultative Committee). In this procedure, the input test signal consists of two sinewaves, of equal magnitude, at relatively closely spaced frequencies, such as 9.5kHz and 10kHz, or 19kHz and 20kHz. If these are passed through a nonlinear system, sum and difference frequency IM products will, once again, be generated, which will result in spurious signals at 500Hz and 19.5kHz, or 1kHz and 39kHz, respectively, for the test signal frequencies proposed above. However, in this case, the measurement technique for determining the magnitude of the IM difference-frequency signal is quite simple, in that all that is needed is a steep cut low-pass filter which will remove the high frequency input test tones, and allow measurement of the low frequency spurious signal

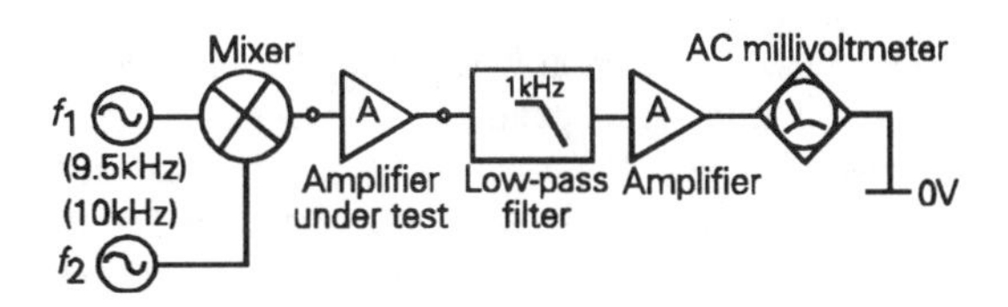

Figure 17.31 *The CCIF intermodulation distortion measurement procedure*

voltages. A suitable circuit layout is shown in Figure 17.31.

In audio engineering practice, the value of IM testing, as, for example, in the case illustrated above, is that it allows a measurement of system nonlinearity at frequencies near the upper limit of the system pass-band, where any harmonic distortion products would require the presence of frequency components beyond the pass-band of the system. Because these harmonics would be beyond the system pass-band they would not be reproduced, nor would they be seen by a simple THD meter, although the presence of nonlinearities would still result in the generation of spurious (and signal muddling) IM difference tones which were within the pass-band.

Spectrum analysis

The second common method of measuring harmonic and other waveform distortions is by sampling the output signal from the system under test by a metering system having a tunable frequency response, so that the measured output voltage will be proportional to the signal component over a range of individual output frequencies. The advantage of this technique is that if the system under test generates a series of harmonics of the fundamental frequency, but there is also an inconveniently large amount of mains hum and circuit noise, which would reduce the accuracy of reading of a simple notch-type THD meter, the relative amplitudes of the individual harmonics – as well as hum voltages – can be individually isolated and measured. In use, the gain of the analyser amplifier is adjusted to give a 100% (0dB) reading for the fundamental (input test signal) frequency, and the relative magnitudes of all the other output signal components are then read off directly from the display. This type of analysis can also be used to reveal the presence of IM distortion products resulting from two or more input test tones.

The background noise level will depend on the

narrowness of the frequency response of the frequency selective circuitry in the analyser, and this will generally depend on the frequency sweep speed chosen, with a lower resolution for a fast traverse-speed oscilloscope display than for an output on a more slowly moving paper chart recorder.

Spectrum analysers of this type are used at radio frequencies as 'panoramic analysers', to show the presence and magnitudes of incoming radio signals in proximity to the chosen signal frequency, as well as for assessing the freedom from hum and spurious signal harmonics in an oscillator or amplifier design.

A typical circuit layout for a spectrum analyser using the superhet system is shown, schematically, in Figure 17.32. In this, the signal to be analysed, and the output from a swept frequency local oscillator signal are separately fed to a low distortion mixer, typically using a diode ring modulator layout, and the composite sum and difference frequency signals are then amplified by an intermediate frequency gain stage, and converted into a unidirectional voltage by a logarithmic response detector circuit. If the frequency of the local oscillator is coupled to the sweep voltage which determines the position of the 'spot' on the horizontal trace of the oscilloscope, or the position of the pen on a paper chart recorder, there will then be a precise and reproducible relationship between measurement frequency and spot or pen position, so that the oscilloscope graticule or the recorder chart paper can then be calibrated in terms of signal magnitude and signal frequency.

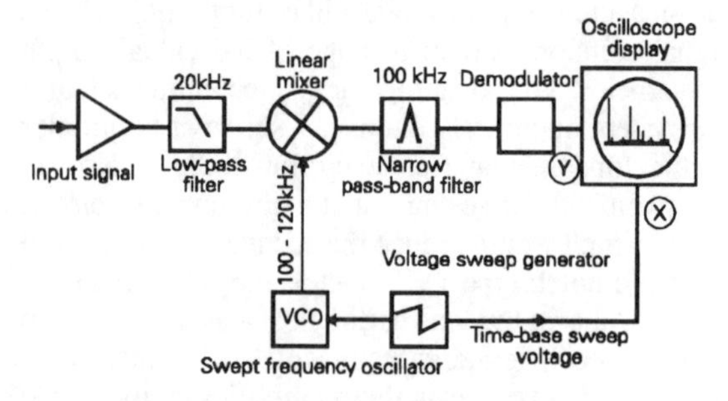

Figure 17.32 *Schematic layout of spectrum analyser*

A typical spectrum analyser display for a high quality audio amplifier, showing mains hum components, as well as the test signal and its harmonics, is shown in Figure 17.33.

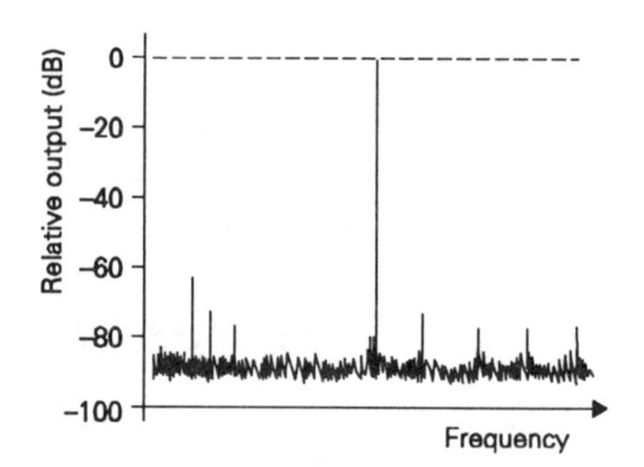

Figure 17.33 *Typical spectrum analyser display*

Transient distortion effects

While the discussion above has been confined to the types of waveform distortions which will affect steady-state (i.e. sinusoidal) waveforms, for many circuit applications it is important to know of the waveform changes which have occurred in 'step function' or rectangular wave signals, as illustrated in Figure 17.34. Since these types of abrupt change in a pre-existing voltage level will imply, from Fourier analysis, an infinite series of harmonics (mainly odd-order – such as 3rd, 5th, 7th and so on – in square-wave or rectangular-wave signals), any signal handling system with less than an infinitely wide pass-band will result in a distortion of the input waveform. This may be just a rounding-off of the leading edge of the signal waveform – due to inadequate HF bandwidth, or 'overshoot' or 'ringing', as illustrated in Figure 17.34, due to the introduction of spurious high frequency components, or just due to errors in the relative phase of the composite signal harmonics.

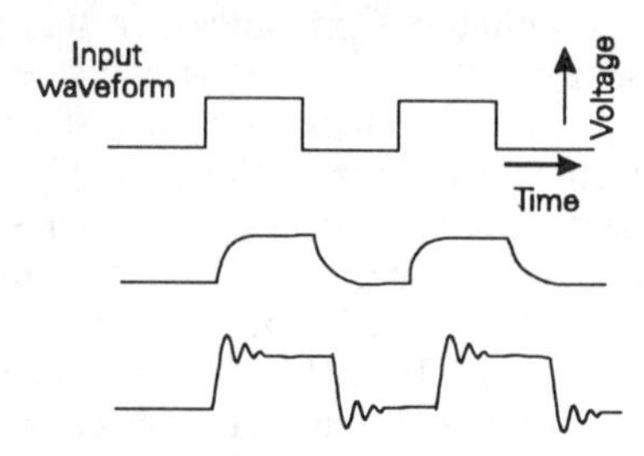

Figure 17.34 *Typical transient distortion effects*

It is difficult to derive numerical values, from simple instrumentation, which relate to the nature of any of these rectangular waveform signals, except 'rise time', overshoot magnitude, 'settling time' – to some specified accuracy level – or slewing rate (the maximum available rise-time, where this is limited by some aspect of the internal system design), characteristics also shown in Figure 17.35. These effects are most easily seen on an oscilloscope display, and if it has calibrated 'X' and 'Y' axis graticules, their magnitudes in time or voltage, may be approximately established.

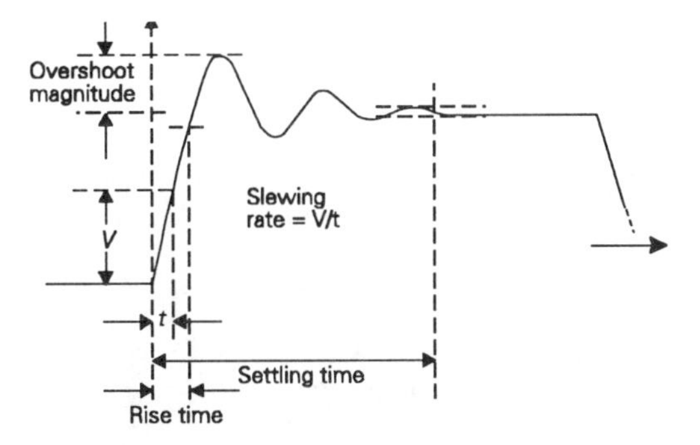

Figure 17.35 *Quantifiable transient response characteristics*

Oscilloscopes

The oscilloscope is one of the most useful and versatile analytical and test instruments available to the electronics engineer. Instruments of this type are available in a wide range of specifications, and at a similarly wide range of purchase prices.

In its most basic form, this type of instrument consists of a cathode ray tube in which the electrons emitted by the heated cathode are focused onto the screen to form a small point of light, and the position of this 'spot' can be moved, in the vertical, 'Y-axis', and horizontal, 'X-axis', directions, by deflection voltages applied to a group of four 'deflection plates' mounted within the body of the cathode ray tube. In normal use, the plates controlling the horizontal movement of the spot are fed with a voltage derived from a linear sawtooth waveform generator, giving an output voltage, as a function of time, which is of the kind shown in Figure 17.36. The effect of this is to allow signals displayed in the vertical, 'Y' axis, to be seen in a time sequence. A range of sawtooth waveform repetition frequencies is normally provided, giving a range of deflection speeds, which will provide cm/s, cm/ms, or cm/μs, calibrations.

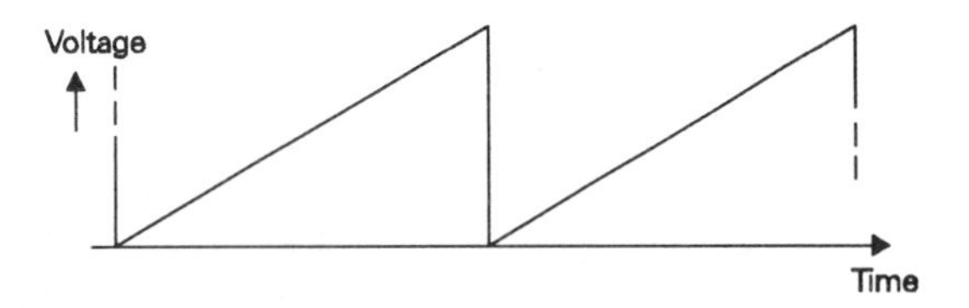

Figure 17.36 *Output sweep voltage applied to oscilloscope X-deflection plates*

The 'X' deflection (time-base) scan speeds are controlled by a range switch, allowing a switched choice in a 1:3:10:30:100:300, type of sequence, but it is also necessary for there to be a variable time-base scan speed control to allow the choice of intermediate values between these set choices of time-base frequency.

An important feature of any time-base sweep generator system is that it should allow the 'X' scan to be synchronized with the periodicity of the signal to be displayed, where this is repetitive, so that the waveform pattern will appear to be stationary along the 'X' axis. The more versatile and effective this synchronization facility is, the easier it will be to use the scope.

A typical time-base sweep voltage generator circuit is shown, schematically in Figure 17.37. It is normal practice in better quality instruments to provide both a time-base sawtooth waveform output, for the control of other instruments, such as spectrum analysers or frequency modulated oscillators (wobbulators), and also a time-base switch position – sometimes designated 'XY' – which will allow the injection of an external signal into the 'X' amplifier circuit. This facility will allow the oscilloscope to be used for measurements such as the relative phase of two synchronous waveforms, and also for determining the repetition frequency of an unknown signal, by means of 'Lissajoux figures', of which some simple patterns are shown in Figure 17.38.

A separate signal input to a low distortion wide-bandwidth voltage gain stage provides the vertical, 'Y' axis, deflection of the oscilloscope 'trace'. In most modern instruments this will be a DC amplifier design, but with a switched capacitor in series with the input circuit to allow the DC component of an input composite signal to be blocked. This would be essential in cases where the presence of a DC offset voltage would otherwise prevent the full range of amplifier gain being used, such as when it was desired to examine a

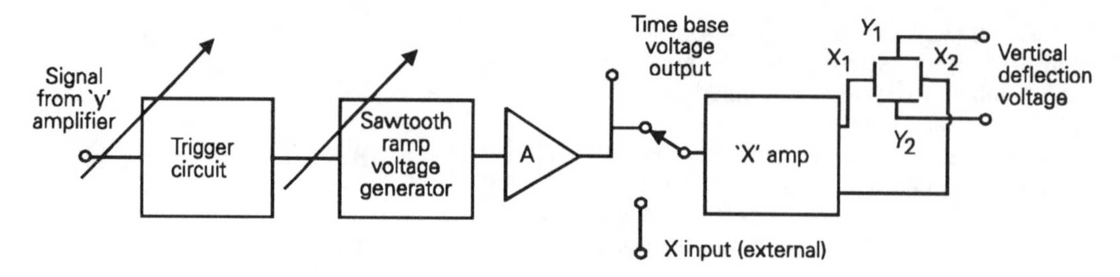

Figure 17.37 *Schematic layout of time-base generator circuit*

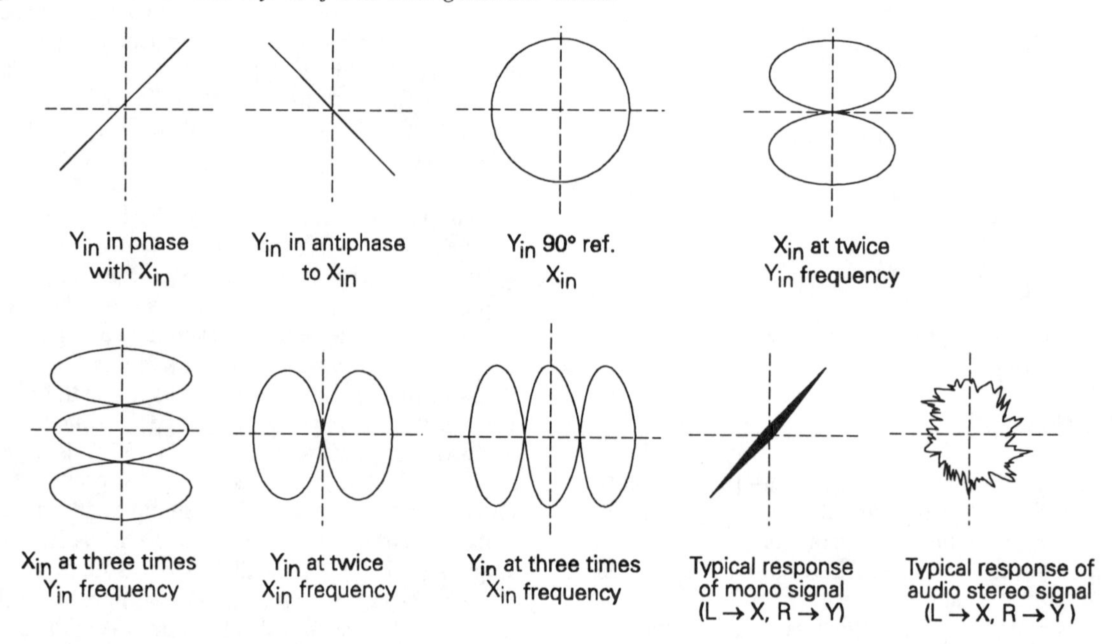

Figure 17.38 *Typical Lissajoux figures*

relatively low magnitude AC signal present at the same time.

Typical 'Y' amplifier specifications offer gain/frequency characteristics which are substantially flat from DC to perhaps 10, 20, 50, or 100MHz, with some specialist scopes giving upper –3dB turn-over frequencies as high as the GHz range. In the lower bandwidth designs, preset switched input sensitivity values ranging from 1mV/cm to, perhaps, 300V/cm are offered.

Multiple trace displays

The enormous usefulness of being able to to display two or more signals, simultaneously, so that one could, for example, with a dual trace instrument, look at the input and output waveforms from a signal handling stage, has made muliple beam scopes much more common. The early twin-beam oscilloscopes used a cathode ray tube which had two entirely separate electron gun, focus and 'Y' deflection assemblies, so that the incoming 'Y' axis signals could be displayed entirely independently. However, this type of tube is both expensive and bulky, so that, in modern instruments, a single gun tube is used, and the time base trace is split, electronically, into two or more separate scans, by applying a rectangular waveform voltage signal to the input of the final stage of the 'Y' deflection amplifier. The same signal is also applied to an electronic

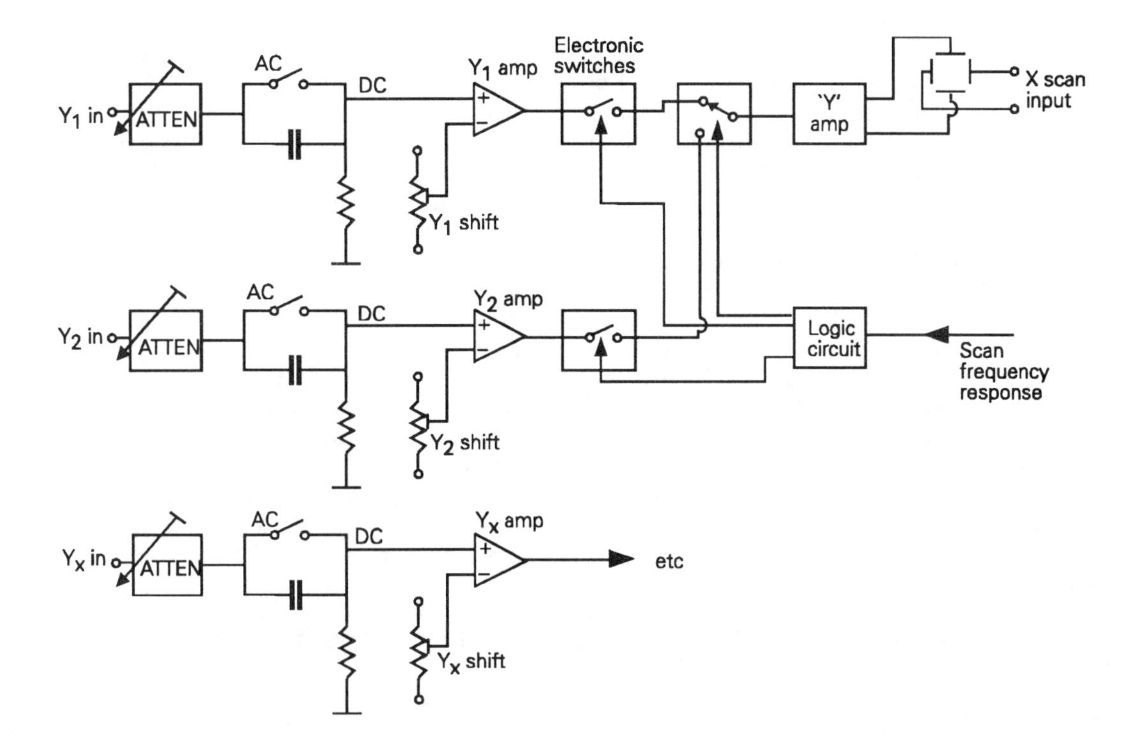

Figure 17.39

switch between the outputs of the 'Y1', 'Y2', 'Y3', etc., amplifiers and the output display driver stage. A schematic layout for this type of multiple trace display system is shown in Figure 17.39.

What happens, in practice, at higher scan frequencies, is that, for a time interval corresponding to the duration of a single 'X' axis sweep, a vertical deflection voltage derived from, say, the 'Y1' front panel 'Y' shift control, combined with the output from the Y1 amplifier itself, are switched to the 'Y' axis output voltage amplifier and deflection plates. During the next 'X' axis sweep, a vertical deflection voltage from the front panel 'Y2' vertical shift control, and the output from the 'Y2' amplifier, are switched to the display system, but the outputs from the 'Y1' amplifier and shift controls are disconnected. Similarly, if there are 'Y3' or 'Y4' input circuits, these will also be used to display their outputs on successive time-base sweeps.

Because of the persistence of vision, at higher time-base speeds, the successive appearances of signals from the various 'Y' inputs will be merged in the perception of the viewer, and will seem to be present simultaneously. Moreover, since the vertical shift voltages are also switched, it is possible to move the sequential 'Y' displays, independently, on the scope screen, exactly as if there were a number of completely independent gun and deflection assemblies within the tube.

At lower sweep frequencies, where the limited persistence of vision would no longer allow a flicker free display, it is normal practice to switch the amplifier outputs and shift voltages at a some high multiple of the scan frequency, so that, if the resolution of each trace were high enough for it to be seen, each trace would consist of a sequence of closely spaced dots.

Storage oscilloscopes

The normal 'real-time' oscilloscope is only able to display the character of a signal, in a given voltage/time sequence, at the moment it happens, and only then if it is accurately repetitive – so that each successive 'X' direction scan repeats and reinforces its predecessors. However, there are instances where the signal which is of interest is sporadic or randomly occurring, and it is wished to examine in detail this particular event, recorded, briefly perhaps, on an externally triggered trace.

In early 'scopes, the only available technique was to use a 'long-persistence' screen phosphor, so that the image of the trace would only fade away fairly slowly. An improvement on this system is to use a type of 'scope design in which one layer of a two layer screen is coated with a phosphor which has a fluorescence threshold somewhat above the energy level provided by the 'writing' beam. The image of the last scan pattern can then be recovered by flooding the screen with electrons from a so-called 'flood' electron gun, and this image will persist, at high brilliance, until the flood gun is switched off, and normal writing continues.

However a whole new family of 'scopes has arisen, to make use of the ability of fast analogue to digital (A–D) converters and high storage capacity digital memory techniques to store and recover digitally encoded signals. Using this technique, an incoming 'Y' axis voltage signal is sampled, at a repetition frequency high enough to give, say, 500 or more voltage sampling points across the 'X' axis scan. This series of discrete signal voltage levels can be stored in random access memory (RAM), cells, from which the digital information can be recovered in its proper time sequence, and decoded, once more, by a digital to analogue. (D–A), converter, to generate an accurate replica of the signal present on the scope at the time when the signal was stored.

The great advantage which this technique offers, by comparison with long-persistence phosphors, or flood gun systems, is that, depending on the resolution of the stored signal – which depends on the number of voltage sampling points, in the 'Y' axis provided by the A–D and D–A converters, and on the number of time sampling points in the 'X' axis – the recovered signal can be examined in greater detail by expanding various points of the vertical and horizontal image scale. Although the shape of the signal will be preserved, the time axis is now an arbitrary one, recostructed from pulses derived from the system clock. Markers indicating the voltage and time characteristics of the original waveform may be superimposed on the recovered signal to simulate the characteristics of a real-time signal.

A schematic layout of a digital storage oscilloscope system is shown in Figure 17.40.

Frequency measurement

Although an approximate indication of signal frequency can be obtained from a Lissajoux figure, on an oscilloscope display, this is an awkward measurement to make, and is usually time consuming. For high frequencies, the standby of the 'amateur' transmitter enthusiast was the 'grid dip' meter, or its transistorized variants, such as that shown in Figure 17.41. In this, the current drawn from the power supply by a low power oscillator is reduced when the oscillator frequency is adjusted to coincide with that of the unknown signal, when this signal is inductively ar capacitatively coupled to the dip-meter oscillator coil.

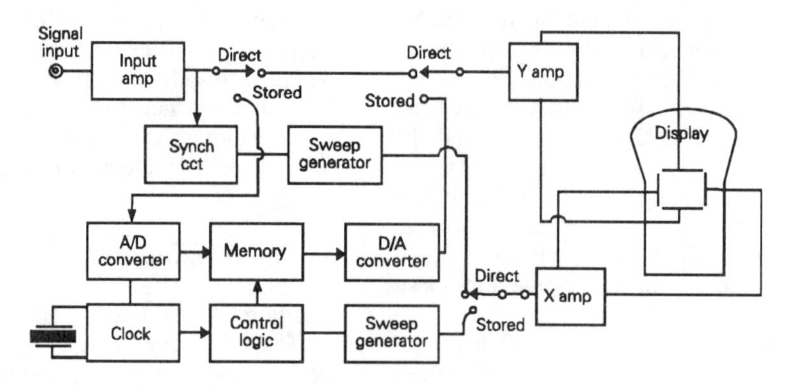

Figure 17.40 *Schematic layout of digital storage oscilloscope*

This, also, is only as accurate as the calibration of the dip-meter tuned circuit.

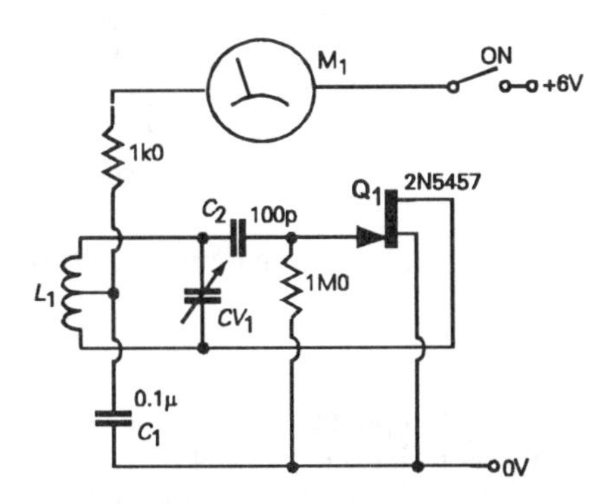

Figure 17.41 *Grid dip frequency meter*

A much better system, and one which is now almost universally used, is the digital display 'frequency counter'. This operates in the way shown, schematically, in Figure 17.42. In this, the incoming signal, after 'conditioning' to convert it into a clean rectangular pattern pulse train, of adequate amplitude, is passed through an electronic switch (gating) circuit, which feeds those pulses received during a precise time interval on to a digital pulse counter system. For convenience in use, the final count is switched to the display through a storage register, to allow the counter to be reset between each measurement interval. The accuracy obtainable is determined by the precision of the setting of the frequency of the crystal oscillator which controls the timing cycle, and by the number of pulses which are allowed through the gating switch, which, in turn, depends on the input frequency. At 1MHz, a gate duration of 0.01 seconds would allow an accuracy of one part in ten thousand, but at 1000Hz, a gating interval, and display refresh periodicity, of once every ten seconds would be needed to achieve the same accuracy.

For high precision frequency measurements where very protracted counting periods would otherwise be necessary, 'frequency difference' meters are usually employed. In these instruments, the difference between the incoming signal frequency, and that of a heterodyne signal derived from an accurate reference frequency is determined. This technique allows accurate frequency measurements to be made in greatly reduced time intervals.

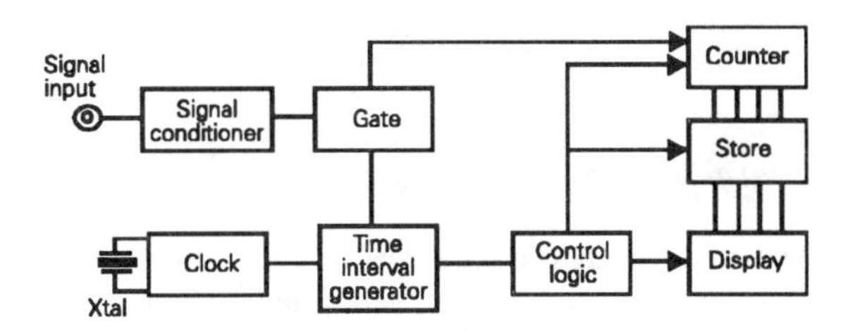

Figure 17.42 *Schematic layout of frequency counter*

Signal generators

The majority of the tests carried out on AC amplifying or signal manipulation circuits require the use of some standardized input signal, in order to determine the waveform shape or amplitude changes brought about by the circuit under test. These input signals are derived from laboratory 'signal generators', and these are available in a range of forms. In low frequency sinewave generators, the major requirements are usually stability of frequency, purity of waveform, and freedom from output amplitude 'bounce', on changes of frequency setting, with the accuracy of frequency or output voltage being relatively less important.

In HF and RF signal generators, precision and stability in the output frequency are essential, as is usually the accuracy of calibration of the output voltage – particularly where the instrument is to be used to determine receiver sensitivity or signal to noise ratio. The purity of the output waveform is usually less important provided that the magnitudes of the output signal harmonics, or other spurious signals, are sufficiently small.

RF signal generators usually make provision for amplitude modulation of the signal, either by an internal oscillator or by an external signal source. This facility is useful to determine the performance and linearity of the demodulator circuit of RF receivers, and may also be helpful in aligning RF tuned circuits.

Other test signals, of square-wave or triangular waveform, are also valuable for test purposes, and these are usually synthesized by specialized waveform generator ICs, and available as wide frequency range 'function generators'. These offer square and triangular output waveforms, as well as a sinewave signal of moderate purity, usually in the range 0.1–0.5% THD. They do have the advantage that they allow a rapid frequency sweep to be made, to check the uniformity of frequency response of the circuit under test.

Beat-frequency oscillators

These instruments derive a relatively high purity, wide frequency range, output signal by heterodyning two internally generated RF signals, as shown in Figure 17.43, and then filtering the output from the mixer stage so that only the difference frequency remains. The major drawback of this technique is that any relative instability in the frequency of either of the RF signals, from which the heterodyne output signal is derived, will be proportionately greater in the resulting difference frequency signal.

Frequency modulated oscillators (wobbulators)

These types of signal generators are almost exclusively used in RF applications, such as in determining the transmission characteristics of the tuned circuits or filter systems which define the selectivity of the receiver, and in aligning these receivers. If the linearity of the modulation characteristics of the oscillator are high enough, it will also allow the distortion characteristics of an FM receiver or demodulator to be measured.

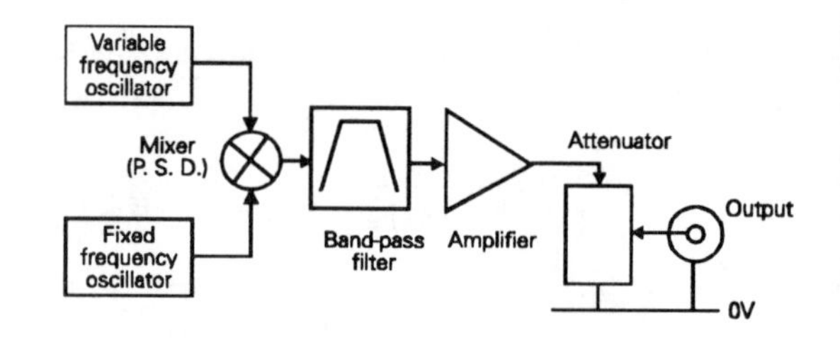

Figure 17.43 *Layout of high frequency BFO (see Figure 10.11)*

Appendix A

Component manufacturing conventions

In the early days of electronics, virtually all components would carry a printed description of the component type and characteristics, usually accompanied by the manufacturer's name, such as 'Wirewound Resistor, 10,000 ohms, Varley Ltd,' or 'Ferranti output transformer, type OPM3, 9:1 ratio'. However, as time passed, components became smaller, and it became increasingly difficult either to affix, or to read, the description on, for example, a half-watt carbon rod resistor, so alternative, non-numerical, descriptive codes were adopted, on an international basis (which, incidentally, also helped the export of components). Of these systems, the earliest was the resistor colour code shown in Table A.1.

Resistor colour codes

Under this system, the value of a resistor, in ohms, would be identified by a three-colour sequence. In the UK, this sequence would be 'body', 'tip', 'dot', where the first two numbers would be as read, while the third, the dot, would be a multiplier denoting the number of zeros following the introductory number.

A modification of this system adopted in the US, and followed by the UK manufacturers, extended the dot into a coloured ring, around the whole of the resistor body, so that the multiplier, which is the most significant digit, would not be concealed from view by awkward component orientation. This system changed fairly rapidly into the now conventional labelling style, in which these colour codes are displayed as a ring sequence of three or more coloured bands, painted near one end of the resistor body, as shown in Figure A.1.

A recent extension to this sequence, included in this list, is the use of silver and gold bands as fractional multipliers to permit coding of very low value resistors.

Colour coding is also used to show the possible error in the actual value of the resistor (termed the tolerance), in which a fourth (or fifth) colour band is added to the bands which specify the resistor value. Using this code, a resistor with a simple three-band sequence, reading from the band nearest the end, of red, red, orange, would denote a 22kΩ resistor, and the absence

Table A.1 *The resistor colour code system*

Colour	Number	Tolerance
Black	0	–
Brown	1	1%
Red	2	2%
Orange	3	
Yellow	4	
Green	5	
Blue	6	
Violet	7	
Grey	8	
White	9	
Gold	(x 0.1)	5%
Silver	(x 0.01)	10%
No colour		20%

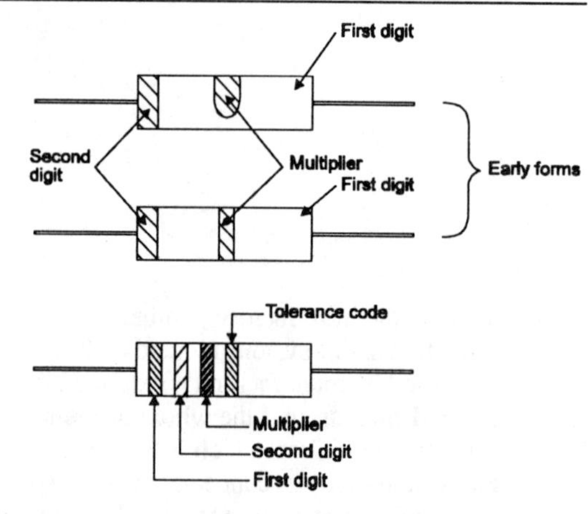

Figure A.1 *Resistor colour coding systems*

of a tolerance specification would imply a +/– 20% possible inaccuracy in its actual value.

A resistor coded yellow, violet, yellow, silver, would mean a 470kΩ component, +/– 10% tolerance, and a brown, black, brown, gold, banding would mean a 100ohm , +/– 5% value.

Unfortunately the situation has become more complicated by the increasing use of close tolerance resistors, in which the resistor value is indicated by a four band code, with an added red (+/– 2%), or brown (+/– 1%) fifth band giving the resistor accuracy specification. By this system, a brown, red, black, brown, red, code sequence would denote a (120 + one zero) 1200ohm , 2% tolerance component, in place of the rather easier to read brown, red, red, silver coding which would be used for a 5% resistor. The close tolerance components can also be rather awkward to decode if the manufacturers have placed the bands too symmetrically on the resistor body.

Preferred value systems

It is clearly impracticable either to manufacture or to stock an infinitely wide range of component values, so a so-called 'preferred' range of values has been adopted, internationally, and it is sensible for circuit designers to choose values for their designs which fall within one or other of these ranges, of which the two most common are listed in Tables A.2 and A.3.

Table A.2 *Preferred values: 5% sequence*

10	27	68	and so on, by multiples or
12	33	82	sub-multiples of ten (e.g.
15	39	100	1.5 ohms or 470,000
18	47	120	ohms) to cover the
22	56	. .	available range.

Table A.3 *Preferred values: 1% sequence*

0	20	39	75	and so on
11	22	43	82	
12	24	47	91	
13	27	51	100	
15	30	56	10	
16	33	62	. .	
118	36	68	. .	

Each of these numerical sequences offers an approximately constant ratio increment from any given value to the next greater value in the series, and the steps are chosen to suit the accuracies of the components in question.

In the case of resistors, the range of values quoted will usually be available in the range 10 ohms to 1 megohm. Manufacturing difficulties tend to limit the precision available outside this range.

There is also a 0.1% preferred range, giving incremental values in the sequence: 100, 102, 105, 107, 110, 113, 115, 118, 121, 124, 127, 130, 133, 137, 140, ... and so on, but these are very expensive and neither widely used nor widely available.

Capacitor coding systems

The same colour/number equivalence coding method has also been used for small value capacitors, particularly ceramic types, where the number denotes the capacitance in picofarads. In the tantalum bead capacitors of the type shown in Figure A.2, a similar system is used to show the capacitance in microfarads. The dot is a multiplier using the factors shown below, and an additional colour band, around the connecting wires, shows the working voltage, using the code shown in Table A.4.

Table A.4 *Tantalum bead capacitor coding*

Colour	Working voltage	Multiplier
White	3V	x0.1
Yellow	6.3V	--
Black	10V	x1
Green	16V	--
Blue	20V	--
Grey	25V	x0.01
Pink	35V	--
Brown	--	x10
Red	--	x100

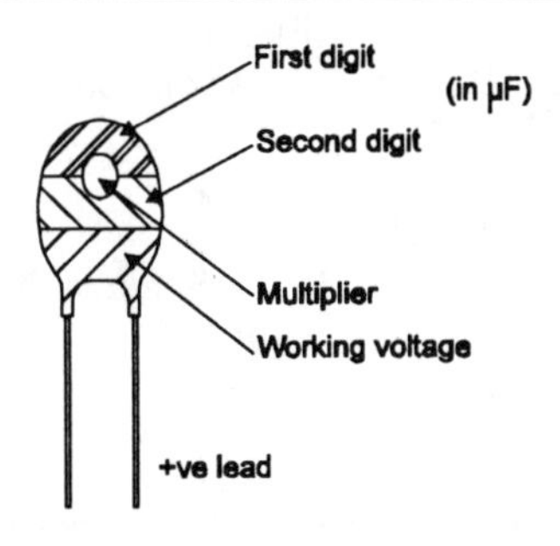

Figure A.2 *Tantalum bead capacitor colour code system*

More recently, the use of colour coding for capacitors has increasingly been abandoned in favour of printed legends, even on very small sized components, where, again, in tantalum bead capacitors the number will show the capacitance in microfarads, while ceramic bead and polystyrene film capacitors will show the value in pF. The tolerances are as shown in the code system of Table A.5.

In the case of ceramic disk capacitors, a 3-digit code, such as '103' is used (as a numerical equivalent of the resistor colour code, 'brown, black, orange') to denote a 10,000 pF capacitor, or '471' to denote a 470pF unit. This will usually be followed by the tolerance code letter. Also, in small value capacitors the letter 'R' is used to show the position of a decimal point, so that '47RJ' would mean a 47pF, +/–5% component, and a '3R3D' code would mean a 3.3pF +/–0.5pF capacitor (see Figure A3).

Table A.5 *Capacitance tolerance coding (EIA system)*

Code	Tolerance
C	+/– 0.25pF
D	+/– 0.5pF
F*	+/– 1pF or +/– 1%
G*	+/–2pF or +/– 2% (whichever is the greater)
J*	+/–5%
K*	+/–10%
L	+/–15%
M*	+/–20%
P	–0 +100%

* This alphabetical tolerance coding system is also used on non-colour coded resistors.

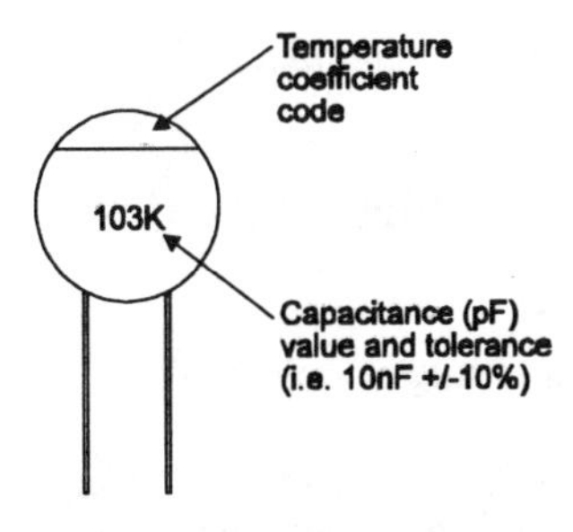

Figure A.3 *Ceramic disk capacitor coding system*

If there is a paint band across the top of the device, this will indicate the temperature coefficient, using the colour code shown in Table A.6.

As in tantalum bead capacitors, the working voltage is indicated in colour coded capacitors, such as the 'radial' lead polyester types, by an additional colour band around the bottom of the component, as shown in Figure A.4. In the case of wound film/foil capaci-

Table A.6 *Ceramic capacitor temperature coefficient coding*

Colour code	Temperature coefficient
Black	'NP0'
Brown	–ve 30ppm/°C
Red	–ve 75ppm/°C
Orange	–ve 150ppm/°C
Yellow	–ve 220ppm/°C
Green	–ve 330ppm/°C
Blue	–ve 470ppm/°C
Violet	–ve 750ppm/°C
Grey/white	
Red/violet	+ve 100ppm/°C

tors, a black or coloured band at one end indicates the lead connected to the outer foil, information which may be necessary in RF applications.

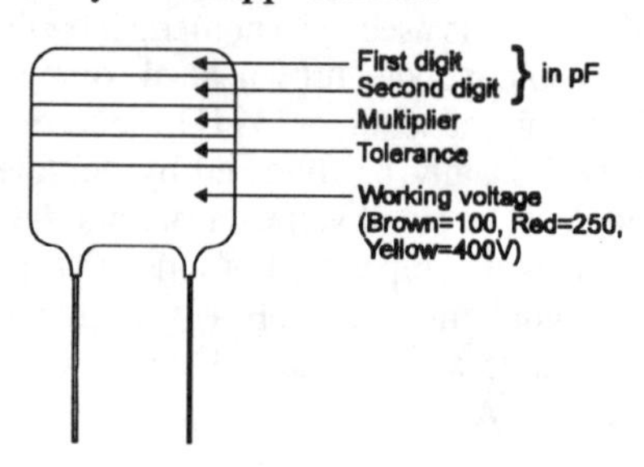

Figure A.4 *Radial lead type polyester coding system*

Semiconductor labelling

Manufacturers' prefixes

There can be differences in specification between apparently identical, and pin compatible, semiconductor devices – especially in the case of linear ICs and output power transistors – so, when replacement is necessary, it is prudent to use devices from the same source. The manufacturers' identification prefixes are shown in Table A.7.

Suffixes generally refer to guaranteed operating temperature range and package types, as shown in Tables A.8 and A.9.

Table A.7 *Semiconductor manufacturers' prefixes*

Prefix	Manufacturer
ACF, AY, GIC, GP	General Instrument (GI)
AD, CAV, AS, HDM	Analog Devices (AD)
AM	Advanced Micro Devices (AMD)
ADC, DH, DM, DS, LFT, LF, LM, NH	National Semiconductor (NS)
BX, CX	Sony
CA, CD, CDP	RCA/GE
CM, HV	Supertex
CMP, DAC, MAT, MUX, OP, PM, REF, SMP	Precision Monolithics (PMI)
D, DF, DG, SI	Siliconix
F, μA, μL	Fairchild
H, HA, HI	Harris
HA, HD, HG, HL, HM, HN	Hitachi
ICH, ICL, ICM	Intersil
IR	International Rectifier
ITT	ITT
LT, LTC, LTZ	Linear Technology
MC	Motorola
N, NE, S, SE	Signetics (Philips)
SG	Silicon General
SL, SP	Plessey
SN, TL, TLC, TMS	Texas Instruments
T, TA, TC, TMM, TMP	Toshiba
UCN, UCS, UDN, UDS, UHP, ULN, ULS	Sprague
XR	Exar
Z, ZD, ZN, ZT, ZTX	Ferranti

Table A.8 *Guaranteed operational temperature range*

Suffix	Temperature range
1. (Harris only)	–55°C – +200°C
M. (2 for Harris, 54 for TTL)	–55°C – +125°C
I.	–25°C – +85° C
C. 5 for Harris, 74 for TTL)	0° – +70°C

Table A.9 *IC package description*

Manufacturer	Metal can TO99 TO100 8pin 10pin	Plastic DIL 8pin	14pin	16pin	Ceramic DIL 8pin	14pin	16pin	Power Plastic TO92	TO220
Advanced Micro Devices	–	P	P	P	D	D	D	–	
Analog Devices	J								
Fairchild	H	T	P	P	R	D	D	W	U
Intersil	K								
ITT	–	N	N	N	D,J	D,J	D,J	---	–
Harris	2*	3*	3*	3*	1*	1*	1*	–	–
Motorola	H,G	P	P	P	V	L	L	P	T
National Semiconductors	H,G	N	N	N	J	J	J	Z	T
Precision Monolithics	J	P	P	P	Z	Y	Q	–	–
Raytheon	H	DN	DB	MP	DE	DC	DD	S	
Signetics	H	N,V,N,E	F,A,N,H	B,NJ	FE	FH	FJ		
Siliconix	A	J	J	J	K	K	K		
Sprague	H	M	A	A	H	H	H	Y	Z
RCA	T								
Texas Instruments	H	P	N	N	JG	J	J	LP	KC
	mainly linear ICs	mainly commercial or industrial types			mainly military or industrial types			transistor types	

In the package description codes, one or more letters will be placed immediately after the temperature code letter, except in the case of the Harris devices, where numbers are used (as shown by an asterisk, and will precede the type number).

Discrete semiconductor type numbering

Transistor type designations are normally a good bit simpler than those used for ICs, since they do not usually carry a prefix identifying the maker, or a suffix specifying one or more package forms, or a working temperature range. The package type is usually implied by the type number of the transistor. Also, unless they are very widely used types, such as the BC109 (which is in a TO18 case), or the BC212 (which is in a TO92(f) encapsulation), they will only be available from one, or maybe two, manufacturers.

The 'BC' type designation, quoted for the two transistor types noted above, is one of the European 'Pro Electron' type classifications listed in Table A.10, which actually give a description of the general type of the device by means of the letters used. The 'JEDEC' listing employed in the USA, which uses prefixes such as '1N', '2N' and '3N', only refers to the time when the device type was registered with the US military authorities, so the only thing which can be inferred from the number is that a '2N5089' is a much more recent device than a '2N930'.

There is, however, some degree of type identification in the prefix, in that '1N' refers to diodes or rectifiers, '2N' refers to bipolar junction or field-effect devices, and '3N' refers to MOSFETs. United States designated transistors (and ICs) are invariably 'second-sourced', which means that the devices are available, to an identical specification, from two or more manufacturers, whereas devices with a Pro Electron code may sometimes be supplied by only one manufacturer, which can be inconvenient.

Figure A.10 *The Pro Electron type classification*

First letter	Second letter	Third letter (if any)	Number
A = Germanium	A = small signal diode	Not usually significant	Manufacturers' catalogue number
B = Silicon	B = varicap or rectifier diode		
C = Gallium arsenide	C = small signal transistor		
	D = power transistor		
	E = point contact diode		
	F, G, L = high frequency transistor		
	R = special purpose device		
	S = switching diode or transistor		
	T = thyristor or triac		
	U = high voltage transistor		
	X = (same as B)		
	Y = power rectifier		
	Z = zener or avalanche diode		

In the case of small signal junction transistors, the final letter or letters in the type number are used to indicate the pin connections, as shown in Figure A.5, and the small-signal current gain (h_{fe}) measured at 1mA collector current, where

$$A = 40 - 120$$
$$B = 150 - 460$$
$$C = 270 - 800$$

while the absence of a final letter indicates that the current gain range has not been determined.

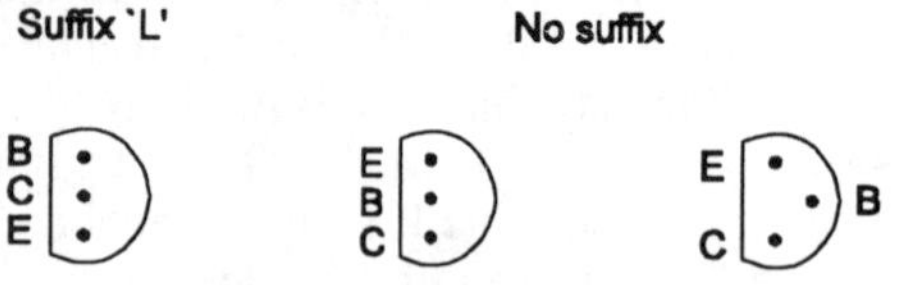

Suffix 'L'

No suffix

Figure A.5 *TO–92 plastic encapsulated transistors (pin view)*

The Japanese Electronic Industry Association (JEIA) classification

This follows the specification laid down in the Japanese Industrial Standard, JIS-C-7012. In this all semiconductor devices have a five group identification sequence. Of these, the first group, usually a single number, indicates the number of active connections, minus one, so that a diode will have a prefix '1', a bipolar transistor, '2', and a MOSFET, with a substrate connection, will carry a number '3'. The second group identifies its registration, and for JEIA registered devices will be 'S'.

The third group will be a letter indicating device type, as set out in Table A.11.

Table A.11 *Semiconductor device classification*

Letter	Type and polarity
A	HF PNP bipolar transistor
B	LF PNP bipolar transistor
C	HF NPN bipolar transistor
D	LF PNP bipolar transistor
E	Thyristor. P-gate
G	Thyristor. N-gate
H	N-base Unijunction trigger device
J	P-channel FET/MOSFET
K	N-channel FET/MOSFET
M	Triac

The fourth group of numbers/letters indicates the time of registration, as in the 'JEDEC' sequence, and does not convey device characteristic information, and the final letter (A, B, etc.) indicates whether this is an improved version of an existing JEIA registered device.

Base connections

Some of the commonly used pin connections for transistors and linear ICs are shown in Figure A.6.

Diode/rectifier polarity marking

The cathode (positive end), of a semiconductor diode is invariably indicated by a paint band around one end of the device. There is not usually any significance in the actual colour used, which will depend on the body colour of the device.

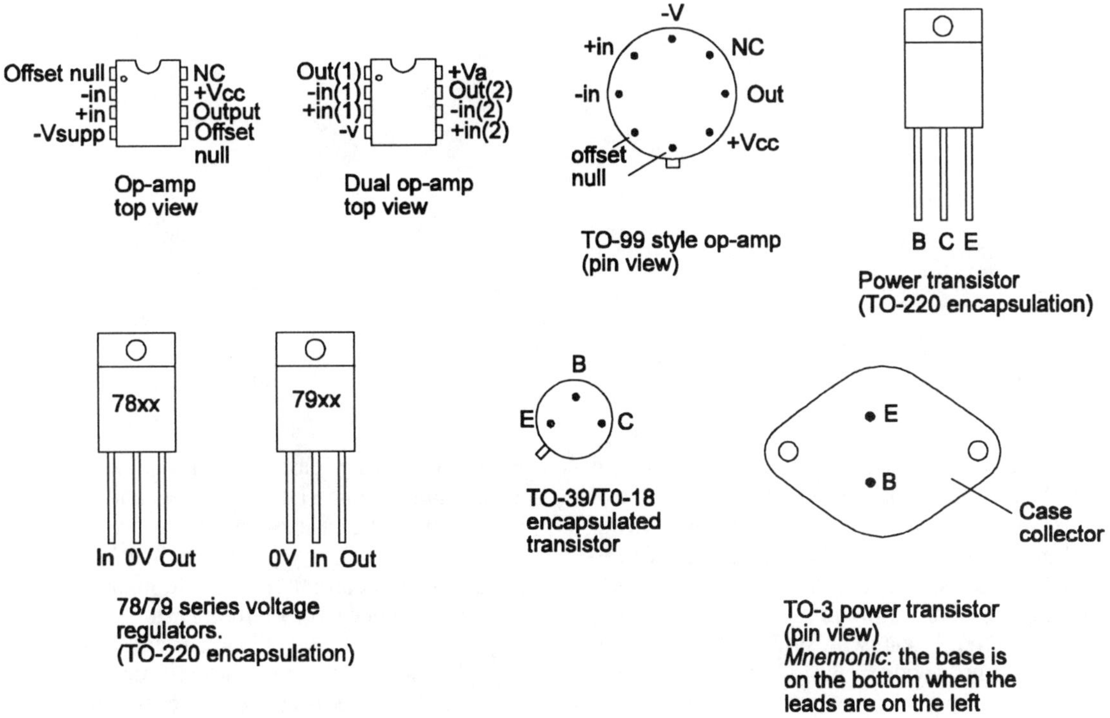

Figure A.6 *Some popular transistor and IC base connections*

Appendix B

Circuit impedance and phase angle calculations

It is frequently required, in electronic circuit design, to introduce some modification in the gain/frequency response of the system. This can be done by inserting a combination of resistors and capacitors, or of resistors and inductors, into the signal, or possibly the feedback, path of an amplifier circuit. This is a very powerful technique, and, with sufficient ingenuity in the circuit design, a wide variety of types of frequency response characteristic can be generated. However, in order to be able to do this with some degree of precision, it is necessary to be able to carry out a few, relatively simple, calculations on the characteristics of networks containing resistors in combination with capacitors and inductors, and these calculations must take into account the phase shifts between the applied voltage and the current flow which arise when inductors or capacitors are present in such networks.

Consider, for example, the case in which an alternating current is passed through a series combination of a resistor and a capacitor, or a resistor and an inductor. In both cases, the voltages developed across the two components will be 90° out of phase with each other, as I have shown graphically in Figures B.1a and B.1b.

In the first of these cases, because the voltage across a capacitor depends on the charge held between its plates, and it takes a finite time for a capacitor to charge or discharge, the voltage 'lags' in phase in relation to the current flowing through it, or to the voltage developed across the resistor – which is always in phase with its internal current flow.

In the case of the resistor/inductor combination, shown in Figure B.1b, the opposite condition is true, since the voltage across the inductor arises as a result of the 'back EMF' produced by changes in the current flow – this is an inherent characteristic of inductance, in which any change in current gives rise to a voltage, and this voltage is such as to oppose the change in current flow which gave rise to it – and the voltage developed across the inductor will 'lead' in phase in relation to the current flow through it.

Earlier in the book, I have noted that the impedance of a capacitor (Z_C) is related to its capacitance and the operating frequency by the equation

$$Z_c = 1/(2\pi fC)$$

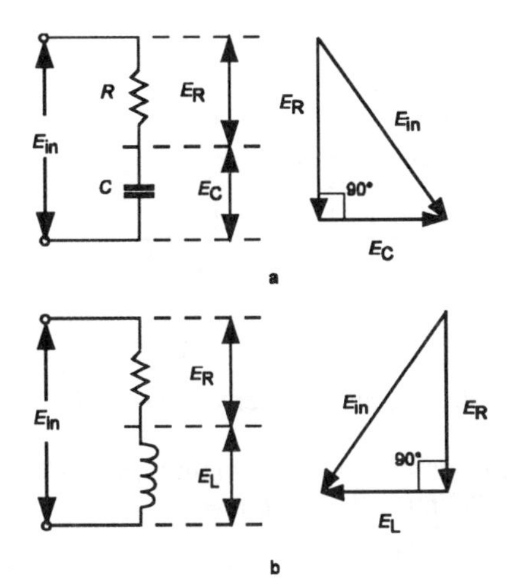

Figure B.1 *Phase angle relationship in RC and RL networks*

Similarly, the impedance of an inductor (Z_L), is defined by the relationship:

$$Z_L = 2\pi fL$$

where f is the frequency of the applied signal, and C and L are the values of the capacitor or the inductor in farads and henries, respectively. However, because of the effects of the relative phase shifts in circuits containing such 'reactive' components, any simple calculation, in which the values of Z_C or Z_L obtained from these equations were treated as numerically equivalent to a similar value of resistance, would give incorrect results.

There is, fortunately, a simple mathematical device by which the phase shifts in such circuits can be taken into account in circuit calculations, and this is the use of the operator 'i' or 'j', which is numerically $\sqrt{-1}$. Pure mathematicians call this i, to denote the fact that it is an 'imaginary' number (since all real numbers give positive values when they are squared). Since electrical engineers have chosen to use the symbol i to denote current, we use the symbol j instead, and its presence implies a 90° phase shift. The use of this operator is justified by the following argument. In a DC system, the opposite of a positive voltage is a negative one. In an AC system, where the applied voltage is continuously oscillating from positive to negative, the opposite of an instantaneous positive potential (and it is convenient to refer to such AC voltages as '*E*' to distinguish them from DC potentials, +/–'*V*'), is the same potential half a cycle (180°) later, when it has swung from positive to negative. A 180° phase shift in an AC system therefore has the effect of mutiplying the initial potential by –1, provided, of course, that the signal is truly sinusoidal. If we have a pair of *RC* (or *LC*) networks in series, each of which produces a phase shift of either +90° (or –90° depending on the way it is connected), the combined effect of the two (the one multiplied by the other), would be to give a 180° phase shift, equivalent to multiplying the input voltage by –1. Each of the separate circuits which produced the +/–90° phase shift must, therefore, have had the effect of multiplying the input voltage by $\sqrt{-1}$, which is exactly the effect which would have been given by the factor j. We can therefore use j in our calculations as the mathematical equivalent of a 90° phase shift.

A further common bit of notational shorthand in circuit equations is the use of the symbol 'ω' (Omega in Greek) to represent '$2\pi f$' – sometimes described as the frequency in 'radians per second' – since these terms nearly always occur grouped together.

The true impedance of a capacitor is not, therefore, $Z_C = 1/(2\pi fC)$, but $Z_C = 1/(j2\pi fC)$, and that of an inductor, $Z_L = j2\pi fL$. In shorthand form these equations become

$$Z_C = 1/(j\omega C) \quad \text{and} \quad Z_L = j\omega L$$

The effect of these phase shifts can be shown graphically, as in Figure B.1, as a 'vector diagram' in which the j term denotes the right angled limb of the triangle, and the input signal is represented by the hypotenuse. Such a graphical display also allows us to derive some further useful bits of information.

In the case of the simple series RC network, shown in Figure B.1a, the circuit impedances can be represented in the manner shown in Figure B.2a, in which the vertical and horizontal limbs represent the resistance and capacitative impedances, R and $1/j\omega C$, respectively. It is unnecessary to write the j symbol in the drawing, in that this is implicit in the position of the w*C* vector at right angles to the *R* limb.

From the theorem of Pythagoras, the length of the hypotenuse, h in Figure B.2b, is $\sqrt{(a^2 + b^2)}$, and the angle Θ is such that $\tan \Theta = b/a$, or, more conveniently, $\Theta = \tan^{-1} b/a$, a calculation which most 'scientific' pocket calculators will do very quickly.

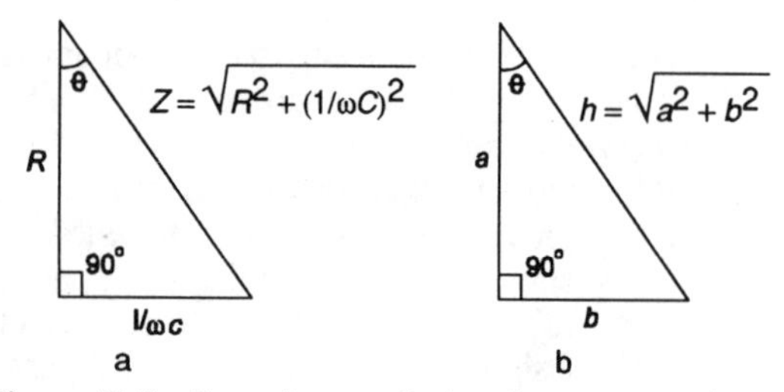

Figure B.2 *Impedance relationships in an RC network*

Returning to the impedance diagram of Figure B.2a, the impedance of the network is therefore

$$Z_{in} = \sqrt{R^2 + \left\{\frac{1}{\omega C}\right\}^2}$$

The angle Θ, which is the phase error between the voltage applied to the network and the current flowing through it (which is the same as the voltage developed across the resistor) is $\tan^{-1}(1/\omega CR)$. It can be inferred from this relationship that, if either R or C were very large, the phase error would be very small indeed.

As noted above, we can identify the phase-shifting effects of Cs and Ls in circuit networks by incorporating the term j in their impedance equations, and we can derive the resultant impedance and phase angle characteristics of these 'complex' networks, mathematically, by sorting out the terms with, and without these j symbols, and treating them by the normal rules of algebraic equations. This procedure works equally well, however many Ls, Cs and Rs there are in the network; if there are a lot, it just becomes more complicated to calculate.

Practical network impedance calculations

Resistor–capacitor networks

In the case of the network shown in Figure B.3a, the resistance of the circuit would be

$$R = \frac{a \cdot b}{a + b}$$

If we substitute R and $1/j\omega C$ for the terms a and b, the

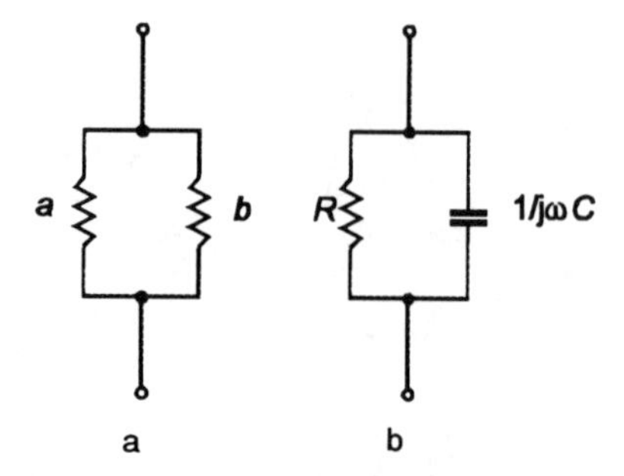

Figure B.3 *Simple RC parallel network*

equation for the impedance of the parallel RC network shown in Figure B.3b will be

$$Z = \frac{R \cdot (1/j\omega C)}{R + (1/j\omega C)}$$

We can simplify this equation by multiplying the top and bottom lines by $j\omega C$, which converts it into the much more manageable form:

$$Z = \frac{R}{1 + j\omega CR}$$

The next mathematical dodge is to get rid of the j term in the bottom line of the equation so that we can divide it into two separate parts representing the in-phase and 90°, 'quadrature', components.

This is done by means of the relationship

$$(a + b)(a - b) = a^2 - b^2$$

When the symbol j appears in these terms, the resultant sign is changed because $j^2 = -1$, with the result that

$$(a + jb)(a - jb) = a^2 + b^2$$

in which, conveniently, the j terms have disappeared.

We can therefore multiply the top and bottom of an equation containing a j term in the denominator by some term of the general form of $a - jb$, which eliminates the j terms from this position. This leaves two separate fractions, one containing the in-phase and the other containing the quadrature components.

Treating the original equation which we derived for Figure B.3b

$$Z = \frac{R}{1 + j\omega CR}$$

in this manner, we get

$$Z = \frac{R}{1+(\omega CR)^2} - \frac{j\omega CR}{1+(\omega CR)^2}$$

which allows us to calculate both the impedance and the phase shifts introduced by the circuit.

The type of network shown in Figure B.4b is a very versatile one, in that, as it stands, it is a useful 'step' network, giving the type of transmission vs. frequency response shown in Figure B.5, while if $R_2 = 0$, it is a simple HF attenuator circuit.

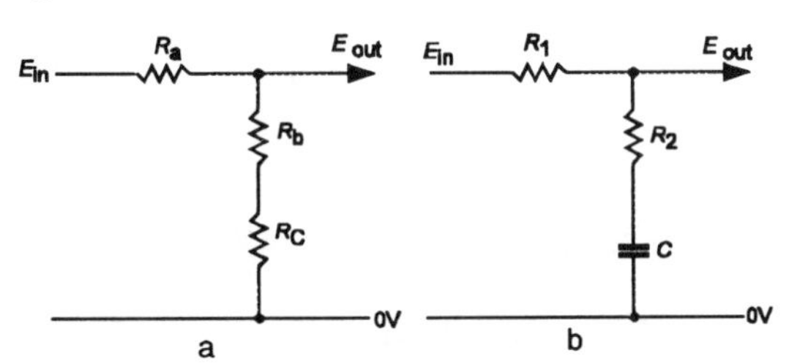

Figure B.4 *Attenuation of an RRC network*

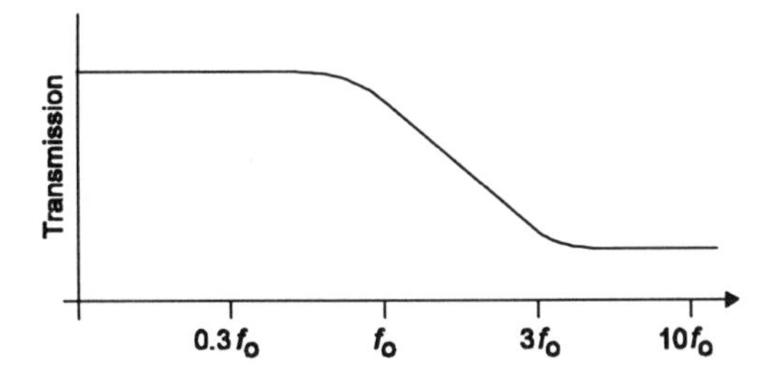

Figure B.5 *Transmission characteristics of an RRC step network*

Taking the case of the simple resistor network of Figure B.4a, the transmission of this circuit would be

$$\frac{E_{out}}{E_{in}} = \frac{R_b + R_c}{R_a + R_b + R_c}$$

By analogy with this, the equation defining the transmission of Figure B.4b will be

$$\frac{E_{out}}{E_{in}} = \frac{1/j\omega C + R_2}{R_1 + 1/j\omega C + R_2}$$

which can be simplified, using the technique shown above, to give the relationship

$$\frac{E_{out}}{E_{in}} = \frac{1 + j\omega CR_2}{1 + j\omega C(R_1 + R_2)}$$

By the same process used above, the in-phase and quadrature components can be extracted, as

$$\frac{E_{out}}{E_{in}} = \frac{1 + \omega^2 C^2 R_2(R_1+R_2)}{1+[\omega C(R_1+R_2)]^2} - \frac{j\omega CR_1}{1+[\omega C(R_1+R_2)]^2}$$

If we make $R^2 = 0$, this equation simplifies to

$$\frac{E_{out}}{E_{in}} = \frac{1}{1+\omega^2 C^2 R_1^2} - \frac{j\omega CR_1}{1+\omega^2 C^2 R_1^2}$$

which is the transmission characteristic of the type of network shown in Figure B.6-3.

From the examples shown in Figures B.2, we can find the actual transmission factor, in numerical tems, by taking the square root of the sum of the squares of the real and imaginary parts of this equation, and the phase angle of the output is given by

$$\tan^{-1}\left\{\frac{\text{quadrature}}{\text{in–phase}}\right\}$$

It is always worthwhile in any calculation of this type to do a quick check on the accuracy of ones arithmetic by substituting some limit values for the components. For example, in the equations above, consider the effect of making the value of $C = 0$. This causes the imaginary part of the transmission equation for both Figure B.6-3 and B.6-5 to disappear, and the real part to become

$$E_{out}/E_{in} = 1$$

which is what we would expect.

On the other hand, if we were to postulate that the value of C was exceedingly large, the first of these examples would give

$$E_{out}/E_{in} = R_2/(R_1 + R_2)$$

and the second would give

$$E_{out}/E_{in} = 0$$

which, again, is what we would expect.

Even in the absence of a desk-top computer, modern programmable scientific pocket calculators make the task of calculating the characteristics of such *RC* networks relatively easy, once the labour of working out the maths has been done, and to save the reader some of this effort, I have shown, in Figure B.6, some of the more commonly used *RC* network arrangements, together with their impedance and transmission equations.

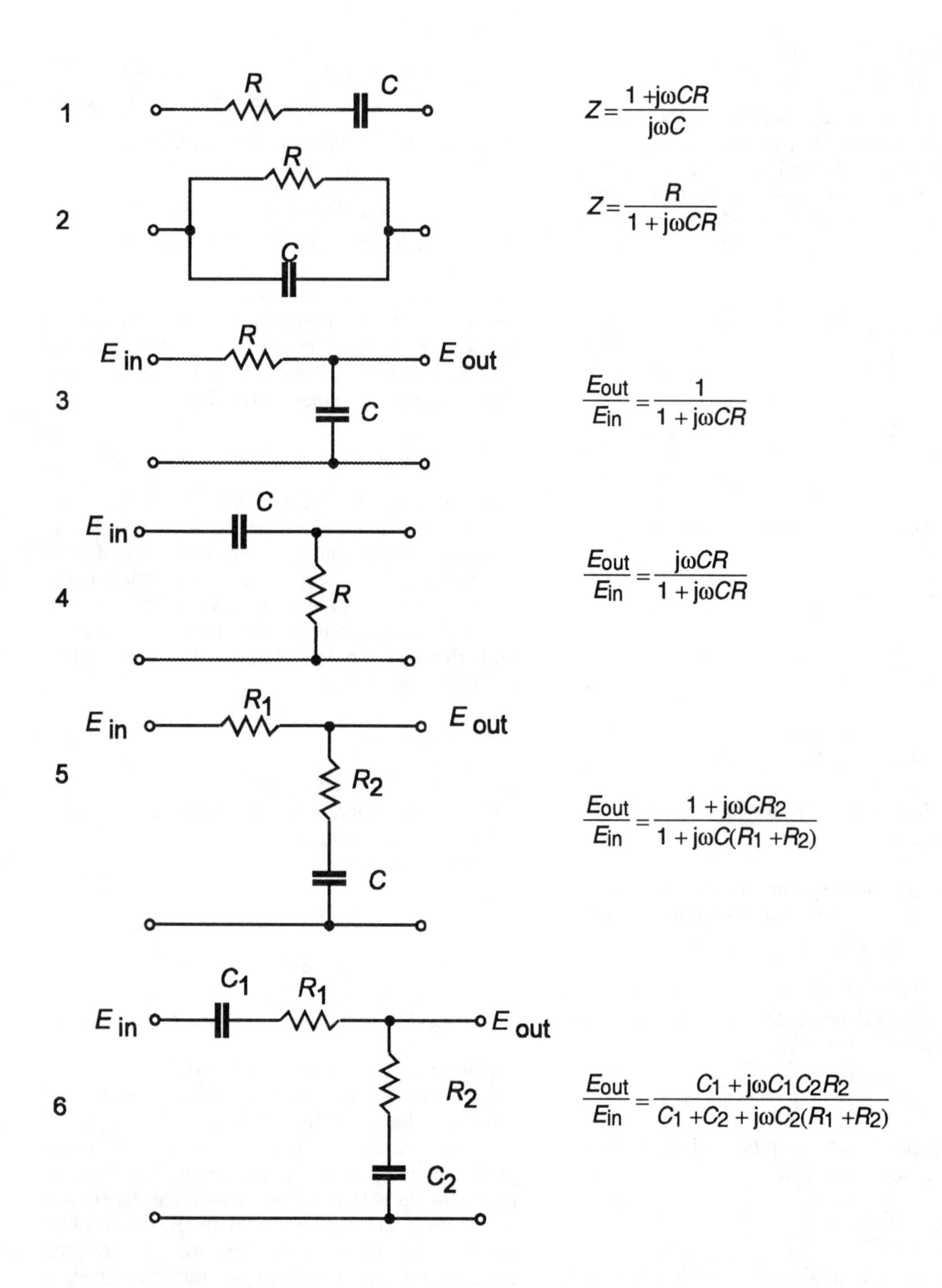

Figure B.6 *Some commom RC network equivalents*

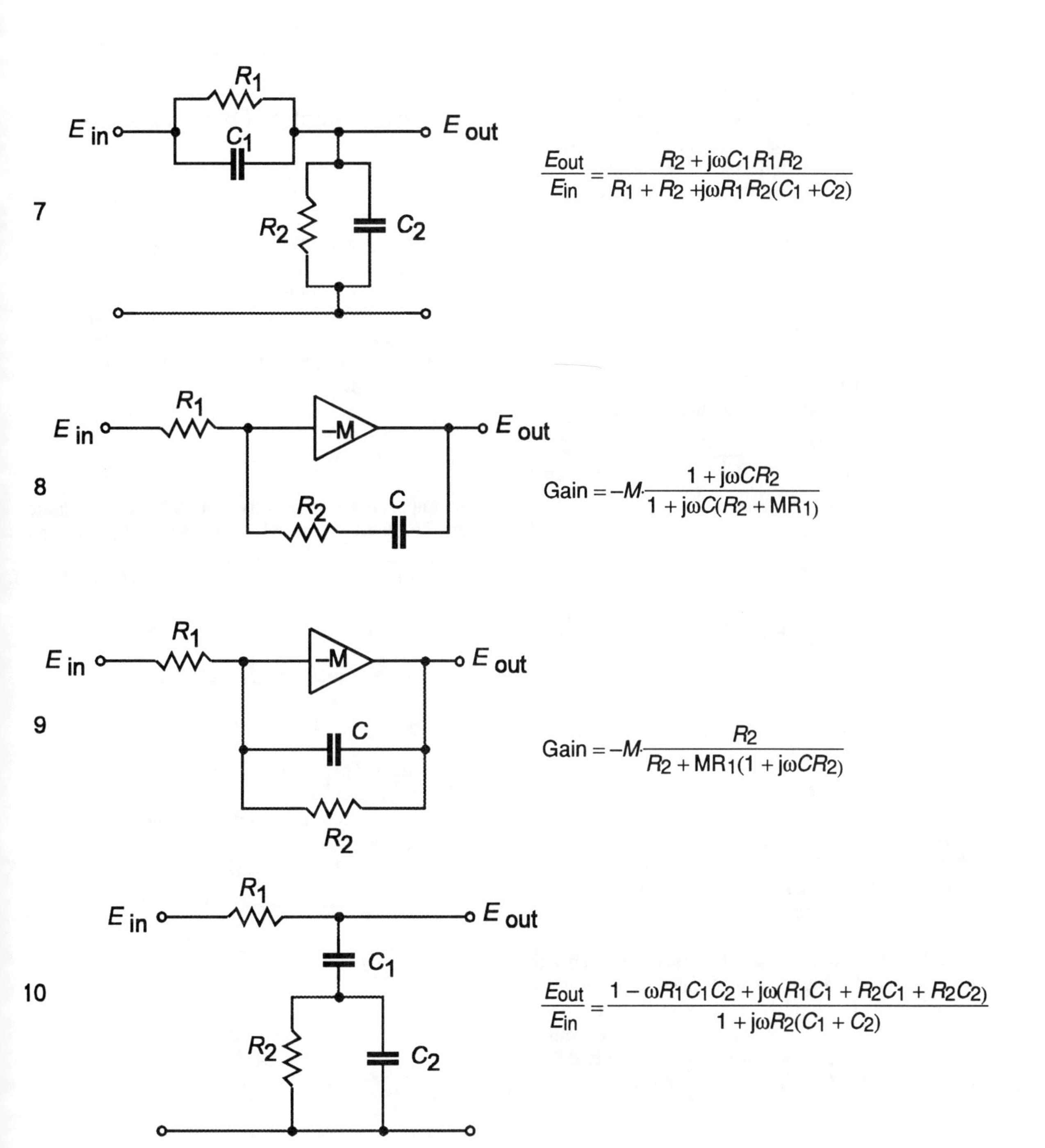

Figure B.6 continued

Resistor–inductor networks

The basic technique for calculating the characteristics of these is the same as that used for RC networks, except that the term $j\omega L$ is used in place of $1/j\omega C$ in the equations. For example, the LR networks shown in Figures B.7 a and b, have the transmission factors of

$$E_{out}/E_{in} = j\omega L/(R + j\omega L)$$

and

$$E_{out}/E_{in} = R/(R + j\omega L).$$

respectively. These can be broken down into the real (in-phase) and imaginary (quadrature) components as

$$\frac{E_{out}}{E_{in}} = \frac{(\omega L)^2}{R^2 + (\omega L)^2} + \frac{j\omega LR}{R^2 + (\omega L)^2}$$

and

$$\frac{E_{out}}{E_{in}} = \frac{R^2}{R^2 + (\omega L)^2} + \frac{j\omega LR}{R^2 + (\omega L)^2}$$

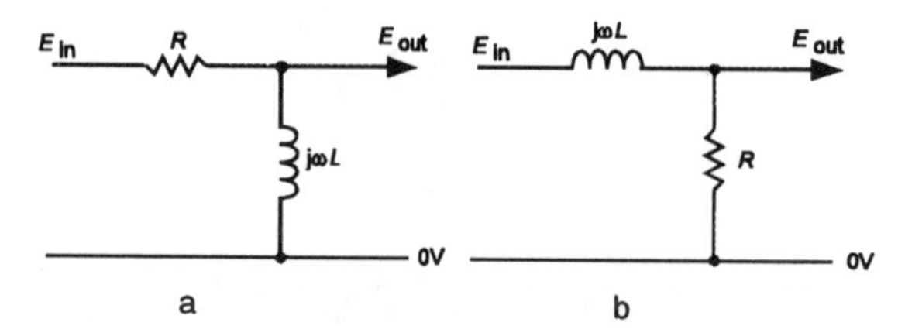

Figure B.7 *RL networks*

Once again, the validity of these formulae can be checked by substituting some limit values. For example, if L = 0, then, in the first case, $E_{out}/E_{in} = 0$ and in the second case, $E_{out}/E_{in} = 1$ which are obviously correct answers.

In all of the examples quoted, it is possible to change the equations for any one kind of network to those for a simpler one by simply removing an R, or a C or an L (by making their values equal to zero). For example, if we take the equation for the network of Figure B.6-7, by removing the Cs ($C_1 = C_2 = 0$) we get

$$E_{out}/E_{in} = R_2/(R_1 + R_2)$$

which is what we would expect, or by deleting C_1 ($C_1 = 0$) we will arrive at the equation for a type 6-3 network but with a resistor (R_2) across its output.

Practical examples of unusual circuit effects

Although the general effect of many RC and RL networks can be guessed from simple inspection, even though the precise turn-over points in the frequency response, or the gain or attenuation may not be known, there are several combinations of these components which give rather unusual effects, which might not easily be guessed but which can be demonstrated by calculation. The first of these is the LC series circuit shown in Figure B.8.

The LC series circuit

The impedance of this is the sum of the two separate parts i.e. $Z = 1/j\omega C + j\omega L$. If we multiply the second of these terms by ($j\omega C/j\omega C$), which is numerically equal to 1, we can rearrange the equation in the form

$$Z = \frac{1 - \omega^2 LC}{j\omega C}$$

This has the curious characteristic that if

$$\omega^2 LC = 1 \quad \text{then} \quad Z = 0$$

This condition is met when $\omega^2 = 1/LC$ or $f = 1/(2\pi\sqrt{(LC)})$

This frequency is termed the resonant frequency of the circuit, often denoted by f_o, and, at this frequency, the impedance of this network is zero.

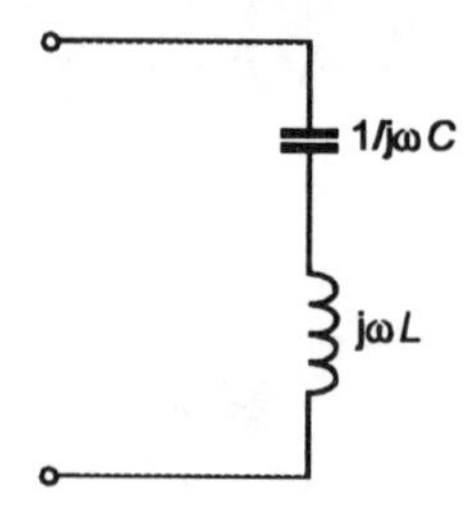

Figure B.8 *Series resonant LC network*

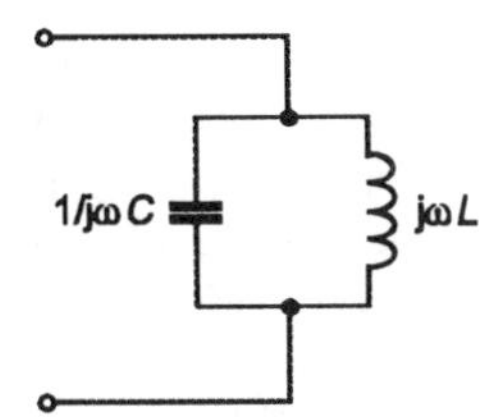

Figure B.9 *Parallel resonant LC network*

The LC parallel resonant circuit.

The circuit arrangement shown in Figure B.9 is basically that of Fig B.6-2, but with the R replaced by jw$L$. The impedance equation can then be written, by analogy, as

$$Z = \frac{j\omega L}{1 - \omega^2 LC}$$

and when $\omega^2 LC = 1$, which is to say when $f = 1/(2\pi\sqrt{LC})$, the denominator becomes zero, and the impedance of the network is infinitely high. This condition is termed 'parallel resonance'.

The Wien network

This interesting and useful circuit arrangement, shown in Figure B.10, is basically a network of the kind shown in Figure B.6-1, in series with a network of the kind shown in Figure B.6-2. In its most commonly used form, both capacitors and both resistors have the same values.

Since we have already worked out the impedance

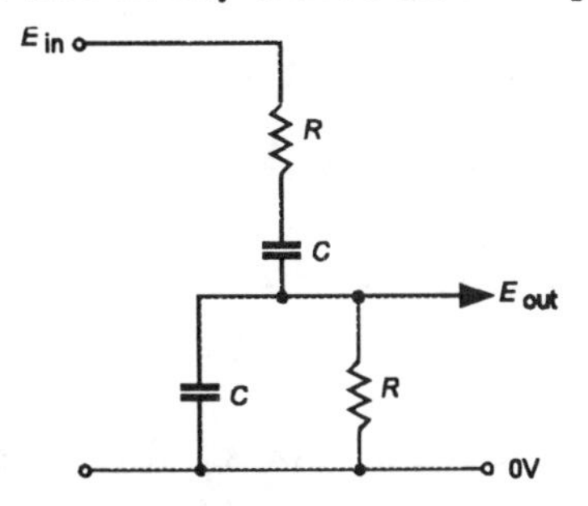

Figure B.10 *Wien network*

characteristics of both of these networks, we can write down the transmission characteristics, using the familiar ratio, derived from a similar resistance network

$$E_{out}/E_{in} = a/(a + b)$$

By fitting in the terms for the resistors and capacitors in place of a and b we get the rather clumsy looking equation

$$\frac{E_{out}}{E_{in}} = \frac{R/(1 + j\omega CR)}{R/(1 + j\omega CR) + (1 + j\omega CR)/j\omega C}$$

We can make this a bit more manageable by multiplying both top and bottom lines by jωC, which gives us

$$\frac{E_{out}}{E_{in}} = \frac{j\omega RC/(1 + j\omega CR)}{j\omega CR/(1 + j\omega CR) + 1 + j\omega CR}$$

which simplifies to

$$\frac{E_{out}}{E_{in}} = \frac{j\omega CR}{1 - (\omega CR)^2 + 3j\omega CR)}$$

and when $(\omega CR)^2 = 1$, which is, of course, when $(\omega CR) = 1$

$$E_{out}/E_{in} = j\omega CR/3j\omega CR = 1/3$$

with no imaginary (quadrature) terms left. But, as we have seen above, $\omega CR = 1$ when $f = 1/(2\pi CR)$, and at this frequency the output signal will be in phase with the input, and the transmission of the network will be 1/3.

Since the condition for continuous oscillation in a feedback circuit is that the signal fed back should be in phase with the input, and that the loop gain is unity, this condition would be met, in the Wien bridge oscillator circuit shown in Figure B.11, at the unique frequency, $f_o = 1/(2\pi CR)$, when the amplifier gain is 3, a possibility which is exploited in a wide range of commercial instruments.

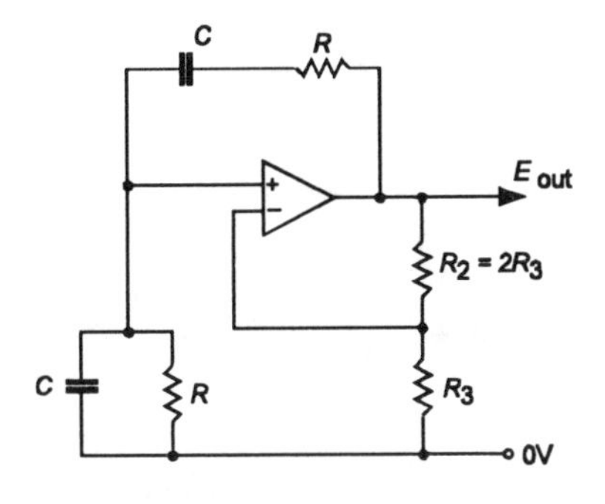

Figure B.11 *Wien bridge oscillator circuit*

The Sallen and Key filter circuit

In Chapter 8, I looked at a range of filter circuits, which could be constructed by the use of one or more gain blocks in conjunction with suitable *RC* networks, as a more elegant alternative, at low frequencies, to the use of *RL*, *RC* or *RLC* attenuator networks.

One of the most popular of these designs is the Sallen and Key filter, which can be assembled from two resistors and two capacitors, in combination with a single op. amp., as shown in the circuit layouts of Figure B.12, to provide a relatively steep-cut (–12dB/octave) high-pass or low-pass filter, with a Butterworth type of filter characteristic, as shown in Figures B.13a and B.13b.

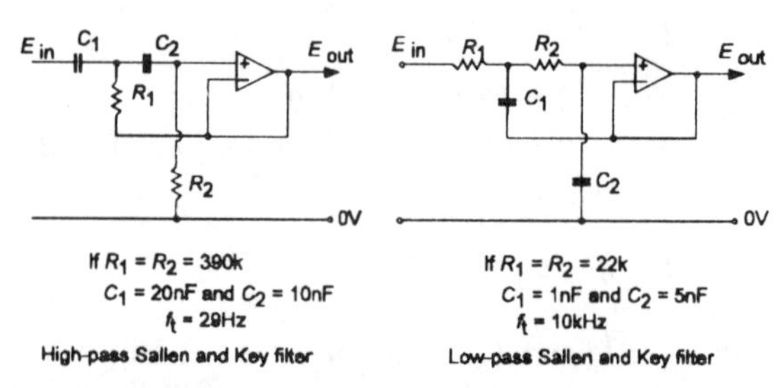

Figure B.12

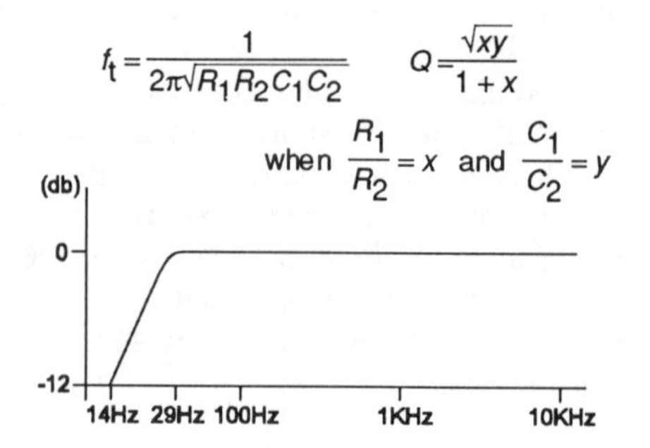

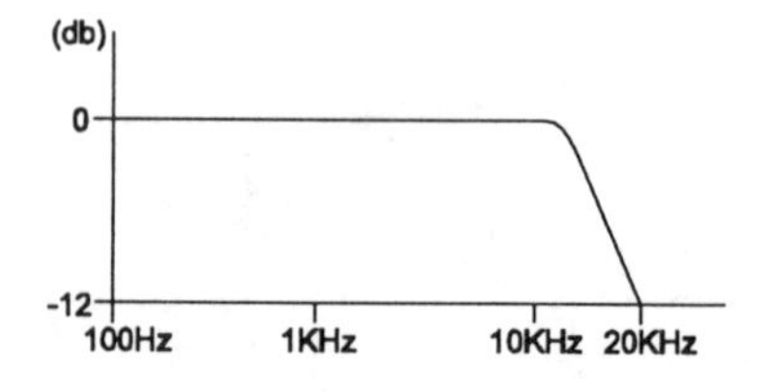

Figure B.13 *Transmission of Sallen and Key filters shown in Figure B.12*

The way this type of filter works can be explored by the use of the j operator, in exactly the same way as the previous circuit analysis. The simplest approach to this is to set up a schematic layout containing impedance blocks, $Z_1 - Z_4$, and an ideal (infinitely high input impedance, zero output impedance), unity gain block, as shown in Figure B.14. If we consider the operation of the circuit in terms of the input and output voltages, and the current flows, i_1 and i_2, and we say that the voltage at the input to the gain block is identical to that at its output, we can derive the following relationships

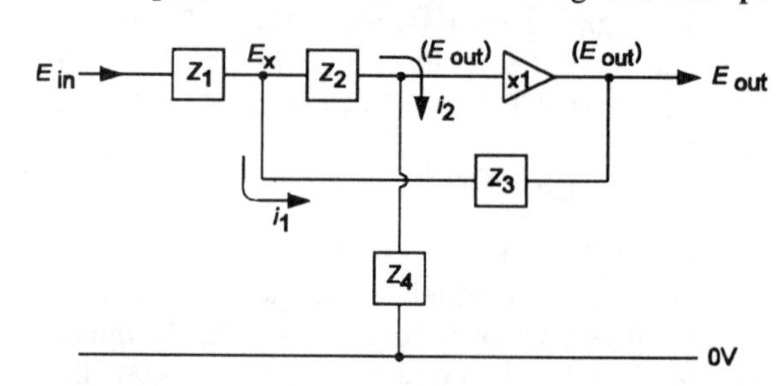

Figure B.14 *Impedance block equivalence of Sallen and Key filter*

$$E_{in} = E_{out} + (i_1 + i_2) \cdot Z_1 + i_2 \cdot Z_2 \qquad (1)$$

but $E_{out} = i_2 . Z_4$, therefore;

$$i_2 = E_{out}/Z_4 \quad \text{also,} \qquad (2)$$

$$i_1 = (E_x - E_{out})/Z_3 \quad \text{and,}$$

$$i_2 \cdot Z_2 = (E_x - E_{out}) \quad \text{Therefore,}$$

$$i_1 = i_2 . Z_2/Z_3 \qquad (3)$$

So, from (1) and (3), we get,

$$E_{in} = E_{out} + i_2 \cdot Z_1 \cdot Z_2/Z_3 + i_2 \cdot Z_2 \qquad (4)$$

and, from (2) and (4), we get:

$$E_{in} = E_{out}(1 + Z_1 Z_2/Z_3 Z_4 + Z_1/Z_4 + Z_2/Z_4) \qquad (5)$$

Therefore

$$\frac{E_{out}}{E_{in}} = \frac{1}{1 + Z_1/Z_4 + Z_2/Z_4 + Z_1 Z_2/Z_3 Z_4}$$

We can now fit in *R*s and 1/jw*C*s in place of these

impedance blocks to get practical circuit equations for the high-pass and low-pass Sallen and Key filter designs.

In the case of the low-pass filter shown in Figure B.12a, where

$$Z_1 = R_1 \qquad Z_3 = 1/j\omega C_1$$

$$Z_2 = R_2 \qquad Z_4 = 1/j\omega C_2$$

we then get the formula

$$\frac{E_{out}}{E_{in}} = \frac{1}{1 + j\omega C_2(R_1 + R_2) - \omega^2(C_1C_2R_1R_2)} \qquad (6)$$

We can deduce several things from this, by simple inspection. For example, when $f = 0$, (i.e. $\omega = 0$), $E_{out}/E_{in} = 1$ (which is to say that the circuit gives unity gain at DC or very low frequencies). Similarly, if the frequency is very high (ω very large) the $-\omega^2$ term in the denominator is the dominant factor and the gain is very small indeed. Since at very low frequencies, there are no significant j terms, the phase shift of the circuit is near zero. The same is true at high frequencies, but, in this case, the negative sign implies a –180° phase shift.

Also, when $\omega^2 (C_1C_2R_1R_2) = 1$, the denominator in this equation is at its smallest, and E_{out}/E_{in} is therefore at a maximum. This condition will occur when $\omega^2 = 1/C_1C_2R_1R_2$, or when $f_o = 1/(2\pi C_1C_2R_1R_2)$ which gives the turn-over frequency of the filter. At this frequency the output of the circuit will be:

$$1/j\omega C_2(R_1 + R_2)$$

which we can call the Q of the circuit. Since, at this frequency, the output can be expressed as $-j/C_2(R_1 + R_2)$, the output will be at quadrature with the input, and in this case the negative sign therefore means that the phase shift will be –90° .

A similar analysis can be made for the high-pass S&K filter, by replacing the Z terms with Rs and $1/j\omega C$s, which gives us the performance equation

$$\frac{E_{out}}{E_{in}} = \frac{\omega^2(R_1R_1C_1C_2)}{\omega^2(R_1R_2C_1C_2) - 1 - j\omega R_1(C_1 + C_2)} \qquad (7)$$

Once again, the output will be at a maximum when

$$\omega^2 = 1/(R_1R_2C_1C_2),$$

or when

$$f_o = 1/(2\pi\sqrt{(R_1R_2C_1C_2)})$$

In this case, at high frequencies, when the ω^2 term dominates this equation, the output will be very close to unity, while at very low frequencies (ω very small) the output will tend to zero.

A further useful mathematical manipulation which can be done with both the low-pass and high-pass filter circuits is to represent the ratios of the Rs and Cs by symbols. Say, for example, we call $R_1/R_2 = x$, and $C_1/C_2 = y$, so that $R_1 = xR_2$, and $C_1 = yC_2$, and suppose we use the term ω_o to define the frequency at which $\omega^2(R_1R_2C_1C_2) = 1$. Then, in the case of the low-pass filter, equation (6)

$$\omega_o^2 = 1/(C_2R_2)^2xy \quad \text{or} \quad \omega_o = 1/C_2R_2\sqrt{xy}$$

and the middle term, $j\omega C_2(R_1 + R_2)$, in the denominator of the equation becomes

$$j\omega C_2R_2(1 + x)$$

If we now express the frequency applied to the circuit as a fraction of ω_o, then equation (7) becomes

$$\frac{E_{out}}{E_{in}} = \frac{1}{1 + j\omega/\omega_o \cdot (1 + x)/\sqrt{xy} - (\omega/\omega_o)^2} \qquad (8)$$

which means that the gain, Q at f_o, (when $\omega/\omega_o = 1$), will be

$$Q = \frac{\sqrt{xy}}{1 + x}$$

A similar process can be used for the high-pass filter layout.

This type of calculation therefore allows us to determine the turn-over frequency of the system, its gain over a wide range of frequencies, and the Q of the system at f_o. For an optimally flat performance from a second order filter of this type, the Q should be $\sqrt{2}/2$.

Conclusions

The use of the j operator as a means of simulating, mathematically, the effect of the phase shift introduced into the system by an inductor or a capacitor, in networks which contain these elements, allows useful performance calculations to be made on these circuits, and allows the reasons for their behaviour to be explained. In addition, if one has some means of performing repetitive calculations, and even a programmable pocket calculator may be adequate, it becomes practicable to determine both the frequency response and the phase shift in any network for which one has the patience to derive the circuit equation.

There are, indeed, other techniques by which this kind of calculation can be made, but I feel that these mainly demand a higher degree of mathematical skill than that which is needed in the use of the j operator in the relatively simple algebraic or trigonometric manipulations explored above.

Index